Characterisation Methods in Inorganic Chemistry

CHARACTERISATION METHODS in INORGANIC CHEMISTRY

MARK T. WELLER | NIGEL A. YOUNG

OXFORD
UNIVERSITY PRESS

Great Clarendon Street, Oxford, OX2 6DP,
United Kingdom

Oxford University Press is a department of the University of Oxford.
It furthers the University's objective of excellence in research, scholarship,
and education by publishing worldwide. Oxford is a registered trade mark of
Oxford University Press in the UK and in certain other countries

Published in the United States of America by Oxford University Press
198 Madison Avenue, New York, NY 10016, United States of America

British Library Cataloguing in Publication Data

Data available

Library of Congress Control Number: 2017932608

ISBN 978–0–19–965441–3

Printed in Great Britain by
Bell & Bain Ltd., Glasgow

Preface

Inorganic chemistry lies at the forefront of many of the most significant, recent scientific advances across materials and molecular chemistry: examples include new photovoltaic compounds, homogeneous and heterogeneous catalysis, pharmaceuticals, and nanomaterials such as graphene. The discovery and identification of these new functional inorganic compounds has relied on the determination of their structures and properties, achieved through the application of a wide variety of spectroscopic and other analytical techniques. Characterisation methods, when applied in inorganic chemistry, have also been central in developing our understanding of the fundamental properties of many compounds. This includes how different atoms from across the whole periodic table interact with each other and bond to form new molecules and complex structures.

At all degree levels a full understanding of inorganic chemistry has, at its heart, an appreciation of the available characterisation techniques and the methods by which instrumental data are analysed. Thus courses in materials chemistry normally include a description of how a solid compound is characterised through the application of techniques such as crystal diffraction, X-ray absorption spectroscopy, and electron microscopy. Likewise main group, molecular chemistry is often taught in parallel with multinuclear nuclear magnetic resonance (NMR) and vibrational spectroscopy, while transition metal complex chemistry is explained using electronic (ultraviolet–visible) spectroscopy and magnetism. The majority of undergraduate laboratory experiments include an element of characterisation of a synthesised compound requiring analysis of instrumental data. Year 3 and 4 practical projects form a core part of many Master's level degree courses and typically employ a large proportion of the characterisation methods described in this textbook.

This text takes a different and fresh approach to teaching inorganic chemistry and one that is aligned with the way that the subject is taught in many university level courses. It is not the intention of this text to deliver a complete inorganic chemistry course, which would cover the descriptive inorganic chemistry of the elements and the basic theory of bonding and reactions of inorganic compounds. These aspects are covered in major texts, such as Weller, Rourke, Overton, and Armstrong's *Inorganic Chemistry*, OUP, and Burrows et al.'s *Chemistry³*, OUP. In fact this book assumes the fundamental chemistry knowledge covered in those texts and introductory degree level inorganic and physical chemistry.

The basis of this text is also one of problem solving and active participation by the student. Thus the text has a large number of worked examples to illustrate how information can be extracted from experimental analytical data. These example problems are followed by numerous self tests so the reader can apply their newly acquired knowledge and interpret similar experimental spectra or diffraction data. In addition further problems are presented at the end of each main chapter or section; for the core techniques of diffraction, NMR, vibrational spectroscopy, and electronic spectroscopy, covered in Chapters 2–5, these problems focus on the fundamental applications and simpler spectra and diffraction data. In the case of the more specialised techniques, described in the later chapters, the problems cover the whole of the material in the chapter. Chapter 11 contains a number of problems that require the understanding and application of more than one experimental technique.

We recognise that students, and lecturers, also appreciate the ability to practise with additional problems, including more advanced examples where we have not had room to include these in the main text. For this reason, alongside the book, we have developed an Online Resource Centre that includes many additional problems. These exercises comprise many further single- and multiple-technique problems, as well as in-depth, longer, and advanced problems.

The level of the text is aimed at undergraduate courses, starting from Year 1 and taking the material to a reasonably advanced level—Master's and early, non-specialised research project level. The complete text should also be of use for research level students, who may wish to refresh and revise their understanding of a specific technique, or discover the capabilities of the full range of characterisation methods available to the chemist. It was not our intention, in a text of this size, to cover fully the more detailed analysis, which often needs computers, nor the very advanced aspects of the various methods. The individual chapter bibliographies include signposts to more specialised texts.

In summary, we believe that this innovative text and its problem-solving approach to teaching analytical inorganic chemistry reflect current degree level pedagogy and provide the modern skills required of a university or college level course.

We acknowledge separately many of the academics and colleagues who have helped with the writing and construction of this text, and over the very lengthy period it has taken to produce a new book from conception to production. We would also like to thank our families and friends, particularly Rachel Young, for their forbearance and support during the many long hours that writing this text has involved. MTW dedicates this text to his father, John Michael Weller, who passed away during the last few months of writing.

Mark T. Weller, Bath
Nigel A. Young, Hull

About the Book

Characterisation Methods in Inorganic Chemistry and its accompanying Online Resource Centre contain many learning features which will help you to build your understanding of the subject. This section explains how to use these features.

Each chapter follows a clear, structured format, which begins with a brief introduction to the technique and basic theory behind it before moving on to the collection of data and its analysis, what form that data typically takes, and how to interpret it. Numerous worked examples, self tests and problems are provided to make your learning as effective as possible.

The material in Chapters 2–5 is clearly divided into introductory and advanced content, making it flexible enough to be used throughout your undergraduate studies.

Examples

Numerous worked examples provide a more detailed illustration of the chapter material. Each one provides practice with calculations and/or problems, and is usually followed by a self test to help you build your understanding, and to monitor your progress.

EXAMPLE 1.1

Calculate the wavelength of electromagnetic radiation with a frequency of 5.00×10^{14} Hz. Which part of the electromagnetic spectrum contains this frequency?

Notes on good practice

In some areas of inorganic chemistry, the commonly used nomenclature can be confusing or archaic. Notes on good practice are included to help you avoid making common mistakes.

NOTE ON GOOD PRACTICE

Other shorthand ways of writing Miller indices include (*hkl*), (*h k l*), *h, k, l*, and just *hkl*, but these can lead to confusion especially when one Miller index is in double figures, for example, the (10, 7, 8) plane might become (1078).

Problems

Problems can be used to check and revise your understanding and practice what you have learned.

Problems

Many of these problems require knowledge of basic solid-state chemistry and the structures of simple inorganic compounds.

Bibliography

Each chapter lists sources of further information particularly useful for the more advanced applications of the specific technique.

Bibliography

P. van der Heide. (2011) *X-ray Photoelectron Spectroscopy: An Introduction to Principles and Practices*. Chichester, UK: Wiley.

About the Online Resource Centre

The Online Resource Centre which accompanies this book provides a number of useful teaching and learning resources and is free of charge.

The site can be accessed at: **www.oxfordtextbooks.co.uk/orc/weller_young/**

Please note that lecturer resources are available only to registered adopters of the textbook. To register, simply visit **www.oxfordtextbooks.co.uk/orc/weller_young/** and follow the appropriate links.

Student resources are openly available to all, without registration.

For Students

Answers to self tests and problems

There are many self tests and problems in each chapter. You can find the worked answers to these in the Online Resource Centre.

For Registered Adopters of the Book

Figures and tables from the book

Lecturers can find the artwork and tables from the book online in ready-to-download format. These can be used for lectures without charge (but not for commercial purposes without specific permission).

The website also provides a number of **additional datasets, problems and exercises**.

Acknowledgements

MTW would like to thank the many supportive academic staff at the University of Bath, including Mike Whittlesey, Paul Raithby, and, in particular, his most esteemed colleague Mary Mahon.

NAY would like to acknowledge the contribution of staff and students at the University of Hull to the development of the examples and spectra used in this text. Adam Bridgeman (now at Sydney) is thanked for many stimulating discussions, and Timothy Prior and Dave Evans for helpful comments. David Collison of Manchester University is especially thanked for many insightful discussions. Robert Lancashire of the University of the West Indies, Mona, Jamaica, is thanked for access to his Tanabe–Sugano data, and many fruitful discussions.

We thank colleagues at Oxford University Press, Martha Bailes and Jonathan Crowe, among others, for their considerable help and support during the writing of this text.

We are also very grateful to the many (anonymous) expert reviewers of this book who provided detailed suggestions as to the content and correctness of the text. Their suggestions were always of great value in a text with such a wide remit.

Brief Contents

Detailed Contents

CHAPTER 1 Fundamental aspects of characterisation methods in inorganic chemistry

CHAPTER 2 Diffraction methods and crystallography

CHAPTER 5 Electronic absorption and emission spectroscopy

CHAPTER 6 X-ray and photoelectron spectroscopy, electron microscopy, and energy dispersive analysis of X-rays

Acronyms and abbreviations

Acronyms

AAS	atomic absorption spectroscopy
AES	atomic emission spectroscopy
AES	Auger electron spectroscopy
ATR	attenuated total reflectance
CBED	convergent beam electron diffraction
CCD	charged coupled device
CIF	crystallographic information file
CHN	carbon hydrogen nitrogen
COSY	correlated spectroscopy
CP	cross polarisation
CSA	chemical shift anisotropy
CW	continuous wave
DMA	dynamic mechanical analysis
DSC	differential scanning calorimetry
DTA	differential thermal analysis
EDAX	energy dispersive analysis of X-rays
EDS	energy dispersive spectroscopy
EFG	electric field gradient
EI	electron ionisation
EPR	electron paramagnetic resonance
ESCA	electron spectroscopy for chemical analysis
ESEM	environmental SEM
ESI	electrospray ionisation
ESR	electron spin resonance
EXAFS	extended X-ray absorption fine structure
FAB	fast atom bombardment
FID	free induction decay
FT	Fourier transform
FTIR	Fourier transform IR
FWHM	full width at half maximum
GC-MS	gas chromatography mass spectrometry
HPLC-MS	high performance liquid chromatography mass spectrometry
ICP-MS	inductively coupled plasma mass spectrometry
ICR	ion cyclotron resonance
INEPT	insensitive nucleus enhancement by polarisation transfer
INS	inelastic neutron scattering
IR	infrared
IS	isomer shift
ISC	inter-system crossing
IVCT	intervalence charge transfer
LMCT	ligand to metal charge transfer
LSIMS	liquid secondary ion mass spectrometry
LUMO	lowest unoccupied molecular orbital
MALDI	matrix-assisted laser desorption/ionisation
MASNMR	magic-angle spinning NMR
MLCT	metal to ligand charge transfer
MO	molecular orbital
NIR	near infrared
NMR	nuclear magnetic resonance
NPD	neutron powder diffraction
NQR	nuclear quadrupole resonance
PDA	photodiode array
PDF	pair distribution function
PES	photoelectron spectroscopy
PXD	powder X-ray diffraction
QCC	quadrupole coupling constant
qNMR	quantitative NMR
RMM	relative molecular mass
RT	room temperature
SAED	selected area electron diffraction
SEM	scanning electron microscopy
SERS	surface-enhanced Raman spectroscopy
SIMS	secondary ion mass spectrometry
SND	single-crystal neutron diffraction
SOMO	singly occupied molecular orbital
SQUID	superconducting quantum interference device
SSIMS	static SIMS
STM	scanning transmission microscope
SXD	single-crystal X-ray diffraction
TEM	transmission electron microscopy
TGA	thermogravimetric analysis
TIP	temperature-independent paramagnetism
TMS	tetramethylsilane
TOF	time-of-flight
UHV	ultra-high vacuum
UPS	ultraviolet photoelectron spectroscopy
UV-vis	ultraviolet–visible
VSEPR	valence shell electronic pair repulsion
VSM	vibrating sample magnetometer
XANES	X-ray absorption near edge structure
XAS	X-ray absorption spectroscopy
XPS	X-ray photoelectron spectroscopy
XRF	X-ray fluorescence

Abbreviations

A	absorbance
A	hyperfine coupling constant
a, b, and c	lattice parameters
B_0	magnetic induction
C	Curie constant
c	concentration
c	speed of light
E	energy
E	identity operation
E_i	ionisation energy
E_R	recoil energy
F_{hkl}	structure factor of the reflection with the Miller indices h, k, and l
$F(R)$	Kubelka–Munk function
f_j	scattering factor for an atom j located at (x, y, z)
G	free energy
g	g factor, g value
g_j	Landé factor
$G(r)$	pair distribution function
H	applied magnetic field
h	Planck's constant
h, k, and l	Miller indices
I	intensity
I	nuclear spin quantum number
i	inversion centre
J	spin–spin coupling constant
J	total angular momentum
k	force constant
k	wavevector
k_B	Boltzmann constant
L	total orbital angular momentum
l	path length
m	mass
m_I	nuclear magnetic energy level
m_S	electron spin states
m_u	atomic mass constant
N_A	Avogadro's number
p	momentum
P_{xyz}	Patterson function
Q	quadrupole moment
R	reflectivity
S	symmetry coordinate
S	total spin angular momentum

T	temperature
T	transmission
T_C	Curie temperature
T_g	glass transition temperature
T_N	Néel temperature
t	crystallite thickness
v	vibrational quantum number
v	velocity
α	polarisability
Γ	reducible or irreducible representation
γ	gyromagnetic ratio
Δ	crystal field/ligand field splitting parameter
Δ_{oct}	octahedral crystal field/ligand field splitting parameter
Δ_{tet}	tetrahedral crystal field/ligand field splitting parameter
ΔE_M	magnetic dipole interaction
ΔE_Q	quadrupole splitting
δ	chemical shift
δ	isomer shift
ε_{max}	molar absorptivity
ζ	single electron spin–orbit coupling constant
η	asymmetry parameter
θ	Bragg angle
θ	Weiss constant
λ	many-electron spin–orbit coupling constant
λ	wavelength
μ	dipole moment
μ	reduced mass
μ_B	Bohr magneton
μ_{eff}	effective magnetic moment
μ_0	vacuum permeability
v	frequency
v	anharmonic vibrational frequency
$\tilde{v}$	anharmonic vibrational wavenumber
ρ	depolarisation ratio
ρ_{xyz}	electron density at a point with coordinates (x, y, z)
σ	symmetry plane
τ	lifetime
φ	work function
χ_g	specific susceptibility
χ_M	molar susceptibility
χ_{opt}	optical electronegativity
ω	harmonic vibrational frequency
$\tilde{\omega}$	harmonic vibrational wavenumber

Fundamental aspects of characterisation methods in inorganic chemistry

1

1.1 Introduction

The characterisation of a new or unknown compound is fundamental to the advancement of inorganic molecular and materials chemistry. Once a new compound or material has been synthesised in the laboratory, identifying its chemical composition and structure is a necessary stage in understanding the reaction processes that have led to its formation, determining exactly what the reaction product is, and in explaining its properties. This is true both of the teaching laboratory and of the frontiers of research. Normally all new compounds obtained during a research project would undergo rigorous analysis and characterisation prior to publication. As a result, the atomic and electronic structure of every inorganic compound or material has, in general, been determined by applying one or more physical characterisation techniques, such as a spectroscopic method using electromagnetic radiation of various wavelengths or through the diffraction of X-rays. The characterisation techniques available to investigate a compound vary greatly in the type and level of structural and compositional information that they yield. The associated instrumentation differs considerably both in its availability and the experimental time for data collection and analysis. Knowing what characterisation methods are available, how the experimental data are collected and analysed, and what each different technique can tell a chemist about a compound are skills that are relevant across both teaching and research.

The structures of inorganic molecular compounds may be discussed in terms of simple bonding models, including valence shell electronic pair repulsion (VSEPR) theory or molecular orbital diagrams. Bonding models, used to predict and understand the arrangement of atoms in molecules, are usually explained through their application to experimentally determined molecular structures. The structures of these inorganic molecules will have been obtained from experimental methods, such as nuclear magnetic resonance (NMR) or infra-red spectroscopy. As a result, the foundations of these characterisation methods are often taught alongside the fundamentals of bonding in inorganic compounds and descriptive inorganic chemistry. Similarly, the chemistry of simple solids, for example, those of close-packed metals or sodium chloride, requires discussion of their structures, normally obtained through the application of X-ray diffraction techniques. This text focuses on describing and illustrating these characterisation methods; it also uses a problem-solving approach so that the relationship between experimental data and the structures and properties of inorganic compounds can be directly appreciated.

Most of the physical techniques used in contemporary inorganic research rely on the interaction of electromagnetic radiation with matter. The full **electromagnetic spectrum** (Figure 1.1) has been used for the characterisation of inorganic compounds, investigating their atomic and electronic structures, and studying their reactions.

While not all of the techniques described in this text use the electromagnetic spectrum as a probe of the structures of inorganic molecules and materials, four of the main physical methods described in this text, NMR spectroscopy, vibrational spectroscopy, X-ray diffraction, and ultraviolet–visible (UV-vis) spectroscopy, all employ light of a specific range of wavelengths. Other less commonly employed techniques, such as Mössbauer spectroscopy, X-ray absorption spectroscopy (XAS), EPR, and nuclear quadrupole resonance (NQR) spectroscopy, also use portions of the electromagnetic spectrum covering the energy range from gamma rays, with wavelengths of a few picometres, to microwaves, with wavelengths of centimetres to metres. Because the full electromagnetic spectrum is employed to characterise inorganic compounds, the data obtained covers a very large wavelength and, therefore, energy range. The following section describes the wide variation in terminology used by chemists to describe these different energy ranges.

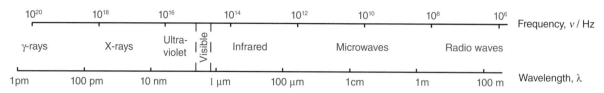

FIGURE 1.1 The electromagnetic spectrum as a function of frequency and wavelength with the main wave types named.

1.2 Units and energy unit conversions

1.2.1 Units of distance and atomic separation

The **SI unit** of length is the metre (m) and wavelengths in the electromagnetic spectrum are normally given, or should be given, in metres or a prefixed metre unit such as nanometres (nm, equal to 1×10^{-9} m) or centimetres (cm, 1×10^{-2} m). However, as the various experimental techniques have developed over the last 150 years, many disciplines have employed their favoured units—which are often not SI units. Thus crystallographers still generally use the **angstrom** with the symbol Å (1 Å = 10^{-10} m = 10^{-8} cm = 10^{2} pm) as a unit of measurement. This unit is convenient because bond lengths between atoms typically lie between 1 and 3 Å and the X-ray wavelength used in crystallography to derive them is usually between 0.3 and 5 Å. A distance such as 1.50 Å is somewhat easier to write, and say, than the equivalent distance in the SI units of pm or nm, that is, 150 pm or 0.150 nm.

1.2.2 Wavelength, frequency, and energy

The wavelength of electromagnetic radiation (λ) is directly related its frequency (ν) through the expression

$$\lambda = \frac{c}{\nu} \qquad \text{Eqn 1.1}$$

that is, wavelength (λ) = speed of light in vacuum (c)/frequency (ν), where $c = 299\,792\,458$ m s^{-1} (3.00×10^{8} m s^{-1} to three significant figures). Frequency is normally expressed in the SI unit hertz (Hz) equivalent to one cycle per second (s^{-1}).

As a result any wavelength of radiation from the electromagnetic spectrum can be restated and quantified in terms of its frequency. Hence, X-rays with a wavelength of 150 pm (to three significant figures) (1.50 Å, 1.50 $\times 10^{-10}$ m) have an associated frequency of

$$1.50 \times 10^{-10}\,\text{m} = 3.00 \times 10^{8}\,\text{m s}^{-1} / \nu$$

$$\nu = 2.00 \times 10^{18}\,\text{Hz (s}^{-1})$$

A similar calculation for a microwave of wavelength 10.0 cm gives $\nu = 3.00 \times 10^{9}$ s^{-1} (3.00 GHz).

EXAMPLE 1.1

Calculate the wavelength of electromagnetic radiation with a frequency of 5.00×10^{14} Hz. Which part of the electromagnetic spectrum contains this frequency?

ANSWER

Using Eqn 1.1 and $c = 300 \times 10^{6}$ m s^{-1}:

$$\lambda = \frac{300 \times 10^{6}\,\text{m s}^{-1}}{5.00 \times 10^{14}\,\text{s}^{-1}} = 6.00 \times 10^{-7}\,\text{m} = 600\,\text{nm}$$

600 nm is in the visible spectrum and is seen as orange light.

SELF TEST

The lowest wavelength light visible to the human eye is 390 nm; what frequency does this correspond to?

The frequency of electromagnetic radiation is directly related to energy through the expression $E = h\nu$ using Planck's constant, $h = 6.626 \times 10^{-34}$ J s. The energy of a single electromagnetic wave may be calculated using this expression so again a single 1.50 Å X-ray has an associated energy of

$$E = 6.626 \times 10^{-34}\,\text{J s} \times 2.00 \times 10^{18}\,\text{s}^{-1}$$

$$= 1.33 \times 10^{-15}\,\text{J (to 3 sig. figs.)}$$

However, as chemists we often deal with moles of a compound so a more convenient energy expression is in J mol^{-1} which gives 7.979×10^{8} J mol^{-1} for X-rays of wavelength 1.50 Å, using Avogadro's number, $N_A = 6.022 \times 10^{23}$ mol^{-1}.

1.2.3 Electronvolts

When used in spectroscopic techniques, such as XAS, X-rays are often produced by accelerating electrons so an alternative unit for describing the energy of electromagnetic radiation is the electronvolt, eV, which is the energy gained by a single unbound electron after acceleration through an electric potential difference of 1 V (equal to 1 joule/coulomb). In addition, the ionisation energies of atoms and molecules are often reported in electronvolts as well as kJ mol^{-1} (e.g. the first ionisation energy of sodium is 5.14 eV = 496 kJ mol^{-1}), as are the separations between energy levels in solids, known as band gaps (the band gap in the semiconductor silicon is given in the literature as 1.1 eV). The charge on an electron is 1.6022×10^{-19} C so 1.000 eV is equivalent to 1.6022×10^{-19} J and a mole of electrons accelerated through 1.0000 V has an energy 96.485 kJ; so 1 eV = 96.485 kJ mol^{-1} (96.485 kJ mol^{-1} is also known as the Faraday constant). Thus, the mole of X-rays of wavelength of 1.50 Å for which we calculated an energy of 7.979×10^8 J (see previous paragraph) can be re-expressed as having an energy of 8269.7 eV, or 8.2697 keV. X-rays of a characteristic wavelength are often generated by bombarding a metal target with high energy electrons accelerated through kV potentials to eject core electrons from the metal atoms. Outer shell electrons falling into the now empty core orbitals produce characteristic radiation energy and this energy can be re-expressed in electronvolts; the energy above which electrons would be absorbed by that metal and fully eject the core electrons is known as an absorption edge in X-ray spectroscopy (Chapter 6).

this is particularly important with calculations involving wavelengths as the values span such a large range of magnitudes, from picometres to kilometres.

SELF TEST

The X-ray absorption edge of Zn is at 9.6659 keV. Calculate the X-ray wavelength equivalent.

The energy separation between the levels which an electron may occupy in a solid is often calculated from theory and expressed in eV. This energy separation, also known as a band gap, can be converted into the wavelength of light that would excite an electron transition between the two energy levels. Thus a band gap of 2.42 eV corresponds to a wavelength of 512 nm derived from the calculation (using the expressions and figures given above) as follows:

$$\lambda(m) = hc/E = (6.626 \times 10^{-34} \text{ J s} \times 2.998 \times 10^8 \text{ m s}^{-1})/(1.602 \times 10^{-19} \text{ J} \times E \text{ (in eV)})$$

$$\lambda(nm) = 1240/E \text{ (eV)}$$

(Note that the value for λ in the first line above is in metres (from the speed of light) and it has been converted to nm in the second expression by multiplication by 10^9.)

1.2.4 Wavenumbers

One other 'energy' unit used by convention in the spectroscopy of inorganic compounds, particularly vibrational spectroscopy, and in some cases electronic absorption spectroscopy, is the **wavenumber**. The 'wavenumber' is defined as the number of waves in unit length $(1/\lambda)$, and is related to the frequency, v, by the expression v/c ($c = 2.998 \times 10^8$ m s^{-1}). The dimensions of wavenumber are 1/length and it is commonly reported in inverse centimetres (cm^{-1}) using centimetres as the unit length. The symbol normally used to represent wavenumber is $\tilde{v}$. Using this formulism, radiation with a frequency of 2.998×10^{13} s^{-1} has a wavenumber given by

$$\tilde{v} = \frac{v}{c} = \frac{2.998 \times 10^{13}}{2.998 \times 10^8} \text{ m}^{-1} = 100\,000 \text{ m}^{-1}$$
$$= 1000 \text{ cm}^{-1} \qquad \textbf{Eqn 1.2}$$

The wavelength of this radiation, λ in cm, is $1/\tilde{v} = 1/1000 = 0.001$ cm, equivalent to 1.000×10^{-5} m, 10 000 nm or 10.000 µm (microns). An absorption at this energy is, therefore, often stated verbally as occurring at 'a frequency of 1000 wavenumbers'; although this is a very widespread practice, it is in fact incorrect as it interchanges frequency (s^{-1}) and wavenumber (cm^{-1}) units.

EXAMPLE 1.2

The shortest wavelength of X-ray generated by bombarding a Mo target with electrons is 0.61978 Å. Calculate the energy (in keV) associated with these X-rays, which corresponds to the Mo absorption edge.

ANSWER

We need to convert a wavelength of 0.61978 Å into eV and we can do this using the conversion process and constants derived above. Thus by combining $E = hv$, $\lambda = c/v$ and Avogadro's number we get

$$E \text{ (J mol}^{-1}) = N_A hc/\lambda \text{ (mol}^{-1} \text{ J s m s}^{-1}/m)$$
$$= \{6.0221 \times 10^{23} \text{ mol}^{-1} \times 6.626 \times 10^{-34} \text{ J s} \times 2.998 \times 10^8 \text{m s}^{-1}\}/ 0.61978 \times 10^{-10} \text{ m}$$
$$= 1.9302 \times 10^8 \text{ J mol}^{-1}$$

Then 1 eV ≡ 96.485 kJ mol^{-1} so $E = 1.9302 \times 10^8$ J mol^{-1} is equivalent to

$$E \text{ (keV)} = 1.9302 \times 10^5 \text{ kJ mol}^{-1}/96.485 \text{ kJ mol}^{-1}/\text{eV}$$
$$= 20.005 \text{ keV}$$

NOTE ON GOOD PRACTICE

As always with calculations it is useful to include units through the calculation to make sure these are consistent—

Table 1.1 summarises the various units used to define the energy of electromagnetic radiation and the conversion factors between them.

EXAMPLE 1.3

A lamp emits violet light with a wavelength of 400.0 nm. What is the energy, calculated in both kJ and eV, of a mole of photons of this radiation?

ANSWER

Initially we need to ensure we are working in base SI units and convert the wavelength to metres and then determine the associated frequency in Hz (s^{-1})

$$400.0 \text{ nm} = 400.0 \times 10^{-9} \text{ m} = 4.000 \times 10^{-7} \text{ m}$$

and the associated frequency can be obtained from $\lambda v = c$:

$$4.000 \times 10^{-7} \text{ m} \times v = 2.998 \times 10^{8} \text{ ms}^{-1}$$
$$v = 7.495 \times 10^{14} \text{ s}^{-1}$$

The energy of a single photon, E_1, is given by $E = hv$:

$$E_1 = (6.626 \times 10^{-34} \text{ J s}) (7.495 \times 10^{14} \text{ s}^{-1})$$
$$E_1 = 4.966 \times 10^{-19} \text{ J}$$

and, therefore, for one mole of photons we multiply by Avogadro's number:

$$E_{mole} = (4.966 \times 10^{-19} \text{ J}) (6.022 \times 10^{23} \text{ mol}^{-1})$$
$$= 299.1 \text{ kJ mol}^{-19}$$

To get to eV we divide by the conversion factor 96.485 kJ mol^{-1}/eV to give 3.100 eV.

SELF TEST

Convert 400 nm into wavenumber units.

1.2.5 NMR energy units

Another technique where the energy units are reported in a different way is in NMR spectroscopy—more detail

of this is given in Chapter 3. A typical transition between energy levels observed in a proton NMR spectrum using a modern spectrometer occurs at an energy with a frequency corresponding to around 300 000 000 Hz (300 MHz). The range of different resonance frequencies observed in inorganic compounds only corresponds to around 3000 Hz so reporting spectra in a more typical energy unit of hertz between 300 003 000 and 300 000 000 Hz would be very cumbersome. Hence, resonance frequencies are usually reported as chemical shifts in parts per million, ppm, relative to a reference material. So for a resonance frequency, v, of 300 000 300 relative to that of a reference, v^o, at exactly 300 MHz the resonance energy is reported via the expression for **chemical shift**, δ, as

$$\delta = \frac{v - v^o}{v^o} \times 10^6 \qquad \text{Eqn 1.3}$$

In this case,

$$\delta = \frac{300 \text{ (Hz)}}{300 \times 10^6 \text{ (Hz)}} \times 10^6 = 1 \text{ ppm}$$

The 'energy scale' in NMR then becomes ppm. Although peak positions are measured in ppm, the separations between resonance peaks are normally stated in hertz; see Chapter 3 for more details.

1.3 The electromagnetic spectrum and spectroscopy

Figure 1.2 shows a more complete version of the electromagnetic spectrum range applicable to the study and characterisation of inorganic compounds—from the shortest wavelength gamma rays to radio waves. It also indicates the positions of various spectroscopies described in detail later in this text and the types of atomic and molecular processes associated with them.

TABLE 1.1 Inter conversion factors for electromagnetic radiation energy units (data taken from NIST database)

Unit	eV	cm^{-1}	kJ mol^{-1}	J	Hz	nm*
eV	1	8065.54	96.4853	1.60218×10^{-19}	2.41799×10^{14}	*1239.842*
cm^{-1}	1.23984×10^{-4}	1	1.19627×10^{-2}	1.98645×10^{-23}	2.99792×10^{10}	*10^7*
kJ mol^{-1}	1.03643×10^{-2}	83.5932	1	1.66054×10^{-21}	2.50607×10^{12}	*119627*
J	6.24151×10^{18}	5.03411×10^{22}	6.02214×10^{20}	1	1.50919×10^{33}	*1.98645×10^{-16}*
Hz	4.13567×10^{-15}	3.33564×10^{-11}	3.99031×10^{-13}	6.62607×10^{-34}	1	*2.99792×10^{17}*
nm	*1239.842*	*10^7*	*119627*	*1.98645×10^{-16}*	*2.99792×10^{17}*	*1*

*To convert to and from nm, the initial value needs to be divided into the conversion factor (e.g. 8265.6 eV = 1239.842/0.1500 nm; 500 nm = 10^7/20000 cm^{-1}). For all other units, the conversion factors just need to be used as multipliers to convert the units.

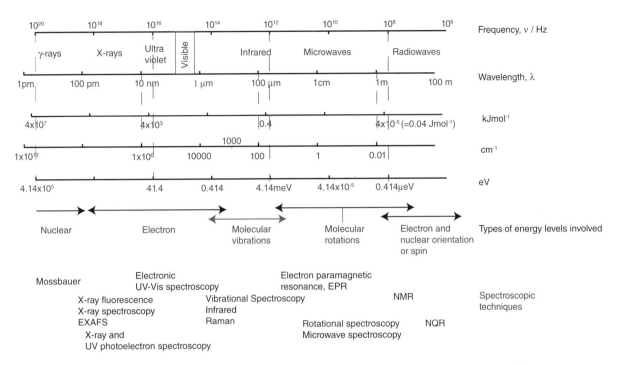

FIGURE 1.2 The electromagnetic spectrum with the main energy units marked and spectroscopic techniques aligned with the appropriate ranges.

1.3.1 The electromagnetic spectrum and transitions between energy levels in atoms

In the various spectroscopic techniques, such as NMR spectroscopy, infrared spectroscopy, and UV-vis spectroscopy, the electromagnetic radiation interacts with the electrons or the atomic nuclei in a compound normally giving up all or some of its energy to the compound. By studying which wavelengths are absorbed and, sometimes, those that are subsequently emitted, every part of a molecule or material can be probed. Thus NMR spectroscopy uses radio waves to excite reorientations of the spins of certain atomic nuclei in a magnetic field and the electromagnetic wavelengths absorbed to produce this reorientation are of relatively low energy. The excitation of an electron generally requires much more energy and spectroscopies associated with this type of process occur in the UV-visible for valence electrons and X-ray region of the spectrum for core electrons.

Some experimental techniques employing electromagnetic radiation in the study of inorganic compounds are not spectroscopic (which implies a 'spectrum' or variation in energy) but rely on elastic scattering where the outgoing electromagnetic radiation is of the same wavelength and energy as the incoming light. Examples include the various diffraction methods, mainly X-ray, where the scattered light from regularly positioned atoms can undergo interference and produce diffraction patterns. Note that the experimental data from such techniques should not be described as a spectrum as it does not involve a change in the energy of the sample, and is usually referred to as either a diffractogram or a diffraction pattern.

1.3.2 The generation of electromagnetic radiation

Table 1.2 summarises the chief sources of the different electromagnetic wavelengths used in spectroscopic techniques. In each case the intensity of the radiation produced coupled with sensitivity of its detection means that the various spectroscopic techniques can typically generate spectra or diffraction patterns in the laboratory in a few seconds to a few hours.

EXAMPLE 1.4

Calculate the wavelength (in nm) of electromagnetic radiation generated in the following electronic transition processes: (a) an electron transferring from a 2p level ($E = 952.3$ eV) to the 1s level ($E = 8979.0$ eV) in a copper atom; (b) a laser emission between two electronic energy levels in a terbium ion separated by $20\,700$ cm^{-1}.

ANSWER

(a) The energy difference between the two levels is 8979.0 – 952.3 eV = 8026.7 eV. We can convert this into nm using the conversion factor in Table 1.1, so 8026.7 eV = 1239.842/8026.7 nm, equivalent to 0.1545 nm or 1.545 Å. This wavelength is in the X-ray region of the electromagnetic spectrum and is generated in an X-ray tube with a copper target and used in X-ray diffraction (Chapter 2).

(b) The conversion factor given in Table 1.1 of 1×10^7 nm/cm^{-1} so 20 700 cm^{-1} is equivalent to $1 \times 10^7/20\,700 = 483$ nm. This is a laser line seen in the emission spectrum of terbium and is in the blue-green part of the visible spectrum.

SELF TEST

In which region of the electromagnetic spectrum is an emission from a neodymium-YAG laser corresponding to a transition between electronic energy levels at 11 502 and 2111 cm^{-1}?

TABLE 1.2 Experimental origins of electromagnetic radiation used in characterisation methods

Radiation	Source
γ-rays	Radioactive nuclei, γ-emitters such as ^{60}Co
X-rays and hard UV	X-ray tube, bombarding metal target with high-energy electrons; synchrotron radiation
Vacuum ultraviolet	Gas discharges (He 21.2 eV, 40.8 eV), Ar, Kr, Xe, synchrotron radiation
UV-vis	Deuterium arc lamp (UV 180–370 nm), tungsten lamp (350–2500 nm)
IR	Heated coiled ceramic filament (Nernst lamp), silicon carbide or nitride, 1–30 μm
Microwaves	Klystron, Gunn diode
Radio waves	Oscillating quartz crystal

1.4 Timescales of characterisation methods

It is useful to consider the 'timescale' of a particular instrumental method as this defines the type of structural information that is extracted from the technique. This is because during the period over which the molecule or compound is sampled by electromagnetic radiation it may undergo changes in geometry due to vibrations or reorientation or indeed in some cases a chemical reaction. Other time dependent factors, such as the lifetime of an excited state formed during its interrogation using electromagnetic radiation, also influence the information extracted. Thus because the periods involved in probing a molecule are different for many of the various experimental methods described in this text we need to define a timescale for each one.

1.4.1 Diffraction

In a typical diffraction experiment using X-rays, the period in which the electromagnetic radiation interacts with an atom that scatters it depends upon the size of the scattering centre. With X-rays the interaction is with the electron cloud and the speed of the X-ray through this is typically a little less than the speed of light in vacuum, as most solids slow down the propagation of light. Given a typical electron cloud dimension of between 1 and 2 Å ($1–2 \times 10^{-10}$ m) this yields an interaction time for an X-ray travelling at 2×10^8 m s^{-1} (two thirds the speed of light in vacuum) of 10^{-18} s. This is much faster than the rate of atomic vibrations, around $10^{13}–10^{14}$ Hz, so the X-ray takes a 'snapshot' of the position of the scattering atom at some fixed point in any vibration. Note that while small changes in atomic positions as a result of vibrations can occur in crystalline solids, molecular rotations can rarely occur in a crystal due to the close proximity of the packed molecules. The diffraction method involves sampling the whole crystal (which will contain typically around 10^{15} molecules) simultaneously using a beam of X-rays and, therefore, all of the instantaneous atomic positions over the duration of the experiment, typically a few hours. Thus atomic positions representing the full range of the vibrations or atomic displacements that occur are obtained (Figure 1.3), and this is built into the structural information that can be extracted from the technique as atomic displacement parameters—see Sections 2.11–2.15 for further discussion.

Recently, with the very high intensity X-ray beams available at synchrotron sources (Section 1.6.5 and Example 1.5), it has become possible to collect X-ray diffraction data very quickly, allowing the structures of

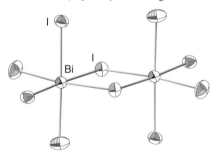

FIGURE 1.3 Representation of the structure of the molecular anion [Bi$_2$I$_{10}$]$^{4-}$ showing atom positions (as defined from the electron density measured in X-ray diffraction) and their associated atomic displacement parameters as ellipsoids; these represent a volume where the atom has a 90% probability of being found.

molecules in electronically excited states, that might last only a few seconds, to be elucidated.

1.4.2 Spectroscopic methods

To determine how X-ray diffraction methods view structures, we only had to consider how long it takes X-rays to interact with the scattering atoms and the length of the experimental data collection. In spectroscopic techniques where electromagnetic energy is typically absorbed (and later re-emitted) the lifetime of the excited state has a profound effect on the structural information extracted and needs to be considered in addition to the electromagnetic wave interaction time and experimental sampling time.

One factor of importance here is the intrinsic **resolution** of the technique. This can be related, in part, to how broad the peaks are in the experimental spectrum compared with the spectral range of the technique. For example, infrared spectra are typically collected over a wavenumber range between 4000 and 400 cm^{-1} (the typical infrared spectral range) and in order to resolve all the different molecular vibrations that might be contributing to a spectrum we would need the individual absorption peaks to be a few wavenumbers wide rather than hundreds of wavenumbers.

Where a molecule absorbs electromagnetic radiation forming an excited state there is a lifetime associated with the excited state before it loses energy and relaxes back to the ground state. This time, denoted the **relaxation time**, is generally, but not universally, short for a large energy difference between the excited and ground state (e.g. with XAS using high energy radiation) and longer where the energy difference is smaller (e.g. with NMR spectroscopy that uses long-wavelength radio waves). With short relaxation times, the Heisenberg uncertainty principle needs to be considered as this tells us that the shorter the time that an excited state exists then the greater the uncertainty in its energy. This relationship is given by the expression

$$\tau \Delta E = h/2\pi \qquad \textbf{Eqn 1.4}$$

where τ is the lifetime of the excited state, ΔE the uncertainty in its energy, and h is Planck's constant. The effect of changing the energy difference between ground state and excited state on the peak widths is also described in Figure 1.4.

Using this expression and employing the relaxation times given in Table 1.3, column 4 allows the uncertainty in the energy of the excited state to be determined and this will define the variability in the absorbed or emitted radiation energy or wavelength; in turn this will then control the width of the line seen in the experimental spectrum (Figure 1.4). Linewidths are also summarised in Table 1.3, column 5. For most experimental techniques this uncertainty in the energy of the excited state has little effect on

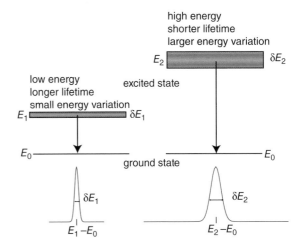

FIGURE 1.4 Energy levels and transitions demonstrating the effect of uncertainty in the energy level of the excited state on peak width and resolution.

the observed experimental spectrum—so in vibrational spectroscopy with an excited state lifetime of ca. 10^{-8} s the typical linewidth is ~10^7 Hz which corresponds to around 5×10^{-4} cm^{-1}, much smaller than the experimental spectral range of 400–4000 cm^{-1}, and also much lower than the widths of the peaks observed in conventional infrared spectra. This also provides a resolution limit, and the highest resolution in specialist commercial instruments is 0.002–0.0035 cm^{-1}. These values should also be compared with the typical measured peaks widths of at least several cm^{-1}. The broader peaks observed experimentally arise because of additional factors that control the vibrational mode transition energy, such as interactions with neighbouring molecules and simultaneous excitation of rotations. Therefore, in infrared spectroscopy relaxation times can be considered as having a negligible effect on the resolution of experimental vibrational spectra.

In solution NMR spectroscopy, the linewidth derived from the uncertainty principle is around 10–0.1 Hz. This can be significant compared to the spectral range for the resonances of some nuclei which are in the range of hundreds to thousands of hertz. For hydrogen, ^{1}H, the value of ~0.1 Hz is relatively small compared with a spectral range of 3000–5000 Hz on modern spectrometers operating at 300–500 MHz. But for other nuclei, and particularly quadrupolar nuclei, which have relaxation times that can be less than a second (1 – 0.1 s in Table 1.3) then resonance widths may be hundreds of hertz and lines so broad they cannot be resolved or even observed above the background.

1.4.3 Introduction to exchange processes

The frequency of the radiation and how long it takes to interact with a molecule are also of importance in controlling the nature of the spectrum obtained. In comparison

TABLE 1.3 Relaxation times for different spectroscopic methods

Spectroscopic technique	Typical radiation and wavelength	Energy of excited state relative to ground state, E_1/Hz	Typical relaxation time/s	Approximate linewidth/Hz
Mössbauer	γ-rays; 10–100 pm	10^{18}	10^{-7}	10^{8}
X-ray absorption, photoelectron spectroscopy	X-rays and hard UV; 0.1–1 nm	10^{17}	10^{-17}	10^{17}
Electronic	UV-vis; 100–1000 nm	10^{16}	10^{-9}	10^{8}
Infrared vibrational	Infrared; 1–100 μm	10^{14}	10^{-6} to 10^{-8}	10^{7}
Rotational	Microwaves; 0.1–10 cm	10^{11}	10^{4}	10^{-4}
NMR and NQR	Radio waves; 1–10 m	10^{8}	10	10^{-1}

with X-ray diffraction, where interaction times are around 10^{-18} s and an almost instantaneous snapshot of the atom positions is obtained (though these are averaged because the experiment samples a large number of atoms over a long experimental period), other techniques use much lower frequency electromagnetic radiation. In NMR spectroscopy different nuclear environments may be separated by just a few J mol^{-1} so the molecule containing the two environments may rearrange very rapidly (particularly at and above room temperature), millions of time per second, that is, with a frequency of MHz. At NMR spectroscopic frequencies which have similar frequency values to the rearrangement process, this can mean that rather than observing the two environments as being distinct an average environment is detected

experimentally. The effect of this **fluxionality** in the molecule is discussed in more detail in Chapter 3 and the example of PF$_5$ described in the next paragraph. A further factor where lifetimes of excited states are long, maybe seconds or longer, is that the molecule can undergo a transformation or reaction in the excited state before the experimental observation of the emitted electromagnetic radiation.

Differences in the electromagnetic radiation frequencies employed by the various spectroscopic techniques that can be used to study a particular compound can result in seemingly contrasting, or even contradictory, information on its structure. One classical example of this is in the study of trigonal bipyramidal AX$_5$ molecules, such as PF$_5$ (Figure 1.5). Techniques which employ relatively high frequency radiation, such as

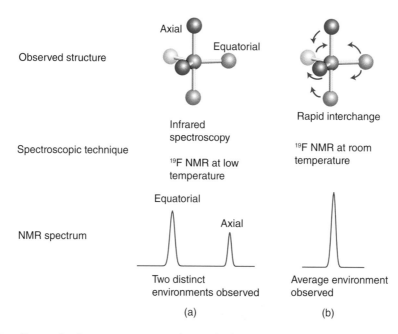

FIGURE 1.5 The effect of intramolecular rearrangements on the signals observed in spectra obtained using two different techniques.

infrared spectroscopy, yield data that distinguish axial and equatorial fluorine atoms as being in different environments. However, on the NMR timescale, with its longer wavelength radiation, the axial and equatorial fluorine atoms exchange positions rapidly so that all fluorine atoms seem to exist in a single type of environment. The rates of exchange between sites depend on temperature and variable temperature NMR techniques can be used to study such phenomena in detail (Section 3.11).

1.5 Fourier transforms in spectroscopy

Normally spectroscopic data from an experimental technique are presented in the form of a spectrum which shows intensity of electromagnetic absorption, transmission or emission as a function of energy—using the energy units described in Section 1.2 applicable to the particular method. Thus infrared spectra are normally presented as percentage transmission on the y axis (abscissa) against wavenumber (on the x axis (ordinate)) between 4000 cm^{-1} and typically 400 cm^{-1} (Figure 1.6). The obvious way of collecting such a spectrum would be to measure the level of radiation absorbed as the energy is scanned across the relevant wavenumber or energy range. This is effectively what is done with UV-vis spectroscopy where light sources that generate a continuous range of energies between 200 and 900 nm are used; a **monochromator** selects specific wavelengths and the detector measures how much light is absorbed at each selected wavelength scanning across the full energy range. Similar wavelength or frequency scanning methods were originally used in all spectroscopic techniques, including infrared spectroscopy and NMR. However, such a method of data collection is relatively inefficient as only a very small proportion of the light

produced by the source is selected by a monochromator at any one time.

A more efficient method of collecting spectroscopic data, now widely used in infrared and NMR spectroscopies, is the **Fourier transform (FT)** technique. This approach involves irradiating the sample under investigation with the full range of energies and discovering which stimulate the molecules under study. Because the excited molecules can emit energy at the same and lower frequencies than those at which they absorb it, the approach is to analyse the complete emission spectrum and find what energies or frequencies are present—this analysis is known as Fourier transform spectroscopy. Fourier transform methods are much faster than scanning methods for collecting experimental spectra as all resonance energies are determined simultaneously rather than one at a time with scanning methods. This technique of analysing the emission spectrum is used in **Fourier transform nuclear magnetic resonance** (FT-NMR, Section 3.3); in **Fourier transform infrared spectroscopy** (FT-IR, Chapter 4) a variant of this analytical method is employed that measures the spectrum produced by an infrared light source with and without the sample present.

One easy way to understand the process is to consider a system which when imparted with energy emits it at only a single frequency. An analogy often used here is a bell. Hitting a bell, that is, giving it energy, results in the emission of sound energy and normally at only a single resonance frequency or note. For a collection of bells, hitting them will result in the simultaneous emission of sound made up of all the resonance frequencies of the different bells combined. Similarly in NMR spectroscopy when a molecule is excited with a pulse of energy it will emit frequencies corresponding to all the possible nuclear magnetic transitions, and their associated energies, in the molecule. Once this emitted spectrum of frequencies is collected it becomes a question of how to deconvolute all the overlapping frequencies present. This is what the Fourier transform method does.

With a single bell or vibration a single frequency is emitted which slowly dies away in intensity and we can represent this emitted radiation as a single decaying sine wave over time (t) (Figure 1.7). The frequency of this sine wave is v_1 so we can fit this emission line with a single frequency sine wave. If we change the representation of the emitted radiation to the frequencies present then we obtain the spectrum in Figure 1.7. This is generally called transforming from a time domain (the sine wave with intensity as a function of time) to the frequency domain, that is, a Fourier transform (FT). This process is named after the French mathematician and physicist, Joseph Fourier, and an FT is just a

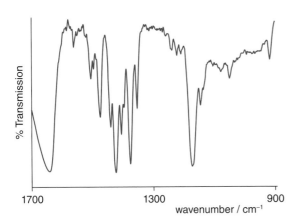

FIGURE 1.6 Part of a typical infrared spectrum of percent transmission versus energy in wavenumber.

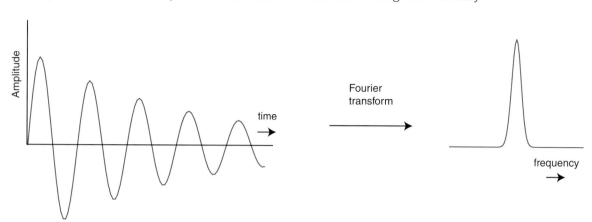

FIGURE 1.7 A decaying sine wave signal and its Fourier transform.

mathematical process to convert any domain to its re-
ciprocal domain.

Now consider the situation where two different bells
are struck simultaneously and produce sine wave sounds
at frequencies v_1 and v_2 with initial amplitudes, I_1 and
I_2, and which both decay slowly. The emitted frequencies and their amplitudes combine together to produce
the more complex wave shown in Figure 1.8. If a Fourier
transform is carried out on this time dependent spectrum
then the two frequencies and their relative intensities can
be extracted.

Typically a molecule has many nuclear magnetic resonance frequencies—maybe tens or hundreds for a complex molecule. However, a Fourier transform can still be
undertaken on the emission spectrum from such a molecule to extract all the component resonance frequencies;
see, for example, Figure 1.9 where four resonances are extracted from a complex decay spectrum. This method of
obtaining all the absorption frequencies in a range of energies is known as Fourier transform spectroscopy and,
as mentioned previously, is the most commonly used
technique in modern NMR spectrometers. Whilst IR
emission spectroscopy can use the above methodology,
the vast majority of IR spectrometers produce absorption

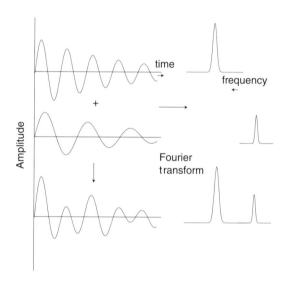

FIGURE 1.8 Fourier transformation of two combined sine
waves into two resonance frequencies.

spectra. However, FT techniques are used to collect the
emission spectrum of the source with and without the
sample, and when these are compared and analysed a
conventional transmission spectrum can be obtained.

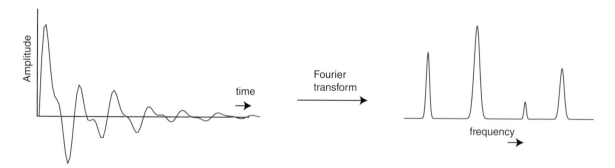

FIGURE 1.9 Fourier transform of a complex wave pattern into four resonance frequencies.

1.6 Experimental considerations—information, experimental times, and sample requirements

Given the wide range of characterisation methods available to the inorganic chemist it can be quite bewildering as to how to choose the appropriate method of analysis. To a large degree this depends on the information required: a simple infrared spectrum will be sufficient to show the presence of a carbonyl group in a molecule, while an NMR spectrum or single crystal X-ray diffraction pattern would give much more information concerning the molecular structure and geometry. Other considerations include how much material is available (micrograms or grams), whether the material needs to be recovered after the experimental technique, and the experiment duration; for example, how stable is the compound, and is the instrument available for minutes or days? Each of the following chapters is dedicated to a particular technique or group of techniques and provides a lot more detail on its capabilities and limitations in characterising an inorganic compound. Here this information is summarised so that comparisons between the different techniques can be made.

1.6.1 Quantities required and experimental times

The sensitivities of most analytical methods have increased many-fold since their invention, which has allowed data to be collected in much shorter times and from much smaller samples. This has, in turn, facilitated investigations of structures and molecular conformations as a function of environment, such as temperature and pressure, as the multiple data sets required can be collected on reasonable timescales. While there is a great variation in instrument capabilities, and they are improving all the time, Table 1.4 summarises the typical experimental times and typical quantities required for the main analytical methods. These are very much generalisations and collecting an NMR spectrum from a low abundance nucleus may take many hours rather than minutes. Some techniques are very quick both in terms of sample preparation and data collection. For example, collecting an infrared spectrum in the range 4000–600 cm^{-1} on a modern Fourier transform infrared spectrometer fitted with an attenuated total reflectance (ATR) system requires just a few milligrams of sample to obtain good statistical quality spectra in 30 s or less. Conversely a single crystal X-ray diffraction experiment would require crystallisation and selection of a suitable crystal and data collection of several hours or perhaps days.

Experimental times also depend very much on the data quality required and factors such as the **signal-to-noise**

and **resolution** requirements will determine how long a data collection will take. A quick qualitative analysis as to whether a characteristic absorption is present in an infrared spectrum or reflection seen in a powder diffraction pattern may take a few seconds or minutes. Conversely, to obtain accurate peak intensities for quantitative analysis or determine whether a weak absorption from a minor component is present or not will require data with a much higher signal-to-noise ratio and, therefore, longer experimental times. In general while the signal increases linearly with data collection time the noise increases with the square root of time: therefore quadrupling data collection times increases the signal-to-noise ratio by a factor of 2. The ability to reduce substantially experimental times by using Fourier transform techniques has been described in Section 1.5. The required resolution of a technique, for example, how broad and well-separated the observed absorptions are in a spectrum or diffraction maxima are in a diffraction pattern, will also strongly impact on the experiment time. While it is very dependent on the specific characterisation method, in many cases instrument data collection parameters can often be adjusted to increase resolution at the expense of longer experimental times.

1.6.2 Sample form and recovery

The form (solid, liquid, solution, or gas) in which a material needs to be presented for a specific characterisation method also needs to be understood. Infrared spectroscopy can be readily undertaken on solids or liquids; UV-vis spectroscopy is normally undertaken using solutions but with appropriate instrumental accessories can be applied to solids. An additional consideration which is often relevant when deciding how to analyse an unknown compound is whether the experimental method is destructive or not. This can be crucial when only a small amount of material is available—perhaps after a long synthesis process or when undertaking forensic analysis of an unknown. This assessment can also be taken in respect of how much material is consumed in a destructive analytical technique. Table 1.4 summarises whether an analytical technique is normally destructive, and whether the compound has to be dissolved in a solvent prior to collecting the instrumental data or whether the analysis can be undertaken on the pure solid, liquid or, very rarely, gaseous compound.

1.6.3 Availability of instrumentation

While the availability and cost of an experimental method may not be a primary consideration in choosing whether to use it to investigate a particular sample it is useful to have some idea of how easy it is to access typical analytical instrumentation. This provides some idea of whether

TABLE 1.4 Sample form, quantities required, and experimental times for the major characterisation methods; whether the sample can be recovered after analysis is also indicated

Technique	Form required (solid or solution) and amount	Destructive or non-destructive	Typical experimental data collection time for routine analysis
Infrared spectroscopy	Solid, powder or suspension or 'mull' in liquid 1–100 mg	Non-destructive if ATR used; if KBr pellet then need to reclaim Difficult to recover suspended solid	10–300 s
UV-vis spectroscopy	0.1 M solution or 100 mg solid	Solid can be recrystallised from solution Non-destructive, but need to reclaim from $BaSO_4$ if diluted	5 min
NMR (solution and solid state)	Normally solution In 1 ml of solvent ^1H 1–10 mg or ^{13}C 10–50 mg or 1 g solid with magic angle spinning (MASNMR)	Solid can be recrystallised from solution Non-destructive	5–100 min
Powder X-ray diffraction	Solid 5–1000 mg	Non-destructive	10–100 min
Single crystal X-ray diffraction	Crystal ~10–100 µm in each dimension	Non-destructive	1–20 h
Thermal analysis—thermogravimetric, differential scanning calorimetry	Solid 10–100 mg	Destructive	30–300 min
X-ray photoelectron spectroscopy	Solid, 10–200 mg Surface area >50 × 50 µm^2 Total sample sizes up to 10 cm	Non-destructive, but challenge to get back after exposure to UHV	1–10 min
Mass spectrometry	10–100 µl of 1 mg/ml solution Solid with ablation	Destructive	1–10 min
Scanning electron microscopy/EDAX	Solid Samples often need coating with conducting layer	Non-destructive Potentially destructive	5–20 min once sample prepared and in microscope
Transmission electron microscopy	Samples need preparation as thin films	Destructive	5–60 min once sample prepared and in microscope

the instrumentation and technique might be available to undergraduates in a teaching laboratory, research groups or only at specialist national or international research facilities. Also the cost, and therefore accessibility, of equipment can vary enormously depending on the quality of data it provides and how quickly experiments can be done. Table 1.5 summarises the availability of instrumentation for each of the characterisation techniques described in this text. Thus IR and UV-vis spectrometers are commonly available in undergraduate teaching laboratories and small research laboratories in industry, while powder and, particularly, single crystal diffractometers and NMR spectrometers are core instrumentation but

normally only at the research level. Instrumentation such as photoelectron spectroscopy, high resolution transmission and scanning electron microscopes are more typically available as central facilities across universities, while very high intensity synchrotron X-ray and neutron sources exist only as national or international facilities.

1.6.4 Choosing a suitable technique

The type and level of information required from a characterisation method varies considerably—this may range from identifying whether a specific functional group is present to the determination of the full molecular

TABLE 1.5 Availability of instrumentation

Technique	Typical availability in a university or specialist research facility
Infrared and Raman spectroscopy	Undergraduate teaching laboratory
UV-vis spectroscopy	Undergraduate teaching laboratory
NMR (solution and solid state (MASNMR))	Low resolution instruments in undergraduate teaching laboratory, higher resolution in departmental research equipment. MASNMR at national research facility
Powder X-ray diffraction	Departmental research equipment; national research facility for rapid and specialised experiments
Single crystal X-ray diffraction	Departmental research equipment; national research facility for rapid and specialised experiments
Neutron diffraction and scattering	National research facility
Thermal analysis—thermogravimetric, differential scanning calorimetry	Departmental research equipment
X-ray photoelectron spectroscopy	Departmental or national facility research equipment
XANES/EXAFS	National research facility
Mass spectrometry	Departmental research equipment
Scanning electron microscopy	University or national research facility
Transmission electron microscopy	University or national research facility

structure and accurate bond lengths. Generally, an increasing level of information comes at the price of longer experiments (Table 1.4), more costly instrumentation (Table 1.5), and more skilled chemists to interpret the data obtained. Tables 1.6, 1.7, and 1.8 summarise the types of characterisation method an inorganic chemist may wish to undertake to determine the structure of a molecule or material and some indication of the type and level of information obtained.

1.6.4.1 Spectroscopic methods

Table 1.6 summarises the main spectroscopic techniques used for the analysis of inorganic compounds. Each of these spectroscopic techniques uses a different electromagnetic wavelength to produce a spectrum (i.e. a range of energies) of absorption or emission probabilities or intensities. A typical spectroscopic experimental data set therefore has the form shown in Figure 1.6 with a series of absorptions or resonances across the range of energies being probed.

1.6.4.2 Chemical, compositional, and physical property analysis methods

Table 1.7 summarises the main techniques used to determine the composition of a compound and these fall into two main types. The first group, such as mass

spectrometry, CHN (carbon hydrogen nitrogen) analysis, and thermogravimetric analysis, involve breaking up the compound and undertaking a mass analysis (either at the atomic or bulk compound scale) of the atoms, ions, and molecules produced. The second group, such as atomic absorption spectroscopy, X-ray photoelectron spectroscopy, and energy dispersive analysis of the X-rays produced in an electron microscope, use the absorption or emission of characteristic radiation by the elements present. The analysis of this characteristic radiation can be qualitative to determine the elemental composition or quantitative by analysis of the intensity of the radiation absorbed or emitted.

1.6.4.3 Diffraction and imaging methods

Table 1.8 summarises the major techniques for structure determination and for the determination of the particle shape and morphology. Because of the need for a level of atomic order in the compound they are mainly applicable to solids and in particular crystalline solids, which have long range ordering of atoms and molecules.

1.6.5 **Research facilities and instrumentation**

The vast majority of the characterisation techniques covered in this text are available using instrumentation found in a typical undergraduate teaching laboratory or

TABLE 1.6 Spectroscopic techniques and derived information

Technique (acronyms)	Energy levels involved and electromagnetic wavelength where appropriate	Outline of derived information
Mössbauer spectroscopy	Transitions between nuclear energy states, γ-rays $1-100 \times 10^{-12}$ m. ^{57}Fe uses 14.4 keV, 0.086 nm	Oxidation state, spin state (if iron) and the chemical, electronic, and magnetic environment of certain nuclei, especially iron
X-ray fluorescence (XRF) and X-ray absorption (XAS, XANES, EXAFS) and emission (EDX) spectroscopies	Transitions from and between core electron energy levels, X-rays, $10^{-11}-10^{-9}$ m	Chemical analysis. Oxidation state, local chemical environment, nanoparticle size
Photoelectron spectroscopy (PES): X-ray (XPS) and ultraviolet (UPS), Auger spectroscopy	Ionisation and transitions from and between core and near-core electron energy levels. 10^{-8} ('soft' X-ray)–10^{-7} ('vacuum' UV) m	Oxidation state, electronic structure and distribution; local environment; nature of occupied energy levels (bonding etc.)
UV-vis spectroscopy, near-infrared spectroscopy (NIR)	Transitions between electronic energy levels 200–1500 nm	Electronic structure, energy levels and symmetries
Infrared spectroscopy (IR) Raman spectroscopy Surface-enhanced Raman spectroscopy (SERS)	Changes in vibrational energy in various molecular modes—bond stretches and bends. Directly measured at 1–25 μm (10 000–400 cm^{-1}) in electromagnetic spectrum (IR) or excited and measured using frequency shift of high intensity visible light (Raman). In SERS molecules adsorbed on surfaces given enhanced signal and sensitivity	Fingerprinting of functional groups, estimation of bond strengths; determination of molecular symmetry and shape
Far infrared spectroscopy/terahertz spectroscopy	Changes in vibrational energy in various molecular modes. 25–1000 μm (400–10 cm^{-1}, 10–0.3 THz)	Low energy modes in molecules, for example, torsional and librational modes, vibrations involving heavy atoms
Microwave spectroscopy Rotational spectroscopy	Molecular rotations excited through absorption of electromagnetic radiation. 100 μm–10 mm (100–1 cm^{-1}, 3000–30 GHz)	Rotational energy levels, shapes of simple molecules
Electron spin (paramagnetic) resonance spectroscopy (ESR/EPR)	Excitation between electron spin energy levels in species with unpaired electrons 3.3 cm; 8.3 mm (X-band, 9 GHz; Q-band, 36 GHz)	Chemical environment of species in solids and molecular species trapped in frozen liquids/solutions
Nuclear magnetic resonance (NMR) Magic angle spinning NMR (MASNMR)	Excitation and reorientation of nuclear spins, (300 MHz, 1 m; 500 MHz, 60 cm)	Comprehensive and detailed information on chemical environment for NMR active nuclei. Molecular structures
Nuclear quadrupole resonance (NQR)	Excitation of energy levels associated with the reorientation of a quadrupolar nucleus in an electric field; 150 cm–30 m	Chemical environment in solids
Inelastic neutron scattering (INS)	Inelastic scattering of neutrons from solids or liquids	Low energy separation electronic energy levels, vibrational and rotational modes. Very low energy modes. Tunnelling and translation energies

as Chemistry Department equipment. These methods generally employ sources of electromagnetic radiation that are compatible with the space available and finances of a teaching university. At the research project level there is the demand for greater sensitivity, faster experiments, and smaller samples, which comes with more powerful radiation sources, and also for specialist radiation such as neutron beams. These specialist, often extremely intense, electromagnetic radiation and particle (neutron and electron) sources have been built as dedicated research facilities catering for the needs of a large international community of scientists in characterising their different compounds. This section introduces the main research facility types and their applications.

TABLE 1.7 Chemical and physical property analysis techniques

Technique (acronyms)	Basis of technique	Outline of derived information
Mass spectrometry	Separation of ions according to mass/charge ratio	Molecular mass from parent ion, and chemical configuration from fragmentation patterns
CHN (carbon hydrogen nitrogen) analysis	Compound pyrolysed in oxidising atmosphere, measurement of amounts of CO_2, H_2O, and N_2 produced	Atomic percentages of C, H, and N in compound
Atomic absorption and emission spectroscopies (AAS, AES). Flame emission spectroscopy	Measurement of characteristic atomic absorption or emission lines in volatilised sample. Strength of absorption proportional to amount present	Quantitative percentage of chemical elements in a compound down to ppb or ppt levels if used in ICP-AES or ICP-MS mode
Scanning electron microscopy (SEM) plus analysis of emitted electromagnetic radiation (EDX/EDAX/EDS)	Excitation of electrons in material leads to the emission of electromagnetic radiation	Chemical analysis for elements (heavier than boron)
Thermogravimetric analysis (TGA)	Sample heated on balance and weight loss as a function of temperature measured	Decomposition temperature, water content
Differential thermal analysis (DTA) Differential scanning calorimetry (DSC)	Sample heated on balance and temperature of sample relative to an inert standard measured. Sample heated on balance and heat needed to keep sample temperature the same as an inert standard measured	Melting point, phase changes, decomposition temperature Quantitative information on heat changes associated with melting and phase transitions
Dilatometry	Measure volume changes in a sample as a function of temperature	Expansion coefficients of solids, powders, pastes and liquids

TABLE 1.8 Diffraction and imaging methods

Technique (acronym)	Basis of technique	Outline of derived information
Single-crystal X-ray diffraction (SXD)	Elastic scattering of X-rays by single crystals to give a diffraction pattern with measured intensities	Complete structure determination; accurate and precise bond lengths
Powder X-ray diffraction (PXD)	Elastic scattering of X-rays by microcrystalline powders to give a diffraction pattern with measured intensities	Fingerprinting solids, phase identification of unknowns, basic crystallographic information, for example, lattice parameters, structure refinement
Single-crystal neutron diffraction (SND) Neutron powder diffraction (NPD)	As SXD and PXD but using neutrons (with similar wavelengths) instead of X-rays	As SXD and PXD but with ability to study light atoms positions and distinguish isoelectronic species, magnetic structure
Pair distribution function (PDF) analysis	Elastic scattering of X-rays or neutrons from non-crystalline or partially crystalline solids and from liquids and solutions	Distances between atoms in amorphous and low crystallinity solids and liquids—crystallinity length scale
Electron diffraction—gas phase	Beam of single energy electrons scattered by gas phase molecules	Interatomic distances in simple molecules
Scanning electron microscopy (SEM)	Optical imaging using electron beams for high magnification	Particle morphology and measurement in the range 100 nm–10 µm
Electron diffraction from crystals	Diffraction patterns formed by elastically back-scattered electrons in electron microscope	Lattice parameters and cell symmetry information

(Continued)

TABLE 1.8 Diffraction and imaging methods (*Continued*)

Technique (acronym)	Basis of technique	Outline of derived information
Transmission electron microscopy (TEM)	Imaging of a solid by analysis of a high energy electron beam that has passed through a very thin specimen	Imaging of nanoparticles, heavy atom resolution images of crystalline solids, defects in crystalline solids, electron diffraction patterns and structure refinement
Optical microscopy, polarised light, heated stage	Visible light microscope using polarising filters	Crystal shape, crystal perfection and domains in solids and liquid crystals, observation of phase changes

1.6.5.1 Synchrotrons and synchrotron radiation

Accelerating electrons moving near the speed of light emit electromagnetic radiation so electrons guided around a circular path using a series of magnets emit intense light in their direction of travel (Figure 1.10a). A synchrotron is an electron storage ring which constantly accelerates them by changing their direction of travel and, therefore, velocity as they move around the ring. Modern synchrotrons, known as third generation sources, have circumferences of over 500 m and with the electron current used (300 mA) generate light 10 billion times brighter than the

Sun's rays reaching the Earth. This light is produced right across the electromagnetic spectrum (Figure 1.10b) but typically is brightest at X-ray and harder UV wavelengths (0.1–100 nm). The advantages of using synchrotron radiation to characterise inorganic compounds and materials derive from its high brilliance, allowing very small samples to be studied or experiments to be undertaken in much shorter times. This in turn allows high throughput experiments where numerous samples, or many small areas within a single sample, can be characterised rapidly. More complex structures also become more amenable to study so that synchrotron radiation is widely used to

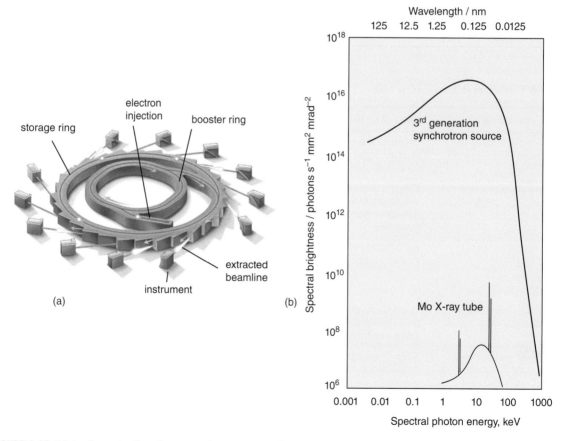

FIGURE 1.10 (a) A schematic of synchrotron radiation source. (b) Comparison of the brilliance of X-ray wavelength generation of a modern synchrotron source and a laboratory X-ray tube.

study bioinorganic compounds including some enzymes and proteins. The sample environment can be varied allowing studies as a function of a large temperature range or pressure within reasonable timescales of a few hours. The wide range of energies produced by a synchrotron, when combined with the use of monochromators, allows the variation of the radiation wavelength incident on a sample and thereby UV-vis and X-ray spectroscopies. Analytical techniques that use X-rays and short-wavelength UV light include powder and single-crystal X-ray diffraction (Chapter 2) and X-ray absorption spectroscopies (Chapter 6).

EXAMPLE 1.5

The brightness of a typical monochromated (single wavelength) laboratory X-ray source is 10^{11} photons s^{-1} mm^{-2} mrad^{-2} while a modern synchrotron source produces X-rays with a brightness of 10^{16} photons s^{-1} mm^{-2} mrad^{-2}. For a crystal of dimensions $100 \times 100 \times 100$ μm^3 the collection of a complete X-ray diffraction pattern on a laboratory instrument took 12 hours. Estimate how long a data collection would take on a synchrotron source.

ANSWER

The ratio of brightness is $10^{16}/10^{11} = 10^5$ so assuming data collection and read out take the same time then $12 \times 60 \times 60$ s$/10^5 = 0.4$ s! In practice experimental times would be considerably longer due to the time taken to read out the data from the detectors.

SELF TEST

If 2 hours of synchrotron X-ray beam time was available what size crystal could be studied assuming zero read out time from the detectors?

1.6.5.2 Laser facilities

Lasers are widely used in laboratory analytical instrumentation as an intense source of electromagnetic radiation, for example, in Raman spectroscopy (Chapter 4). Laser radiation can also be pulsed producing short bursts of very intense light that are particularly useful for time resolved studies of molecular behaviour and of short lived excited states. For example, rather than just studying the ground electronic state of a material it becomes possible with intense laser light to excite sufficient electrons into an excited state. The material's physical and electronic structure, in this short lived excited state, can be probed by techniques such as X-ray diffraction or electronic spectroscopy.

1.6.5.3 High resolution electron microscopy

Low to moderate power and resolution electron microscopes are inexpensive enough to form core instrumentation within university and industrial research departments. However, very high power microscopes, such as scanning transmission microscopes (STMs) which are able to image individual atoms, cost in excess of £1M and need specialist housing such as very low vibration, low extraneous magnetic field, and temperature and humidity controlled buildings.

1.6.5.4 Neutrons

Neutrons are a very useful probe of structure and dynamics in inorganic compounds. Diffraction of neutrons provides information not available from X-ray diffraction techniques (Sections 2.16.1 and 2.17.1) and inelastic scattering of neutrons is an important technique in the study of atomic and molecular vibrational and rotational motion. However, laboratory based neutron sources are extremely weak. Hence, to undertake neutron scattering experiments of inorganic compounds and materials a specialist facility is required. Such purpose-built neutron scattering research facilities have been built since the 1970s and are now an important component in the portfolio of analytical methods available for the more advanced investigation of inorganic compounds. Two methods of generating high flux neutron beams are available, nuclear reactors and spallation sources, though most recent developments have focused on spallation sources.

EXAMPLE 1.6

Neutron energies and wavelengths. Calculate the wavelength associated with a neutron travelling at 1000 m s^{-1}.

ANSWER

The energies of neutrons used for both diffraction and spectroscopy experiments cannot be calculated using the expressions used in Section 1.2 for electromagnetic radiation as they have mass. Instead the energy, and associated wavelength, of a neutron depends on its velocity and can be calculated from the de Broglie relationship:

$$\lambda = \frac{h}{p} = \frac{h}{mv}$$

where λ is the neutron wavelength, p is its momentum, m is its mass, $1.6749286 \times 10^{-27}$ kg, v its velocity, and h is

Planck's constant. Thus for a neutron travelling at 1000 m s^{-1} we have

$$\lambda = 6.626 \times 10^{-34} \text{ J s}/(1.675 \times 10^{-27} \text{ kg} \times 1000 \text{ m s}^{-1})$$

$$\lambda = 3.956 \times 10^{-10} \text{ m (395.6 pm)}$$

This wavelength corresponds to those of X-rays in the electromagnetic spectrum and can be used in diffraction experiments.

SELF TEST

Calculate the kinetic energy of a neutron moving at 1000 m s^{-1}. What type of molecular energies (electronic, vibrational or rotational) does this correspond to?

1.6.5.4.1 Nuclear reactor-based neutron sources

The fission of a uranium nucleus following impact by a single neutron produces on average three neutrons. While the majority of nuclear reactors have been designed to produce electrical power it is possible to design and build a research nuclear reactor to produce intense neutron beams. The neutrons are produced continuously and the energies required for a specific experimental technique can be obtained and selected by a combination of moderators (that slow the neutrons down) and monochromators (that select specific energies or wavelengths).

1.6.5.4.2 Spallation neutron sources

In a spallation neutron source a neutron beam is produced by directing a beam of fast-moving protons at a heavy metal target, typically tantalum, tungsten or uranium. The impact of the proton beam chips ('spallation') neutrons from the metal nuclei. The proton beam is produced in a synchrotron similar to that used to accelerate electrons (Section 1.6.5.1) though in this case packets of hydride (H$^-$) ions are accelerated before being stripped of electrons to produce protons. Because the proton beam is pulsed, typically at 50 Hz, the neutrons are also generated in pulses. While the individual packets of neutrons produced by a spallation neutron source are very bright they are only produced for a short time. However, unlike reactor based sources, which lose a lot

of neutrons through processes such as monochromation, a large portion of the neutron pulse can be used in an experiment making the timescale of the experiments at the two sources similar.

1.7 **Layout of the book**

The layout of this book reflects the way in which many university courses teach inorganic chemistry. The four chief techniques used by chemists to characterise a new compound or material are NMR spectroscopy, infrared spectroscopy, UV-vis spectroscopy, and X-ray diffraction. For each of these main techniques, a major chapter has been allocated in this text and divided into two: the first half describing the basis of the method and analysis of data from relatively simple compounds, the latter half involving more complex data collection and analysis strategies. The early parts of each chapter generally cover material and data analysis which might be covered in the first two years of a university level course while the later sections would cover material typically from advanced topics in years 3 and 4 and in Master's level courses.

The later chapters of the book cover more specialised techniques which have more limited applications in analysing inorganic compounds. For example, they can only be applied to certain types of compound or material, or they provide only limited information on a compound's structure. These later chapters also cover chemical analysis techniques such as elemental analysis and thermal analysis methods where the analysis of the data is relatively straightforward and does not need the underlying theory required and taught for the main four techniques.

While each of these topics is covered in an individual chapter it should be noted that in the analysis of an unknown many of these methods will be used together to provide a full understanding of the structure. Thus a molecular compound would typically be characterised using NMR (and for all NMR active nuclei present), single crystal X-ray diffraction, and infrared spectroscopy as well as chemical analysis, for example, CHN analysis. A new solid material may well be characterised by powder X-ray diffraction, infrared spectroscopy, UV-vis spectroscopy, as well as various thermal analysis methods.

Problems

1.1 The wavelength of the main line in the sodium atomic spectrum line is 589 nm. What are the frequency and the wavenumber for this line? What is the energy of one photon of this wavelength?

1.2

  (a) A laser emits light with a frequency of 4.12×10^{14} s^{-1}. Calculate the energy of one photon.

  (b) If the laser emits a pulse containing 2.0×10^{18} photons of this radiation, what is the total energy of that pulse?

  (c) If the laser emits 1.3×10^{-2} J of energy during a pulse, how many photons are emitted?

1.3 What is the wavelength of an electron moving with a speed of 5.97×10^6 m s^{-1}? ($m_e = 9.11 \times 10^{-31}$ kg)

1.4 The structure of $[Cu(OH_2)_6](BrO_3)_2$ determined using single X-ray diffraction at room temperature shows the copper atom to have a regular copper environment with six equal Cu–O bond lengths while a study using X-ray absorption spectroscopy shows a distorted environment with two long and four shorter Cu–O distances. Describe a possible reason for these observations.

1.5 What experimental characterisation methods could be used to investigate the following:

  (a) Non-destructive identification of a small amount of an unknown metal oxide in a forensic investigation?

  (b) Determining whether N–H or N–D bonds vibrate at a higher frequencies (wavenumbers)?

  (c) Whether a sample is dimethylsilane $((CH_3)_2SiH_2)$ or ethylsilane $(C_2H_5SiH_3)$.

Bibliography

A.D. Burrows, J. Holman, A. Parsons, G. Pilling, and G. J. Price. (2013) *Chemistry³*. 2nd ed. Oxford: Oxford University Press.

M.T Weller, T. L. Overton, J. A. Rourke, and F.A. Armstrong. (2014) *Inorganic Chemistry*. 6th ed. Oxford: Oxford University Press.

D.W.H. Rankin, N.W. Mitzel, and C.A. Morrison. (2013) *Structural Methods in Molecular Inorganic Chemistry*. Chichester, UK: Wiley.

J. Kenkel. (2013) *Analytical Chemistry for Technicians*. 4th ed. Boca Raton, FL: CRC Press.

G.M. Lampman, J.A. Vyvyan, G.S. Kriz, and L. Pavia. (2012) *Introduction to Spectroscopy*. 5th ed. Cengage India.

J.M. Hollas. (2004) *Modern Spectroscopy*. 4th ed. Chichester, UK: Wiley.

2 Diffraction methods and crystallography

FUNDAMENTALS: BASIC DIFFRACTION METHODS AND CRYSTALLOGRAPHY

2.1 Introduction to diffraction

Diffraction techniques, particularly those using X-rays, are the most important structure determination methods available to the chemist. X-ray diffraction has been responsible for ascertaining the structures of over two million different crystalline compounds. Diffraction methods provide a complete description of the structure of a crystalline compound, including the positions of atoms, ions and molecules and the bond lengths and bond angles between them. This derived structural information can be interpreted in terms of sizes of atoms and how they bond and pack together in a solid, and its analysis allows chemists to explain many of the physical properties of a compound or material. X-ray diffraction methods may also be used to investigate the purity, particle size, and composition of powders (polycrystalline materials made up of many very small crystals) making them useful tools in quality control and forensic applications.

Diffraction methods involve the study of crystalline solids, which have a long range, regular arrangement of atoms. These techniques are not generally applicable to amorphous solids, glasses, liquids, or solutions; though even for amorphous materials, where there is some local ordering in the arrangement of the atoms, some structural information can be extracted. Diffraction is normally a non-destructive analytical method, in that the sample remains unchanged after the data collection and it may be analysed subsequently by using a different technique.

Diffraction occurs when waves are scattered by an object in their path, giving rise to interference. X-rays are scattered elastically (with no change in energy) by the electrons in atoms, and diffraction can occur for a periodic array of scattering centres separated by distances similar to the wavelength of the radiation (about 100 pm, 1 Å), as exist in a crystal. Thus, an X-ray beam impinging on a crystalline compound with an ordered arrangement of atoms will produce a set of diffraction maxima, termed a **diffraction pattern** (Figure 2.1a). Other types of radiation or particles, including electrons or neutrons, moving with velocities such that their associated wavelengths, λ (from the de Broglie relationship, $\lambda = h/mv$), are similar to interatomic separations, can also be diffracted in a similar way to X-rays.

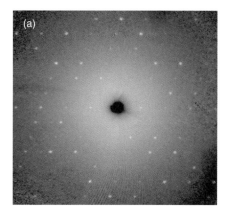

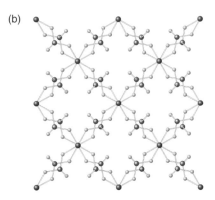

FIGURE 2.1 (a) An X-ray diffraction pattern obtained by the scattering of X-rays from a single crystal, diffraction maxima are seen as the white spots on an image plate: each spot originates from the X-rays scattering from regularly spaced atoms, as in the crystalline material structure shown in (b).

The intensity and position of a diffraction maximum in a diffraction pattern depends on the exact atomic structure, and this is defined by the positions and types of the atoms in a crystal. The collection and analysis of X-ray diffraction data allows crystallographers to determine how the atoms had to be arranged in order to produce the diffraction pattern (Figure 2.1b).

There are two principal X-ray diffraction techniques: the **powder method**, in which the material being studied is in a polycrystalline or powdered form, and **single-crystal diffraction**, where the sample is a single crystal with dimensions of several tens of micrometres or larger. Before the diffraction data obtained from each of these techniques can be interpreted some of the basic concepts in crystallography need to be introduced.

2.2 Unit cells, lattices, and the description of crystal structures

A **unit cell** of the crystal is an imaginary parallel-sided region (a 'parallelepiped') from which the entire crystal can be built up by purely translational displacements;[1] unit cells so generated fit perfectly together with no space excluded. The angles (α, β, γ) and lengths (a, b, c) used to define the size and shape of a unit cell are the **unit**

---

[1] A translation exists where it is possible to move an original figure or motif in a defined direction by a certain distance to produce an exact image. In this case a unit cell reproduces itself exactly by translating it parallel to a unit cell edge by the unit cell parameter.

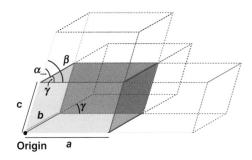

FIGURE 2.2 The unit cell parameters, angles (α, β, γ) and lengths (a, b, c), defining the size and shape of a unit cell.

cell parameters (the '**lattice parameters**'); the angle between a and b is denoted γ, that between b and c is α, and that between a and c is β (Figure 2.2). Unit cells may be chosen in a variety of ways but it is generally preferable to choose the smallest cell that exhibits the greatest symmetry.

The relationship between the lattice parameters in three dimensions as a result of the symmetry of the structure gives rise to the seven **crystal systems** (Table 2.1 and Figure 2.3). The symmetry present in the different crystal systems ranges from the highest symmetry cubic, with all the lattice parameter lengths identical and the angles all 90°, to triclinic which has no relationship between the lengths nor angles. All the ordered structures adopted by compounds belong to one of these crystal systems. However, most of those described in this chapter, which deals with simple compositions and stoichiometries, belong to the higher symmetry cubic, tetragonal, and hexagonal systems.

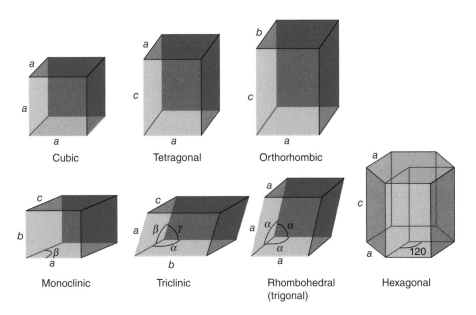

FIGURE 2.3 The seven crystal systems showing relationships between lattice parameters.

TABLE 2.1 The crystal systems

Unit cell dimensions	Crystal system	Example compounds
$a = b = c,\ \alpha = \beta = \gamma = 90°$	Cubic	NaCl, CsCl, MgAl$_2$O$_4$, C$_{60}$K$_3$
$a = b \neq c,\ \alpha = \beta = \gamma = 90°$	Tetragonal	K$_2$NiF$_4$, TiO$_2$, BaTiO$_3$ (298 K)
$a \neq b \neq c,\ \alpha = \beta = \gamma = 90°$	Orthorhombic	YBa$_2$Cu$_3$O$_7$
$a \neq b \neq c,\ \alpha = \gamma = 90°\ \beta \neq 90°$	Monoclinic	KH$_2$PO$_4$
$a \neq b \neq c,\ \alpha \neq \beta \neq \gamma \neq 90°$	Triclinic	Many molecular compounds
$a = b \neq c,\ \alpha = \beta = 90°\ \gamma = 120°$	Hexagonal	LiNbO$_3$
$a = b = c,\ \alpha = \beta = \gamma \neq 90°$	Trigonal/rhombohedral	BaTiO$_3$ below −80°C

EXAMPLE 2.1

Use Figure 2.4 to identify the crystal systems of the structures of (a) CsI and (b) α-Hg.

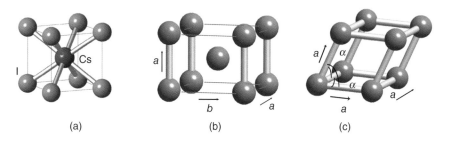

(a)　　　　　(b)　　　　　(c)

FIGURE 2.4 Structures of (a) CsI, (b) α-Hg, and (c) β-Hg.

ANSWER

By looking for the relationships between unit cell dimensions and angles it is possible to assign the correct crystal system to a structure. Thus a structure which propagates or has identical translations along three directions at right angles to each other (orthogonal directions) will be cubic. In the structure of

(a) CsI the unit cell repeat along the three orthogonal directions is the same distance so CsI has a cubic unit cell. In the structure of (b) α-Hg the repeat directions are all at right angles but while two of the repeats are identical in length the third is different so the unit cell here has $a = b \neq c,\ \alpha = \beta = \gamma = 90°$ and is tetragonal.

SELF TEST

Identify the crystal system of β-Hg shown in Figure 2.4(c).

A **lattice** is defined as an array of equivalent points, **lattice points**, in one, two, or, more normally for inorganic compounds, three dimensions. The lattice provides no information on the actual positions of atoms or molecules in space but shows the translational symmetry of the material by locating equivalent positions. The environment of an atom placed on any one of these lattice points would be identical to that placed on any other lattice point. The simplest illustration of this is a

one-dimensional lattice consisting of an infinite series of equally spaced points along a line (Figure 2.5a). The distribution of other lattice points about any randomly selected lattice point is identical. A structure may be built up by associating an atom with a lattice point, and, thereby, each and every lattice point, as in the chain of sodium atoms in Figure 2.5(b). In two dimensions a lattice consists of a grid of equivalent points, as the example in Figure 2.5(c).

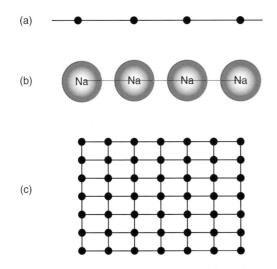

FIGURE 2.5 (a) A one-dimensional lattice of equidistant points. (b) A one-dimensional structure, a chain of sodium atoms, formed by associating a sodium atom with each lattice point of (a). (c) A two-dimensional lattice.

In three dimensions four lattice types are found. The simplest lattice type is known as **primitive**, given the symbol **P**, and a unit cell with a primitive lattice contains a single lattice point. For a primitive lattice the only purely translational symmetry is that of the unit cell. Figures 2.6(a) and 2.6(b) show examples of primitive lattices; lattice points are normally placed at the corners of the unit cell parallelepiped but a single lattice point inside the unit cell is an alternative description Figure 2.6(c).

For the remaining three lattice types, as well as the translational symmetry of the unit cell, there is additional translational symmetry within the unit cell. A second lattice type is body centred, which is given the symbol **I**, and an example of a body centred unit cell is shown in Figure 2.7. This shows lattice points at the cell corners and, for body centring, the additional lattice point is at the cell centre with fractional coordinates (½, ½, ½). Note that the coordinate system (x, y, z) corresponds to the unit cell parameters a, b, c

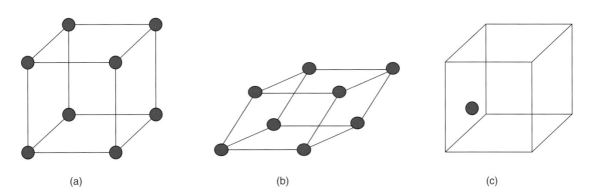

FIGURE 2.6 Two primitive lattices from different crystal systems: (a) cubic and (b) triclinic; (c) is also a primitive (cubic) lattice but the chosen lattice point is not at the unit cell origin.

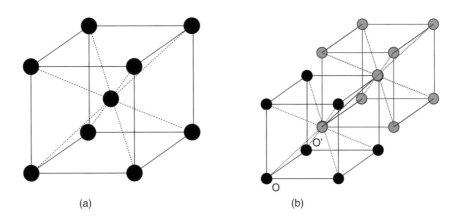

FIGURE 2.7 (a) A body centred (**I**) unit cell. (b) The lattice point at the cell centre in (a) is in an identical environment to those on the corners; this can be seen if the origin of the lattice is repositioned from O to O′.

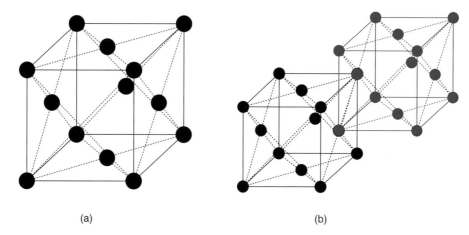

(a)   (b)

FIGURE 2.8 (a) A face centred (*F*) unit cell. (b) The lattice point at the face centre in (a) is in an identical environment to those on the corners if the origin of the lattice is repositioned.

so the position (½, ½, ½) is halfway along each of the *a*, *b*, and *c* lattice parameter directions. This means that if an atom is placed on a general position within a body centred unit cell with fractional coordinates (*x*, *y*, *z*) the lattice will generate a second atom at the position with fractional coordinates (*x* + ½, *y* + ½, *z* + ½). This atom at (*x* + ½, *y* + ½, *z* + ½) will have surroundings identical to those of the atom at (*x*, *y*, *z*). Further atoms introduced into this lattice type at (*x*′, *y*′, *z*′) will have an equivalent atom at (*x*′ + ½, *y*′ + ½, *z*′ + ½), thereby, building up the complete crystal structure with the additional translation symmetry.

A three-dimensional lattice which has lattice points at the centre of all the unit cell faces, as well as at the corners, is known as face centred and given the symbol *F* (Figure 2.8). The additional translational symmetry in this case consists of the three elements +(½, ½, 0), +(0, ½, ½), and +(½, 0, ½), hence for an atom on a general site (*x*, *y*, *z*) three additional identical atoms will be generated within the unit cell with the coordinates (*x* + ½, *y* + ½, *z*), (*x*, *y* + ½, *z* + ½), and (*x* + ½, *y*, *z* + ½).

Finally a lattice which has points in just *one* of the faces is also known as face-centred but given the symbol *A*, *B*, or *C* (Figure 2.9). A *C*-type lattice refers to the case where the additional translational symmetry places lattice points at the centres of the faces delineated by the *a* and *b* directions as well as at the origin. In fractional coordinate terms, in addition to the general site (*x*, *y*, *z*) for a *C* centred lattice, a second site is generated at (*x* + ½, *y* + ½, *z*). The *A* and *B* face-centred lattices are obtained

in an identical manner but the additional lattice points occur in the *bc* and *ac* planes, respectively. However, the *A* and *B* descriptions are not normally used, as redefinition of the *a*, *b*, and *c* directions will produce the *C* centred description.

The lattice types describe the pure translational symmetry of the structure and show equivalent positions within unit cells related by translational symmetry. It should be noted that the lattice type does not provide any information on the total number of atoms or molecules in the unit cell. In the case of a material crystallising with a primitive cubic lattice there may be more than one atom or molecule of a particular kind within the unit cell, but their positions in terms of their environments will be quite different.

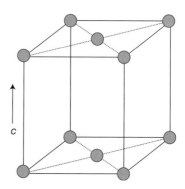

FIGURE 2.9 A cell centred in only one face, usually designated as *C* centred.

EXAMPLE 2.2

Identify the lattice type of the structures shown in Figures 2.10(a), (b), and (c).

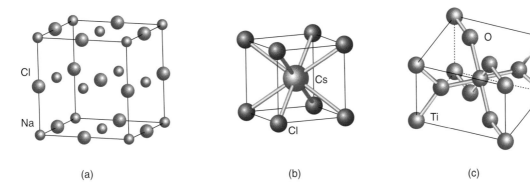

(a)   (b)   (c)

FIGURE 2.10 The structures of (a) NaCl, (b) CsCl, and (c) TiO_2 (rutile); cations are blue and anions are black in each structure.

ANSWER

By inspecting a unit cell of a crystal structure all the translational symmetry present can be identified and the presence of additional lattice points determined. This allows the lattice type of the unit cell to be assigned.

(a) The unit cell of NaCl is shown with sodium ions at the corners of the unit cell and these ions also occur at the centres of each face; chloride ions are at sites halfway along the cell edges and at the cell centre. Thus the additional translational elements, present for both ion types, are $(x + \frac{1}{2}, y + \frac{1}{2}, z)$, $(x, y + \frac{1}{2}, z + \frac{1}{2})$, and $(x + \frac{1}{2}, y, z + \frac{1}{2})$ so the lattice type is face-centred, **F**. Note that the structure of NaCl is often described as 'face-centred cubic' denoting the lattice type and crystal system of this compound.

(b) The unit cell of CsCl only has the translational symmetry of the unit cell. The ion at the centre of the unit cell is different from the one on the corner so there is no $(+\frac{1}{2}, +\frac{1}{2}, +\frac{1}{2})$ translational symmetry and the lattice type is not **I**. Therefore, the lattice type is primitive, **P**.

(c) The TiO_2, rutile, unit cell is tetragonal and consideration of the Ti ions shows that for these ions there is $(+\frac{1}{2}, +\frac{1}{2}, +\frac{1}{2})$ symmetry with ions on the corners and in the centre of the unit cell. However, consideration of the oxide ions shows that applying the $(+\frac{1}{2}, +\frac{1}{2}, +\frac{1}{2})$ translational

symmetry element to these does not arrive at another oxide ion in an identical environment (see Figure 2.11). Remembering that the lattice type's translational symmetry must apply to the whole structure and all the atom sites present means that the only translational symmetry remaining for TiO_2 is that of the unit cell so the lattice type is primitive, **P**.

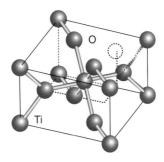

FIGURE 2.11 The TiO_2, rutile, unit cell is tetragonal and the lattice type is primitive. Translational symmetry of $(+\frac{1}{2}, +\frac{1}{2}, +\frac{1}{2})$ exists for the Ti ions but is absent for the oxide ions as shown by the vacant site at the position shown by the broken circle.

SELF TEST

Identify the lattice types of (a) α-Hg and (b) β-Hg shown in Figures 2.4(b) and (c).

2.3 **Lattice planes and Miller indices**

In a regular array or grid of points in two dimensions it can be seen that in certain directions the lattice points form lines (Figure 2.12a). Lattice points that form a regular array in three dimensions may be connected by two-dimensional **lattice planes** (Figure 2.12b). Normally these lattice planes are shown for a single unit cell (Figure 2.12c) but the planes extend infinitely in each direction. Each plane is a representative member of a family of parallel equally spaced planes; a possible lattice plane for a primitive cubic unit cell is shown in Figures 2.12(c) and (d).

These lattice planes are labelled using **Miller indices**. For a three-dimensional unit cell, three indices are required and designated conventionally h, k, and l; the Miller indices for a particular family of planes are usually written (h, k, l) where h, k, and l are integers, positive, negative, or zero.

> **NOTE ON GOOD PRACTICE**
> Other shorthand ways of writing Miller indices include (hkl), $(h\ k\ l)$, h, k, l, and just hkl, but these can lead to confusion especially when one Miller index is in double figures, for example, the (10, 7, 8) plane might become (1078).

In the case of a three-dimensional structure, the derivation of Miller indices may be illustrated by considering the two adjacent lattice planes from the same set which cut the a, b, and c axis at the unit cell origin and at the least distance along the various cell directions. The Miller indices of this family of planes are given by the reciprocals of the fractional intercepts along each of the cell directions. For example, using the lattice plane shown in Figure 2.13(a), the intercepts are at $\frac{1}{2} \times a$, $\frac{1}{1} \times b$, and $\frac{1}{3} \times c$. The Miller indices of this plane, one of a set of equally spaced parallel planes, are thus (2, 1, 3) (Figure 2.13b). The intercept along the unit cell a direction, at a/h, gives rise to h, along

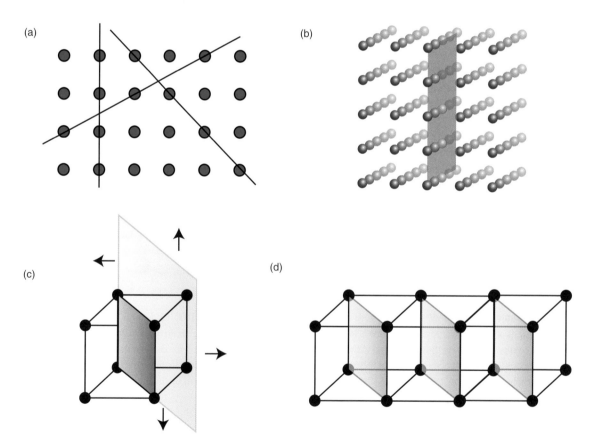

FIGURE 2.12 (a) A two-dimensional regular array of atoms showing their arrangement into lines in certain directions. (b) A plane of atoms in a three-dimensional array. (c) A possible lattice plane for a primitive cubic cell also showing that it extends infinitely. (d) Each plane is a representative member of a family of parallel equally spaced planes.

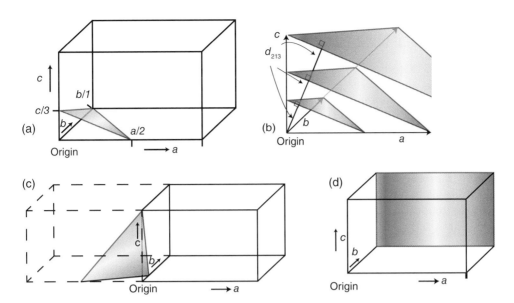

FIGURE 2.13 (a) Construction of a plane with Miller indices (2, 1, 3) and the distance d_{213}, the distance to the origin and (b) adjacent planes with Miller indices (2, 1, 3). (c) Construction of a plane with a negative Miller index, the (−2, 4, 1) plane. (d) Construction of a plane with a zero Miller index, the (0, 1, 0) plane.

b, at b/k, gives k, and along c, at c/l, gives l. For planes that have a negative Miller index the intercept on the axis is in the negative direction when referred to the unit cell origin (Figure 2.13c). In this case, the (−2, 4, 1) plane intercepts the a axis at $−a/2$. For planes which are parallel to one of the unit cell directions the intercept is at infinity and, therefore, the Miller index for this axis is $1/\infty = 0$. Figure 2.13(d) shows an example of a (0, 1, 0) plane for an orthorhombic unit cell.

The separation of the planes is known as the d-spacing and is normally denoted d_{hkl}. From Figure 2.13(a) it follows that this is also the perpendicular distance from the origin to the nearest plane. It is also the separation between all adjacent planes in the family with the Miller indices (h, k, l) (Figure 2.13b). The relationship between d-spacing and the lattice parameter can be determined geometrically but is dependent upon the crystal system (Section 2.4).

EXAMPLE 2.3

Drawing planes with defined Miller indices

For an orthorhombic cell of dimensions $5 \times 3 \times 4$ Å, draw planes with the Miller indices (a) (2, 1, 1), (b) (−4, 1, 1), and (c) (1, 1, 0).

ANSWER

Given a set of Miller indices for a specified unit cell it is possible to draw one example of the plane and mark the d-spacing for this plane using the fact that the plane cuts the unit cell directions at a/h, b/k, and c/l. Figure 2.14 shows these planes and their construction.

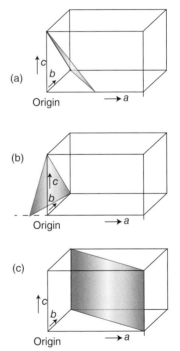

FIGURE 2.14 Diagrams showing (a) (2, 1, 1), (b) (−4, 1, 1), and (c) (1, 1, 0) planes.

SELF TEST

For the same orthorhombic cell of dimensions $5 \times 3 \times 4$ Å, draw planes with Miller indices (a) (2, −1, −2) and (b) (4, 0, 0).

2.4 Relationship between Miller indices and *d*-spacing

Trigonometry may be used to derive the relationship between the *d*-spacing for any plane and its Miller indices. It is perhaps easiest to demonstrate this calculation using a two-dimensional case and then logically extend this to three dimensions; also by considering a crystal system with all cell angles as 90° the calculation is simplified. We can undertake the calculation in two dimensions, using simple trigonometry, by considering a plane with one Miller index as zero, that is, an (h, 0, l) plane, as shown in Figure 2.15.

Hence, from Figure 2.15 it follows that:

$$\sin\varphi = \frac{d_{h0l}}{(a/h)} \quad \cos\varphi = \frac{d_{h0l}}{(c/l)} \qquad \text{Eqn 2.1}$$

The trigonometric expression $\sin^2\varphi + \cos^2\varphi = 1$ allows these two expressions to be combined giving

$$d_{h0l}^2 = \frac{1}{\left(\dfrac{h}{a}\right)^2 + \left(\dfrac{l}{c}\right)^2} \qquad \text{Eqn 2.2}$$

and, therefore,

$$\frac{1}{d_{h0l}^2} = \left(\frac{h}{a}\right)^2 + \left(\frac{l}{c}\right)^2 \qquad \text{Eqn 2.3}$$

Extension of this expression to a three-dimensional case with right angles between the lattice directions, an orthorhombic system, produces the general expression (for convenience the notation d_{hkl} has been shortened to d):

$$\frac{1}{d^2} = \frac{h^2}{a^2} + \frac{k^2}{b^2} + \frac{l^2}{c^2} \qquad \text{Eqn 2.4}$$

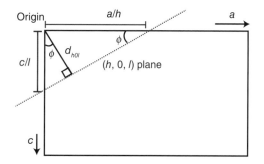

FIGURE 2.15 Construction used to derive the relationship between planes with Miller indices (h, 0, l) and d_{h0l} in an orthorhombic crystal system.

This expression (Eqn 2.4) can be used for other crystal systems with orthogonal axes, that is, cubic and tetragonal, by incorporating the appropriate relationships between a, b, and c. So for a cubic crystal system where $a = b = c$:

$$\frac{1}{d^2} = \frac{h^2}{a^2} + \frac{k^2}{a^2} + \frac{l^2}{a^2} \qquad \text{Eqn 2.5}$$

which can be simplified to

$$d^2 = \frac{a^2}{(h^2 + k^2 + l^2)} \qquad \text{Eqn 2.6}$$

For all crystal systems, with unit cell angles α, β, γ not equal to 90°, the relationships giving the *d*-spacing, *d*, are summarised in Table 2.2.

EXAMPLE 2.4

Calculations involving *d*-spacings

Using the appropriate expressions given in Table 2.2, calculate *d*-spacings for the (1, 0, 1) and (−1, 0, 1) planes in an orthorhombic cell with dimensions $a = 3$ Å (300 pm), $b = 5$ Å (500 pm), and $c = 4$ Å (400 pm) and compare the values. Carry out the same calculation for a monoclinic cell with the same *a*, *b*, and *c* but with $\beta = 110°$. Show diagrammatically, using unit cells viewed down the *b* axis, the positions of these planes in both the orthorhombic and the monoclinic crystal systems. Explain why for a monoclinic crystal system $d_{101} \neq d_{-101}$.

ANSWER

Substituting the data into the formula gives the following value for an orthorhombic unit cell:

$$d_{101} = d_{-101} = 2.40 \text{ Å}$$

Obviously, with the squares of the Miller indices only in the formula, the negative indices produce the same value as the positive ones; see Figure 2.16(a). Using the expression for a monoclinic unit cell (Table 2.2), the following *d*-spacings are obtained:

$$d_{101} = 1.956 \text{ Å} \quad \text{and} \quad d_{-101} = 2.755 \text{ Å}$$

with the '*hl*' term in the expression then negative Miller indices make a difference. This is illustrated by the unit cell in projection in Figure 2.16(b).

TABLE 2.2 Expressions for d-spacings in the different crystal systems

Crystal system	Expression for d (d_{hkl}) in terms of lattice parameters and Miller indices
Cubic	$\dfrac{1}{d^2} = \dfrac{h^2 + k^2 + l^2}{a^2}$
Tetragonal	$\dfrac{1}{d^2} = \dfrac{h^2 + k^2}{a^2} + \dfrac{l^2}{c^2}$
Orthorhombic	$\dfrac{1}{d^2} = \dfrac{h^2}{a^2} + \dfrac{k^2}{b^2} + \dfrac{l^2}{c^2}$
Hexagonal	$\dfrac{1}{d^2} = \dfrac{4}{3}\left(\dfrac{h^2 + hk + k^2}{a^2}\right) + \dfrac{l^2}{c^2}$
Monoclinic	$\dfrac{1}{d^2} = \dfrac{1}{\sin^2\beta}\left(\dfrac{h^2}{a^2} + \dfrac{k^2 \sin^2\beta}{b^2} + \dfrac{l^2}{c^2} - \dfrac{2hl\cos\beta}{ac}\right)$
Triclinic	$\dfrac{1}{d^2} = h^2\dfrac{b^2c^2\sin^2\alpha}{V^2} + k^2\dfrac{a^2c^2\sin^2\beta}{V^2} + l^2\dfrac{a^2b^2\sin^2\gamma}{V^2}$ $+2hk\dfrac{abc^2(\cos\alpha\cos\beta - \cos\gamma)}{V^2} + 2kl\dfrac{a^2bc(\cos\beta\cos\gamma - \cos\alpha)}{V^2}$ $2lh\dfrac{ab^2c(\cos\gamma\cos\alpha - \cos\beta)}{V^2}$

where V is the volume of the unit cell given by $V = abc(1 - \cos^2\alpha - \cos^2\beta - \cos^2\gamma + 2\cos\alpha\cos\beta\cos\gamma)^{\frac{1}{2}}$

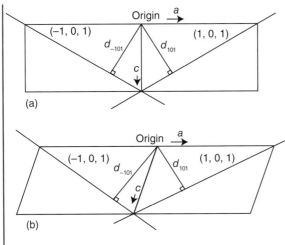

(a)

(b)

FIGURE 2.16 (a) The equivalence of d_{101} and d_{-101} in an orthorhombic unit cell and (b) the non-equivalence of d_{101} and d_{-101} in a monoclinic unit cell.

SELF TEST

By consideration of the expressions for d-spacing values in Table 2.2 determine for which crystal systems the d-spacing values of the (1, 1, 1) and (−1, −1, −1) planes will be equal.

2.5 Scattering of X-rays by crystalline solids: Bragg's law and the Bragg equation

X-rays are scattered by electrons and in a crystalline solid the electron distribution necessarily forms a regular array corresponding to the atom positions. The scattered X-ray waves from a periodic arrangement of scattering points will undergo interference effects leading to diffraction. This is illustrated in Figure 2.17 in two dimensions for the scattering of a planar wave at two adjacent points—each equivalent to an electron cloud. The scattered X-rays, in two dimensions, have circular wavefronts and in certain directions the peaks of these waves coincide to give an interference maximum.

The scattering of X-rays from the atoms in a crystalline solid is more easily illustrated by considering a set of parallel lattice planes containing those atoms, as shown in Figure 2.18. The points marked A and D in successive planes would represent centres of electron density in the crystal that scatter the X-rays. This construction can be used to work out the relationship between the separation of the planes (the d-spacing introduced in Section 2.3), the wavelength of the X-rays, λ, and their angle of incidence to the plane, θ.

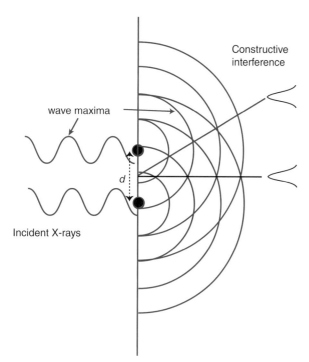

FIGURE 2.17 The scattering of planar incident X-ray wave at two points separated by a distance, d; in a crystalline solid these would represent two atoms. Constructive interference, where the scattered wave peaks (shown as semicircles) coincide, gives rise to a diffraction maximum.

As drawn in Figure 2.18, the impinging X-ray beam (shown in blue as two waves that are in phase) will be scattered, respectively, from the points A and D in neighbouring planes. The scattered X-rays (shown as dotted blue waves) will undergo constructive interference if the additional distance travelled by the X-ray photon scattered from D (compared with that scattered from A) is an integral number of wavelengths. This additional path difference, marked as the distance BD + DC, will depend on the lattice spacing or d_{hkl}, where (h, k, l) are the Miller indices for the planes under consideration, and will also be related to the angle of incidence of the X-ray beam, θ. Using trigonometry in the triangles ABD and ADC yields BD = DC = $d\sin\theta$. Therefore, to get constructive interference the following relationship between θ, d_{hkl}, and λ can be derived:

$$\text{path difference} = \text{BD} + \text{DC} = 2 \times d_{hkl} \times \sin\theta = n\lambda$$

where n is an integer and λ is the X-ray wavelength. The expression

$$2d\sin\theta = n\lambda \qquad\qquad \textbf{Eqn 2.7}$$

is known as the **Bragg equation** or **Bragg's law** and it relates the measured angles of the diffraction maxima to the d-spacings. The angle at which diffraction occurs is known as the **Bragg angle**. Note that the total angle through which the beam is turned relative to the incident beam path by the diffraction process is 2θ, twice the Bragg angle, and this is the angle measured in an X-ray diffraction experiment, the **diffraction angle**. The diffraction maxima in X-ray diffraction patterns are normally referred to as 'reflections'; an X-ray scattering from the (2, 1, 3) plane is termed the (2, 1, 3) reflection in diffraction pattern. This terminology, while incorrect in terms of the process involved in origin of the maxima, which is diffraction, is so widely used that it is the accepted notation.

In general, n in the Bragg equation is always taken as unity or first-order diffraction; the reason for this is illustrated in Figure 2.19. Rewriting the Bragg equation as $\lambda = 2 \times (d_{hkl}/n) \times \sin\theta$ and consideration of Figure 2.19 shows that the higher order diffraction maximum, with $n = 2$, is equivalent in representation to diffraction from a set of planes with half the separation. Hence, the second-order diffraction from, for example, the (1, 0, 0) plane is identical to the first-order diffraction from the (2, 0, 0) plane. In general, it is always possible to select lattice planes such

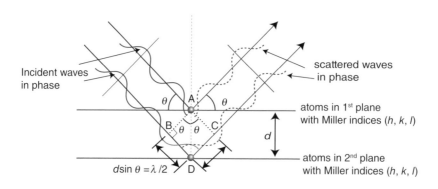

FIGURE 2.18 Derivation of Bragg's law.

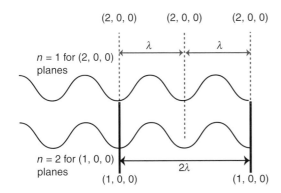

FIGURE 2.19 Equivalence of diffraction of $n = 2$ for the $(1, 0, 0)$ reflection and $n = 1$ for the $(2, 0, 0)$ reflection.

that n is unity. Note that slightly off the Bragg angle, θ, the diffracted X-ray beams from two successive planes will only be slightly out of phase. However, in a crystal with a large number of planes, scattering from each successive plane will be further and further out of phase. Over the whole crystallite this leads to destructive interference at angles a little away from the Bragg angle so the diffraction maxima are sharp with small width—typically around 0.1°.

In a crystalline material, an infinite number of lattice planes with different Miller indices exist and each set of planes will have a separation defined by d_{hkl}. All the various d-spacings possible can be obtained by consideration of Table 2.2 for a specific crystal system. Each of the d-spacings possible (for a set of lattice parameters and integer choice for h, k, and l) can, in principle, through the Bragg equation give rise to a diffraction maximum at the measured diffraction angle, 2θ. The positions of these diffraction maxima, in terms of the Bragg angle, θ, can be obtained by combining the Bragg equation with the expression for the d-spacing for a particular crystal system and eliminating d_{hkl}.

Thus, for the cubic crystal system combining Bragg's law $2d\sin\theta = n\lambda$ with the appropriate relationship between d-spacings and the Miller indices,

$$\frac{1}{d^2} = \frac{h^2 + k^2 + l^2}{a^2},$$

gives, setting $n = 1$,

$$\sin^2\theta = \frac{\lambda^2}{4a^2}(h^2 + k^2 + l^2) \qquad \textbf{Eqn 2.8}$$

The scattering of an X-ray beam by a crystal will thus give rise to a large number of reflections whose positions (defined by θ) are determined by lattice parameters, the Miller indices, and the X-ray wavelength. Hence, by studying the diffraction of X-rays by crystalline solids, structural information, such as the lattice parameters, can be obtained.

EXAMPLE 2.5

Calculations using the Bragg equation

Calculate the diffraction angle, 2θ, of an X-ray with wavelength 1.5406 Å for:

(a) a set of planes with d-spacing 2.100 Å;

(b) the $(1, 1, 1)$ and $(2, 2, 2)$ planes of a cubic crystal structure with lattice parameter 5.400 Å.

ANSWERS

(a) Using the Bragg equation (Eqn 2.7) and inserting the values $\lambda = 1.5406$ Å and $d = 2.100$ Å, setting $n = 1$, gives $\theta = 21.52°$, $2\theta = 43.04°$.

(b) Using Eqn 2.8 and the values given yields for the $(1, 1, 1)$ planes $\sin^2\theta = 0.0610$ and $2\theta = 28.61°$ and for the $(2, 2, 2)$ planes $\sin^2\theta = 0.2442$ and $2\theta = 59.23°$. Note that larger h, k, l values lead to smaller d-spacing values and higher diffraction angles.

SELF TEST

Calculate diffraction angles using $\lambda = 1.5406$ Å for the $(1, 0, 0)$ and $(0, -1, 0)$ planes of a cubic crystal structure of lattice parameter 6.31 Å.

2.6 The powder X-ray diffraction experimental method

2.6.1 Sources of X-rays

The standard source of X-rays is an 'X-ray tube' (Figure 2.20) which consists of an evacuated tube made from glass or ceramic and a metal end plate. Electrons are produced by passing a current through a wire filament and these are then accelerated through a high electrical potential, typically 30–60 kV, onto a water-cooled metal target block. Some of the high velocity electrons cause ejection of core electrons from the atoms of the metal target. An electron from a higher orbital drops down in energy to replace the core orbital vacancy so produced and in the process it emits radiation of a defined frequency and wavelength ($E = h\nu = hc/\lambda$); for most transition metals this radiation is in the X-ray region of the electromagnetic spectrum (with $\lambda \sim 1$ Å or 100 pm); see Section 1.2. Several transitions are possible so the output of radiation is labelled K, L, M, etc. to represent different principal quantum numbers of the shell from which the electron was ejected (K represents the 1s level) with a sub-script α, β, etc. to define the principal quantum number of the higher energy shell (Figure 2.21).

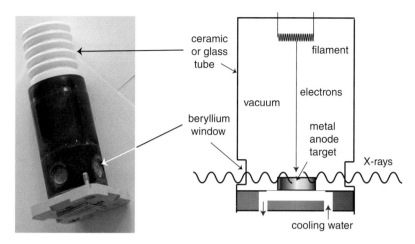

FIGURE 2.20 A photograph of an X-ray tube and a schematic showing its key components.

The X-ray spectrum from the target consists of a series of intense sharp maxima which are superimposed on a background of radiation ('bremsstrahlung') resulting from the electrons slowing down as they enter the metal target (Figure 2.22).

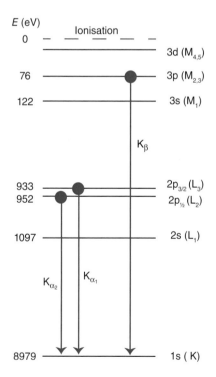

FIGURE 2.21 Possible quantised electron transitions contributing to the K-lines in the X-ray spectrum of copper. These are labelled K, L, M depending on the principal quantum number of the final electron energy level.

To undertake diffraction experiments a single (or near single) X-ray wavelength is needed. This wavelength, usually the most intense, can be selected from the rest of the X-ray radiation using filters that only allow X-rays of energies above a certain energy to pass, or by exploiting the Bragg equation in a primary monochromator. In the latter case the X-ray beam emerging from the tube is directed on to a single crystal of known structure (often germanium, silicon or graphite) suitably oriented so that the desired wavelength satisfies the Bragg equation at the appropriate angle to the sample. The most commonly used X-ray tube target materials are copper and molybdenum, which give characteristic X-rays of wavelengths 1.5406 Å (Cu K_{α_1}) and 0.71073 Å (Mo K_α), respectively. Note that the copper X-ray wavelength can be given to fewer significant figures or as the $K_{\alpha_1}/K_{\alpha_2}$ average, K_α, 1.5418 Å.

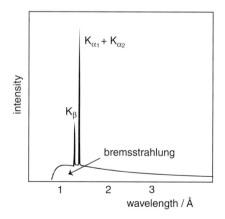

FIGURE 2.22 The K-line X-ray spectrum produced from bombarding a copper target with high-energy electrons.

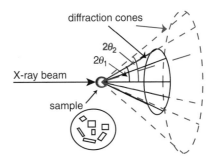

FIGURE 2.23 Formation of diffraction cones produced by X-rays scattered from a polycrystalline sample.

EXAMPLE 2.6

Use the Bragg equation to calculate the monochromator angle required to select copper Cu $K_{\alpha 1}$ radiation using the (1, 1, 1) plane of a germanium crystal ($d = 3.266$ Å).

ANSWER

We need to find the correct angle using the Bragg equation $2d\sin\theta = n\lambda$. Putting $n = 1$, $\lambda = 1.5406$ Å, and $d = 3.266$ Å in this expression gives $\theta = 13.64°$ ($2\theta = 27.28°$).

SELF TEST

Calculate the monochromator angle with the same germanium crystal needed to select Mo K_α radiation.

The intensity of the X-ray beam produced by an X-ray tube is limited by the need to cool the static target, as much of the energy from the impacting electrons is turned into heat. To cool the target more effectively 'rotating anode' X-ray tubes can be used, yielding an order of magnitude gain in intensity.

Much more intense X-ray beams are generated by synchrotron sources (Chapter 1, Section 1.6.5). Synchrotron X-ray sources allow diffraction data to be collected several orders of magnitude faster than is possible with an X-ray tube. An additional advantage of a synchrotron X-ray source derives from that fact that X-rays are produced over a wide range of energies. This allows any wavelength over the X-ray portion of the electromagnetic spectrum to be selected (using a monochromator) for an experiment.

2.6.2 Data collection

To detect an X-ray beam that has been diffracted, according to Bragg's law, from a particular lattice plane, the orientation of the X-ray source, crystal, and detector must be correct. In single-crystal X-ray diffraction, each of these diffraction maxima will appear as a spot on a detector, so

that the diffraction pattern consists of an array of spots each one corresponding to scattering from a plane with a different set of Miller indices in the single crystal (see Figure 2.1).

A powder or polycrystalline sample contains thousands of very small crystallites, typically 0.1–10 μm in dimension and these particles will adopt, randomly, a full range of possible orientations. An X-ray beam striking a polycrystalline sample can be diffracted in all possible directions as governed by the Bragg equation and the d-spacing for any h, k, l values present in the sample. The effect of this is that each lattice spacing permitted for the crystal system will give rise to a cone of diffraction, as shown in Figure 2.23. In fact, each cone consists of a set of closely spaced dots each one of which represents diffraction from a single crystallite within the powder sample; with a very large number of crystallites these dots join together to form the cone. Figure 2.24 shows the experimental image produced on a photographic film placed behind a powder sample exposed to a monochromatic X-ray beam.

2.6.3 Experimental methods

In order to obtain powder X-ray diffraction (PXD) data in a form useful for analysis, the positions of the various diffraction cones need to be determined. This can be achieved by using either photographic film or a detector sensitive to X-rays.

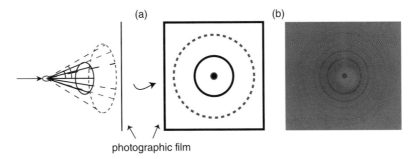

FIGURE 2.24 (a) Schematic of the image produced on a photographic film placed perpendicular to an X-ray beam diffracted by a powder sample. (b) An experimental photograph of a powder diffraction pattern.

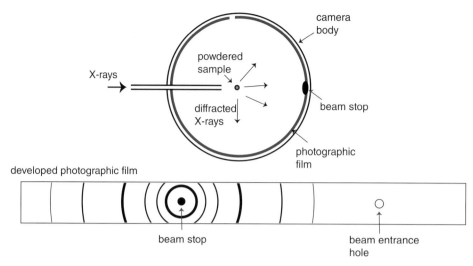

FIGURE 2.25 Powder X-ray diffraction data obtained using a Debye–Scherrer camera.

2.6.3.1 Photographic methods

The Debye–Scherrer camera (Figure 2.25), while no longer used in modern laboratories, provides a simple illustration of the PXD experimental method. A strip of X-ray sensitive film is placed inside a cylindrical camera which has, at its centre, the sample. The X-ray beam enters through one side of the cylinder and is collimated onto the powder sample which is normally mounted by coating a glass fibre or filling a narrow glass capillary. The sample is often rotated around the axis of the fibre in order to present as many orientations of the crystallites as possible to the beam. The diffraction cones cut the film at various diffraction angles, 2θ, where θ is the Bragg angle. Once the film is developed and laid flat, a typical powder diffraction pattern, as shown in Figure 2.25, is obtained. Each arc on the film represents a different d-spacing in the sample. At low angles the lines have distinct curvature as the diffraction cone angle is small and a large part of the cone's curve is intersected; at angles

nearer 90° only a small section of the cone intersects the film and the line curvature is less marked. The position of the undiffracted beam can be seen as the dark area surrounding the shadow of the beam stop. From the radius of the camera and the distance along the film from the beam stop position, the diffraction angle, 2θ, can be determined for each of the lines.

2.6.3.2 The powder diffractometer

The powder diffractometer uses an X-ray detector, typically a scintillation counter, to measure the positions of the diffracted X-rays. X-rays reaching the scintillation detector material (typically sodium iodide doped with thallium) generate a pulse of visible light whose intensity is measured. Scanning the detector around the sample along the circumference of a circle cuts through the diffraction cones at the various diffraction maxima (Figure 2.26a). The intensity of the X-rays detected as a function of the detector angle, 2θ, for a typical material is shown in

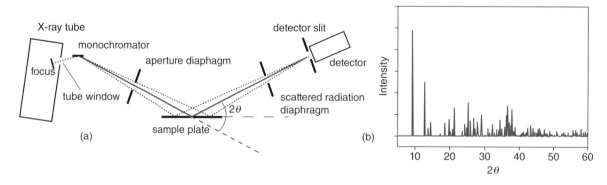

FIGURE 2.26 The operation of a powder diffractometer. (a) Schematic of the diffractometer with the detector arm rotated around the diffraction angle with the X-rays scattered from the flat sample plated (rotated at half the rate of the detector) and, thereby, focused on to the detector circle. (b) A typical diffraction pattern obtained on a powder X-ray diffractometer.

Figure 2.26(b). The reflection geometry frequently used for diffractometers is useful in that it is to some degree focusing, leading to narrow peaks which can be reasonably well resolved from each other. This can be seen in Figure 2.26(a), where the sample acts like a mirror and provided the geometry of the equipment is correctly designed the diffracted beam can be focused on to the detector.

2.7 Applications of powder X-ray diffraction

Table 2.3 summarises some of the main applications of PXD. The technique can provide information that ranges from the rapid identification of an unknown solid through to a full structure determination if very high quality data are available.

TABLE 2.3 Summary of the applications of powder X-ray diffraction

---

Identification of solid, crystalline unknowns

Determination of sample purity

Qualitative and quantitative solid phase analysis

Construction of phase diagrams, determining the degree of a solid state reaction

Thermal expansion and phase changes

Particle size determination

Refinement of lattice parameters and determination of lattice type

Structure refinement

Ab initio structure determination

2.7.1 Phase identification

As described in Section 2.5, the 2θ positions of the reflections observed in the PXD pattern depend on the d-spacings of the lattice planes. These are in turn defined by the lattice parameters of the unit cell adopted by the compound. The intensities of the reflections depend on the positions of the atoms in the unit cell and how strongly they scatter X-rays. As every compound crystallises with a unique unit cell, and arrangement of atoms within that unit cell, it produces a unique PXD pattern. Thus, the PXD pattern obtained experimentally can be used as the characteristic 'fingerprint' of a compound. In a mixture of crystalline compounds each phase present contributes to the whole powder diffraction pattern. Its contribution to the pattern is in proportion to the

amount present and how strongly the elements present in that phase scatter X-rays.

Powder X-ray diffraction data, collected experimentally or calculated from known structures, have been compiled into large databases. For example, the **Powder Diffraction File** managed by the International Centre for Diffraction Data contains several hundred thousand entries. Comparison of an experimental powder diffraction pattern with all those in a database can be carried out using a computer in a few seconds and the phases present identified. These computer programs can also resolve diffraction patterns from mixtures into the contributions from several component phases. Therefore, the PXD technique provides a quick, nondestructive method of identifying an unknown or mixture of unknowns.

It should be noted that this method will only work for crystalline materials. An amorphous material does not have the long range structural order to produce sharp diffraction maxima in the diffraction pattern. Materials that are amorphous, including glasses and some plastics and polymers, produce broad, background features in the diffraction pattern. The amount of crystalline material that can be identified in a mixture of phases will depend on the quality of the diffraction data and how well each component present scatters X-rays. On a typical laboratory powder X-ray diffractometer a few per cent of most phases can be identified in a mixture and this is of importance in determining the purity of polycrystalline materials. With very high intensity synchrotron X-ray sources, a fraction of a per cent of a phase can sometimes be observed as a contributor to the overall diffraction pattern. How strongly a material scatters X-rays, and the complexity of the unit cell, are important factors to consider when assessing whether diffraction from it will be observed in the overall diffraction pattern and in quantitative analysis (Section 2.7.5). In an equimolar mixture of Li_2O, SnO_2, and Li_2SnO_3 the peaks from Li_2O will be much weaker than those of the other two components because low atomic number elements, lithium and oxygen in this case, scatter X-rays much more weakly than electron-rich tin. Materials with small, high-symmetry unit cells, such as cubic and hexagonal, have relatively few diffraction peaks in their patterns compared with materials adopting large, triclinic unit cells. The scattering from high symmetry materials is, therefore, concentrated into a few peaks that are easier to observe experimentally. To illustrate this, consider the diffraction pattern from a 50:50 mixture of salt, NaCl (cubic, unit cell volume 180 Å^3) and washing soda, $Na_2CO_3.10H_2O$ (monoclinic, unit cell volume 1270 Å^3); peaks from the cubic sodium chloride phase will be much stronger than those of $Na_2CO_3.10H_2O$.

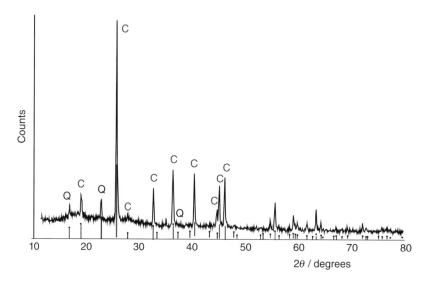

FIGURE 2.27 The PXD pattern obtained from the 'witch bottle' sample with peaks derived from calcite (C) and quartz (Q) marked.

2.7.2 Forensic applications and phase identification of unknowns

The non-destructive nature of PXD and its ability to collect data from a few milligrams of an unknown make it a valuable technique in forensic science and archaeology. As an example, Figure 2.27 shows the powder diffraction pattern obtained from a solid found inside a 'witch bottle', a small glass bottle, dating from the sixteenth and seventeenth centuries, traditionally buried in doorways and hearths to act as a charm. The PXD peaks from the solid were quickly assigned to silicon dioxide (quartz) and calcium carbonate (calcite) and the material was identified as a mixture of sand and chalk (the latter being the major component as it gives rise to the strongest reflections). Similar applications in forensic science include the identification of the pigments present in a sample of paint and the determination of the presence and level of Epsom salts, $MgSO_4$, which is often used, at various concentrations, to dilute methamphetamine street drugs.

2.7.3 Determination of phase purity, following reactions to completion, and phase diagrams

The synthetic route to many complex solid state compounds involves heating together the component oxides for several days until the reaction is complete. At the end of the reaction the reactants will be totally consumed and a single-phase product generated. For example, the spinel oxide $MgAl_2O_4$ can be obtained by reacting MgO

with Al_2O_3 at 1400°C for several days. The progress of this reaction can be monitored using PXD. When the reactants are initially mixed the PXD pattern will consist of the diffraction peaks from just the two reactants. As the reaction proceeds, the diffraction peaks of the spinel product will appear at the expense of the characteristic reflections of MgO and Al_2O_3; see Figure 2.28. Eventually only peaks from $MgAl_2O_4$ are observed in the PXD pattern and the reaction can be deemed as being complete. This is with the caveat that because very weak peaks from poorly scattering materials might not be observed experimentally, a few per cent of reactant could still be present.

Phase diagrams represent which distinct new solid structures exist between two or more simpler components, such as metals or binary oxides. Thus the phase diagram between MgO and Al_2O_3 would show the existence of the spinel phase $MgAl_2O_4$ formed at a 1:1 molar ratio of these oxides, but no other complex oxide (as $MgAl_2O_4$ is the only ternary phase that exists in this system). As a result, the systematic mapping of a phase diagram can be achieved using PXD. In such a mapping process, the reaction products obtained from different reaction mixture stoichiometries are analysed using PXD. The patterns generated show the presence or not of the various phases that might exist and thus be formed across the complete composition range. Their levels can also be determined quantitatively (see Section 2.7.5) from the peak intensities. For example, the reaction product from a 2:1 molar mixture of MgO and Al_2O_3 would have a PXD pattern showing reflections from MgO and $MgAl_2O_4$ in a 1:1 intensity ratio.

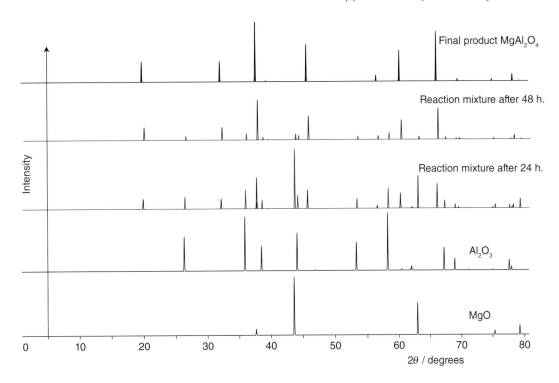

FIGURE 2.28 Powder X-ray diffraction patterns obtained from monitoring an equimolar reaction between MgO and Al_2O_3 to synthesise $MgAl_2O_4$. Pure phases are shown in black and the reactants in the part-complete reaction material are shown in blue.

2.7.4 **Phase changes and thermal expansion**

Many crystalline materials undergo a phase change as a function of some external environmental parameter, for example, temperature or pressure. This process will involve a rearrangement of the atom positions in material, the adoption of a new unit cell, and concomitant changes to the PXD pattern. Therefore, the PXD technique provides a method for investigating phase behaviour of a material as a function of variables such as temperature and pressure. Equipment that facilitates changes in sample environment, including a furnace, cryostat, humidity controller, or pressure cell, can be mounted on the powder diffractometer and data collected continuously as a function of temperature, etc. As well as changes in structure, the size of the unit cell will usually change with temperature—normally an expansion with increasing temperature. A larger unit cell will mean increasing d-spacings and as a result the diffraction maxima will slowly shift to lower diffraction angles.

Figure 2.29 demonstrates how phase changes and thermal expansion can be observed using powder diffraction techniques. In this figure a series of powder diffraction patterns, collected at different temperatures, is viewed from above and shaded so that the diffraction peaks appear dark above the background. The diffraction angle, 2θ, is on the x-axis (with data presented for 2θ values between 65° and 90°) and the increasing temperature on the y-axis.

The diffraction maxima are observed as darker lines and with increasing temperature (e.g. between 200 and 300 K) these can be seen to shift to lower 2θ values, indicating larger d-spacings (from the Bragg equation, $2d\sin\theta = n\lambda$) in the material as it expands. These values can be analysed to obtain a thermal expansion coefficient for a material—including different values along each of the different unit cell directions. At 165 K there is an abrupt change in the diffraction pattern with one set of reflections disappearing and another appearing at different angles. This represents an abrupt change in structure,

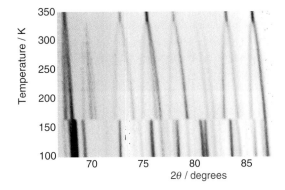

FIGURE 2.29 A stacked plot of PXD patterns collected as a function of temperature. Diffraction maxima are shown as dark lines.

a 'phase change', and the formation of a new atomic arrangement. A similar transition happens at 330 K but in this instance the phase change is gradual with one structure (in this case adopting a tetragonal unit cell below 330 K) slowly transforming into a closely related one above 330 K (with a cubic unit cell).

2.7.5 Quantitative analysis and quality control

The PXD technique determines whether a crystalline phase is present and also, from the reflection intensities, the proportion of that phase present in a mixture (subject to the considerations discussed in Section 2.7.1 regarding the effects of atomic scattering power and unit cell complexity on the overall scattered X-ray intensity from a phase). This leads to many applications of PXD in analytical chemistry and quality control. Accurate quantitative analysis of a complex mixture of different phases may be achieved by using standards and standard mixtures. An example of this application is in the analysis of multicomponent mixtures of clay minerals. These can contain, for example, kaolinite, quartz, muscovite mica, and various feldspars, and PXD data can be analysed to determine the levels of each component to a precision of ±0.1%. Similarly in the pharmaceutical and agrochemical industries, commercial products may contain different polymorphs or the hydrated forms of the active ingredient. The levels of each of these components can be determined rapidly from PXD data.

2.8 Powder X-ray diffraction data analysis

2.8.1 Indexing powder diffraction patterns from cubic systems

The relationship between Bragg angle and the lattice parameter is obtained by combining the Bragg equation with the expression for d-spacing in terms of lattice parameter and Miller indices. In the cubic crystal system the following expression was obtained in Section 2.5 (setting $n = 1$):

$$\sin^2\theta = \frac{\lambda^2}{4a^2}(h^2+k^2+l^2) \qquad \text{Eqn 2.9}$$

which can be rewritten as

$$\sin^2\theta = A(h^2+k^2+l^2) \qquad \text{Eqn 2.10}$$

where A is a constant as both a and λ are fixed values. The $\sin^2\theta$ values for each of the various reflections can be

readily obtained from their diffraction angles, 2θ. These will all be related to each other through the common multiplier A as h, k, and l can only take integer values. Note that this will only be true for cubic systems; the more complex expressions for the d-spacings in terms of the lattice parameters in the other crystal systems means that more than one multiplier will be required. This argument may also be reversed in that if all the reflections in the powder diffraction pattern can be related through their $\sin^2\theta$ values by a single, common multiplier then the crystal system must be cubic. Once A is determined then each reflection can be assigned its Miller indices, a process known as '**indexing**' or '**indexation**'. For example, the reflection with the multiplier $1 \times A$ must have $(h^2+k^2+l^2) = 1$ so this is the reflection from the $(1, 0, 0)$ plane (or an equivalent since for a cubic system the $(0, 1, 0)$, $(0, 0, 1)$, $(-1, 0, 0)$, etc. planes all have identical d-spacings).

The process involved in indexing data from a cubic system and obtaining the lattice parameter is illustrated in Table 2.4. The $\sin^2\theta$ values are tabulated and they can then be inspected for the common multiplier, A. The best way of accomplishing this is to divide through all the $\sin^2\theta$ values by the $\sin^2\theta$ value for the first reflection, and then inspect the values obtained. If these are all integers then we have identified A and, as we will see, the first reflection is probably from the $(1, 0, 0)$ plane, with $A = 1$. If the ratios are not integers then the values obtained are normally obvious improper fractions such as 1.333 (4/3) or 2.500 (5/2); in these cases the first reflection is not from the $(1, 0, 0)$ plane and the multiplier of A needs to be adjusted (in the cases given previously division by three and two respectively) so that all reflections have an integer ratio between their $\sin^2\theta$ values (3A:4A and 2A:5A for the examples above). More examples of indexing PXD data of this form are given in Section 2.9.2.

The multiplier of A for each peak is the sum of the squares of the Miller indices and by considering the values that $(h^2+k^2+l^2)$ may take, for all possible h, k and l integer values, each reflection may be assigned its Miller indices. Thus, a multiplier value of $6 \times A$ is derived from $2^2 + 1^2 + 1^2$ and is the reflection from the $(2, 1, 1)$ plane. Table 2.4 completes this analysis for a powder diffraction data set from a cubic material where the first reflection has Miller indices $(1, 0, 0)$.

Note that there are some values which $h^2 + k^2 + l^2$ cannot take. For instance it is impossible to choose the integer values of the Miller indices such that the sum of their squares equals 7 or 15. This means that there can be no peak in the powder diffraction pattern corresponding to 7A or 15A. Once the data have been indexed, derivation of the lattice parameter is readily undertaken. Any peak may be chosen and the appropriate Miller indices,

TABLE 2.4 Indexing of the powder diffraction data from a cubic material ($\lambda = 1.5418$ Å $= 154.18$ pm)

2θ	$\sin^2\theta$	Ratio	Miller indices
22.983	0.03969	1.00	(1, 0, 0)
32.729	0.07938	2.00	(1, 1, 0)
40.372	0.11907	3.00	(1, 1, 1)
46.962	0.15876	4.00	(2, 0, 0)
52.908	0.19845	5.00	(2, 1, 0)
58.418	0.23814	6.00	(2, 1, 1)
68.595	0.31752	8.00	(2, 2, 0)

wavelength, and $\sin^2\theta$ values for that reflection substituted back into Eqn 2.9 to give a. For example, from Table 2.4, and working in with the X-ray wavelength in angstroms,

$$0.31752 = (1.5418^2/4a^2)\times(8)$$

$$a = 3.8695\text{Å} \ (= 386.95 \text{ pm})$$

In general, if all lines have been measured with the same precision of 2θ then the use of a high angle line will produce a more accurate lattice parameter. In practice, a computer-based, least squares minimisation, refinement technique, fitting all the reflection positions simultaneously, is used to obtain the best value for the lattice parameter.

TABLE 2.5 Indexing table for Example 2.7

2θ	$\sin^2\theta$	Ratio	Miller indices
23.74	0.04231	1.00	(1, 0, 0)
33.82	0.08460	2.00	(1, 1, 0)
41.74	0.1269	3.00	(1, 1, 1)
48.58	0.1692	4.00	(2, 0, 0)
54.77	0.2116	5.00	(2, 1, 0)
60.51	0.2539	6.00	(2, 1, 1)
71.15	0.3385	8.00	(2, 2, 0)
76.21	0.3808	9.00	(3, 0, 0)/(2, 2, 1)
81.15	0.4231	10.00	(3, 1, 0)
86.03	0.4654	11.00	(3, 1, 1)
90.89	0.5078	12.00	(2, 2, 2)
95.74	0.5500	13.00	(3, 2, 0)
100.64	0.5923	14.00	(3, 2, 1)
110.73	0.6770	16.00	(4, 0, 0)
116.01	0.7193	17.00	(4, 1, 0)/(3, 2, 2)

The lattice parameter can be determined from any reflection but using the highest angle one:

$$\sin^2\theta = \frac{\lambda^2}{4a^2}(h^2 + k^2 + l^2)$$

$$0.7193 = \frac{1.5418^2}{4a^2}(17)$$

So $a = 3.748$ Å

EXAMPLE 2.7

Index the following PXD data collected from a primitive cubic system using $\lambda = 1.5418$ Å and calculate the lattice parameter.

2θ 23.74 33.82 41.74 48.58 54.77 60.51 71.15 76.21 81.15 86.03 90.89 95.74 100.64 110.73 116.01

ANSWER

Draw up a table in the same form as Table 2.4, as shown in Table 2.5. Determine the $\sin^2\theta$ values and divide through all values by the smallest of these; if the first reflection is the (1, 0, 0) reflection then this $\sin^2\theta$ value is equal to **A** and all other reflections should be an integer multiple of this. This is the case here and the ratios determined are equivalent to $(h^2 + k^2 + l^2)$. Some simple arithmetic then allows all values of h, k, and l to be assigned.

SELF TEST

Index the following PXD data collected from polonium ($\lambda = 1.5418$ Å) and calculate the lattice parameter.

2θ 26.512 37.845 46.803 54.596 61.695 68.345 80.870 86.933

2.9 Lattice type and systematic absences

2.9.1 The origin of systematic absences

In indexing the powder patterns in the previous section it has been assumed that all the possible reflections are observed, that is, scattering from each of the different lattice planes is sufficiently intense to contribute to the diffraction profile. As a result the reflection from the plane with Miller indices $(1, 0, 0)$ should be the first line observed in the diffraction pattern and the subsequent reflections will be $(1, 1, 0)$, $(1, 1, 1)$, etc. This is normally so for a primitive lattice, but for body-centred and face-centred lattices restrictions occur on the values that h, k, and l may take if the reflections are to have any intensity. This results in certain reflections not being observed in the powder diffraction pattern and these are known as **systematic absences**. The origin of these absences can be illustrated with regard to Figure 2.30(a) and (b). Consideration of diffraction from the $(1, 0, 0)$ plane of a cubic material with a body-centred lattice shows that X-rays diffracted at the Bragg angle from the $(1, 0, 0)$ planes (the faces of the cube) will be, by definition, in phase. However, halfway between the $(1, 0, 0)$ planes, as a result of the body centring, there is

an identical plane of atoms shifted in the x, y, and z directions by $(+\frac{1}{2}, +\frac{1}{2}, +\frac{1}{2})$. X-ray diffraction from this plane of atoms, which is a $(2, 0, 0)$ plane, will have a path-length difference relative to that from the $(1, 0, 0)$ planes exactly $\lambda/2$ out of phase, and will, therefore, destructively interfere with it. In a body-centred lattice there is one lattice point in the $(2, 0, 0)$ plane for every lattice point in the $(1, 0, 0)$ plane and overall there will be total destructive interference for the $(1, 0, 0)$ reflection. Note that it does not matter where the atoms or ions actually are in the unit cell, it is the fact that they always occur in pairs (at (x, y, z) and at $(x + \frac{1}{2}, y + \frac{1}{2}, z + \frac{1}{2})$) generated by the body-centring that causes the destructive interference.

Extension of these considerations to other possible values of h, k, and l in a body-centred lattice shows that for a reflection to be present then $h + k + l$ must be even. Alternatively, this may be expressed as the condition $h + k + l = 2n + 1$ (where n is an integer) is the systematic absence for I-centred lattices. Similar considerations for a face-centred cubic lattice show the following restrictions on h, k, and l for a reflection to be observed: $h + k = 2n$, $k + l = 2n$, and $h + l = 2n$. Figure 2.30(c) shows the position of the interleaving plane for the absent $(1, 0, 0)$ reflection in a cubic **F** lattice. This restriction may also be expressed in the terms h, k, and l must be all odd or all even for a reflection to be present.

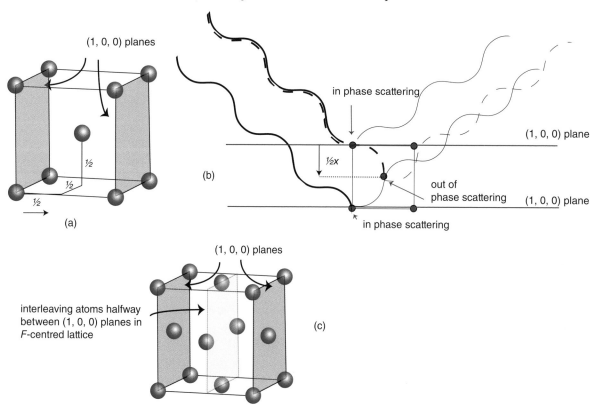

FIGURE 2.30 The origin of systematic absences. (a) $(1, 0, 0)$ planes in a body-centred cubic lattice showing equivalent atom positions at corners and body centre. (b) Out of phase scattering from an atom halfway between $(1, 0, 0)$ planes. (c) Interleaving atoms between $(1, 0, 0)$ planes in an **F**-type cubic lattice.

EXAMPLE 2.8

Demonstrate using a diagram that the (1, 1, 0) reflection will be absent for a face-centred cubic lattice.

ANSWER

In a projection diagram of a cubic **F** lattice (see Figure 2.31) the planes with Miller indices (1, 1, 0) are separated by an interleaving plane of atoms which will scatter X-rays $\lambda/2$ out of phase and cause destructive interference.

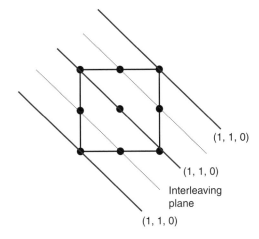

FIGURE 2.31 The origin of the (1, 1, 0) systematic absence for a face-centred cubic lattice.

SELF TEST

Show that for a face centred cubic lattice the (1, 1, 1) plane has no interleaving planes and will, therefore, be present in a PXD pattern.

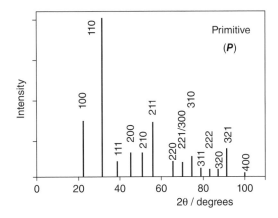

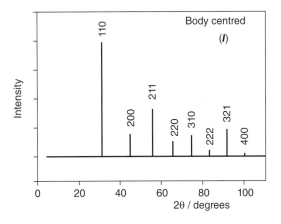

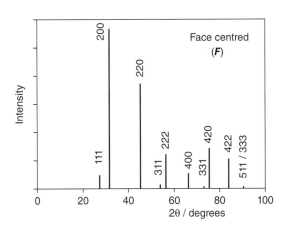

FIGURE 2.32 Typical forms of the diffraction patterns obtained from **P**, **I**, and **F** cubic lattices.

Using these systematic absence conditions for **I** and **F** lattice types we can determine which reflections should be seen in the PXD patterns and Table 2.6 summarises this for small integer values of h, k, and l.

The form of the observed patterns for the three cubic lattice types reflects these systematic absences, as shown in Figure 2.32. As well as being able to predict which reflections can be present in a PXD pattern, the systematic absences found in an experimental PXD pattern can now be analysed to identify the lattice type.

2.9.2 Indexing data from *I*- and *F*-centred lattices

The existence of systematic absences can be a complicating factor when attempting to index cubic systems with **I** and **F** lattice types as the first observed reflection in the powder diffraction pattern will not be the (1, 0, 0) reflection. This was assumed in the indexing process described in

Section 2.8.1. However, the process used previously can easily be modified to cover **I** and **F** lattice types.

2.9.2.1 Indexing an *F*-centred lattice ($\lambda = 1.5418$ Å)

The first stage of the indexing process, as before in Section 2.8.1, is to calculate $\sin^2\theta$ values for all the reflections (Table 2.7, column 2) from the 2θ values (column 1).

TABLE 2.6 Reflection conditions imposed by lattice type

Miller indices			Lattice type and condition for reflection to be seen		
			Primitive, *P* All possible values of *h*, *k*, and *l*	Body centred, *I* *h* + *k* + *l* = 2*n*	Face centred, *F* *h*, *k*, *l* all odd or all even
1	0	0	✓	✕	✕
1	1	0	✓	✓	✕
1	1	1	✓	✕	✓
2	0	0	✓	✓	✓
2	1	0	✓	✕	✕
2	1	1	✓	✓	✕
2	2	0	✓	✓	✓
2 2 1/3 0 0			✓	✕	✕
3	1	0	✓	✓	✕
3	1	1	✓	✕	✓
2	2	2	✓	✓	✓
3	2	0	✓	✕	✕
3	2	1	✓	✓	✕
4	0	0	✓	✓	✓

We then inspect the ratios between these values by dividing through by the $\sin^2\theta$ value of the first reflection. This gives column 3. As these values are not integers the first reflection cannot be from the (1, 0, 0) plane. However, all values in column 3 can be converted to integers by multiplying through by 3 giving the true integer ratios shown in column 4. The reflections can now be assigned Miller indices, column 5. The first peak is now assigned as the reflection from the (1, 1, 1) plane. All the reflections, once indexed, have (*h*, *k*, *l*) all odd or all even, showing that this material is face-centred cubic.

TABLE 2.7 Indexing table for a face-centred cubic lattice

2θ	$\sin^2\theta$	Apparent ratio	True ratio, $h^2 + k^2 + l^2$	(*h*, *k*, *l*)
27.377	0.0560	1	3	(1, 1, 1)
31.701	0.0746	1.33	4	(2, 0, 0)
45.444	0.1492	2.66	8	(2, 2, 0)
53.871	0.2052	3.66	11	(3, 1, 1)
56.482	0.2239	4	12	(2, 2, 2)
66.221	0.2984	5.33	16	(4, 0, 0)
73.070	0.3544	6.33	19	(3, 3, 1)
75.298	0.3731	6.67	20	(4, 2, 0)
83.996	0.4477	8	24	(4, 2, 2)

The lattice parameter can be calculated by substituting appropriate values into Eqn 2.9, and using the high angle (4, 2, 2) reflection gives *a* = 5.644 Å.

2.9.2.2 Indexing an *I*-centred lattice (λ = 1.5418 Å)

In a body-centred cubic system, the systematic absences can also be misleading in terms of the initial stages of indexing. The first stage of the indexing process, as before in Section 2.8.1, is to calculate $\sin^2\theta$ values for all the reflection 2θ values (Table 2.8, columns 1 and 2) and inspect the ratios between these values by dividing through by the $\sin^2\theta$ value of the first reflection. This gives column 3. The $\sin^2\theta$ values of the first lines will be found to have the ratios 1:2:3:4:5:6:7:8. The presence of a line with a ratio of 7 points to these lines originating from a body-centred system as there are no values of *h*, *k*, and *l* which square and sum together to give 7 (column 5), even for a primitive lattice type. The common multiplier must be adjusted, by being halved, for these lines to give the ratios 2:4:6:8:10:12:14:16 (column 4), corresponding to the Miller indices (1, 1, 0), (2, 0, 0), (2, 1, 1), etc. (column 6). The line with initial common ratio 7 becomes ratio 14 and can be indexed as (3, 2, 1). With this indexing all reflections have *h* + *k* + *l* = 2*n*, corresponding to the systematic absence condition for a body-centred cubic lattice. The lattice parameter can, once again, be calculated by substituting appropriate values into Eqn 2.9, and using the high angle (2, 2, 0) reflection gives *a* = 3.232 Å.

TABLE 2.8 Indexing table for a body-centred cubic lattice

2θ	$\sin^2\theta$	Apparent ratio	True ratio	Apparent hkl	True hkl
40.39	0.1192	1.00	2	(1, 0, 0)	(1, 1, 0)
58.40	0.2380	1.99	4	(1, 1, 0)	(2, 0, 0)
73.63	0.3591	3.00	6	(1, 1, 1)	(2, 1, 1)
87.50	0.4782	4.00	8	(2, 0, 0)	(2, 2, 0)
101.0	0.5954	4.99	10	(2, 1, 0)	(3, 1, 0)
115.5	0.7153	6.00	12	(2, 1, 1)	(2, 2, 2)
131.8	0.8333	6.98	14	???	(3, 2, 1)
154.1	0.9498	7.97	16	(2, 2, 0)	(4, 0, 0)

EXAMPLE 2.9

Index the following PXD data (2θ in degrees, $\lambda = 1.5418$ Å) obtained from a crystalline sample of C_{60} (fullerene) and determine the lattice type and lattice parameter.

2θ	10.85	12.53	17.76	20.86	21.79
	25.22	27.52	28.25		

ANSWER

The process, summarised in Table 2.9, follows that described for indexing PXD data from a material adopting a face-centred cubic lattice. By appropriate choice of a common ratio between $\sin^2\theta$ values the reflections can be assigned Miller indices shown in column 5; as all reflections have (h, k, l) all odd or all even the lattice type is **F**.

TABLE 2.9 Indexing table for Example 2.9

2θ	$\sin^2\theta$	Apparent ratio	True integer ratio	Miller indices
10.85	0.008938	1.00	3	(1, 1, 1)
12.53	0.01191	1.332	4	(2, 0, 0)
17.76	0.02383	2.666	8	(2, 2, 0)
20.86	0.03277	3.667	11	(3, 1, 1)
21.79	0.03572	3.996	12	(2, 2, 2)
25.22	0.04766	5.332	16	(4, 0, 0)
27.52	0.05658	6.329	19	(3, 3, 1)
28.25	0.05955	6.663	20	(4, 2, 0)

Using a high-angle line and substituting back into Eqn 2.9 gives $a = 14.13$ Å.

SELF TEST

Determine the diffraction angle of the next highest angle diffraction peak, above the (4, 2, 0) reflection, expected to be seen for crystalline C_{60} in its PXD data.

EXAMPLE 2.10

X-rays are scattered by electrons in crystalline compounds, which means that atoms and ions with similar electron configurations scatter X-rays to an almost identical degree—and therefore 'look' the same to an X-ray. All the potassium halides crystallise with the sodium chloride (rock salt) structure. Explain the following observation.

The powder X-ray diffraction patterns of KF, KBr, and KI are indexed on face-centred cubic lattices of dimensions $2(r_+ + r_-)$ while that of KCl can be indexed on a primitive cubic lattice of dimensions half that, $(r_+ + r_-)$. (r_+ and r_- refer to the ionic radii of the K^+ and the halide ion, respectively.)

ANSWER

Because K^+ and Cl^- are isoelectronic they scatter X-rays to almost an identical extent, that is, X-rays 'see' these two ions as being almost identical. Consideration of the rock salt structure (Figure 2.33) shows that assigning all the sites occupied by cations and anions as identical produces a new repeating unit which has a primitive cubic lattice with half the repeat of the original face-centred cubic unit cell. The other anion–cation pairs are not isoelectronic so the observed PXD patterns are representative of the true face-centred cubic structure.

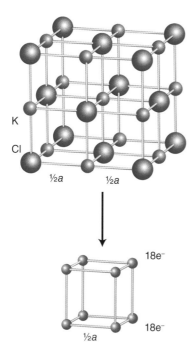

18e⁻

18e⁻

½a

FIGURE 2.33 The structure of KCl and the pseudo-primitive unit cell of half the dimensions produced when considering K⁺ and Cl⁻ as isoelectronic, 18e⁻, species that scatter X-rays to an almost equivalent degree.

SELF TEST

Which of the Group II sulfides, MgS, CaS, SrS, or BaS, which all adopt the rock salt structure, will display a PXD pattern that can be indexed using a primitive lattice type?

2.10 Extracting simple structural information using PXD data from cubic materials

Once the PXD data from a cubic material have been indexed, the lattice parameter and type can also be obtained. This information, together with a basic knowledge of the most common simple structure types, often allows the extraction of useful information such as interatomic distances.

The simplest structure types are often those of metallic elements many of which adopt structures with face-centred cubic (fcc, *F* lattice) and body-centred cubic (bcc, *I* lattice) lattice types (Figure 2.34). In both these cases the shortest M–M distance can be determined and, hence, the metal radius.

EXAMPLE 2.11

Calculating a metal radius

The PXD pattern of tungsten metal (W) can be indexed using a cubic lattice parameter of 3.187 Å and the systematic absences show the lattice type to be body centred. The PXD data from iron metal at 1000°C (γ-Fe) was indexed with a fcc lattice type and a lattice parameter of 3.660 Å. Calculate metallic radii for tungsten and iron.

ANSWER

The structure types for the cubic metal unit cells, bcc and fcc, are shown in Figure 2.34. Tungsten metal has the

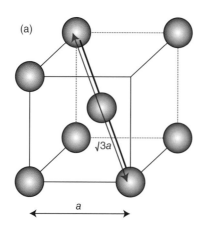

(a)

√3a

a

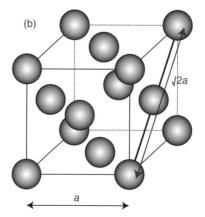

(b)

√2a

a

FIGURE 2.34 The common cubic metal unit cells, (a) body-centred cubic and (b) face-centred cubic, showing how the distances between metal atoms can be derived knowing the lattice parameter.

structure type shown in (a), and iron that in (b). For (b), application of Pythagoras's theorem to a cube face shows that the face diagonal is $\sqrt{2} \times a$ (the lattice parameter) and the Fe–Fe distance is $\sqrt{2}a/2 = 2.588$ Å. So the iron metallic radius is 1.294 Å. For tungsten metal in (a), application of Pythagoras's theorem to a cube body-diagonal shows that it equals $\sqrt{3} \times a$. So the W–W distance is $\sqrt{3}a/2 = 2.760$ Å. So the tungsten metallic radius is 1.380 Å.

SELF TEST

The structure of solid C_{60} is face-centred cubic arrangement C_{60} molecules. Use the lattice parameter for solid C_{60} derived in Example 2.9 to obtain an estimate for the size of a single C_{60} molecule.

Similar calculations of interionic distances can be undertaken for simple ionic solids, such as the Group 1 halides or the Group 2 oxides and halides. These compounds often adopt cubic unit cells, including the well-known rock salt (NaCl) structure and the fluorite structure (CaF_2). Table 2.10 summarises some important cubic structure types.

TABLE 2.10 Lattice types of common cubic structures

Structure type/ name	Lattice type	Stoichiometry and typical compositions
Rock salt	Face-centred	AX: NaCl, MgO
Zinc blende/ sphalerite	Face-centred	AX: ZnS, CuCl, CdS, GaP
Fluorite/antifluorite	Face-centred	AX_2: CaF_2, ZrO_2, Li_2O
Caesium chloride	Primitive	AX: CsCl, TlI, CuZn
Perovskite	Primitive	ABX_3: $SrTiO_3$

EXAMPLE 2.12

Structural information from PXD data from a simple salt

The PXD pattern ($\lambda = 1.5418$ Å) of TlBr at 298 K showed reflections at the following positions:

2θ 22.313 31.761 39.161 45.533 51.272
     56.582 66.360 70.967

A PXD data set collected at 200 K showed reflections at the following positions:

2θ 23.366 27.044 38.619 45.628 47.780
     55.762 61.274 63.045 69.883

Interpret these data and determine the shortest Tl–Br distance at the two temperatures.

ANSWER

At 298 K the data can be indexed using the method described previously as follows:

2θ 22.313 31.761 39.161 45.533 51.272
     56.582 66.360 70.967

Miller indices (1, 0, 0) (1, 1, 0) (1, 1, 1) (2, 0, 0)
     (2, 1, 0) (2, 1, 1) (2, 2, 0) (3, 0, 0)/(2, 2, 1)

There is no systematic absence, so the lattice type is primitive (**P**); the lattice parameter can be obtained as 3.9842 Å. A primitive cubic structure type for a 1:1 salt means that TlBr is likely to adopt the CsCl structure type, P-type lattice, at room temperature (Table 2.10). The shortest Tl–Br distance in the CsCl structure type is $\sqrt{3}/2 \times a = 3.450$ Å.

At 200 K the data can be indexed using the method described previously, Section 2.9.2.1, as follows:

2θ 23.366 27.044 38.619 45.628 47.780
     55.762 61.274 63.045 69.883

Miller indices (1, 1, 1) (2, 0, 0) (2, 2, 0) (3, 1, 1)
     (2, 2, 2) (4, 0, 0) (3, 3, 1) (4, 2, 0) (4, 2, 2)

Miller indices are all odd or all even so the lattice type is face centred (**F**); the lattice parameter can be obtained as 6.5940 Å. A face-centred cubic structure type for a 1:1 salt means that TlBr at 200 K is likely to adopt the NaCl structure type (Figure 2.10a) at room temperature (the zinc blende structure is also a possibility though unlikely for the large ions Tl^+ and Br^-). The shortest Tl–Br in the NaCl structure type is $a/2 = 3.297$ Å.

SELF TEST

The transition temperature between the two polymorphs of TlBr occurs at 232 K. Sketch a figure, similar in form to that of Figure 2.29, but over the angular range $2\theta = 10$–$60°$ showing the expected evolution of the diffraction pattern ($\lambda = 1.5418$ Å) of TlBr between 200 and 300 K.

2.11 The single-crystal X-ray diffraction technique—introduction

In single-crystal X-ray diffraction, a beam of X-rays impinges on a single crystal, and undergoes diffraction producing a pattern of spots with each spot corresponding to a different Miller index set, (h, k, l), for the crystal. As with PXD, the pattern produced was historically collected using a photographic film

FIGURE 2.35 Part of a typical single-crystal X-ray diffraction pattern; each spot represents diffraction from a unique set of Miller indices, (h, k, l).

though, these days, scintillation counters or, more commonly, image plates are used. As the crystal can be rotated and the detector aligned to various positions relative to the crystal all the possible reflections from the one crystal can be measured in terms of their angle of diffraction and intensities. Figure 2.35 shows part of a single-crystal X-ray diffraction pattern. The major advantages of single-crystal X-ray diffraction over the powder method are that for complex structures many thousands of reflection positions and their intensities can be measured. Analysis of these data normally leads to a full structure determination with atom positions, so that interatomic distances (bond lengths) and angles can be extracted. While it is possible to extract structures from powder diffraction data (see Section 2.16.4), this is normally a lengthy and difficult process and limited in terms of the complexity of structure that can be analysed.

Single-crystal X-ray crystallography has three key stages which involve producing and selecting a suitable crystal, data collection, and data analysis. Producing and finding a suitable single crystal can be problematic for some compounds, though with modern diffractometers, using image plate detectors, a crystal of the order of 50 μm in all dimensions is sufficiently large. Smaller crystals, down to 10 μm in one or two directions, can often be studied on modern laboratory equipment particularly if the material contains good X-ray scatterers, that is, heavy, electron-rich atoms, such as most metals. Using synchrotron radiation X-ray sources the study of much smaller crystals down to a few microns or less in each direction becomes possible.

In the second stage, data collection, the crystal is mounted in the X-ray beam and initially the unit cell is determined. Using an image plate detector around 10–30

diffraction images are collected adopting different crystal and detector positions. These yield angular positions for several tens or hundreds of reflections. Computer programs then analyse these data to determine the lattice parameters and, therefore, the crystal system. A full data collection then follows. For an image plate detector this involves gradually rotating the crystal in the X-ray beam around all its various axes and also rotating the detector around the crystal. Under these experimental conditions, every possible reflection will be aligned onto the detector at some point and the individual intensities that arise can be measured. A full data set typically contains thousands or tens of thousands of reflections and is effectively a list of Miller indices and their diffraction intensities.

The final step in a crystal structure determination involves working back from the diffraction peak intensities to obtain the positions of the atoms in the unit cell. The basis of this process is explained in the following sections. This analysis may also draw on additional chemical information such as the composition of the compound and its density. The level of detail in the crystallographic model obtained depends to some degree on the complexity of compound being studied. **Small-molecule crystallography,** which covers a lot of inorganic chemistry including coordination compounds, many materials, and organometallic chemistry, concerns structures that contain up to several hundred non-symmetry related atoms in the unit cell. For these structures it becomes possible to discern the exact position of every atom. For more complex structures, which may have many thousands of atoms in a very large unit cell (typically over 100 000 Å^3; these systems include complex macromolecules, such as proteins and dendrimers), the determined structures have historically been less well resolved and only the overall distribution of electron density is determined so that the positions of, say, chains of atoms are extracted. However, more recent advances with high-intensity X-ray sources, such as synchrotrons (Chapter 1, Section 1.6.5), have allowed atomic positions to be found for many bioinorganic, metalloprotein structures, which often adopt very large unit cells containing thousands of atoms.

2.12 Reflection intensities and structure factors in X-ray diffraction

In order to understand the basis of the method by which single-crystal X-ray diffraction data may be analysed to produce the arrangement of atoms from its origins, a useful analogy may be made with the way in which we see objects, particularly using a microscope. This process can also be thought of as 'back-transforming' the diffraction data to obtain a regular arrangement of scattering centres, the atoms, which produced the diffraction pattern. Large objects are seen by the eye because a portion of any light falling on them is scattered and some of this light is collected by the eye's lens and focused to produce an image on the retina. As light consists of waves, which have a phase and an intensity, the image of an object on the eye is a representation of the way in which the object scattered light, that is, information on the nature of the object is carried in the intensity of the observed light and its phases. This analogy can be extended to smaller objects as viewed under a microscope, where the lens collects some of the scattered visible light from a microscopic object and, through refraction, produces an image. To image even smaller objects much shorter wavelengths of electromagnetic radiation are required—hence the use of X-rays (wavelengths around 0.5–2.5 Å) to study interatomic spacings (with bond lengths typically between 1 and 3 Å).

The fundamental basis of the analysis method in single-crystal X-ray diffraction involves reconstructing the positions of the atoms that caused the diffraction using the measured reflection intensities. The mathematical relationship between an object and the scattering pattern it produces is a **Fourier transform** (Section 1.5), so the intensities of the diffraction pattern and the arrangement of atoms (and their associated electrons as the scattering centres) in the unit cell of the crystal structure are related to each other by a Fourier transformation: the diffraction pattern is the Fourier transform (FT) of the electron density, and the electron density is the FT of the diffraction pattern.

Each scattered X-ray contributing to the diffraction pattern is defined by two numerical values, its amplitude denoted by $|F|$ and the phase of the diffracted wave, φ. Figure 2.36 illustrates how these values are associated with different waves and shows $|F|$, the height of the wave, and the phase, the offset from a defined origin.

The amplitude and phase of a diffracted wave can be represented by a single complex number in the form

$$F = |F|e^{i\varphi} \qquad \textbf{Eqn 2.11}$$

So that for each reflection with Miller indices we have

$$F_{hkl} = |F_{hkl}|e^{i\varphi_{hkl}} \qquad \textbf{Eqn 2.12}$$

F_{hkl} is known as the structure factor of the reflection with the Miller indices h, k, and l. In the X-ray diffraction experiment, the intensity of a reflection, I_{hkl}, is measured. This is related to the structure factor through the expression

$$I_{hkl} = C_{hkl}F_{hkl}{}^2 \qquad \textbf{Eqn 2.13}$$

where C_{hkl} is a calculable constant dependent on the sample absorption and various instrumental factors. The structure factor for any reflection, F_{hkl}, is the sum of all the scattering of an X-ray beam by all the electrons in the crystal (Figure 2.37).

The crystal is constructed from numerous identical repeating units (the unit cells) and the scattering factor of a reflection with Miller indices (h, k, l) is the sum of the scattering from all electrons in the unit cell. A rigorous mathematical analysis of this scattering summation (an integration) gives the expression

$$F_{hkl} = \int_{unit\ cell} \rho_{xyz}\, e^{2\pi i(hx+ky+lz)}\, dV \qquad \textbf{Eqn 2.14}$$

where ρ_{xyz} is the electron density at a point with co-ordinates (x, y, z) in the unit cell. The structure factor (amplitude and phase) for reflection hkl is obtained by taking the value of the electron density at each point in the unit cell, ρ_{xyz}, multiplying it by the complex number

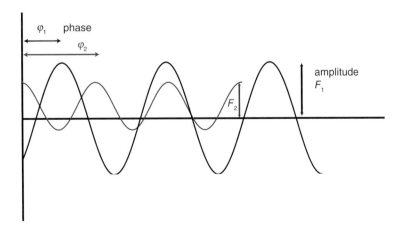

FIGURE 2.36 Two examples of scattered waves defined by their amplitudes, F_1 and F_2, and their phases, φ_1 and φ_2 relative to the origin.

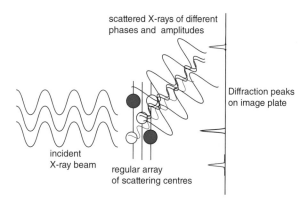

FIGURE 2.37 Scattering of a monochromatic X-ray beam produces a diffraction maximum whose structure factor (amplitude and phase), F_{hkl}, is the summed result from all the scattered waves.

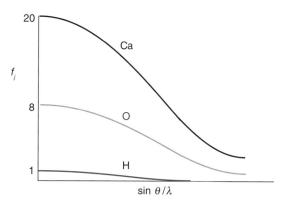

FIGURE 2.38 The structure factor, f_j, shown as a function of $\sin \theta / \lambda$ for Ca ($Z = 20$), O ($Z = 8$), and H ($Z = 1$).

$e^{2\pi i(hx + ky + lz)}$, which describes the phase, and summing (integrating) these values over the whole cell volume V. Undertaking this calculation using the continuous function ρ_{xyz} across all points in the unit cell would be very cumbersome, but we know that electrons are associated with atoms. Therefore, it is easier to assign the electrons to the points in the unit cell (coordinates (x, y, z)) where the atoms are located and with the number of electrons at that point equivalent to the number on that atom (its atomic number). The above expression for the structure factor then simplifies to

$$F_{hkl} = \sum_j f_j e^{2\pi i(hx+ky+lz)}$$

$$= \sum_j f_j [\cos 2\pi(hx + ky + lz)$$

$$+ i \sin 2\pi(hx + ky + lz)] \qquad \text{Eqn 2.15}$$

where f_j represents the **scattering factor** for an atom j located at (x, y, z); it defines how well an atom scatters X-rays as a function of Bragg angle. This equation (Eqn 2.15) can be seen to represent the scattering shown pictorially in Figure 2.37, that is, the total structure factor for a reflection with the Miller indices (h, k, l) is the sum of all scattered wave amplitudes (determined through the terms f_j) and their phases (determined through the complex functions $\cos 2\pi(hx + ky + lz) + i \sin 2\pi(hx + ky + lz)$) from all the atoms in the unit cell.

As X-rays are scattered by electrons the value of scattering factor, f_j, is directly proportional to the number of electrons in an atom or ion. At the Bragg angle $\theta = 0°$, f_j can be taken as the atomic number of an atom. However, as the electron cloud around a nucleus has dimensions of a similar magnitude to the wavelength of X-rays (1–2 Å) the scattering from different parts of the electron cloud will not be perfectly in phase except in the straight-on

direction, $\theta = 0°$. This effect increases as a function of scattering angle. Therefore the scattering that takes place from an atom, that is, the scattering factor for that atom, is a function dependent on angle, decreasing with increasing θ (or $\sin \theta$) (see Figure 2.38).

If a crystal structure is known, as defined by the unit cell dimensions and where the atoms are located within it, the calculation of F_{hkl} can be carried out to reproduce the diffraction pattern obtained experimentally from that crystal. This can be done for every possible reflection with Miller indices (h, k, l) to produce a set of calculated structure factors, each with an amplitude $|F_{hkl}|$ and a phase φ_{hkl}.

EXAMPLE 2.13

Calculating a structure factor

Use Eqn 2.12 to calculate $|F_{hkl}|$ for the (2, 1, 0) reflection from CsCl which adopts a primitive cubic unit cell with Cs on the cell corner and Cl at the cell centre. Take f_{Cs} as 41 and f_{Cl} as 13 for the diffraction angle, 2θ, of the (2, 1, 0) reflection.

ANSWER

The unit cell can be described with Cs on (0, 0, 0) and Cl on (½, ½, ½). As this structure is centrosymmetric and for every atom at (x, y, z) there is another one at $(-x, -y, -z)$, the structure factor expression (Eqn 2.15) can be simplified (as $\sin(-x) = -\sin(x)$ these parts of the equation cancel out, that is, the phase for a centrosymmetric structure is always 0 or π) to

$$F_{hkl} = \sum_j f_j \cos 2\pi(hx + ky + lz)$$

(Note that the calculation needs to be undertaken using radians as the structure factor expression is derived using these units.)

$$F_{210} = f_{Cs} \cos 2\pi \left[(2 \times 0) + (1 \times 0) + (0 \times 0) \right]$$
$$+ f_{Cl} \cos 2\pi \left[(2 \times \tfrac{1}{2}) + (1 \times \tfrac{1}{2}) + (0 \times \tfrac{1}{2}) \right]$$
$$F_{210} = f_{Cs} \cos 0 + f_{Cl} \cos 3\pi = f_{Cs} - f_{Cl} = 41 - 13 = 28$$

SELF TEST

(a) Calculate $|F_{hkl}|$ for the (2, 0, 0) reflection for CsCl using $f_{Cs} = 45$ and $f_{Cl} = 15$.

(b) Repeat the calculation of $|F_{hkl}|$ for the (2, 1, 0) reflection but for CsI instead of CsCl and using $f_I = 41$. Comment on your answer.

While the calculation of the diffraction pattern structure factors for a known crystal structure can be carried out using the procedure just discussed, the reverse process is more difficult. This is because it is the intensities of the reflections that are measured in the diffraction pattern. As we have seen, the measured intensity is proportional to the structure factor squared, so

$$I_{hkl} \propto F_{hkl}{}^2 \qquad\qquad \textbf{Eqn 2.16}$$

Taking the square root of a measured intensity will give the magnitude of the structure factor $|F_{hkl}|$ but provides no information on its phase. Working out the phase associated with each reflection is a key stage in the analysis of single-crystal X-ray diffraction data which we will discuss in the following section as part of the full method of structure solution.

2.13 Structure solution from single-crystal X-ray diffraction data

The previous sections have described some of the mathematical basis of the single-crystal X-ray diffraction method and the relationship between X-ray scattering by electrons and the diffraction pattern produced. We now tackle how the data sets are obtained and the important stages involved in their analysis to produce an accurate crystal structure. Figure 2.39 summarises the stages involved in this process as a flowchart indicating what information is extracted at each stage; each of these is then dealt with in more detail. While computers and sophisticated software are essential, nowadays, to undertake a full structure determination, an outline

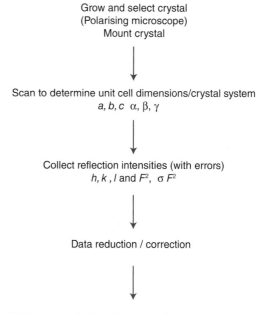

Grow and select crystal
(Polarising microscope)
Mount crystal

Scan to determine unit cell dimensions/crystal system
a, b, c α, β, γ

Collect reflection intensities (with errors)
h, k, l and F^2, σF^2

Data reduction / correction

Initial structure solution - Patterson, direct, or other methods

Structure refinement, Fourier methods to complete model

Check solution, bond lengths, composition
Produce crystallographic information file (cif)

FIGURE 2.39 The stages involved in determining a crystal structure using single-crystal X-ray diffraction.

of the processes involved in this analysis is provided in this section.

2.13.1 Obtaining suitable single crystals

The sample must be a single crystal, in which all the unit cells are identical and are aligned in the same orientation so that they scatter to give a clear diffraction pattern consisting of well-defined maxima or spots. While the outward appearance of a regular shape is often a useful indicator of crystal quality some crystals with poorly formed crystal faces are internally of high regularity and some amorphous glassy materials may look like single crystals. Normally the quality of a crystal is investigated by quickly collecting some initial portions of the diffraction pattern and checking the reflections are well-defined spots.

The larger the crystal the more intense the diffraction pattern so, in general, larger crystals will provide better quality data. However, as X-rays are also absorbed, particularly by the heavier elements in a crystal, then corrections to the data need to be made, which can become problematic for very large crystals. Also, as the dimensions of the incident X-ray beam are less than a millimetre and it is essential to bathe the crystal in a uniform beam this also puts an upper limit on the crystal size. Generally crystals of around 0.05–0.5 mm in each dimension are most suitable for single-crystal X-ray diffraction experiments, though with the most modern instrumentation and synchrotron X-ray sources crystals as small as 1–10 μm can be studied. Such crystals usually need to be examined and handled under a microscope. A microscope with polarising filters provides a useful test of the quality of a crystal. Polarised light passing through a crystal will be optically rotated. Because of this optical rotation polarised light that has passed through a perfect single crystal is normally fully extinguished by an analysing polarising filter. A 'single crystal' which transmits only partly extinguished, polarised light may have fault planes running through it or be constructed of many smaller crystals and so be unsuitable for single-crystal X-ray diffraction analysis.

2.13.2 **Instrumentation and collection of diffraction data**

Once a single crystal is selected it is mounted on a 'pip', a fine glass fibre or polymer grid, which will hold it firmly in the X-ray beam. Apart from the sample itself, no crystalline material should be in the X-ray beam, so the crystal can be glued (most glues are fully amorphous) to a fine glass fibre. Alternatively, the crystal can be manipulated in an oil, and so coated, and then attached to a glass fibre or thin plastic sample mount. The sample is then cooled on the diffractometer, typically to around 100–150 K, which freezes the oil to an amorphous solid that holds the crystal in place. The oil also provides a protective coating around the crystal which allows air sensitive samples to be handled and studied. The crystal mounted on the glass fibre or plastic sample mount is then attached to the goniometer head (Figure 2.40). The single-crystal diffraction experiment involves rotating the crystal in the beam during exposure and the goniometer allows the lateral adjustments required to position the crystal accurately and with a precision of a few microns. The goniometer head also ensures centring of the crystal in the X-ray beam and facilitates rotations of the crystal during data collection to an accuracy of a hundredth of a degree.

The diffraction conditions represented by the Bragg equation are stringent, and will be satisfied for only very few reflections for a given, general, orientation of

FIGURE 2.40 A goniometer head on which a single crystal is mounted and rotated during single-crystal data collection.

a stationary crystal bathed in an X-ray beam. For a fixed crystal orientation, the initial pattern recorded on X-ray image plate will only show a few spots. In order to bring more lattice planes into diffracting positions the crystal must be rotated in the X-ray beam. Recording the whole of the diffraction pattern involves rotating the crystal around its various axes, in small angular amounts, and collecting the diffracted intensity at a particular point or several reflection intensities on an image plate. The most widely used type of diffractometer has three rotation axes for the crystal and one for the detector, known as **a four-circle diffractometer**. With a point X-ray detector the reflection positions and intensities are observed one at a time. Modern X-ray detectors or image plates act like photographic film. They can record over a large area and are position-sensitive, allowing the intensities and positions of multiple reflections to be recorded simultaneously and thereby shortening data collection times.

2.14 **Solving the structure from single-crystal diffraction data**

Having collected the diffraction data its analysis involves 'working back' to a structural model of the atom locations in the unit cell which gave rise to the unique diffraction pattern. This process is a reverse Fourier transform. The equation for the electron density at a point (x, y, z), ρ_{xyz}, is

$$\rho_{xyz} = \frac{1}{V} \sum_{hkl} \left| F_{hkl} \right| e^{i\phi_{hkl}} e^{-2\pi i (hx + ky + lz)} \qquad \textbf{Eqn 2.17}$$

Whereas a simplified expression for F_{hkl} was given as Eqn 2.15, the full expression for F_{hkl}, known as the **structure factor expression**, includes additional terms

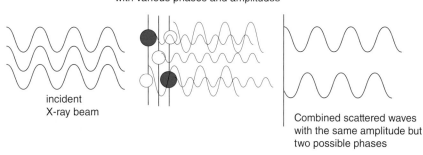

X-rays scattered by different atoms
with various phases and amplitudes

incident
X-ray beam

Combined scattered waves
with the same amplitude but
two possible phases

FIGURE 2.41 The scattering of X-rays by atoms in a crystal gives rise to diffraction maxima whose intensity (amplitude) can be measured but whose phase cannot be determined directly.

which, for example, allow for the local displacement of atoms caused by thermal motion and non-fully occupied sites (site occupancy factors). The amplitudes $|F_{hkl}|$ can be measured as $(I_{hkl})^{\frac{1}{2}}$ but the phases of the reflections ($e^{i\varphi}$ in the above equation) are not known and the calculation of ρ_{xyz} cannot be carried out directly. This can be seen from Figure 2.41 in that summing the scattered X-rays from several atoms in the unit cell gives a resultant wave whose amplitude can be measured but the phase of which could be any value. Therefore some method of determining or assigning phases is needed. Of the various methods used, two are, by far, the most common and important, the Patterson method and direct methods.

The Patterson synthesis works best for structures containing one or just a few atoms with significantly more electrons than the rest ('heavy atoms'), such as in metal complexes. It allows likely positions for the heavy atom(s) in the unit cell to be determined and these can then be used to assign initial reasonable values to the phases of the reflections. Consideration of Eqn 2.15 shows that F_{hkl} will be dominated by f_j(heavy atoms) as they are much larger than terms with f_j(light atoms) so phases calculated using just these values are likely to be correct. Subsequent structure refinement and identification of the lighter atom sites allows more accurate phasing of the reflections and the full crystal structure to be extracted.

The direct method process is more appropriate for 'equal atom' structures, such as organic compounds where the majority of the scattering atoms have similar atomic numbers (C, O, and N). This uses statistical methods to propose reasonable possibilities for the phases of a set of reflections allowing initial electron density maps to be calculated.

2.14.1 The Patterson synthesis

While the value of $|F_{hkl}|\,e^{i\phi}$ is not known the quantities $|F_{hkl}|^2$ are determined experimentally as they are directly related to the reflection intensities, so the function

$$P_{xyz} = \frac{1}{V} \sum_{hkl} |F_{hkl}|^2 \, e^{-2\pi i(hx+ky+lz)} \qquad \textbf{Eqn 2.18}$$

where P_{xyz} is the Patterson function, can be calculated precisely from the experimental data. Undertaking this calculation for a large number of points (x, y, z) in the unit cell will produce values of P_{xyz} that can be plotted to produce a **Patterson map** (Figure 2.42). The Patterson map is a map of vectors between pairs of atoms in the structure and heights of the peaks in the map are proportional to the scattering factors of those two atoms multiplied together. Thus if there are two atoms in the unit cell with coordinates A_1 at (x_1, y_1, z_1) and A_2 at (x_2, y_2, z_2), respectively, then there will be peaks in the Patterson map at $(x_1 - x_2, y_1 - y_2, z_1 - z_2)$ and $(x_2 - x_1, y_2 - y_1, z_2 - z_1)$ representing the vectors A_1 to A_2 and A_2 to A_1. Of course with many atoms in the unit cell there will be a large number of vectors and, therefore, peaks in the Patterson map. This is why the technique is most applicable to structures that contain a few heavy atoms, as the strongest peaks in the map will correspond to the few vectors between these heavy atoms. Each of the strong peaks in the Patterson map can be associated with a vector between two heavy atoms and as a result the likely coordinates, (x, y, z), of the heavy atoms determined. Figure 2.42 illustrates the origin of the information in a Patterson map for a simplified structure that contains three atoms, one at a defined origin $(0, 0, 0)$ and two at (x_1, y_1, z_1) and (x_2, y_2, z_2).

Part of the known structure of a crystalline material is shown in Figure 2.42(a). The possible interatomic vectors have been plotted in Figure 2.42(b) and there will also be a 'vector' between an atom and itself $(0, 0, 0)$. In the Patterson map the vectors will have identical directions to those on the crystal structure relative to the origin of the map and end on maxima in the Patterson function (P_{xyz}); the heights of these peaks in the map are proportional to the product of the atomic numbers of the two atoms involved. The positions of these maxima are recorded using the coordinate system (u, v, w) to represent

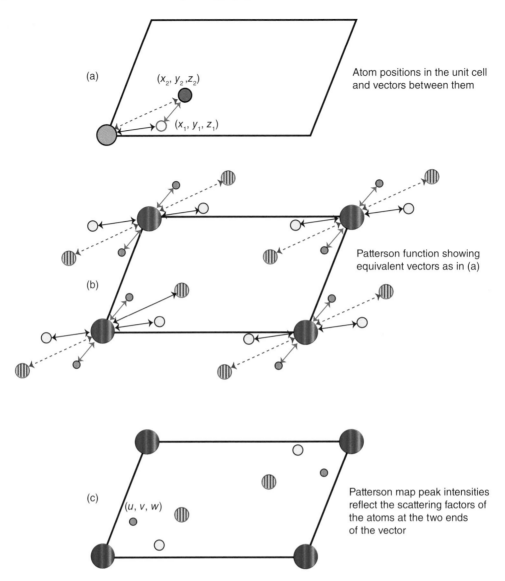

FIGURE 2.42 The origin of the peaks observed in a Patterson map and its interpretation. (a) Three atom positions marked with shaded circles; the radius of the circle represents the magnitude of the scattering factor for that atom. The grey-shaded atom at the origin is the 'heavy atom'. (b) The possible vectors between these atoms and (c) these vectors (u, v, w) plotted on the same unit cell (with the coordinate system $(u, v, w) \equiv (x, y, z)$) and their heights represented by the size of the circle radius.

the differences between the coordinates of each pair of atoms in the original crystal structure. So one maximum in the Patterson map will be at (u, v, w) with $u = x_1 - x_2$, $v = y_1 - y_2$, and $w = z_1 - z_2$. Note that at the origin, and the equivalent positions at the corners of the Patterson cell, there is always the highest peak maximum, as this corresponds to the interatomic vectors of each atom with itself, coordinates $(0, 0, 0)$. In this simple case, with just three atoms and one heavy one at the origin, interpreting the Patterson map would be relatively simple as the strongest vectors (u, v, w) also correspond to the next heaviest atom positions, as $(x_1 = y_1 = z_1 = 0)$. See Example 2.14 for a further simple example of interpreting a Patterson map.

EXAMPLE 2.14

Analysis of a Patterson map

The compound $Zr(OH)_2(NO_3)_2.5H_2O$ crystallises with a unit cell that contains two zirconium atoms whose positions are related through a centre of symmetry so they can be denoted as being at (x, y, z) and $(-x, -y, -z)$. The largest peaks in the Patterson map were calculated using single-crystal X-ray diffraction data from $Zr(OH)_2(NO_3)_2.5H_2O$, including one at $(u, v, w) = (0.896, 0.886, 0.472)$. Determine the coordinates of the zirconium atoms in the structure.

ANSWER

With the centre of symmetry, one vector present in the Patterson map will be between (x, y, z) and $(-x, -y, -z)$, that is, $(2x, 2y, 2z)$, so we can assign the strong peak at $(0.896, 0.886, 0.472)$ to these values. Hence, we can calculate the position of the zirconium atoms as near $(0.448, 0.443, 0.236)$ and the symmetry related position $(-0.448, -0.443, -0.236)$.

Once the heavy atom positions in the unit cell have been located from the Patterson map it is then possible to calculate values for the amplitudes and phases of the reflections using just those atom positions. These calculated values should show some correspondence to the actual measured intensities and for some reflections their true phases, especially as the heavy atoms are the main contributors. The rest of the structure can then be determined, as discussed in Section 2.14.3, using Fourier electron density difference maps.

With complex structures containing many tens of atoms with similar scattering powers the Patterson map will contain many hundreds of vectors so be very difficult to interpret; in such case direct methods (Section 2.14.2) are used to phase reflections.

2.14.2 Direct methods

Direct methods involve trial and error techniques to assign phases to the strongest reflections observed in the diffraction pattern. The way in which these phases are assigned is not totally random and they exploit constraints or statistical correlations between the phases that result from the fact that the electron density at each point in the unit cell, ρ_{xyz}, must be a positive real number. These trial assignments and calculations, carried out using computers, can then be assessed as to how well they fit the experimental data and for recognisable structural features, for example, reasonable interatomic distances of 1–2 Å between peaks in an electron density map. From the electron density maps produced a significant portion of the structure is often apparent and the rest of the structure can then be determined, as discussed in Section 2.14.3.

2.14.3 Completing the structure

Once a portion of the structure, that is, the part of the electron density map corresponding to positions of some of the atoms, has been determined using Patterson or direct methods it becomes possible to calculate a set of $|F_{hkl}|^{calc}$ and associated phases, φ_{hkl}, using the structure factor expression (Eqn 2.15). These values can be compared with the experimental values $|F_{hkl}|^{obs}$ and a convenient

way of doing this is by calculating a **residual** or **R-factor** which measures the overall difference between the two:

$$R = \frac{\sum_{hkl} \left| \left| F_{hkl}^{obs} \right| - \left| F_{hkl}^{calc} \right| \right|}{\sum_{hkl} \left| F_{hkl}^{obs} \right|}$$ **Eqn 2.19**

Ideally this residual should tend to zero as the structural model gets closer to replicating the true electron density at all points in the unit cell. This difference between the observed and calculated structure factors can also be used to improve the structural model and find sites of electron density not located in the initial Patterson or direct methods analysis. This normally involves calculation of a difference **Fourier map** by which a Fourier transformation is carried out using the values of $|F_{hkl}|^{obs} - |F_{hkl}|^{calc}$ and phases, which are calculated from the partially right $|F_{hkl}|^{calc}$, to generate the electron density map. This electron density map will not be exact as the phases assigned using $|F_{hkl}|^{calc}$ will not all be correct but it will show regions of unassigned electron density and to a large degree how much density needs to be added at that point (and, therefore, the atom type that is missing), as shown in Figure 2.43.

Atoms can then be assigned to these sites and further calculations of $|F_{hkl}|^{calc}$ undertaken leading to a hopefully reduced residual (Eqn 2.19) and a model closer to reality as the set of calculated reflection intensities becomes more aligned with the experimental one. Iterative application of this procedure should eventually lead to a complete structural model. Further refinement of the structural model involves calculating and applying small shifts to atom positions to minimise further the residual,

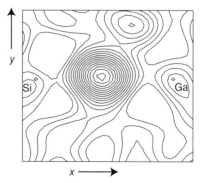

FIGURE 2.43 Part of a two-dimensional Fourier difference map with unassigned electron density shown as contour lines. In this structure silicon and gallium atom positions have already been identified at the positions marked. A large 'difference peak' can be seen between Si and Ga indicating that an atom (in this case an oxygen atom) needs to be added to the structural model at that position.

as well as allowing for other features of the structural model, such as atom displacements caused by thermal motion. In this way the proposed model approaches the true electron density distribution in the unit cell and the *R*-factor diminishes. A good structural model will yield values for the *R*-factor of less than 0.05 demonstrating only a few per cent difference between the experimental structure factors and the calculated ones. Obviously the model should also make chemical sense with reasonable distances between the atoms. However, where the electron density associated with an atom is low, as it is for hydrogen, it may not be possible to locate these scattering centres. This is especially true where hydrogen is present in combination with strong X-ray scatterers, such as transition metal ions, that will dominate the X-ray scattering making it hard to locate these weakly scattering atoms. Normally, hydrogen positions can be inferred because the nature of the structural unit present is known, for example, an acetate ligand will have three methyl hydrogen atoms. If the exact hydrogen positions are of interest, other techniques, such as neutron diffraction, normally have to be employed (see Sections 2.16.3 and 2.17.1).

EXAMPLE 2.15

Analysis of a Fourier map

The zirconium atom position in the structure of $Zr(OH)_2(NO_3)_2.5H_2O$ was determined in Example 2.14. Calculation of a difference Fourier map, $|F_{hkl}|^{obs} - |F_{hkl}|^{calc}$, using this zirconium position produced the map section (at $z = 0.24$) shown in Figure 2.44, which is centred near to the zirconium atom position marked. Determine the coordinates of two of the oxygen atoms in the structure directly coordinated to zirconium.

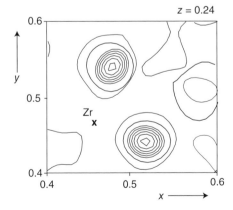

FIGURE 2.44 A difference Fourier map for $Zr(OH)_2(NO_3)_2.5H_2O$.

ANSWER

The difference Fourier map shows two clear peaks close to the zirconium atom which are likely to be oxygen atoms directly coordinated to it. Reading from the map, the approximate positions of these peaks and, therefore, the oxygen atoms are near (0.52, 0.44, 0.24) and (0.48, 0.54, 0.24).

2.15 Crystallographic information from single-crystal X-ray diffraction

At the end of a successful structure analysis the position of all the atoms in the unit cell will have been determined. This structural information also includes the crystal system, lattice type, additional translation symmetry, atomic displacement parameters (which provide information on the vibrational motion of the atoms) and also, for disordered structures, site occupancy factors. This information may be used to view the structure and also extract bond lengths and angles and so examine the interactions and bonding present. The structural information is readily presented and stored using a **crystallographic information file** (CIF). The key elements of a CIF are summarised in Figure 2.45.

The structure determination of a compound is usually an unambiguous characterisation of its composition and geometry. Techniques such as nuclear magnetic resonance (NMR, Chapter 3) and vibrational spectroscopy (Chapter 4) when applied to small molecules provide useful information on molecular geometry, types of site and their relative positions in a molecule but not detailed structural information, such as bond distances and angles. Also for larger molecules interpretation of NMR data often requires some knowledge of the molecular geometry and composition in assigning resonances. For this reason many journal publications require a single crystal structure determination to confirm the identity of a new compound.

CIFs are stored in databases which allow their easy retrieval and comparison with any putative new structure or compound. The most notable of these is the Cambridge Structural Database (CSD) which has over 750 000 entries. A subset of this is the Inorganic Chemical Structure Database (ICSD), around 160 000 entries, which contains all inorganic compounds (i.e. it omits those containing an organic unit; hence organometallic compounds are found in the CSD). With all this structural information in the databases it becomes possible to analyse it to understand structure and bonding trends across series of inorganic compounds. Some of the systematics that have been investigated in this way for inorganic compounds include bond lengths as a function of composition and ligand size, and the strengths of hydrogen bonds.

Compound composition, unit cell dimensions, and symmetry

Atom coordinates

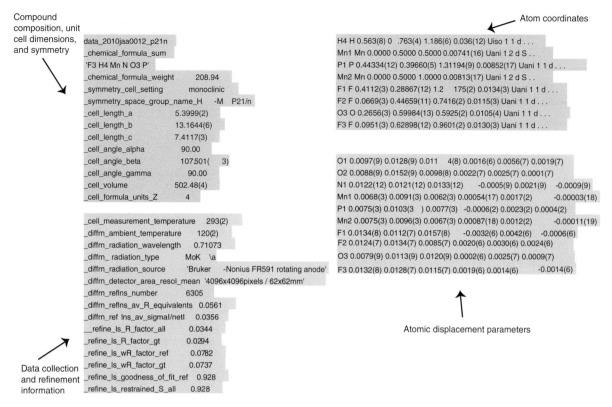

```
data_2010jaa0012_p21n
_chemical_formula_sum
  'F3 H4 Mn N O3 P'
_chemical_formula_weight        208.94
_symmetry_cell_setting          monoclinic
_symmetry_space_group_name_H  -M   P21/n
_cell_length_a            5.3999(2)
_cell_length_b            13.1644(6)
_cell_length_c            7.4117(3)
_cell_angle_alpha         90.00
_cell_angle_beta          107.501(   3)
_cell_angle_gamma         90.00
_cell_volume              502.48(4)
_cell_formula_units_Z     4

_cell_measurement_temperature    293(2)
_diffrn_ambient_temperature      120(2)
_diffrn_radiation_wavelength     0.71073
_diffrn_radiation_type        MoK  \a
_diffrn_radiation_source       'Bruker   -Nonius FR591 rotating anode'
_diffrn_detector_area_resol_mean '4096x4096pixels / 62x62mm'
_diffrn_reflns_number          6305
_diffrn_reflns_av_R_equivalents  0.0561
_diffrn_ref lns_av_sigmaI/netI   0.0356
__refine_ls_R_factor_all         0.0344
_refine_ls_R_factor_gt          0.0294
_refine_ls_wR_factor_ref         0.0782
_refine_ls_wR_factor_gt          0.0737
_refine_ls_goodness_of_fit_ref  0.928
_refine_ls_restrained_S_all      0.928
```

Data collection and refinement information

```
H4 H 0.563(8) 0  .763(4) 1.186(6) 0.036(12) Uiso 1 1 d . . .
Mn1 Mn 0.0000 0.5000 0.5000 0.00741(16) Uani 1 2 d S . .
P1 P 0.44334(12) 0.39660(5) 1.31194(9) 0.00852(17) Uani 1 1 d . . .
Mn2 Mn 0.0000 0.5000 1.0000 0.00813(17) Uani 1 2 d S . .
F1 F 0.4112(3) 0.28867(12) 1.2   175(2) 0.0134(3) Uani 1 1 d . . .
F2 F 0.0669(3) 0.44659(11) 0.7416(2) 0.0115(3) Uani 1 1 d . . .
O3 O 0.2656(3) 0.59984(13) 0.5925(2) 0.0105(4) Uani 1 1 d . . .
F3 F 0.0951(3) 0.62898(12) 0.9601(2) 0.0130(3) Uani 1 1 d . . .


O1 0.0097(9) 0.0128(9) 0.011    4(8) 0.0016(6) 0.0056(7) 0.0019(7)
O2 0.0088(9) 0.0152(9) 0.0098(8) 0.0022(7) 0.0025(7) 0.0001(7)
N1 0.0122(12) 0.0121(12) 0.0133(12)    -0.0005(9) 0.0021(9)   -0.0009(9)
Mn1 0.0068(3) 0.0091(3) 0.0062(3) 0.00054(17) 0.0017(2)    -0.00003(18)
P1 0.0075(3) 0.0103(3   ) 0.0077(3)  -0.0006(2) 0.0023(2) 0.0004(2)
Mn2 0.0075(3) 0.0096(3) 0.0067(3) 0.00087(18) 0.0012(2)    -0.00011(19)
F1 0.0134(8) 0.0112(7) 0.0157(8)   -0.0032(6) 0.0042(6)   -0.0006(6)
F2 0.0124(7) 0.0134(7) 0.0085(7) 0.0020(6) 0.0030(6) 0.0024(6)
O3 0.0079(9) 0.0113(9) 0.0120(9) 0.0002(6) 0.0025(7) 0.0009(7)
F3 0.0132(8) 0.0128(7) 0.0115(7) 0.0019(6) 0.0014(6)    -0.0014(6)
```

Atomic displacement parameters

FIGURE 2.45 An example of a crystallographic information file (CIF) annotated with key blocks of information.

With the crystallographic information available in a CIF, it is possible to draw a representation of the crystal structure using various software packages. These representations are useful for showing molecular geometries and also the interactions between neighbouring molecules. Information on the thermal motions of atoms may also be included in these diagrams; this is done by drawing a thermal ellipsoid whose axes' lengths represent the probability of the atom being at a position along that axis. One of the original, and still commonly used, software packages for drawing such diagrams is **ORTEP** (Oak Ridge Thermal Ellipsoid Program) and a typical molecular representation drawn using this package is shown in Figure 2.46.

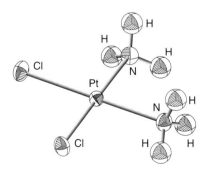

FIGURE 2.46 An ORTEP diagram of cisplatin, cis-$PtCl_2(NH_3)_2$.

Problems

Many of these problems require knowledge of basic solid-state chemistry and the structures of simple inorganic compounds.

2.1 Identifying crystal systems and lattice types

2.1.1 Explain what is meant by the term lattice. What are the lattice types of the following structure types?

(a) CsCl   (b) NaCl   (c) cubic ZnS

2.2 Drawing lattice planes and Miller indices

2.2.1 Draw a diagram to represent an orthorhombic unit cell of dimensions 3 Å × 5 Å × 4 Å. Mark on it (or copies of it) lattice planes with Miller indices

(1, 2, 4)   (1, 0, −1)   (0, 0, 1)

Mark the distance corresponding to d_{124} on the appropriate diagram.

2.3 d-Spacing calculations and Miller indices

2.3.1 For an orthorhombic unit cell of dimensions $a = 3$, $b = 5$, and $c = 8$ Å, draw on a diagram (or diagrams) of the unit cell, planes with the Miller indices

(2, 1, 2) and (4, 2, 4)

Use Eqn 2.4 to calculate d-spacing values for the (2, 1, 2) and (4, 2, 4) planes. How are these related?

2.4 Bragg equation calculations

2.4.1 A powder diffractometer uses the reflection from the (1, 1, 1) plane of a cubic germanium crystal ($a = 5.658$ Å) as a monochromator. Calculate the experimental diffraction angle, 2θ, from this plane for X-rays of wavelength 2.23 Å.

2.5 Systematic absences. Use an X-ray wavelength of 1.5418 Å.

2.5.1 Li_2O adopts the antifluorite structure with an **F**-type cubic lattice. The lattice parameter of Li_2O is 4.61 Å. What will be the 2θ value of the first reflection observed in its powder X-ray diffraction pattern?

2.5.2 Diamond has an **F**-type cubic lattice. What will be the Miller indices of the first five allowed reflections observed in its X-ray diffraction pattern?

2.5.3 The first reflection in the X-ray diffraction pattern of CsCl is found at $2\theta = 21.55°$. Calculate the lattice parameter of CsCl.

2.5.4 Draw a unit cell of iron metal, which adopts a face-centred cubic arrangement of metal atoms at high temperature. Mark on this diagram two adjacent planes with the Miller indices (1, 1, 0). Hence explain why the (1, 1, 0) reflection is absent in the powder X-ray diffraction pattern of iron metal.

2.6 Indexing of powder X-ray diffraction (PXD) data

2.6.1 The PXD pattern ($\lambda = 1.5418$ Å) collected from a sample of radium (Ra) metal showed reflections at the following values of 2θ (degrees)

24.45 34.85 43.04 50.12 56.53 62.50
68.15 73.60 78.89

Index the data; determine the lattice type and the lattice parameter.

Heating radium in fluorine gas afforded the product **A** containing 14.39% F whose powder diffraction pattern could be indexed using a cubic lattice parameter of 6.381 Å with systematic absences indicative of an **F**-centred lattice. Describe a likely structure for **A** and calculate the shortest Ra–F distance.

2.6.2

(a) Cadmium selenide and cadmium sulfide form a complete series of compounds with the general composition $CdS_{1-x}Se_x$. This material is used as a pigment whose colour varies from yellow at $x = 0$ to red at $x = 1$. Both CdS and CdSe adopt the cubic ZnS structure, which can also be described as Cd^{2+} ions occupying half the tetrahedral holes in a cubic close-packed arrangement of sulfide ions.

  (i) Draw the structure of CdS clearly labelling and marking the atom positions.

(ii) Given the following ionic radii

  S^{2-} 164 pm Se^{2-} 182 pm Cd^{2+} 67 pm,

  calculate the expected lattice parameters for CdS and CdSe.

(iii) Sketch a diagram showing how the lattice parameter of $CdS_{1-x}Se_x$ would be expected to vary as a function of x.

(iv) A sample of orange paint from a crime scene was analysed by powder X-ray diffraction ($\lambda = 1.5418$ Å) and showed reflections at the following values of 2θ (degrees)

  28.47 32.99 47.35 56.18 58.92 69.21

  Index these data; determine the lattice type and the lattice parameter.

(v) Hence determine the composition of the pigment in the orange paint assuming it to be from the $CdS_{1-x}Se_x$ solid solution.

2.6.3

(a) The ionic radii (Å) relevant to the caesium halides (CsF, CsBr, and CsI) may be taken as

  Cs^+ 1.88 F^- 1.29 Br^- 1.96 I^- 2.20.

  (i) Use radius ratio rules to predict the structure type adopted by each of these three caesium halides.

  (ii) Use the ionic radii to calculate values for the lattice parameter of each structure.

  (iii) Describe the three lattice types known for the cubic crystal system and state the systematic absences expected for each in their powder X-ray diffraction data.

  (iv) Predict the position of the first reflection observed in the powder X-ray diffraction patterns from CsF, CsBr, and CsI. Take $\lambda = 1.54$ Å.

  (v) The first reflection observed in the powder X-ray diffraction pattern of CsI is measured at 26.73°. Explain any difference between this value and the value calculated in (iv).

2.7 Structure factor calculations

2.7.1 Silver metal adopts a simple **F**-centred cubic structure with atoms at the coordinate positions (0, 0, 0), (½, ½, 0), (½, 0, ½), and (0, ½, ½). The structure is centrosymmetric so the structure factor expression simplifies to

$$F_{hkl} = \sum_j f_j \ \cos 2\pi(hx + ky + lz)$$

Calculate a value of the structure factor, in terms of f_{Ag}, for the (1, 0, 0), (1, 1, 0), and (1, 1, 1) reflections. Comment on your answer in respect of the systematic absences seen for **F**-centred lattices.

2.8 Structure solution methods

2.8.1 Decide whether Patterson or direct methods would be the most appropriate starting point for the structure solution of the following compounds.

(a) $NaSiAlO_4$ (b) $PbCO_3$ (c) $AgSnI_3$

ADVANCED DIFFRACTION TECHNIQUES

2.16 Powder X-ray diffraction

2.16.1 Analysis of PXD data from tetragonal and hexagonal systems

The indexing of powder X-ray diffraction (PXD) data from cubic crystal systems was covered in Section 2.8 and, because there is only a single lattice parameter, all the reflections had $\sin^2\theta$ values that were related through a single common ratio. The combination of the expression for the d-spacings in a tetragonal or hexagonal unit cell (Table 2.2) with the Bragg equation gives rise to expressions which contain terms based upon the two lattice parameters, a and c. Replacement of constant terms $\lambda^2/4a^2$ and $\lambda^2/4c^2$ by A and C, respectively, gives the following expressions:

$$tetragonal \ \sin^2\theta = A\ (h^2 + k^2) + C\ l^2 \qquad \textbf{Eqn 2.20}$$

$$hexagonal \ \sin^2\theta = A\ (h^2 + hk + k^2) + C\ l^2 \quad \textbf{Eqn 2.21}$$

where A and C incorporate the wavelength and the appropriate lattice parameter. For these crystal systems, $\sin^2\theta$ values obtained from the powder diffraction will be related through two multipliers. By consideration of the possible values which the Miller indices can take, it is feasible to derive the relationships between the various $\sin^2\theta$ values. This is done for tetragonal and hexagonal systems in Table 2.11.

Assigning these ratios to particular peaks by inspection or trial-and-error can be quite difficult; it would normally be done nowadays using computers. However, the method can be illustrated for a tetragonal system by consideration of a system where there is only a slight distortion from a cubic unit cell, that is, the lattice parameters a and c are very similar. In this case the relationship between the reflections seen in the powder diffraction pattern of a tetragonal material to

those of a cubic phase is shown schematically in Figure 2.47. The (1, 0, 0) reflection of a cubic material is made up of diffraction from the (1, 0, 0), (0, 1, 0), (0, 0, 1), (−1, 0, 0), (0, −1, 0), and (0, 0, −1) lattice planes which all have an identical d-spacing. The number of planes which contribute towards a reflection is termed the **multiplicity** and in this case the multiplicity is 6. The $\sin^2\theta$ values for all these reflections are identical as they depend only on $(h^2 + k^2 + l^2)$ and a single factor common ratio, A, is needed during indexing.

In a tetragonal material with slightly different a (= b) and c lattice parameters, where c > a, the two constants A and C are needed but these will be very similar. Hence, the $\sin^2\theta$ values for the (1, 0, 0) and (0, 0, 1) reflections will diverge only marginally and these maxima will be observed at only slightly different 2θ values in the powder diffraction pattern. In Figure 2.47, the pair of peaks near 22° can be assigned as deriving from the (1, 0, 0) reflection of the cubic material and these two reflections indexed as (0, 0, 1) and (1, 0, 0). Lattice planes contributing to the (1, 0, 0) peak will be the (1, 0, 0), (0, 1, 0), (−1, 0, 0) and (0, −1, 0), giving a multiplicity of 4; those contributing to the (0, 0, 1) reflection will be the (0, 0, 1) and (0, 0, −1), giving a multiplicity of 2. The multiplicity is reflected in the peak intensities which have a 2:1 ratio.

In this material, which is metrically slightly-distorted from cubic, assignment of Miller indices to all the other reflections becomes straightforward. Once the (1, 0, 0) and (0, 0, 1) reflections have been indexed, values of the constants A and C can be obtained using Eqn 2.20 and the other observed $\sin^2\theta$ values assigned using the possible ratios shown in Table 2.11. Again, following full indexing of the data, the values of A and C may be used to calculate the a and c lattice parameters. For the tetragonal compound in Figure 2.47, $BaTiO_3$, the values obtained are $a = 3.995$ Å (399.5 pm) and $c = 4.034$ Å (403.4 pm).

TABLE 2.11 The $\sin^2\theta$ relationships in tetragonal and hexagonal systems, $h, k, l \leq 2$

R	(1, 0, 0)	(0, 0, 1)	(1, 1, 0)	(1, 0, 1)	(1, 1, 1)	(2, 0, 0)	(0, 0, 2)	(2, 0, 1)	(1, 0, 2)	(2, 1, 0)	(2, 1, 1)	(2, 0, 2)
Tetragonal	A	C	2A	A + C	2A + C	4A	4C	4A + C	A + 4C	5A	5A + C	4A + 4C
Hexagonal	A	C	3A	A + C	3A + C	4A	4C	4A + C	A + 4C	7A	7A + C	4A + 4C

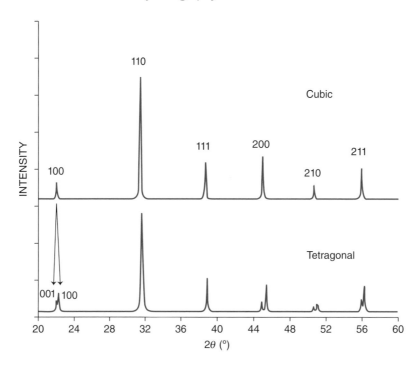

FIGURE 2.47 Diffraction patterns from cubic and tetragonal materials, where $a \sim c$ and $c > a$ for the tetragonal material and a is similar in the two materials. The six lines contributing to the multiplicity of the cubic phase $(1, 0, 0)$ reflection resolve into two reflections of relative intensities 2:4.

EXAMPLE 2.16

Index the following pattern from chromium metal assuming the crystal system is hexagonal ($\lambda = 1.5418$ Å) and determine the lattice parameters.

2θ 20.02 38.17 40.69 43.47 57.02 62.86 68.99
75.99 81.68 83.30 85.17

ANSWER

The $\sin^2\theta$ values can be calculated as follows:

0.0302 0.1069 0.1209 0.1371 0.2278 0.2719
0.3207 0.3789 0.4276 0.4417 0.4579

These values can now be inspected for the relationships in Table 2.11. So dividing through by the first $\sin^2\theta$ value (= C) shows the third reflection to be $4C$, the sixth $9C$—so these could be the $(1, 0, 0)$, $(2, 0, 0)$, and $(3, 0, 0)$ or the $(0, 0, 1)$, $(0, 0, 2)$, and $(0, 0, 3)$ reflections, most likely the latter as we do not have the ratio $3A$ which should appear for the $(1, 1, 0)$ reflection. Dividing through by the second $\sin^2\theta$ (A) shows ratios $3A$ with line 7 and $4A$ with line 9 which implies these are the $(1, 0, 0)$, $(1, 1, 0)$, and $(2, 0, 0)$ reflections. We now have values for A and C and so can associate ratios and index the remainder of the lines as follows:

0.0302	0.1069	0.1209	0.1371	0.2278	0.2719	0.3207	0.3789	0.4276	0.4417	0.4579
C	A	$4C$	$A+C$	$A+4C$	$9C$	$3A$	$A+9C$	$4A$	$3A+4C$	$4A+C$
(0, 0, 1)	(1, 0, 0)	(0, 0, 2)	(1, 0, 1)	(1, 0, 2)	(0, 0, 3)	(1, 1, 0)	(1, 0, 3)	(2, 0, 0)	(1, 1, 2)	(2, 0, 1)

Substituting the values for A and C back into Eqn. 2.21 gives $a = 2.358$ Å and $c = 4.436$ Å.

SELF TEST

Calculate the expected position of the $(1, 1, 1)$ reflection from chromium metal, a reflection not observed in the experimental data.

2.16.2 **Crystallite size from PXD data**

In order to observe sharp diffraction maxima in a PXD pattern, the crystallites need to be of sufficient size to ensure that slightly away from the diffraction angle 2θ maximum, destructive interference occurs. The requirement is for the presence of a sufficiently large number of parallel planes so that summation of the diffracted waves, which are only slightly out of phase between two successive planes, will eventually over tens or hundreds of planes produce destructive interference as the scattering becomes progressively more out of phase (Section 2.5). In materials with very small crystallites and, therefore, significantly fewer planes, the scattered X-rays at angles slightly away from the Bragg angle do not totally destructively interfere. This leads to some diffracted intensity at angles slightly away from the diffraction angle, 2θ, maximum and as a result broadening of the diffraction peaks as seen in Figure 2.48.

The **Scherrer formula** relates the thickness of the crystallites, that is, the number of planes, to the breadth of peaks in the PXD pattern:

$$t = \frac{0.9\lambda}{\sqrt{B_M^2 - B_S^2}\ \cos\theta} \qquad \text{Eqn 2.22}$$

where t is the crystallite thickness, λ is the X-ray wavelength, θ is the Bragg angle (half the measured diffraction angle, 2θ, in *degrees*), and B_M and B_S are the width in *radians* of diffraction peaks of the sample and of a standard

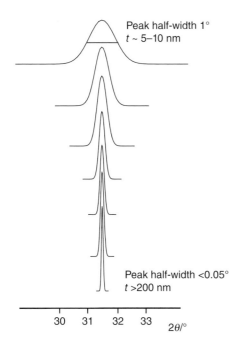

FIGURE 2.48 The observed broadening of a PXD peak half-width as a function of crystallite size.

Peak half-width 1°
$t \sim$ 5–10 nm

Peak half-width <0.05°
t >200 nm

30 31 32 33
$2\theta/°$

sample at half height (the half-widths of the peaks). The standard peak width is obtained from a highly crystalline sample (with a diffraction peak at a similar diffraction angle to that of the sample under investigation) and represents the instrumental broadening. However, the correction due to instrumental broadening is often very small, especially for nanoparticles (1–100 nm in size), so the Scherrer formula is often used in the simpler form:

$$t = \frac{0.9\lambda}{B_M\ \cos\theta} \qquad \text{Eqn 2.23}$$

The Scherrer method of determining crystallite size from powder data is of wide application. For materials which are of plate-like habit, for example, clays, the size of the crystallites in the different directions, the a, b, and c lattice parameter directions, may be ascertained by consideration of the peak widths of the various $(h, 0, 0)$, $(0, k, 0)$, and $(0, 0, l)$ reflections.

EXAMPLE 2.17

Application of the Scherrer formula

The $(1, 1, 1)$ reflection, at $2\theta = 39.2°$ with $\lambda = 1.5418$ Å radiation, from gold nanoparticles formed by reducing a solution of $HAuCl_4$ was found to have a half-width of 1.00°, while a reflection obtained from a high crystallinity silica sample at a similar diffraction angle had a half-width of 0.060°. Calculate the size of the gold nanoparticles.

ANSWER

The half-width angles need to be converted to radians for input into Eqn 2.22. $B_M = 1/57.296 = 0.0174$ radians and $B_S = 1.047 \times 10^{-3}$ radians. We can use the Scherrer formula and note that as the wavelength is given in Å the particle size will be derived in the same units. Also ensure that the angle is the Bragg angle (half the diffraction angle) and is in degrees.

$$\begin{aligned}
t &= \frac{0.9\lambda}{\sqrt{B_M^2 - B_S^2}\ \cos\theta} \\[2mm]
&= \frac{0.9 \times 1.5418}{\sqrt{0.0174^2 - 0.001047^2}\ \cos(39.2/2)} \\[2mm]
&= 84.8\ \text{Å}\ \ (= 8.48\ \text{nm})
\end{aligned}$$

SELF TEST

Calculate the size of a gold nanoparticle that gives rise to a $(1, 1, 1)$ reflection of width 0.5°.

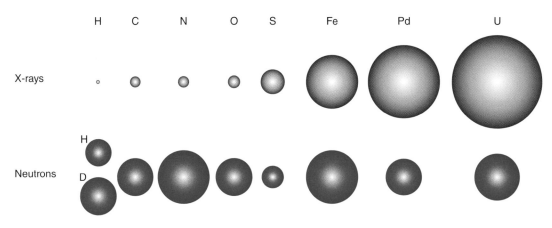

FIGURE 2.49 Relative scattering factors of selected elements for X-ray and neutron diffraction methods.

2.16.3 **Neutron powder diffraction**

Diffraction occurs from crystals for any particle moving with a velocity such that its associated wavelength (through the de Broglie relation, $\lambda = h/mv$) is comparable to the separations of the atoms or ions in the crystal. Neutrons in thermal equilibrium at room temperature travel with velocities of ~2200 m s^{-1}. These have associated wavelengths of the order of 100–200 pm (1–2 Å) and, hence, can undergo diffraction by crystalline compounds. The generation of neutron beams that are intense enough for diffraction work is performed at national or international facilities (Section 1.6.5). Neutron beams of the appropriate wavelength are produced by 'moderating' (slowing down) neutrons generated either in nuclear fission reactors or through a process known as spallation, in which neutrons are chipped off the nuclei of heavy elements by an accelerated beam of protons.

The instrumentation used for collecting data and analysing single-crystal or neutron powder diffraction (NPD) patterns is often similar to that used for X-ray diffraction, particularly for reactor sources of neutrons. In an analogous manner to a laboratory powder X-ray diffractometer a single wavelength can be selected from the as-generated neutron beam using a monochromator. This monochromated beam is directed onto a sample around which the detector can travel to collect neutrons diffracted at all Bragg angles. The scale of instrumentation is much larger because neutron beam fluxes are much lower than those of laboratory X-ray sources. For example, a typical sample size for NPD will be several grams, about 10–100 times larger than is needed for PXD. Data collection times for NPD data are normally of the order of several hours rather than tens of minutes for the X-ray method. Investigation of an inorganic compound with this technique is, therefore, much less routine and as a result its application is limited to systems where studies using X-ray diffraction fail to determine all the features of a structure.

The advantages of neutron diffraction stem from the fact that neutrons are scattered by nuclei rather than by the surrounding electrons. As a result, neutrons are sensitive to structural parameters that often complement those obtained using X-rays. In particular, the scattering is not dominated by the heavy elements. This can be a problem with X-ray diffraction for many inorganic compounds, which contain both heavy and light elements. For example, locating the position of a light element, such as hydrogen or lithium, in a material that also contains lead can be impossible with X-ray diffraction, as almost all the electron density is associated with the lead atoms. In contrast, with neutrons, the scattering from light atoms is often similar to that of heavy elements (Figure 2.49).

Where the light atoms present in a compound contribute significantly to the overall scattering the intensities in the diffraction pattern change markedly; see Figure 2.50.

Another area of application of neutron diffraction is in distinguishing some near isoelectronic species. In X-ray scattering, pairs of elements that neighbour each other in the periodic table, such as O and N (see Figure 2.49) or Cl and S, are nearly isoelectronic and scatter X-rays to a similar extent; therefore they are hard to differentiate in a crystal structure that contains them both. However, the elements in these pairs do scatter neutrons to very different extents, N 50% more strongly than O, and Cl about four times more strongly than S, so distinguishing these atoms can be more credibly achieved using neutron diffraction data. Example 2.18 (and its self test) demonstrate how the near neighbours and isoelectronic species K^+ and Cl^-, and Fe and Co, can be distinguished using neutron diffraction.

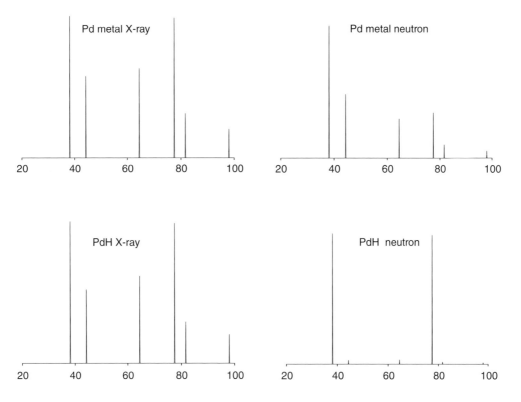

FIGURE 2.50 The PXD and NPD patterns of Pd and PdH. Both materials adopt a face-centred cubic unit cell of dimensions 4.085 Å with PdH having the rock salt structure and all the X-ray and neutron diffraction patterns show the characteristic patterns and systematic absences for face-centred cubic, see Figure 2.32. The addition of hydrogen to Pd metal has almost no effect on the intensity of the reflections in the PXD patterns due to the low scattering from this atom. In the NPD patterns larger changes in the reflection intensities occur, as the hydrogen scattering power is similar to that of palladium.

EXAMPLE 2.18

Neutron powder diffraction from KCl

Describe how the NPD pattern obtained from KCl (rock salt structure type) would differ from the PXD pattern in terms of observed reflections. The relative neutron scattering powers of potassium and chlorine are 3.7 and 9.6 compared with X-ray scattering values (f_j) of 18 for both atom types. Consider your answer in relation to that given for Example 2.10.

ANSWER

The effect of having isoelectronic scatterers, K^+ and Cl^-, in the rock salt structure was discussed in Example 2.10. With neutrons the nuclei of K^+ and Cl^- can be readily distinguished and both the unit cell and lattice type reflected in the NPD pattern of KCl (Figure 2.51) are those of the rock salt structure type with $a = 5.43$ Å and *F*-centring systematic absences (h, k, l are all odd or all even for the reflection to be present).

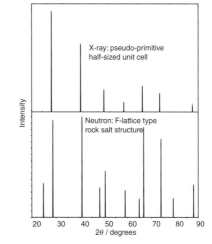

FIGURE 2.51 Comparison of the powder diffraction patterns obtained from KCl with X-rays and neutrons of the same wavelength.

SELF TEST

The structure of the alloy FeCo is shown in Figure 2.52. Predict which reflections would be observed in (i) the PXD pattern from FeCo and (ii) the NPD pattern from FeCo. Take the X-ray scattering powers of the neighbouring elements Fe and Co to be identical and the relative neutron scattering powers as 9.45 (Fe) and 2.49 (Co).

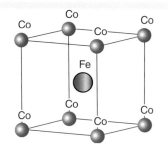

FIGURE 2.52 The structure of FeCo with ordered Fe and Co positions.

2.16.4 **PXD intensities and full diffraction profile analysis**

As we have seen under single-crystal X-ray diffraction, the peak intensities of the reflections in an X-ray diffraction experiment are determined by the atom types and their distribution in the unit cell; measurement of these intensities allows the X-ray crystallographer to work out the structure. In the PXD experiment, because all the reflections occur along the single 2θ axis, for any non-simple structure the reflections overlap, particularly at high 2θ values. This peak overlap is particularly serious for low symmetry crystal systems, monoclinic and triclinic, and large unit cells, >1000 Å^3. As a result, except for very simple structures, extracting individual reflection intensities, essential for a structure determination, becomes very difficult.

Useful structural information may still be obtained from powder diffraction patterns using a technique known as **profile fitting**, often called the **Rietveld method**

(named after the Dutch crystallographer Hugo Rietveld). In this method, a trial structure (in which likely positions, in the unit cell, are assigned to the different atoms present) is used to calculate intensities for the various reflections. These intensities may then be combined with the various factors controlling the form of the powder pattern, for example, lattice parameters and radiation wavelength, to generate a complete 'calculated' powder pattern or 'profile' based on the trial structure. Comparison of this profile with the experimental one can then be undertaken and adjustment of the trial structure parameters carried out in order to obtain the best fit between the experimental and calculated powder patterns (Figure 2.53). Because the whole pattern is fitted, peak intensity information from overlapping reflections is still extracted in this method of analysis. This then provides reasonably accurate information on the atom positions in the unit cell.

As stated above, a trial structure is normally required before the technique can be employed though considerable advances are being made in full structure solution

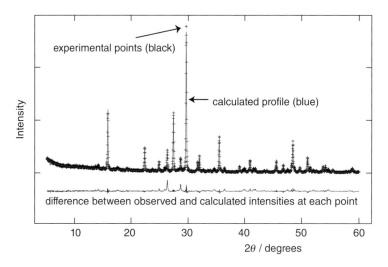

FIGURE 2.53 A fit between the experimental and calculated PXD profiles as generated from a Rietveld structure refinement analysis.

from powder diffraction data. Generally this requires very high quality powder diffraction data, such as those obtained at synchrotron sources, with excellent resolution so that any overlap of the reflections is minimised. In this way a sufficient number of reflection intensities can be determined with confidence to allow the application of the structure determination methods described for single-crystal diffraction data (which use just reflection (h, k, l) and intensity values). Once a basic structural model is obtained the profile analysis method, described previously, can be applied to extract and refine the complete structure.

2.17 Single-crystal diffraction

2.17.1 Single-crystal neutron diffraction

Neutron diffraction (Section 2.16.3) may also be applied to single crystals, that is, **single-crystal neutron diffraction**, though the benefits and limitations of NPD are reproduced. For the accurate determination of hydrogen positions in inorganic compounds the method is unrivalled. Where location of the precise and accurate position of hydrogen atoms is an important structural feature, such as in some organometallic hydride compounds and in many ferroelectrics containing OH groups, the technique is often used. Specific examples include the bonding of hydrogen in complexes such as $OsClH_3(PPh_3)_3$ (Figure 2.54) and understanding strong hydrogen bonds and proton displacement in compounds such as KH_2PO_4. However, due to the low neutron fluxes available in comparison with X-ray beams, the single crystals have to be much larger, typically a few cubic millimetres (such crystals are around ten thousand times larger in volume than

those needed for X-ray diffraction studies), and preparing high quality crystals of this size can be problematic. Even with a large crystal, experimental data collection times on a high intensity neutron source are at least an order of magnitude longer compared with laboratory X-ray experiments, typically taking several days.

2.17.2 Electron density and X-n maps

In the analysis of single-crystal X-ray diffraction (Sections 2.11–2.15), the electron density that scatters the X-rays was associated with the atomic nuclei positions and assumed to be spherical. Such analysis is normally sufficient if just the overall structure needs to be determined. However, the true distribution of electrons in a compound is not perfectly spherical around all the atoms. For example, there is a build-up of electron density between nuclei in bonds and also at positions adopted by lone pairs. The observed X-ray diffraction data is the FT of the electron density, so the inverse FT of high quality diffraction data produces a detailed scattering density map—that is, for X-rays, a distribution of electron density throughout the crystal structure. An example of such a map is shown in Figure 2.55.

While both X-rays and neutrons probe the structure of a compound, because they are scattered by different centres (electrons and nuclei, respectively) they provide different crystallographic information. Overall because to a large extent electrons in the unit cell are associated with nuclei, and form approximately spherical clouds around them, the structures determined using the two diffraction techniques have the same major features. If, however, the distribution of electrons is not spherically symmetric around the atoms then a careful comparison

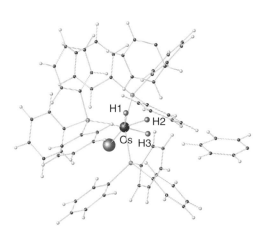

FIGURE 2.54 The structure of $OsClH_3(PPh_3)_3$ determined using single-crystal neutron diffraction data showing the hydrogen positions adjacent to the heavy osmium atom.

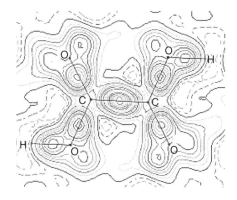

FIGURE 2.55 A detailed X-ray scattering density map of oxalic acid, $HO_2C–CO_2H$, with atomic electron scattering deducted; the residual electron density in the bonds and associated with the lone pair on oxygen atoms of the carbonyl groups can be seen. (Data from Rigaku promotional literature, www.rigaku.com/en/products/smc/rapid/app006.)

FIGURE 2.56 An X-n map of the formate anion in Li(HCOO). H_2O. The additional electron density in the C–O and C–H bonds can be seen from the contour lines. (Data from J.O. Thomas, R. Tellgren, and J. Almhöf, *Acta Crystallogr. B* 31 1946 (1975).)

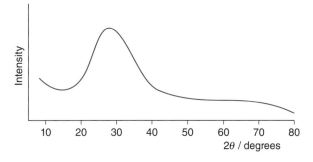

FIGURE 2.57 The PXD pattern obtained from amorphous SiO_2 ($\lambda = 1.5418$ Å).

of an X-ray determined electron density map with a neutron determined nuclear position map can give detailed insights into exact electron distributions. In this way features such as lone pairs and bonding electron distributions can be produced in an **X-n map** (Figure 2.56).

2.17.3 Amorphous materials, total diffraction, and pair distribution function analysis

Many solid materials do not have a perfect crystalline structure with long range order of atoms over thousands of unit cells but adopt structures in the solid that have only local order. A classic example is a glass, such as silicon dioxide, which has been rapidly cooled from the melt, amorphous SiO_2. In such a solid perfectly repeating atom positions across multiple unit cells do not exist, but some local structural elements do occur throughout the material. For example, in amorphous SiO_2, silicon, wherever it is in the glassy structure, will be coordinated to four oxygen atoms at a distance of around 1.6 Å; further away, the next shell of atoms is likely to be silicon, via bent Si–O–Si bonds, at about 2.8–3.1 Å distant from the original silicon atom. X-ray scattering from these regularly spaced (at short range, <5 Å) atoms gives rise to broad diffraction maxima. The PXD pattern ($\lambda = 1.5418$ Å) of amorphous SiO_2 (Figure 2.57) shows the local Si–O–Si separations as a broad peak, half-width around 4°, centred on $2\theta \sim 30°$ (corresponding to a *d*-spacing of 3.0 Å).

In other material structures, repeating unit cells of the same size exist, but the exact arrangement of part of the crystal structure within a unit cell may be random and differ in neighbouring unit cells. One example of this

behaviour is the orientation of cyanide groups in AuCN. This material, while adopting a perfectly ordered alternating arrangement of gold and cyanide ions, may have the cyanide ions randomly orientated as C–N or N–C along one crystallographic direction. The local orientation of the cyanide groups also affects the exact gold atom position depending on whether it is coordinated to carbon or nitrogen, so there are small variations in its position throughout the material. The X-ray scattering from samples with some elements of long range repeating order but local disorder, typically consist of some broadened Bragg reflections plus **diffuse scattering** which appears as broad features underneath some of the Bragg diffraction maxima (Figure 2.58).

The full interpretation of such data can be achieved through pair distribution function (PDF) analysis. The PDF function, $G(r)$, is the probability of finding a neighbouring atom at a distance *r*. It can be calculated by considering all scattering centres in the sample and how well the atoms and neighbours scatter X-rays (or neutrons in a neutron PDF function). The PDF function of a perfectly crystalline solid will consist of sharp peaks extending to large values of *r*, as the long range order of repeating unit cells ensures that exact atom separations exist to large distances (Figure 2.59a). At very large distances these peaks overlap strongly and the function $G(r)$ tends to constant value. In disordered and amorphous solids all possible atom separations will again be represented as peaks in the $G(r)$ function. At short distances (<2–4 Å) distinct peaks representing local co-ordinations are observed in $G(r)$ though these are broadened. At larger distances the disorder spreads out the atom to atom separations and the $G(r)$ function decays more rapidly towards a constant value (Figure 2.59b). Experimentally, $G(r)$ can be obtained from the total diffraction pattern intensity, all Bragg peaks and diffuse background, by an FT. Computer modelling packages are available to allow fitting of a structural model, that includes local atom positional information, to the full PXD experimental data (Figure 2.59a).

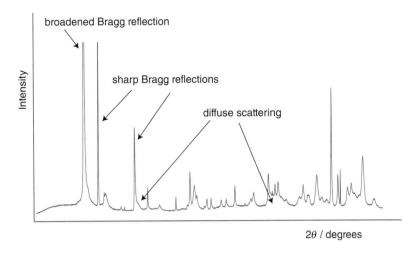

FIGURE 2.58 A PXD pattern from a material with partial long range order and local disorder showing peak broadening and diffuse scattering.

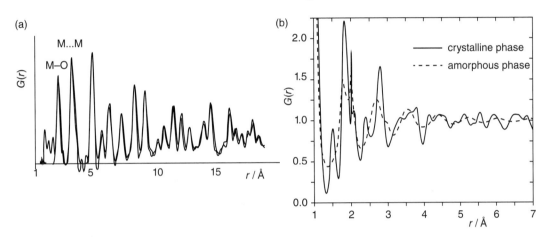

FIGURE 2.59 (a) A typical radial distribution function, $G(r)$, for a crystalline metal oxide. Peaks in $G(r)$ show interatomic separations with nearest neighbour M–O and next nearest M...M distances marked at around 2 and 3.5 Å, respectively. (b) Comparison of $G(r)$ functions for crystalline and amorphous phases. (Data from N.C. Karayiannis, K. Foteinopoulou, and M. Laso, *Int. J. Mol. Sci.* **14** 332 (2013).)

Bibliography

M.T Weller, T. L. Overton, J.A. Rourke, and F.A. Armstrong. (2014) *Inorganic Chemistry*. 6th ed. Oxford: Oxford University Press.

Powder diffraction

R.E. Dinnebier. (2008) *Powder Diffraction: Theory and Practice*. Cambridge: Royal Society of Chemistry.

V. Pecharsky and P. Zavalij. (2008) *Fundamentals of Powder Diffraction and Structural Characterization of Materials*. 2nd ed. New York: Springer.

D.L. Bish and J.E. Post. (1989) *Modern Powder Diffraction*. Washington, DC: Mineralogical Society of America.

R. Jenkins and R.L. Snyder. (1996) *Introduction to X-ray Powder Diffractometry*. Chichester: Wiley.

Single-crystal diffraction

W. Clegg. (2015) *X-Ray Crystallography* (Oxford Chemistry Primers). 2nd ed. Oxford: Oxford University Press.

S.K. Chatterjee. (2010) *X-ray Diffraction: Its Theory and Applications*. 2nd ed. Prentice-Hall of India Pvt Ltd.

C. Hammond. (2015). *The Basics of Crystallography and Diffraction* (International Union of Crystallography Texts on Crystallography). 4th ed. Oxford: Oxford University Press.

A.J. Blake, J.M. Cole, J.S.O. Evans, P. Main, S. Parsons, D.J. Watkin, and W. Clegg. (2009) *Crystal Structure Analysis: Principles and Practice* (International Union of Crystallography Texts on Crystallography). Oxford: Oxford University Press.

Nuclear magnetic resonance (NMR) spectroscopy

FUNDAMENTALS

3.1 Introduction to NMR spectroscopy

Nuclear magnetic resonance (NMR) is the most powerful and widely used spectroscopic method for the investigation of the molecular structures of compounds, dissolved in solution or as pure liquids. It yields information about the shape and symmetry of molecules. For many common atom types NMR spectroscopy defines the different environments present and their numbers. Analysis of NMR spectra often allows the full molecular configuration of a molecule to be determined (Figure 3.1).

The technique can also provide useful information about the rate and nature of ligand interchange in fluxional molecules and can be used to follow reactions and so determine their mechanistic pathway. As well as fully defining the arrangement of the atoms in small molecules, the technique has aided the definition of the structures of large molecules, such as proteins and enzymes with molar masses of over 100 000 atomic mass units. Often the

information complements the descriptions obtained from diffraction techniques (Chapter 2) in defining molecular geometry, though in some cases molecular conformations in solution can differ from those in the crystalline state. While diffraction techniques usually produce accurate interatomic distances and angles, NMR studies generally yield more limited information on the separation of nuclei, such as which atoms are bonded or close neighbours in a molecule. Multinuclear NMR spectroscopy is an excellent technique for the investigation of many important atom types of interest in inorganic chemistry and can be used to characterise compounds containing hydrogen, carbon, fluorine, phosphorus, silicon, and boron. Many transition metal nuclei, such as those of platinum, rhodium and silver, can also be easily studied provided the compounds do not have unpaired electrons. Other important elements are much less amenable to study using the method, that is, oxygen, nitrogen, and sulfur, though even for these elements NMR studies using specific isotopes (which may be enriched) have proved very valuable. Specialised NMR

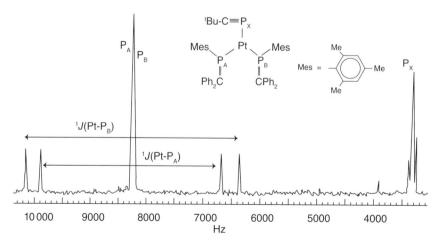

FIGURE 3.1 The 28.7 MHz ^{31}P {^{1}H} NMR spectrum used to define the structure of the molecule shown in the inset. (Data from J.F. Nixon, *Chem. Commun.* 930 (1983).) Analysis of the NMR spectrum allows the different phosphorus environments, P_A, P_B, and P_X, to be identified.

techniques can also be applied to the study of some solids, including materials such as silicates, fullerides (C_{60} derivatives), and bioinorganic metal–protein complexes. The technique is normally non-destructive and the sample can be recovered from solution after the resonance spectrum has been collected, if it is required.

3.2 The basis of the NMR technique

Many spectroscopic techniques involve measuring the energy required, usually input as electromagnetic radiation, to excite a system between two defined energy levels (Chapter 1). In NMR spectroscopy a variant of direct electromagnetic excitation is employed, whereby the two energy levels, the ground state and an excited state of the molecule, are brought into **resonance** with the electromagnetic radiation frequency and there is a constant transfer of energy between them. In the case of NMR, the separation in energy of the two levels involved is controlled by the application of an external magnetic field. The energy levels involved and the resonance conditions for different nuclei of interest in inorganic chemistry are discussed in the following sections.

3.2.1 Nuclear magnetic moments

Fundamental particles have different **spin quantum numbers**. The electron (e^-), proton (^1H), and neutron (^1n) all have spin (designated by s (for electrons) or I (for protons and neutrons)) of ½, while a photon of electromagnetic radiation has $I = 1$. When a charge is associated with this spin, as with a proton forming part of an atomic nucleus, it gives rise to a magnetic moment. The magnetic moment is related to the spin by a factor known as the **gyromagnetic ratio**, γ. The gyromagnetic ratio, which is a constant for a fundamental particle or a specific nuclear isotope (e^-, ^1H, ^7Li, ^{19}F, etc.), is the ratio of its magnetic moment to its spin angular momentum.

Nuclei, which consist of a combination of neutrons and protons, may have an associated **magnetic moment**.

This is not always the case as in some combinations of protons and neutrons, such as in ^{16}O, the individual proton and neutron spin values sum to give a nuclear $I = 0$. The nuclear magnetic moment is an intrinsic property of a particular nucleus and its value depends on its exact composition in terms of the number of protons and neutrons. The NMR technique depends on energy transitions involving a nuclear magnetic moment, when it is situated in an external magnetic field, so it is useful to be able to identify which nuclei can be studied using NMR spectroscopy. In order to undertake NMR spectroscopy a nucleus with non-zero spin, $I \geq ½$, is needed; nuclei with even atomic numbers and even mass numbers, such as ^{12}C and ^{16}O, have $I = 0$ and are invisible in NMR spectra.

Every nuclear isotope that is composed of a different number of protons and neutrons will have its own spin, I. The empirical rules relating the number of protons (equivalent to the atomic number, Z), the number of neutrons (which can be calculated from the mass number and Z), and the spin quantum number (I) are shown in Table 3.1.

EXAMPLE 3.1

Predicting nuclear spin quantum numbers and NMR activity

On the basis of their mass and atomic numbers, assign the nuclear spin quantum number category (I as zero, half integer, or integer) to the following isotopes: (i) ^{106}Pd, (ii) ^{125}Te, and (iii) ^{138}La.

ANSWER

Using Table 3.1 and determining the number of protons and neutrons present for each nucleus of interest we can define the spin category to which the nucleus belongs. (i) For ^{106}Pd, $Z (= p) = 46$ and $n = 106 - 46 = 60$ giving an even number of protons and an even number of neutrons, so $I = 0$. (ii) For ^{125}Te, $Z = 52$ and $n = 73$; this isotope has an even number of

TABLE 3.1 Nuclear spin quantum numbers arising from odd and even combinations of protons and neutrons.

Number of protons, p	Number of neutrons, n	Spin (I)	Examples
Even	Even	0	^{12}C ($p = 6$, $n = 6$) $I = 0$
Even	Odd	½, ³⁄₂, ⁵⁄₂, etc.	^{13}C ($p = 6$, $n = 7$) $I = ½$
Odd	Even	½, ³⁄₂, ⁵⁄₂, etc.	^{19}F ($p = 9$, $n = 10$) $I = ½$
			^{23}Na ($p = 11$, $n = 12$) $I = ³⁄₂$
Odd	Odd	Integral, 1, 2, etc.	^{14}N ($p = 7$, $n = 7$) $I = 1$
			^{10}B ($p = 5$, $n = 5$) $I = 3$

protons and an odd number of neutrons so I will have a half integer value. (The value of $I = \frac{1}{2}$ for this nucleus, though this cannot be deduced from Table 3.1). (iii) For ^{138}La, $Z = 57$ and $n = 81$, with an odd number of protons and an odd number of neutrons so I is predicted to be an integer spin (in practice $I = 5$).

SELF TEST

Determine a spin category for a ^{139}La nucleus.

In summary all isotopes with an odd atomic number and/or an odd atomic mass will have a nuclear spin, Thus a hydrogen nucleus, ^{1}H, a proton, has $I = \frac{1}{2}$, while a deuterium nucleus, ^{2}H (D), consisting of a proton and a neutron has $I = 1$. Tables 3.2 ($I = \frac{1}{2}$ nuclei) and 3.3 ($I \geq 1$ nuclei; column 2 gives the I value for the specific isotope) summarise the key NMR parameters for a number of important nuclei that can be studied by NMR spectroscopy. In general, nuclei with $I = \frac{1}{2}$ provide the best quality NMR spectra and structural information on a molecule, though many nuclei with $I \geq 1$ also yield useful NMR

TABLE 3.2 Key NMR data for spin $I = \frac{1}{2}$ nuclei . (Adapted from the Bruker Almanac 1991; also available at www.bruker.com/almanac)

Isotope[a]	Natural abundance[b] /%	Gyromagnetic ratio, $\gamma/10^7$ rad s^{-1} T^{-1}	Frequency ratio referenced to ^{1}H = 100 MHz	Reference compound and conditions	Relative receptivity compared to ^{1}H[c]
^{1}H	99.9885	26.752	100.000	Me$_4$Si 1% in CDCl$_3$	1.000
^{3}H (T)	–	28.535	106.663	Me$_4$Si-tritiated	–
^{3}He	0.000134	–20.379	76.181	He gas	5.75×10^{-7}
^{13}C	1.11	6.728	25.145	Me$_4$Si 1% in CDCl$_3$	1.76×10^{-4}
^{15}N	0.368	–2.712	10.136	Pure MeNO$_2$	3.84×10^{-6}
^{19}F	100	25.181	94.094	CCl$_3$F	0.832
^{29}Si	4.683	–5.319	19.807	Me$_4$Si 1% in CDCl$_3$	3.68×10^{-4}
^{31}P	100	10.839	40.480	85% H$_3$PO$_4$	6.65×10^{-2}
^{57}Fe	2.19	0.868	3.237	Fe(CO)$_5$ in C$_6$D$_6$	7.39×10^{-7}
^{77}Se	7.58	5.121	19.071	Pure Me$_2$Se	5.26×10^{-4}
^{89}Y	100	–1.316	4.920	Y(NO$_3$)$_3$ in D$_2$O	1.18×10^{-4}
^{103}Rh	100	–0.847	3.156	Rh(acac)$_3$ in CDCl$_3$	3.12×10^{-5}
^{107}Ag	51.82	–1.088	4.047	AgNO$_3$ in D$_2$O	3.44×10^{-5}
^{109}Ag	48.12	–1.252	4.653	AgNO$_3$ in D$_2$O	4.86×10^{-5}
^{111}Cd	12.80	–5.705	21.202	Me$_2$Cd	1.22×10^{-3}
^{113}Cd	12.26	–5.961	22.179	Me$_2$Cd	1.34×10^{-3}
(^{115}Sn)	0.34	–8.801	32.718	Me$_4$Sn	1.21×10^{-4}
(^{117}Sn)	7.61	–9.589	35.632	Me$_4$Sn	3.44×10^{-3}
^{119}Sn	8.59	–10.031	37.290	Me$_4$Sn	4.44×10^{-3}
(^{123}Te)	0.89	–7.059	26.169	Me$_2$Te	1.57×10^{-4}
^{125}Te	6.99	–8.511	31.549	Me$_2$Te	2.21×10^{-3}
^{129}Xe	26.44	–7.452	27.856	XeOF$_4$	5.60×10^{-3}
^{169}Tm	100	–2.137	7.99	–	5.66×10^{-4}
^{171}Yb	14.3	4.735	17.699	Yb(C$_5$Me$_5$)$_2$/THF	7.81×10^{-4}
^{183}W	14.28	1.128	4.218	1M Na$_2$WO$_4$ in D$_2$O	1.04×10^{-5}

(Continued)

TABLE 3.2 Key NMR data for spin $I = \frac{1}{2}$ nuclei (*Continued*)

Isotope[a]	Natural abundance[b] /%	Gyromagnetic ratio, $\gamma/10^7$ rad s^{-1} T^{-1}	Frequency ratio referenced to ^{1}H = 100 MHz	Reference compound and conditions	Relative receptivity compared to ^{1}H[c]
^{187}Os	1.64	0.6193	2.282	OsO$_4$ 0.98 M in CCl$_4$	2.0×10^{-7}
^{195}Pt	33.83	5.8385	21.496	Na$_2$PtCl$_6$ 1.2 M in D$_2$O	3.36×10^{-3}
^{199}Hg	16.84	4.846	17.871	Me$_2$Hg	1.00×10^{-3}
^{203}Tl	29.52	15.539	57.123	Tl(NO$_3$)$_3$ in D$_2$O	5.79×10^{-2}
^{205}Tl	70.47	15.692	57.634	Tl(NO$_3$)$_3$ in D$_2$O	0.136
^{207}Pb	22.6	5.626	20.920	Me$_4$Pb	2.07×10^{-3}

[a] Nuclei in parentheses are not the most favourable for NMR studies of the element concerned.
[b] Typical isotopic compositions.
[c] The receptivity is proportional to natural abundance $\times \gamma^3 I(I + 1)$.

TABLE 3.3 Key NMR data for selected quadrupolar, $I \geq 1$, nuclei . (Adapted from the Bruker Almanac 1991; also available at www.bruker.com/almanac)

Isotope	Spin	Natural abundance /%[a]	Gyromagnetic ratio, $\gamma/10^7$ rad s^{-1} T^{-1}	Quadrupole moment, Q /fm^2	Frequency ratio, referenced to ^{1}H = 100 MHz	Reference compound and conditions	Line-width factor	Relative receptivity compared to ^{1}H[b]
^{2}H (D)	1	0.015	4.106	0.2860	15.350	(CD$_3$)$_4$Si	0.41	1.45×10^{-6}
^{6}Li	1	7.42	3.937	−0.0808	14.716	LiCl in D$_2$O	0.033	6.45×10^{-4}
^{7}Li	$\frac{3}{2}$	92.58	10.397	−4.01	38.863	LiCl in D$_2$O	21	0.271
^{9}Be	$\frac{3}{2}$	100	−3.760	5.288	14.051	BeSO$_4$ in D$_2$O	37	1.39×10^{-2}
^{10}B	3	19.9	2.875	8.459	10.744	BF$_3$.Et$_2$O in CDCl$_3$	14	3.95×10^{-3}
^{11}B	$\frac{3}{2}$	80.1	8.585	4.059	32.084	BF$_3$.Et$_2$O in CDCl$_3$	22	0.132
^{14}N	1	99.632	1.934	2.044	7.226	CH$_3$NO$_2$	21	1.00×10^{-3}
^{17}O	$\frac{5}{2}$	0.038	−3.628	−2.558	13.556	D$_2$O	2.1	1.11×10^{-5}
^{21}Ne	$\frac{3}{2}$	0.27	−2.113	10.155	7.894	Ne gas 1 atm	140	6.65×10^{-6}
^{23}Na	$\frac{3}{2}$	100	7.080	10.4	26.452	0.1 M NaCl in D$_2$O	140	9.27×10^{-2}
^{25}Mg	$\frac{5}{2}$	10.0	−1.639	19.94	6.121	11 M MgCl$_2$ in D$_2$O	130	2.68×10^{-4}
^{27}Al	$\frac{5}{2}$	100	6.976	14.66	26.056	1.1 M Al(NO$_3$)$_3$ in D$_2$O	69	0.207
^{33}S	$\frac{3}{2}$	0.76	2.056	−6.78	7.676	(NH$_4$)$_2$SO$_4$	61	1.72×10^{-5}
^{35}Cl	$\frac{3}{2}$	75.78	2.624	−8.165	9.798	0.1 M NaCl in D$_2$O	89	3.58×10^{-3}
^{37}Cl	$\frac{3}{2}$	24.22	2.184	−6.435	8.155	0.1 M NaCl in D$_2$O	55	6.59×10^{-4}
^{39}K	$\frac{3}{2}$	93.26	1.250	5.85	4.666	0.1 M KCl in D$_2$O	46	4.76×10^{-4}
^{43}Ca	$\frac{7}{2}$	0.135	−1.803	−4.08	6.730	0.1 M CaCl$_2$ in D$_2$O	2.3	8.68×10^{-6}
^{45}Sc	$\frac{7}{2}$	100	6.508	−22.0	24.291	0.06 M Sc(NO$_3$)$_3$ in D$_2$O	66	0.302
^{47}Ti	$\frac{5}{2}$	7.44	−1.511	30.2	5.638	TiCl$_4$	290	1.56×10^{-4}
^{49}Ti	$\frac{7}{2}$	5.41	−1.511	24.7	5.639	TiCl$_4$	83	2.05×10^{-4}

(Continued)

TABLE 3.3 Key NMR data for quadrupolar, $I \geq 1$, nuclei (*Continued*)

Isotope	Spin	Natural abundance /%[a]	Gyromagnetic ratio, $\gamma/10^7$ rad s^{-1} T^{-1}	Quadrupole moment, Q /fm^2	Frequency ratio, referenced to ^{1}H = 100 MHz	Reference compound and conditions	Line-width factor	Relative receptivity compared to ^{1}H[b]
^{51}V	7/2	99.750	7.046	−5.2	26.303	VOCl$_3$	3.7	0.383
^{53}Cr	3/2	9.501	−1.515	−15.0	5.652	K$_2$CrO$_4$ in D$_2$O	300	8.63×10^{-5}
^{55}Mn	5/2	100	6.645	33.0	24.789	KMnO$_4$ in D$_2$O	350	0.179
^{59}Co	7/2	100	6.332	42	23.727	K$_3$[Co(CN)$_6$] in D$_2$O	240	0.278
^{61}Ni	3/2	1.140	−2.395	16.2	8.936	Ni(CO)$_4$	350	4.09×10^{-5}
^{63}Cu	3/2	69.17	7.111	−22.0	26.515	[Cu(CH$_3$CN)$_4$][ClO$_4$] in CH$_3$CN	650	6.50×10^{-2}
^{65}Cu	3/2	30.83	7.604	−20.4	28.404	[Cu(CH$_3$CN)$_4$][ClO$_4$] in CH$_3$CN	550	3.54×10^{-2}
^{67}Zn	5/2	4.10	1.676	15.0	6.257	1.1 M Zn(NO$_3$)$_2$ in D$_2$O	390	1.18×10^{-4}
^{71}Ga	3/2	39.89	8.182	10.7	30.497	1.1 M Ga(NO$_3$)$_3$ in D$_2$O	150	5.71×10^{-2}
^{73}Ge	9/2	7.73	−0.936	−19.6	30.497	(CH$_3$)$_4$Ge	28	1.09×10^{-4}
^{75}As	3/2	100	4.596	31.4	17.122	0.5 M NaAsF$_4$ in CD$_3$CN	1300	2.54×10^{-2}
^{81}Br	3/2	49.31	7.249	26.2	27.001	0.01 M NaBr in D$_2$O	920	4.91×10^{-2}
^{83}Kr	9/2	11.49	−1.033	25.9	3.847	Kr gas	50	2.18×10^{-4}
^{87}Rb	3/2	27.83	8.786	13.35	32.720	0.01 M RbCl in D$_2$O	240	4.93×10^{-2}
^{87}Sr	9/2	7.00	−1.164	33.5	4.334	0.5 M SrCl$_2$ in D$_2$O	83	1.90×10^{-4}
^{91}Zr	5/2	11.22	−2.497	−17.6	9.296	Zr(C$_5$H$_5$)$_2$Cl$_2$ in CH$_2$Cl$_2$	99	1.07×10^{-3}
^{93}Nb	9/2	100	6.574	−32.0	24.476	K[NbCl$_6$] in CH$_3$CN	76	0.488
^{95}Mo	5/2	15.92	−1.751	−2.2	6.516	2 M Na$_2$MoO$_4$ in D$_2$O	1.5	5.21×10^{-4}
^{99}Tc	9/2	–	6.046	−12.9	22.508	NH$_4$TcO$_4$ in D$_2$O	12	–
^{99}Ru	5/2	12.76	−1.229	7.9	4.605	0.3 M K$_4$[Ru(CN)$_6$] in D$_2$O	20	1.44×10^{-4}
^{105}Pd	5/2	22.33	−1.23	66.0	4.576	K$_2$PdCl$_6$ in D$_2$O	1400	2.53×10^{-4}
^{115}In	9/2	95.71	5.897	81.0	21.912	0.1 M In(NO$_3$)$_3$ in D$_2$O	490	0.338
^{121}Sb	5/2	57.21	6.4435	−36.0	23.931	KSbCl$_6$ in CH$_3$CN	410	9.33×10^{-2}
^{127}I	5/2	100	5.390	−71.0	20.007	0.01 M KI in D$_2$O	1600	9.54×10^{-2}
^{133}Cs	7/2	100	3.533	−0.343	13.116	0.1 M CsNO$_3$ in D$_2$O	0.016	4.84×10^{-2}
^{137}Ba	3/2	11.232	2.992	24.5	11.113	0.5 M BaCl$_2$ in D$_2$O	800	7.87×10^{-4}
^{139}La	7/2	99.91	3.808	20.0	14.125	0.01 M LaCl$_3$ in D$_2$O	54	6.05×10^{-2}
^{181}Ta	7/2	99.988	3.243	317.0	11.989	KTaCl$_6$ in CH$_3$CN	1400	3.74×10^{-2}
^{187}Re	5/2	62.60	6.168	207.0	22.751	0.1 M KReO$_4$ in D$_2$O	1400	8.95×10^{-2}
^{193}Ir	3/2	62.7	0.5227	75.1			7500	2.34×10^{-5}
^{197}Au	3/2	100	0.473	54.7			4000	2.77×10^{-5}
^{209}Bi	9/2	100	4.375	−51.6	16.069	Bi(NO$_3$)$_3$ in D$_2$O	200	0.144

[a] Typical isotopic compositions.
[b] The receptivity is proportional to natural abundance $\times \gamma^3 I(I + 1)$.

data for the characterisation of inorganic compounds. It is worth reiterating that if $I = 0$, as for ^{12}C, ^{16}O, and ^{32}S, the isotope cannot be studied using NMR spectroscopy.

3.2.2 Energy levels for NMR spectroscopy

Placing a nucleus of spin I in an applied magnetic field means that it can take up $2I + 1$ orientations relative to the direction of the field, labelled m_I (Figure 3.2a). Possible values of m_I range from $m_I = I$ to $m_I = -I$, via $I - 1$, $I - 2$, and so on until $-I$ is reached. Each orientation will have a different energy in an applied magnetic field and it is the transitions between these energy levels that the NMR technique exploits. Transitions between m_I levels are subject to the selection rule $\Delta m_I = \pm 1$.

Considering initially a nucleus with $I = \frac{1}{2}$ (Figure 3.2b), such as ^1H or ^{19}F, the energy separation of the two levels, $m_I = +\frac{1}{2}$ (sometimes also labelled α) and $m_I = -\frac{1}{2}$ (sometimes labelled β) is given by

$$\Delta E = \hbar \gamma B_0 \qquad \textbf{Eqn 3.1}$$

where B_0 is the magnitude of the applied magnetic field (more precisely, the magnetic induction in tesla, T), $\hbar = h / 2\pi$, where h is Planck's constant, and γ is the gyro-magnetic ratio of the nucleus. The gyromagnetic ratio, a constant for a specific nuclear isotope, is the ratio of its magnetic moment to its spin angular momentum (values of γ for important $I = \frac{1}{2}$ and $I \geq 1$ nuclei are given in Tables 3.2 and 3.3, respectively). The energetic ordering of the $m_I = \frac{1}{2}$ and $m_I = -\frac{1}{2}$ levels depends on the sign of γ. Note that many NMR texts and the chemistry literature use the symbol H (in amps per metre) for the magnetic field strength. Modern superconducting magnets produce a field, B_0, of between 5 and 25 T and resonance, equivalent to a transition between the $m_I = \frac{1}{2}$ and $m_I = -\frac{1}{2}$ levels, is achieved under these conditions with electromagnetic radiation in the range 200–1000 MHz. This corresponds to the radiofrequency region of the electromagnetic spectrum (Section 1.3). The transition between the $m_I = \frac{1}{2}$ and $m_I = -\frac{1}{2}$ levels is observed as a single peak or resonance in the NMR spectrum (Figure 3.2b).

The initial sections of this chapter deal mainly with $I = \frac{1}{2}$ nuclei, as these are often the easiest to investigate using the NMR technique and provide high-quality spectra that are often the easiest to interpret. Many quadrupolar nuclei (those that have $I \geq 1$) can also provide very useful NMR spectra; Section 3.16 describes some of the factors that control the quality of NMR spectra obtained for these nuclei. An $I > \frac{1}{2}$ nucleus in an external magnetic field will give rise to $2I + 1$ equally spaced m_I

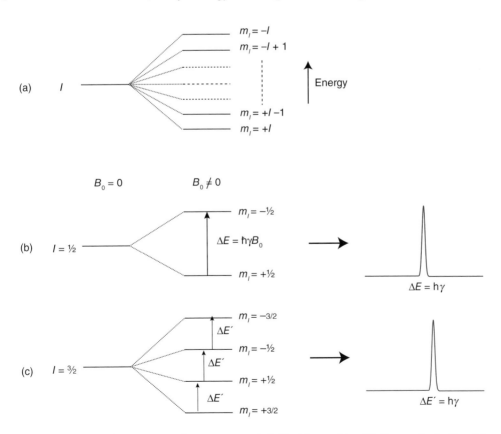

FIGURE 3.2 (a) Splitting of nuclear I into m_I values in a magnetic field with γ positive; (b) diagram for $I = \frac{1}{2}$ and the resulting NMR spectrum; (c) diagram for $I = \frac{3}{2}$ and the resulting NMR spectrum.

levels as shown in Figure 3.2(a) and transitions between these levels will be possible subject to the rule $\Delta m_I = \pm 1$. For an $I = 3/2$ nucleus, such as ^{11}B, four different m_I energy levels form in the external magnetic field, B_0, and three transitions are possible between these levels (Figure 3.2c). Because the energy separations between all the neighbouring m_I levels are the same, the NMR spectrum would consist of a single resonance corresponding to all three possible $\Delta m_I = \pm 1$ transitions.

EXAMPLE 3.2

Calculate the resonance frequency of a ^{1}H nucleus in (a) a 5.000 T field and (b) a 23.4868 T field.

ANSWER

Using $E = \hbar\gamma B_0$ and $\Delta E = h\nu$ gives $\nu = \frac{\gamma B}{2\pi}$ if the correct units of rad s^{-1} T^{-1} are used for γ. From Table 3.2, γ for ^{1}H is 2.6752×10^8 rad s^{-1} T^{-1}, which for $B_0 = 5.000$ T gives $\nu = 2.1289 \times 10^8$ s^{-1}, that is, 212.89 MHz; for $B_0 = 23.4868$ T, $\nu = 1000$ MHz. The resonance frequency is directly proportional to the magnetic field strength and it is common to refer to an 11.744 T magnet as a 500 MHz magnet (the proton resonance frequency), though it should be noted that because different nuclei have different γ values they will resonate at different frequencies for the same field strength.

SELF TEST

Calculate the resonance frequency of a ^{19}F nucleus in an 11.744 T magnetic field using the data in Table 3.1.

3.2.3 Sensitivity of the NMR technique

The energy difference between the $m_I = +\frac{1}{2}$ and $m_I = -\frac{1}{2}$ levels in the applied magnetic field is relatively small, only around 10^{-25} J (for a 10 T field and the proton gyromagnetic ratio). With such a small energy difference, nuclei in a sample may adopt either of the m_I orientations and the distribution of the nuclei (with N_{m_I} representing the number in each m_I level) over these two energy levels, $m_I = +\frac{1}{2}$ and $m_I = -\frac{1}{2}$, will be given by the Boltzmann distribution:

$$\frac{N_{m_I=-\frac{1}{2}}}{N_{m_I=+\frac{1}{2}}} = \frac{N_{\text{upper}}}{N_{\text{lower}}} = e^{\left(-\frac{\Delta E}{k_B T}\right)} \qquad \textbf{Eqn 3.2}$$

where ΔE is the energy difference between the two m_I levels, k_B is the Boltzmann constant (1.3806×10^{-23} J

K^{-1}), and T is the system temperature in kelvin. Using $\Delta E = 10^{-25}$ J and $T = 298$ K this ratio can be calculated as 1.000024. Thus, on average, of 1 000 000 nuclei 500 006 would be in the lower energy m_I state and 499 994 in the higher energy m_I state, that is, the population of the lowest energy level is only marginally more than the higher level. This reduces the number of nuclei that can resonate at any one time, as the resonance signal intensity depends on the excess of nuclei in the ground state. As a result the **sensitivity** of the NMR technique can be quite low as relatively few nuclei in the sample may contribute to the experimental signal. The application of a strong magnetic field increases the separation of the $m_I = +\frac{1}{2}$ and $m_I = -\frac{1}{2}$ levels. Consequently the population difference between them increases leading to an increased signal intensity. Doubling the applied magnetic field doubles ΔE in Eqn 3.2 and from Eqn 3.1 this would change the ratio in the lower to upper m_I states to 1.000049:1, markedly increasing the sensitivity.

A small γ for a nucleus also reduces the sensitivity of the NMR technique, as $\Delta E = \hbar\gamma B_0$, leading to only a small excess of nuclei in the ground state from Eqn 3.2. When studying nuclei with small gyromagnetic ratios increasing the applied magnetic field, B_0, can compensate for the disadvantageous population distribution. This provides an important way of enhancing the quality of NMR spectra for these nuclei. The sample temperature is also important in determining the sensitivity of the NMR experiment. A lower temperature in Eqn 3.2 increases the population of the ground state, with $m_I = +\frac{1}{2}$, and, hence, the NMR experiment sensitivity.

One related factor in the NMR experiment is the **relaxation time** for the nucleus being probed. Under resonance conditions the population of the excited state increases and the signal would be reduced in intensity (and eventually become 'saturated') unless there is a mechanism by which the excited state nuclei can relax back to the ground state and, thereby, re-establish the Boltzmann distribution. The processes by which this relaxation happens, for $I = \frac{1}{2}$ nuclei, involve interaction of the magnetic moment with local random magnetic fields that originate from the rotational and translational motions of the molecule. Different sites for a nucleus in a compound can relax back to the equilibrium distribution with varying relaxation times. Fast relaxation allows, in the NMR experiment, for the resonance condition to be probed quickly and NMR spectra to be collected in seconds or minutes. In this case of fast relaxation to the equilibrium distribution the measured peaks intensities reflect the amount of that nuclear environment in the molecule. In general relaxation times are fast for most important $I = \frac{1}{2}$ nuclei, such as ^{1}H, ^{31}P, and ^{19}F. This means that in the NMR spectra of these nuclei the resonance signal intensity reflects quantitatively the nuclear abundance. If relaxation is slow, often the case for ^{13}C, then quantitative peak intensities cannot be obtained with reasonably short experimental times, as the ground state equilibrium will not have necessarily been established for all the different ^{13}C environments during the NMR experiment.

Another factor that determines the relative, overall sensitivity of the NMR technique is the natural abundance of the nucleus under investigation. For ^{1}H NMR, nearly all the hydrogen nuclei in a molecule will potentially be in resonance (subject to the disadvantageous Boltzmann distribution) at the resonance frequency because the natural abundance of ^{1}H is 99.989%. For some nuclei, such as carbon, the sensitivity of the NMR technique is stated as being 'low'. This is because only 1.07% of carbon atoms in a sample have $I > 0$ (^{13}C) and so would be probed during the collection of the NMR spectrum. The greater majority of carbon atoms, ^{12}C, have $I = 0$ and this nucleus is NMR inactive. The **receptivity** of a nucleus, given relative to ^{1}H = 1.00, is a measure of the overall ease of undertaking NMR spectroscopy on a particular nucleus; values of the receptivities of various nuclei are also given in Tables 3.2 and 3.3 and discussed in Example 3.3.

3.2.4 Energy levels, frequencies, and the NMR nucleus

Each nucleus, including different isotopes of the same element, has a unique value of the gyromagnetic ratio with values for $I = \frac{1}{2}$ nuclei ranging from small values, such as that of ^{187}Os ($\gamma = 0.616 \times 10^{7}$ rad s^{-1} T^{-1}) to that of ^{19}F ($\gamma = 25.18 \times 10^{7}$ rad s^{-1} T^{-1}). This means that resonances for different nuclei will occur at very different frequencies in the same magnetic field, remembering that $\Delta E = \hbar \gamma B_0$. As a result NMR spectra are *specific* to a particular nucleus. For example, the ^{19}F NMR spectrum of OsO_3F_2 will show a resonance from fluorine but it will not contain any resonances due to ^{187}Os; these will be at a very different energy and thus frequency value. Where multiple NMR spectra from a single compound are of interest, for example, ^{187}Os and ^{19}F from OsO_3F_2, the NMR spectrometer will need to be re-tuned to sample very different radiofrequencies to bring the ^{187}Os nuclei and then, in a subsequent experiment, the ^{19}F nuclei into resonance.

EXAMPLE 3.3

Discuss the variation in the receptivities of $I = \frac{1}{2}$ nuclei given in Table 3.2.

ANSWER

While there are a number of factors that influence a nucleus's receptivity the Boltzmann distribution between the two m_I levels is an important one. This in turn depends on γ from Eqns 3.1 and 3.2. In addition the proportion of nuclei of an element that are NMR active, with $I > 0$, will determine how many in a sample could be at resonance. When all factors are taken into account the receptivity is described as being proportional to [natural abundance $\times$ $\gamma^3 I(I + 1)$]. The higher γ the greater the receptivity and the high γ values for nuclei such as ^{19}F and ^{31}P (which are also 100% abundant in Nature) make these nuclei highly amenable to study by NMR. For these nuclei high quality data can be obtained with short experiment times. Alternatively small amounts of samples containing low atomic percentages of fluorine or phosphorus will still provide good quality spectra on reasonable experimental timescales. Conversely it is difficult to study nuclei with low receptivities, such as ^{187}Os (low natural abundance and small γ), where an NMR spectrum collection may often take hours.

SELF TEST

Explain why ^{125}Te is preferred over ^{123}Te for NMR spectroscopic studies of tellurium compounds.

EXAMPLE 3.4

Calculate the resonance frequencies of (i) ^{19}F and (ii) ^{187}Os nuclei in a 5 T field.

ANSWER

Using $v = \frac{\gamma B}{2\pi}$, as in Example 3.2, and from Table 3.2, γ for ^{19}F is 25.18×10^{7} rad s^{-1} T^{-1}. This gives $v = 2.004 \times 10^{8}$ s^{-1} or 200 MHz, while for Os, $\gamma = 0.6193 \times 10^{7}$ rad s^{-1} T^{-1} so $v = 4.901 \times 10^{6}$ s^{-1} equivalent to 4.9 MHz; a difference of approximately 195 MHz! ^{19}F NMR spectra are usually obtained for a specific nucleus over a range of few thousand Hz at most (compare with 195 MHz, so that resonances will occur over the narrow MHz range 194.995–195.005 MHz). Spectra will normally only contain resonances from the nucleus to which the spectrometer has been tuned.

SELF TEST

Calculate the resonance frequencies of ^{203}Tl and ^{205}Tl nuclei in an 11.744 T magnetic field using the data in Table 3.2. Comment on your answer.

As can be seen from the calculation in Example 3.2 the resonance frequency of a nucleus is highly sensitive to the exact magnetic field in which it sits, so that a small change of this value, from say 11.744 T to 11.745 T, will cause a significant shift in the resonance frequency; in this case from 500.026 MHz (11.744 T) to 500.068 MHz (11.745 T), a difference of 42 kHz. The basis of the NMR technique is that each nucleus in a chemical compound experiences a different local perceived magnetic field caused by summing together the external field (B_0), any magnetic fields that exist locally within the molecule, and any magnetic field screening effects (where a magnetic field is generated within the molecule by the external magnetic field so as to oppose it). These locally produced fields and screening effects include those derived from the electrons surrounding a nucleus and any other nearby magnetic nuclei (with $I \geq$ ½). Thus, the resonance frequency of a nucleus can be used as a probe of its environment. Each unique nuclear environment in a compound will experience a slightly different local magnetic field and so give a different resonance frequency. Because it is possible to distinguish very small resonance frequency differences in experimental NMR spectroscopy, the technique is a very sensitive probe of environment.

3.3 Experiment and instrumentation

The essence of the NMR experiment requires the determination of the resonance frequency of a nucleus (which has a magnetic moment) in a compound. Each unique environment of the specific nucleus being probed in the NMR experiment in a compound will experience a slightly different local magnetic field and, thereby, resonate at a slightly different frequency, but within a narrow frequency range. Normally a spectrum will be collected containing all the resonance frequencies of a specific nucleus, for example, a ^{1}H NMR spectrum. Where more than one nucleus with a magnetic moment exists within a compound more than one spectrum would normally be collected, for example, the ^{31}P and ^{19}F NMR spectra of the $[PF_6]^-$ anion. Remember that resonances for different nuclei occur at very different energies or frequencies and so the spectrometer can only investigate one nucleus type at a time before being re-tuned to the next frequency or energy range of interest.

NMR spectra were originally obtained in a **continuous wave (CW) mode**, but this has been superseded by **Fourier transform (FT)** methods (Section 1.5). However, as CW is conceptually easier to understand, it will be considered first. This experiment can be carried out in two ways: the sample is either subjected to a constant radiofrequency and the magnetic field increased (by changing the current flowing in an electromagnet used to generate the field) until the resonances are encountered, or the field is held constant and the radiofrequency range swept. In either mode as the energy range is scanned, either through changes in magnetic field or radiofrequency, each different environment of the nucleus under investigation will come into resonance in turn and the excitation detected as a signal. Figure 3.3 shows a schematic experimental

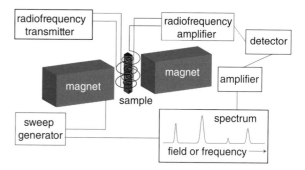

FIGURE 3.3 Schematic of the experimental arrangement of a CW-mode NMR spectrometer and the form of a typical NMR spectrum obtained.

arrangement of a CW-mode NMR spectrometer and the form of a typical spectrum obtained.

In contemporary spectrometers, the energy separations are identified by exciting all the nuclei of interest (i.e. all the protons in ^{1}H NMR) in the sample with a sequence of radiofrequency pulses and then observing the return of the nuclear magnetisation back to equilibrium over time, known as the **free induction decay (FID)**. Fourier transformation (Section 1.5) then converts the time-domain data to the frequency domain with peaks at frequencies corresponding to transitions between the different nuclear energy levels. Figure 3.4 shows the experimental arrangement of an **FT-NMR** spectrometer and a typical spectrum obtained using this method.

Since the earliest NMR experiments, ca. 1950, the magnetic field strengths used in NMR spectrometers have increased enormously from around 0.6 T (equivalent to ~30 MHz for ^{1}H) such that modern superconducting magnets use fields of over 20 T (corresponding to a 1 GHz or 1000 MHz NMR spectrometer). High magnetic field strengths make the technique more sensitive (see Section 3.2.3) by increasing the population difference between the ground and excited states. In addition at high field strengths the spectrum is spread out over a much greater frequency range improving resolution by increasing the separation (in Hz) between the resonance signals.

When referring to different spectrometers such as 200 MHz or 1000 MHz it should be remembered that this denotes the resonance frequency of a proton (^{1}H NMR) in the magnetic field of that spectrometer. Because of the different gyromagnetic ratios different inorganic nuclei have different resonance frequencies (Tables 3.2, column 3, and 3.3, column 6) so a 200 MHz ^{1}H NMR spectrometer is equivalent to a 80.96 MHz ^{31}P NMR spectrometer and a 6.372 MHz ^{103}Rh NMR spectrometer; this is important when comparing NMR data for two different nuclei in the same compound.

^{1}H NMR spectra are normally recorded from the material dissolved in a solvent and the inclusion of signals

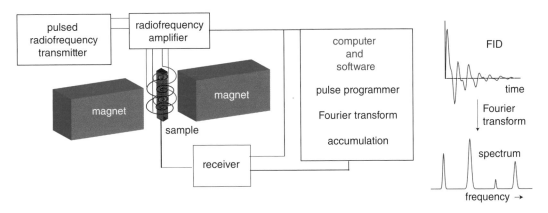

FIGURE 3.4 Schematic of the experimental arrangement of an FT-NMR spectrometer and a typical free induction decay and FT-NMR spectrum obtained using this method.

from the solvent protons is undesirable. Deuterated solvents are normally used, most commonly deuterated chloroform, $CDCl_3$, and deuterated dimethyl sulfoxide, (d^6-DMSO), $(CD_3)_2SO$. By using deuterated solvents a 'deuterium frequency-field lock' can be employed whereby the NMR instrument constantly monitors a D resonance from the solvent and makes changes to the electromagnetic current to ensure the applied B_0 is kept constant. Protein NMR spectra are often recorded using water or mixture of 90% H_2O/10% D_2O as a solvent (and the water ^1H NMR signal suppressed though specialist data collection procedures); the use of some heavy water in the solvent aids the investigation of exchangeable protons in proteins, such as those in amide groups.

Sample tubes are normally spun during the collection of NMR spectra which helps average out any inhomogeneities in the external magnetic field. Historically solvents were supplied with a small amount (typically 0.1%) of tetramethylsilane (TMS) added as an **internal standard**; this allowed calibration of the chemical shifts of each proton in the compound under investigation. TMS shows a single resonance that defines a zero on the NMR spectrum for ^1H NMR (a chemical shift of 0 ppm, see Section 3.4). If the sample needs to be recovered after the experiment TMS is volatile, thus it can usually be easily removed. Modern spectrometers are sensitive enough to reference spectra based on the very weak proton signals derived from residual protonated solvent (e.g. 0.01% $CHCl_3$, in $CDCl_3$) so that deuterated NMR solvents are nowadays normally supplied without added TMS.

3.4 Chemical shifts

If the only magnetic field experienced by a nucleus in a compound under investigation was that of the applied magnetic field B_0 then all NMR spectra would consist of a single line. However, as electrons are charged and have spin and orbital angular momentum they produce small additional, local magnetic fields. These locally induced magnetic fields will slightly modify the magnetic field experienced by a nucleus so that the local magnetic field of each nucleus will reflect its exact environment. Electrons in full atomic orbitals generate a local magnetic field which opposes the applied magnetic field and the nucleus is said to be **shielded**. It is also possible for higher energy excited electronic states, which have unpaired electrons, to mix partially with the ground state and in this case the effect is to reinforce the applied magnetic field B_0 and the nucleus is said to be **unshielded** or **deshielded**. As we have seen (Eqn 3.1), the frequency of an NMR transition depends on the local magnetic field experienced by the nucleus so that each unique chemical environment for a nucleus will give rise to a slightly different resonance frequency in the NMR spectrum.

The intensities of resonances in most NMR spectra reflect the proportion of nuclei in different environments (noting the need for fast relaxation, Section 3.2.3). Therefore, if a compound contains three fluorine nuclei, two of which are shielded to a different extent from the other because of their different environment, the NMR spectrum will contain two resonances with the intensity ratio 2:1.

EXAMPLE 3.5

Identify the number of different chemical environments for fluorine in IF_7, which has a pentagonal bipyramidal shape (Figure 3.5). Hence predict the number of resonances expected in the ^{19}F NMR spectrum of IF_7 and their relative intensities.

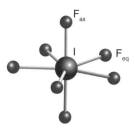

FIGURE 3.5 The pentagonal bipyramidal molecular structure of IF_7.

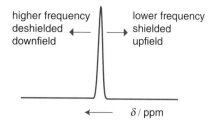

FIGURE 3.6 Summary of the terminology used in defining the relationship between shielding, field, frequency, and δ, the chemical shift.

ANSWER

The pentagonal bipyramidal geometry has two different sites for fluorine, equatorial (five fluorine sites) and axial (two fluorine sites). Each distinct site will experience different levels of shielding and so two resonances would be expected to be observed in the NMR spectrum. The intensities of the resonances reflect the number of nuclei in each different environment and the two resonances would be expected to have signal strengths in the ratio 5:2.

SELF TEST

Predict the number of different fluorine environments and, therefore, the number of ^{19}F resonances in (a) a trigonal bipyramidal molecule, such as PF_5, and (b) an octahedral molecule, such as SF_6.

The magnitude of the shielding effect will depend on the strength of the magnetic field. Since different spectrometers operate with different B_0 values a method of measuring the degree of chemical shielding that is independent of magnetic field is needed. It is expressed in terms of the **chemical shift**, δ, the difference between the resonance frequency of nuclei (ν) in the sample and that of a reference compound (ν°):

$$\delta = \frac{\nu - \nu^\circ}{\nu^\circ} \times 10^6 \qquad \textbf{Eqn 3.3}$$

The chemical shift δ is dimensionless, though common practice is to report it as 'parts per million' (ppm) in acknowledgement of the factor of 10^6 in the definition. Note that the definition of chemical shift by reference to a standard compound (ν°) removes the effect of the applied field, that is, the chemical shift is independent of spectrometer used. The common calibration standard for ^1H, ^{13}C, or ^{29}Si spectra is TMS. When the nucleus is shielded (sometimes also called 'upfield' or 'high-field') relative to the standard then $\delta < 0$, while $\delta > 0$ corresponds to a nucleus that is

deshielded ('downfield' or 'low-field') with respect to the reference. Figure 3.6 summarises the terminology used in defining the relationship between shielding, local field, and resonance frequency.

EXAMPLE 3.6

Reconcile the following descriptions of the NMR spectra (collected at a resolution that only distinguishes nuclear environments) with the geometry of the molecule investigated.

(a) The ^{19}F NMR spectrum of SF_4 (Figure 3.7a) contains two resonances of equal intensity.

(b) The ^{31}P NMR spectrum of the P_4O_9 molecule (Figure 3.7b) consists of two resonances with intensities in the ratio 3:1.

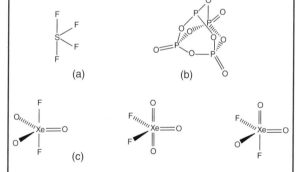

FIGURE 3.7 Molecular structures of (a) SF_4, (b) P_4O_9, and the possible isomers of (c) XeO_3F_2.

ANSWERS

(a) The data indicate that the four fluorine atoms are in two different environments with two atoms in each; this is consistent with the shape of the SF_4 molecule predicted from valence shell electron pair repulsion (VSEPR) theory (Figure 3.7a).

(b) The data require the four phosphorus atoms in the molecule to adopt two different environments in the ratio 3P:1P. This is consistent with the molecular geometry shown in Figure 3.7(b).

SELF TEST

Predict the ^{19}F NMR spectra of the three possible XeO_3F_2 isomers shown in Figure 3.7(c). Is it possible to distinguish all these isomers using ^{19}F NMR?

3.5 Chemical shift and environment

The numerous environmental factors that control the value of the chemical shift include parameters such as oxidation state, charge on an ion or molecular ion, type of bonding (e.g. the degree of covalency in the bonds), and the number of ligands. It is, therefore, difficult to predict chemical shifts. Some of the factors that affect the electron density around a nucleus, and can be expected to modify locally experienced magnetic fields, include oxidation state associated with the nucleus, the overall charge of the species, and the electronegativity of neighbouring atoms. Correspondingly, the chemical shift values yield information on these factors, so by using databases and tabulated chemical shift information it is often possible to assign a specific resonance in an NMR spectrum to a particular type of environment. Some such data are summarised in Tables 3.4, 3.5, and 3.6 for ^{1}H, ^{31}P, and ^{19}F nuclei, respectively. For example, the ^{1}H chemical shift in CH_4 is only 0.1 ppm because the H nuclei are in an environment similar to that in the standard (TMS), but the proton chemical shift observed for GeH_4 (where the hydrogen is bonded to the more electropositive germanium) is $\delta = 3.1$. A hydrogen atom bound to a closed-shell, low-oxidation-state, d-block element from Groups 6–10 (such as in $[HCo(CO)_4]$) is hydride-like, H^-, and generally found to be highly shielded (negative δ), whereas in an oxoacid (such as H_2SO_4) the proton, H^+, is deshielded.

Chemical shift variations for the same element in inequivalent positions within a molecule may also be interpreted with respect to bonding and electron distributions. For instance, in the T-shaped molecule ClF_3 the chemical shift of the unique $^{19}F_{eq}$ nucleus is separated by $\Delta\delta = 125$ ppm from that of the other two F_{ax} nuclei (Figure 3.8). This can be interpreted with respect to the two long $Cl–F_{ax}$ bonds (1.698 Å) and one short $Cl–F_{eq}$ (1.598 Å) and more electron density associated with the F_{ax} fluorine sites shielding these nuclei. A more detailed discussion of the relationship between chemical shift and chemical environment is given under advanced topics in Section 3.12.

TABLE 3.4 Proton, ^{1}H, chemical shift ranges for important environments, relative to TMS

Proton environment type	Typical chemical shift or range/ppm	Proton environment type	Typical chemical shift or range/ppm
Organic C–H/O–H	0 to 12	Main group E–H	–2 to +8
CH_4	0.1	Si-H (SiH_4)	3 to 4 (3.2)
Saturated CH_3	0.9	GeH_4	3.1
Saturated CH_2	1.3	SnH_4	3.85
Saturated CH	1.5	Me_3Pb**H**	7.7
$=CH_2$	4.6 to 5.9	B-H/$[BH_4]^-$	0.5 to 1.5
Aromatic C–H	6 to 8.5	$[AlH_4]^-$	2.75
R–NH_2	1 to 5	$[GaH_4]^-$	3.47
Alkyl C–OH	1 to 5.5	P-H	1.8 to 4
Aromatic –OH	4 to 12	S-H	1.5 to 2.5
η^2-C_2H_2	3 to 4	$[F-H-F]^-$	16.1
η^5-C_5H_5	4 to 5	Transition metal M–H	–5 to –14
η^6-C_6H_6	3 to 7	W–H	–12.3
		Pt–H	–8.8

TABLE 3.5 Phosphorus, ^{31}P, chemical shift data (from Bruker Almanac, 1991), relative to 85% H_3PO_4

Compound P(III)	Chemical shift (ppm)	Compound P(V)	Chemical shift (ppm)
PMe_3	–62	Me_3PO	+36.2
PEt_3	–20	Et_3PO	+48.3
$PMeF_2$	245	$[Me_4P]^+$	+24.4
$PMeH_2$	–163.5	$[PO_4]^{3-}$	+6.0
$PMeCl_2$	+192	PF_5	–80.3
$PMeBr_2$	+184	PCl_5	–80
PMe_2F	+186	$MePF_4$	–29.9
PMe_2H	–99	Me_3PF_2	–158
PMe_2Cl	–96.5	Me_3PS	+59.1
PMe_2Br	–90.5	$[PS_4]^{3-}$	+87
		$[PF_6]^-$	–145
		$[PCl_4]^+$	+86
		$[PCl_6]^-$	–295
		Me_2PF_3	+8.0

TABLE 3.6 Fluorine, ^{19}F, chemical shift data (from Bruker Almanac, 1991), relative to neat $CFCl_3$

Compound	Chemical shift (ppm)	Compound	Chemical shift (ppm)
$CFCl_3$	0.00	SbF_5	−108
CH_3F	−271.9	$[SbF_6]^-$	−109
CF_2H_2	−1436	SeF_6	+55
F_2	+422.92	SiF_4	−163.3
CF_3Cl	−28.6	$[SiF_6]^{2-}$	−127
ClF_3	+116, −4	TeF_6	−57
ClF_5	+247, +412	WF_6	+166
IF_5	+10, +58	XeF_2	+258
IF_7	+170	XeF_4	+438
AsF_3	−40.6	XeF_6	+550
AsF_5	−66	NF_3	147
$[AsF_6]^-$	−69.5	SOF_2	75.68
BF_3	−131.3	SF_6 (dilute)	57.617
$[BF_4]^-$	−163	SF_6 (10% conc.)	57.42
MoF_6	−278	SO_2F_2	33.17
ReF_7	+345	PF_3	−34.0
SF_6	+57.42	PF_5	−71.5
SO_2F	−78.5	POF_3	−90.7
$S_2O_5F_2$	+47.2		

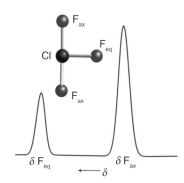

FIGURE 3.8 The ClF_3 molecular shape and its low-resolution ^{19}F NMR spectrum, which shows only the different chemically-shifted, resonance frequencies.

EXAMPLE 3.7

Interpret the ^{1}H chemical shift data for the Group IV hydrides given in Table 3.4 with reference to their All-red–Rochow electronegativities: C (2.50), Si (1.74), Ge (2.02), Sn (1.72), and Pb (1.55).

ANSWER

The chemical shift values show a good correlation (moving to more positive δ values, i.e. deshielded) with decreasing electronegativity (increasingly electropositive) as the Group is descended. This reflects changes in the electron density around the hydrogen nucleus as the bonding moves away from purely covalent in CH_4 to more ionic in Me_3PbH.

SELF TEST

Explain the chemical shift variation in the series Si–H, P–H, and S–H shown in Table 3.4.

3.6 Intensities

Provided sufficient time is allowed during NMR spectrum acquisition for the full relaxation of the observed nucleus (Section 3.2.3), integrated intensities ('integrals') can be used with confidence to aid spectral assignment for most nuclei (note that for some nuclei, e.g. ^{13}C, quantitative signal intensity information is difficult to obtain). For instance, as we have seen in the low resolution ^{19}F NMR spectrum of ClF_3 (Figure 3.8), the relative integrated intensities are 2:1. This pattern is consistent with the T-shaped structure because it indicates the presence of two symmetry-related F nuclei and one inequivalent F nucleus. The ^{19}F resonance intensities distinguish this less symmetrical structure from trigonal planar or pyramidal structures, which would have a single resonance with all fluorine environments equivalent.

EXAMPLE 3.8

Predict low resolution (number of resonances and intensities) NMR spectra for the following species.

(a) PF_5, trigonal bipyramidal, ^{19}F NMR ($I = \frac{1}{2}$, 100%) abundant

(b) BrF_5, square based pyramidal, ^{19}F NMR ($I = \frac{1}{2}$, 100%) abundant

ANSWERS

(a) In PF_5 there are three equatorial fluorine atom positions and two axial ones so the NMR spectrum is predicted to consist of two signals with intensity ratio 3:2.

(b) In BrF_5 there is one axial fluorine atom positions and four in plane fluorine atoms so the NMR spectrum is predicted to consist of two signals with intensity ratio 1:4.

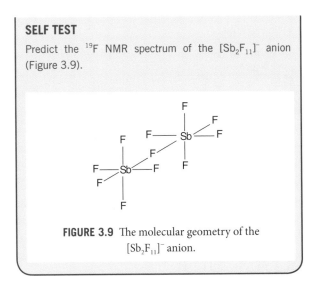

FIGURE 3.9 The molecular geometry of the $[Sb_2F_{11}]^-$ anion.

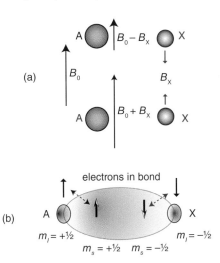

FIGURE 3.10 (a) Schematic of the effect of a neighbouring magnetic nucleus, **X**, magnitude B_x, on the field experienced at nucleus **A**. (b) Magnetisation transfer, spin–spin coupling, facilitated through a bond and the electrons therein.

3.7 First-order spin–spin coupling or scalar coupling

3.7.1 Origins of spin–spin coupling

Spin–spin coupling, sometimes also referred to as scalar coupling, arises where there are other NMR active nuclei (i.e. those having $I \geq \frac{1}{2}$) near to the nucleus under investigation. These neighbouring nuclei, because they have $I \geq \frac{1}{2}$, will have magnetic moments which will alter the field experienced by the nucleus being studied and modify its resonance frequency. The magnetic field of a nearby non-equivalent nucleus adds to, or subtracts from, the externally applied magnetic field (Figure 3.10a), causing small changes in the observed chemical shift. The effect is independent of the applied field and, because it provides information about the neighbouring nuclei, it is very useful in determining relationships between atom positions in a molecule.

3.7.2 First-order spin–spin coupling for two heteronuclear $I = \frac{1}{2}$ nuclei

The mechanism by which spin–spin coupling modifies the NMR spectrum can be illustrated by considering a simple case where two non-equivalent spin ½ nuclei form a bond. An example would be a ^{31}P–^{19}F bond in the compound Z_4PF, where Z is NMR inactive. For a generic **heteronuclear** (two different nuclei, e.g. ^{31}P and ^{19}F) spin–spin coupled system we can label the two different nuclei as **A** and **X** (Figure 3.10a)); specifically in Z_4PF, **A** = ^{31}P and **X** = ^{19}F. We consider the case where the chemical shifts of the **A** and **X** are very different. This is known as **first-order coupling**. Where the chemical shifts are similar, second-order effects can occur and these are covered in Section 3.14.

Considering the **A** nucleus alone initially we would expect a single resonance between the $m_I(\mathbf{A}) = +\frac{1}{2}$ and $m_I(\mathbf{A}) = -\frac{1}{2}$ levels (Figure 3.11a). Note the use of the labels α ($m_I = +\frac{1}{2}$) and β ($m_I = -\frac{1}{2}$) as it helps simplify the notation when discussing relative m_I values on neighbouring nuclei and transitions between them. If the orientation on nucleus **A** is $m_I(\mathbf{A}) = +\frac{1}{2}$ and on nucleus **X** is $m_I(\mathbf{X}) = -\frac{1}{2}$ we can write this state as αβ. NMR transitions between these states can be considered in two ways—either by consideration of the energy of the whole **AX** system (Section 3.7.2.1) or by just considering nucleus **A** (Section 3.7.2.2). Many courses discussing NMR just consider the simpler analysis covered in Section 3.7.2.2.

3.7.2.1 Spin–spin coupling of two $I = \frac{1}{2}$ nuclei: complete system energy level diagram

We now consider the effect of an adjacent **X** nucleus, which has its own possible nuclear magnetic levels $m_I(\mathbf{X}) = +\frac{1}{2}$ (α) and $m_I(\mathbf{X}) = -\frac{1}{2}$ (β); as for the **A** nucleus in the same applied magnetic field, the α state on **X** will lie to a lower energy than the β state. Considering the system as a whole there are four possible configurations of the spins on **A** and **X**, αα, αβ, βα, and ββ. If there is no interaction between the nuclei, these levels lie in the energy order αα (lowest) < αβ < βα < ββ (highest) in Figure 3.11(b), shown as blue horizontal lines. Resonances

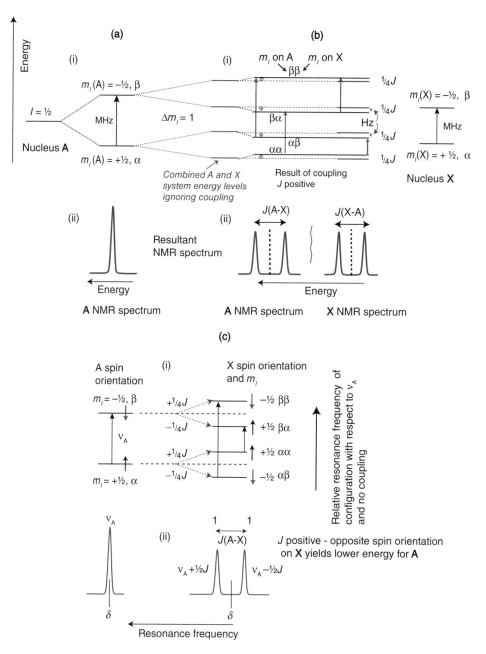

FIGURE 3.11 Energy level diagrams for two interacting nuclei, **A**, ^{31}P ($I = \frac{1}{2}$), and **X**, ^{19}F ($I = \frac{1}{2}$), and the resultant **A** nucleus NMR spectrum. Allowed NMR transitions on **A** are shown in black and those on **X**, are shown in blue. Note that the energy axis is not to scale. (a) (i) The energy level diagram for nucleus **A** from Figure 3.2(b) and (ii) the resultant single resonance NMR spectrum of **A**. (b) (i) Energy levels in the overall **AX** system showing the changes in **A** and **X** energy levels with J positive and allowed NMR transitions on **A** and **X** (ii) the resultant doublet NMR spectra of **A** and of **X**. (c) (i) Energy level diagram for nucleus **A** alone which only considers the relative energies of the states on **A** and how they are modified by interaction with **X** with J positive; note that the *shifts* in energy of the αα, αβ, βα and ββ states mirror those in (b). (ii) Resultant NMR spectra of **A**.

($\Delta m_I = 1$) will occur on **A** (αα $\leftrightarrow$ βα and αβ $\leftrightarrow$ ββ, which are both at the same energy if there is no interaction between nuclei) and **X** (αα $\leftrightarrow$ αβ and βα $\leftrightarrow$ ββ, the same energy for this nucleus with no interaction with **A**) in their specific NMR frequency ranges.

An interaction between the magnetic nuclear spins, spin–spin coupling with a positive coupling constant,

has the effect of raising the whole system αα energy state, lowering αβ and βα, and raising ββ, as drawn in Figure 3.11(b) (i). Using the selection rule $\Delta m_I = 1$ on nucleus **A** shows that two resonances are still possible between the ground and excited state on **A** (αα $\leftrightarrow$ βα and αβ $\leftrightarrow$ ββ) and shown in black in Figure 3.11(b). These resonances are now of different energies and also occur at energies that are different from the case without spin–spin coupling—one to a higher energy/frequency/chemical shift (deshielded/downfield) and one to a lower energy/frequency/chemical shift (shielded/upfield). This has the effect of 'splitting' the original resonance between the $m_I = +\frac{1}{2}$ and $m_I = -\frac{1}{2}$ levels into two resonances, termed **a doublet**, with a separation expressed as the **spin–spin coupling constant**, defined as J, with units in hertz, Hz (Figure 3.11b). J can be either positive or negative and is defined as positive if the energy of **A** is lower when **X** has the opposite spin, as in Figure 3.11.

The advantage of this approach to deriving the energy levels in a coupled **AX** system (Figure 3.11b) is that NMR resonances on both **A** and **X** nuclei can be seen on the same diagram and the same coupling constant, J, can be seen to apply to both nuclei. Transitions between the energy levels on **X** (i.e. $\alpha\alpha \leftrightarrow \alpha\beta$ and $\beta\alpha \leftrightarrow \beta\beta$) are shown in dark blue in Figure 3.11(b) and again occur at different energies from the case with no coupling but with a same separation of J (labelled $J(A-X)$ and $J(X-A)$ in Figure 3.11(b)). The overall effect of these energy level shifts, in both the NMR spectra of **A** and **X** nuclei, is to produce two resonances with a separation of J. Each resonance will be $\frac{1}{2}J$ either side of the uncoupled resonance positions, δ, in both the **A** and **X** NMR spectra.

Note that the energy scale in Figure 3.11 is not in proportion. For heteronuclear **AX** the resonances on **A** and **X** will occur at very different energies and be in two very different regions in the MHz frequency range. The coupling constants are typically in the Hz range so the shifts in energy levels in Figure 3.11(b) are much smaller than the α and β level energy separations in Figure 3.11(a). In the case of $\mathbf{A} = {}^{31}\mathrm{P}$ and $\mathbf{X} = {}^{19}\mathrm{F}$ in $\mathrm{Z_4PF}$ on a 300 MHz spectrometer the ${}^{31}\mathrm{P}$ doublet resonances will be around 121 440 000 Hz, with a separation of around 800 Hz, while the ${}^{19}\mathrm{F}$ resonances will be near 282 282 000 Hz with the same separation.

3.7.2.2. Single nucleus energy level diagram for spin–spin coupling of two $I = \frac{1}{2}$ nuclei

When considering spin–spin coupling on a single nucleus, such as **A**, Figure 3.11(b) can be simplified to show just the shifts in the m_I energy levels that occur for nucleus **A**, caused by interaction with the m_I levels on **X** ($I = \frac{1}{2}$) in **AX**. This is shown in Figure 3.11(c). Note that the ordering of the energy levels is modified in this diagram compared to those in Figure 3.11(b), as we are considering only the energy levels of nucleus **A** and their interaction with the magnetic fields of **X**, rather than the total energy of **A** and **X** nuclei together in the applied magnetic field. Note that the direction of the energy *shifts* in Figures 3.11(b) and (c) are the same.

In this description, for a positive coupling constant J, the β state on **X** will lower the energy of the α state on **A** through the mechanism shown in Figure 3.10; similarly

the β orientation on **X** will increase the energy of the α state on **A**. For the β state on **A**, the $\beta\beta$ interaction increases the energy of this configuration and the $\beta\alpha$ interaction is a lower energy state. The allowed resonances $\Delta m_I = 1$ ($\alpha\alpha \leftrightarrow \beta\alpha$ and $\alpha\beta \leftrightarrow \beta\beta$) occur at different energies giving rise to the expected doublet in the NMR spectrum of the **A** nucleus (Figure 3.10c).

This (less rigorous but simple) analysis of the shifts in m_I energy levels on **A** alone, possible NMR transitions and their probabilities can be extended more generally for $\mathbf{AX}_n$, $n > 1$, where **A** has more than one neighbouring nucleus to which it is coupled. These systems and the resultant spectra are discussed more fully in Section 3.7.4.

3.7.3 Spin–spin coupling nomenclature, coupling strengths, and interaction distances

The coupling constant is often written as $J(\mathbf{A}\text{-}\mathbf{X})$ or $J_{A\text{-}X}$, where **A** is the nucleus being probed and **X** the one to which it is coupled. This avoids confusion where more than one spin–spin coupling interaction exists in a molecule. Coupling constants between nuclei are dependent upon their gyromagnetic ratios and these are specific to a particular nuclear isotope so coupling constants are often written to define the specific nucleus as in $J({}^{31}\mathrm{P}\text{-}{}^{1}\mathrm{H})$. This is particularly important when dealing with elements that have more than one NMR active nucleus, so coupling of fluorine to silver (Table 3.2) would need to define $J({}^{19}\mathrm{F}\text{-}{}^{107}\mathrm{Ag})$ and $J({}^{19}\mathrm{F}\text{-}{}^{109}\mathrm{Ag})$, and the ratio of these two would be in the ratio of the ${}^{107}\mathrm{Ag}$ and ${}^{109}\mathrm{Ag}$ gyromagnetic ratio values. The magnitude of J between two heteronuclear nuclei is proportional to the product of their gyromagnetic ratios. Large coupling constants, 400–1500 Hz, are seen between ${}^{31}\mathrm{P}$ and ${}^{19}\mathrm{F}$ ($\gamma = 10.839$ and 25.181×10^7 rad s^{-1} T^{-1}, respectively), while much smaller ones are seen for coupling to ${}^{187}\mathrm{Os}$ ($\gamma = 0.6193 \times 10^7$ rad s^{-1} T^{-1}); for example, $J({}^{19}\mathrm{F}\text{-}{}^{187}\mathrm{Os})$ in *cis*-$\mathrm{OsO_2F_2}$ is 35.1 Hz.

As noted previously the spin–spin coupling does not depend on the magnitude of the applied field and Hz is used as the unit for J (rather than ppm units normally used for chemical shift). Giving J in Hz also allows values to be compared immediately for heteronuclear systems where NMR spectra are obtained for two different nuclei at two different resonance frequencies. The strength of the coupling, in Hz, between two nuclei remains the same regardless of which nucleus is being observed in an NMR experiment, that is, $J({}^{31}\mathrm{P}\text{-}{}^{19}\mathrm{F}) = J({}^{19}\mathrm{F}\text{-}{}^{31}\mathrm{P})$. This can be seen in Figure 3.11(b) for the ($\alpha\alpha \leftrightarrow \beta\alpha$ and $\alpha\beta \leftrightarrow \beta\beta$) and ($\alpha\alpha \leftrightarrow \alpha\beta$ and $\beta\alpha \leftrightarrow \beta\beta$) resonances. Therefore, provided the correct spectrometer resonance frequencies are used then the $J(\mathbf{A}\text{-}\mathbf{X})$ values determined in NMR experiments probing nucleus **A** can be used to assign the couplings observed in the NMR spectrum of **X**.

Spin–spin coupling between nuclei decreases rapidly with distance and in most cases it is greatest when the two atoms are directly bonded to each other. The magnetic interaction between nearby nuclei is transmitted through space, so the interacting nuclei do not have to be bonded, but it is strengthened where there are bonds between the nuclei. This is because the coupling between the nuclear spins and electronic spins efficiently transmits information on the arrangement of neighbouring nuclear spins (Figure 3.10b). The nomenclature ^{n}J, where n is the number of bonds between the coupled nuclei, is normally used as shorthand to describe the coupling constant between nuclei. So in the molecule Z_4PF the coupling constant would be written as $^{1}J(^{31}P-^{19}F) = 680$ Hz. If there was coupling between the methyl group hydrogen atoms and phosphorus and fluorine then in $(H_3C)_4PF$ these would be denoted $^{2}J(^{1}H-^{31}P)$ and $^{3}J(^{1}H-^{19}F)$, respectively. Sometimes where coupling constants have been measured from NMR experiments and not assigned to specific interaction the nomenclature $J, J', J'' \ldots$ is used for decreasing measured values.

EXAMPLE 3.9

Using data in Table 3.2, (i) calculate the separation in ppm of the $^{1}J(^{31}P-^{19}F)$ 680 Hz doublet in the ^{31}P NMR spectrum of Z_4PF on a 200 MHz ^{1}H NMR spectrometer. (ii) Calculate the same separation in ppm on a 500 MHz ^{1}H NMR spectrometer. (iii) Calculate the separation in ppm of the 680 Hz doublet in the ^{19}F NMR spectrum of Z_4PF on a 200 MHz ^{1}H NMR spectrometer.

ANSWER

We can use and rearrange the expression for chemical shift (in ppm), that is, Eqn 3.3,

$$\delta = \frac{v - v^{\circ}}{v^{\circ}} \times 10^6 \, ,$$

where v° is the spectrometer frequency, and consider two resonances at δ_1 (v_1) and δ_2 (v_2) separated $\Delta\delta$ (Δv which is equivalent to J).

$$\Delta v = v_1 - v_2 = \frac{v_1 - v_2}{v^{\circ}} \times 10^6 = \frac{\Delta v}{v^{\circ}} \times 10^6$$

$$\Delta v = \frac{\Delta v \times v^{\circ}}{10^6}$$

which is equivalent to:

J(Hz) = [chemical shift separation (in ppm)] × [spectrometer frequency (when given as a number of MHz)].

NMR spectrometer frequencies refer to ^{1}H NMR and we need to ensure we use the correct resonance frequency for the nucleus under investigation. (i) From Table 3.2, the ^{31}P resonance frequency equivalent to 200 MHz is 80.96 MHz, so $\Delta\delta(^{31}P) = 680/80.96 = 8.40$ ppm. (ii) On a 500 MHz ^{1}H NMR spectrometer $\Delta\delta = 680/202.40$ ppm = 3.36 ppm.

(iii) The value for $^{1}J(^{19}F-^{31}P)$ will also be 680 Hz and on a 200 MHz ^{1}H NMR spectrometer the ^{19}F equivalent operating frequency from Table 3.2 is $(200 \times 94.094/100) = 188.19$ MHz. Therefore, $\Delta\delta(^{19}F) = 680/188.19$ ppm = 3.61 ppm.

SELF TEST

Calculate the separation in ppm of the $^{1}J(^{19}F-^{31}P)$ 680 Hz doublet in the ^{19}F NMR spectrum of Me_4PF on a 500 MHz ^{1}H NMR spectrometer.

3.7.4 First-order heteronuclear coupling in AX_n systems, $n > 1$

First-order coupling for an **AX** system gives rise to a doublet with a coupling constant, J, given by the separation of the resonances and a chemical shift by their mid-point (Figure 3.12a). When there is more than one nearby $I = \frac{1}{2}$ nucleus, as in AX_n with $n > 1$, the NMR spectrum of **A** will show more complex coupling patterns, for example, Figures 3.12(b) and (c). These spectra will represent the effects of coupling to the various orientations of the nuclear magnetic moments on all the **X** nuclei, and we now consider an AX_2 system in more detail, again with ^{1}J positive.

We take as an example a molecule Z_3PF_2, where Z is a non-NMR active nucleus ($I = 0$), and its ^{31}P NMR spectrum (Figure 3.13a). We only consider, as in Section 3.7.2.2 for **AX** systems, the effect of coupling on the relative energy levels on **A**. We have to consider the possible m_I values on the two bonded equivalent fluorine atom nuclei, F_1 and F_2, and their effect on the magnetic field experienced on phosphorus. Possible orientations of these are $m_I(F_1) = +\frac{1}{2}$ and $m_I(F_2) = +\frac{1}{2}$, $m_I(F_1) = +\frac{1}{2}$ and $m_I(F_2) = -\frac{1}{2}$, $m_I(F_1) = -\frac{1}{2}$ and $m_I(F_2) = +\frac{1}{2}$, and $m_I(F_1) = -\frac{1}{2}$ and $m_I(F_2) = -\frac{1}{2}$, which yields three different values for the sum of $m_I(F)$ (Σm_I) namely +1, 0 (two arrangements), and −1. Each of these three Σm_I values will produce a different local magnetic field at the ^{31}P nucleus, through the mechanism shown in Figure 3.10, so the resonance will be split into a **triplet** with intensities 1:2:1, the intensities reflecting the likelihood of the Σm_I arrangement. The chemical shift for this triplet resonance is again that of the centroid of the multiplet, which in this case is that of the central, most intense peak (Figure 3.12b).

An alternative way of visualising the coupling scheme in an AX_n system is a '**coupling tree**' or **stick coupling diagram**, shown in Figure 3.13(b). In a coupling tree, coupling to other nuclei is added one nucleus at a time. In the case of the $-PF_2$ ($-AX_2$) system the phosphorus couples to one fluorine nucleus to give two 'branches' on the tree—representing the αα and αβ states as before. It then couples to the other fluorine nucleus, with the same coupling constant. This produces three branches on the tree (ααα, ααβ ≡ αβα, and αββ), and three resonances in the NMR spectrum, as two of the branches become

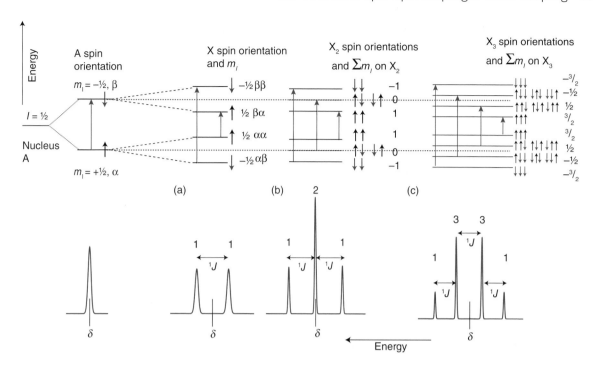

FIGURE 3.12 Energy levels, possible transitions and resultant NMR spectra (a) **AX**, (b) **AX$_2$**, and (c) **AX$_3$**. Relative peak intensities and the measurement of the coupling constant ^1J are shown for each coupling system.

coincident. The central branch, having been arrived at by two routes, is twice as intense in the NMR spectrum.

In general, a multiplet of $2nI + 1$ lines is obtained when a spin ½ nucleus (or a set of symmetry-equivalent spin

½ nuclei) is coupled to n equivalent nuclei of spin I. For $I = ½$ nuclei such as ^{1}H, ^{19}F, and ^{31}P this simplifies to $n + 1$. The ^{29}Si NMR spectrum (5% abundance, $I = ½$) of SiH$_4$ is a quintet arising from coupling to the four equivalent ^{1}H $I = ½$

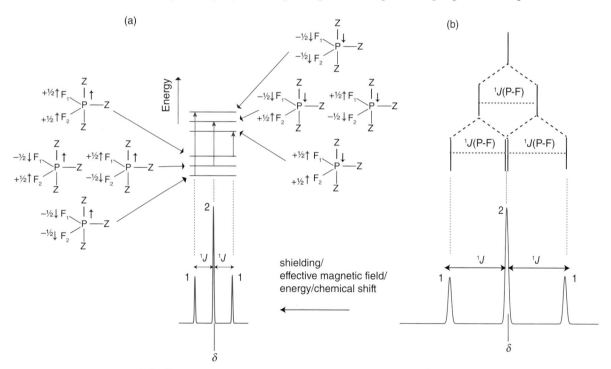

FIGURE 3.13 Origin of the $^1J(^{31}$P-^{19}F) spin–spin coupling triplet for Z$_3$PF$_2$ and the observed ^{31}P NMR spectrum and intensities. (a) Shows how the orientations of m_I on fluorine influence the magnetic field observed at phosphorus. (b) Shows the coupling tree for the –PF$_2$ system in Z$_3$PF$_2$.

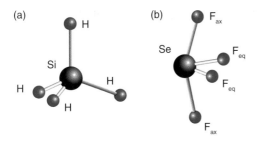

FIGURE 3.14 Molecular structures of (a) SiH_4 and (b) SeF_4.

nuclei in the SiH_4 molecule; see Figures 3.13 right-hand side and 3.14(a). The values of n strictly refers to the number of magnetically equivalent nuclei which must be chemically equivalent in terms of their local environment—as they are in SiH_4; for a more detailed discussion of magnetic equivalence see Section 3.15.

If there is non-equivalence of neighbouring sites, then the coupling constants to these different sites will be different and this will be evident in the NMR spectrum, for example, in SeF_4 (with a 'saw-horse' or 'see-saw' structure), where the fluorine sites are divided into two distinct environments, axial and equatorial (Figure 3.14b). In $^{77}SeF_4$ the couplings from ^{77}Se to the two different fluorine nuclei, $^1J(^{77}Se-^{19}F_{ax})$ and $^1J(^{77}Se-^{19}F_{eq})$, need to be considered in turn. Taking $^1J(^{77}Se-^{19}F_{ax})$ as the largest, then coupling to two F_{ax} sites would produce a triplet; the $^1J(^{77}Se-^{19}F_{eq})$ coupling to two equatorial fluorine atoms would then split each resonance into a further triplet. The ^{77}Se NMR spectrum of $^{77}SeF_4$ would consist of a triplet of triplets.

The treatment of coupling to non-equivalent neighbouring nuclei can cause confusion when compared with the nomenclature used in NMR spectroscopy for many simple organic compounds. For organic compounds n in an $n + 1$ multiplet is used to give the number of coupled nuclei next to the nucleus under observation. This approach works for many organic compounds because such neighbouring

nuclei are usually chemically equivalent (or are very close to chemically equivalent and so may be taken as such). This is not usually the case for inorganic compounds where more complex molecular geometries lead to chemically inequivalent NMR nuclei, as we have seen for SeF_4.

For compounds with larger numbers of equivalent neighbouring coupled nuclei, with $I = \frac{1}{2}$, the intensities and number of resonances in the NMR spectrum can easily be obtained from Pascal's triangle; each of the neighbouring nuclei has a 50% likelihood of having $m_I = +\frac{1}{2}$ or $m_I = -\frac{1}{2}$ (Figure 3.15). Note that this use of Pascal's triangle is only applicable for coupling involving $I = \frac{1}{2}$ nuclei. Pascal's triangle is identical to the branching coupling tree approach described previously, where the numbers represent the total different ways of reaching that point through adding coupled nuclei.

In the example given in Figure 3.15, the fourth row of Pascal's triangle gives the **number of lines** and **their relative intensities** for a quartet obtained when a central nucleus couples to three equivalent neighbours, as in the ^{31}P NMR spectrum of POF_3, see also Figure 3.12(c); the chemical shift, δ, in this case is given by mid-point between the two central resonances. An octahedral species, for example, $^{77}SeF_6$, would exhibit a septet with intensities in the ratio 1:6:15:20:15:6:1 whose peaks would be separated by a constant frequency $J(Se-F)$ and whose chemical shift would coincide with the central most intense resonance.

The $2nI + 1$ lines rule and Pascal's triangle can be applied sequentially to determine coupling patterns in NMR spectra, noting that the strength of spin–spin coupling normally decreases with increasing n in ^nJ. In a molecule such as $H–P(R_3)–F$ that contains the unit $^1H–^{31}P–^{19}F$ the ^{19}F NMR spectrum will contain a resonance that will be split into a doublet by heteronuclear coupling to ^{31}P nucleus, $^1J(F-P)$, and then each component split into a doublet by coupling to ^1H, $^2J(F-H)$; the spectrum is, thus, described as a doublet of doublets (Figure 3.16).

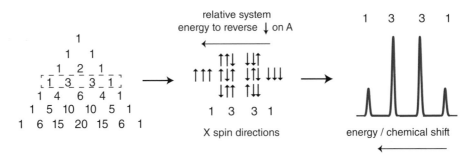

FIGURE 3.15 Pascal's triangle for an AX_n system with each of the neighbouring nuclei, X ($I = \frac{1}{2}$), having a 50% likelihood of $m_I = +\frac{1}{2}$ or $m_I = -\frac{1}{2}$. For the AX_3 system, the eight possible arrangements of spin on the three X nuclei are shown which fall into four system transition energies with the number distribution of 1:3:3:1. This gives rise to four possible NMR transitions of different energies and probabilities with relative intensities of 1:3:3:1.

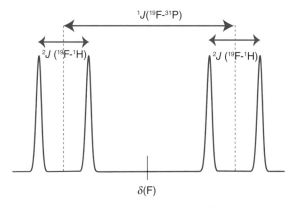

FIGURE 3.16 Heteronuclear coupling in the ^{19}F NMR spectrum of the ^{1}H–^{31}P–^{19}F system giving rise to a 'doublet of doublets'.

3.7.5 Homonuclear coupling

Homonuclear coupling which occurs between nuclei of the same element is detectable when the nuclei are unrelated by the symmetry operations of the molecule. When nuclei are equivalent by symmetry the spin–spin coupling between them is not observed in the NMR spectrum; the origin of this lies in the fact that the transitions are symmetry forbidden. Thus, a single ^{19}F signal is found for a tetrahedron-based molecule such as CF_3I as the three nuclei are equivalent via the threefold axis of the molecule.

Homonuclear coupling can be seen in the ^{19}F NMR spectrum of ClF_3 in Figure 3.17; Cl is effectively NMR inactive. ClF_3 is a 'T-shaped' molecule with two equivalent axial fluorine sites and a distinct equatorial fluorine atom site (Figure 3.17, inset).

The signal ascribed to the two axial F nuclei (^{19}F, $I = \frac{1}{2}$, 100%) is split into a doublet by the single equatorial ^{19}F nucleus, and the latter is split into a triplet by the two axial ^{19}F nuclei. Note that the integrated peak intensities of 2:1, axial : equatorial, reflect the abundance of the two different environments in the molecule. Thus, the pattern of ^{19}F resonances in high resolution NMR spectroscopy readily distinguishes this unsymmetrical structure from trigonal-planar and trigonal-pyramidal structures, both of which would have three equivalent fluorine nuclei and, hence, produce a single ^{19}F resonance.

EXAMPLE 3.10

Predict the features expected in the ^{19}F NMR spectrum of each of the following compounds (^{19}F, $I = \frac{1}{2}$, 100%, ^{31}P, $I = \frac{1}{2}$, 100%; take S, Cl, and I to be NMR inactive).

(i) octahedral SF_6;

(ii) trigonal bipyramidal PCl_3F_2 with (a) two equatorial fluorine atoms, (b) one axial and one equatorial fluorine atom and (c) two axial fluorine atoms.

Interpret the ^{19}F NMR spectra given for the following molecules:

(iii) SF_4—saw-horse structure (Figure 3.18a);

(iv) IF_5—square based pyramid (Figure 3.18b).

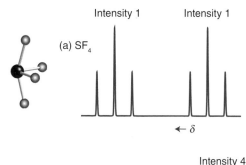

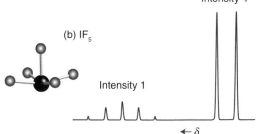

FIGURE 3.18 ^{19}F NMR spectra of (a) SF_4 and (b) IF_5.

ANSWERS

(i) All the fluorine positions are equivalent in an octahedron so no spin–spin coupling occurs and a single resonance is observed.

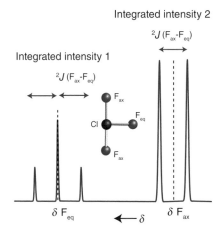

FIGURE 3.17 High resolution ^{19}F NMR spectrum of ClF_3 (inset) showing homonuclear coupling between the axial and equatorial fluorine sites which are non-equivalent.

(ii) (a) The two equatorial fluorine positions are symmetrically equivalent so no homonuclear coupling occurs between them. Heteronuclear coupling will occur to ^{31}P so the spectrum consists of a doublet with splitting $^{1}J(^{19}F-^{31}P)$.
(b) There are two different fluorine environments and each will show heteronuclear coupling to ^{31}P and homonuclear coupling to the non-equivalent fluorine atom. The ^{19}F NMR spectrum will consist of a pair of doublet of doublets.
(c) The two axial fluorine positions are equivalent, so the spectrum will be of the same form as in (a).

(iii) The structure of SF_4 is often described as a 'saw-horse' geometry and derives from a trigonal bipyramid with a lone pair in an equatorial position; see Figure 3.18(a). Thus the fluorine atoms occupy two different environments; two on axial positions and two on equatorial (as referenced to the original trigonal bipyramidal description). At low resolution where spin–spin coupling is not resolved the spectrum will consist of two equal intensity resonances reflecting the two distinct environments. Homonuclear coupling between these two types of fluorine atoms will split (with $^{2}J(F_{eq}-F_{ax})$) each of the resonances into a triplet (coupling to two non-equivalent nuclei) giving rise to a spectrum consisting of a pair of triplets (Figure 3.19a).

(iv) IF_5 adopts a square based pyramidal structure which has a unique fluorine atom at the pyramid's apex and four equivalent fluorine atoms in the base. Following similar reasoning to part (iii), the spectrum will consist of a quintet of overall intensity 1 (the apical site coupled to the four inequivalent basal-plane fluorine atoms, $^{2}J(F_{ap}-F_{basal})$) and a doublet of total relative intensity 4 (the basal plane site coupled to the apical fluorine site, also $^{2}J(F_{ap}-F_{basal})$), as shown in Figure 3.19(b).

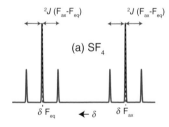

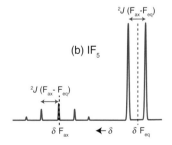

FIGURE 3.19 ^{19}F NMR spectra of (a) SF_4 and (b) IF_5 with assignments.

SELF TEST

Interpret the ^{19}F NMR spectra in Figure 3.20 from (a) $[SF_5]^+$ (S can be assumed to be NMR inactive) and (b) Me_2PF_3 (with both methyl groups equatorial in a trigonal bipyramid; assume no ^3J coupling to hydrogen) in terms of the structure of the species (^{31}P, $I = ½$, 100%).

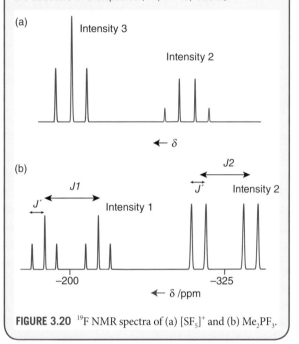

FIGURE 3.20 ^{19}F NMR spectra of (a) $[SF_5]^+$ and (b) Me_2PF_3.

When describing the form of NMR spectra in written terms the data are often given as a chemical shift values with a shorthand description of any multiplets. Thus a doublet is written as (d), a triplet (t), and so on. Complex coupling schemes are described using multiple letters so a doublet of triplets would be (dt). Relative intensity values can also be given to each signal. The spectrum given in Figure 3.20(b) would be summarised as:

δ −200 (dt) relative intensity 1;
δ −325 (dd) relative intensity 2.

3.7.6 First-order coupling to spin $I > ½$ nuclei

Coupling to quadrupolar nuclei with $I \geq 1$ can also be observed in NMR spectra when the quadrupolar nucleus is in a highly symmetric environment. Consider the case of an $I = ½$ nucleus with a single neighbour that has $I = 1$. The $I = 1$ nucleus has possible m_I values of +1, 0, and −1, giving rise through coupling to three energy states (Figure 3.21) for each m_I level on the $I = ½$ nucleus. There are three transitions possible on the observed nucleus, subject to the selection rule $\Delta m_I = \pm1$, and the NMR spectrum of the $I = ½$ nucleus will consist of a triplet (Figure 3.21). These will have equal intensities for all the three resonances (Figure 3.21),

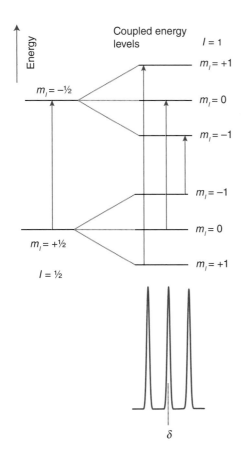

FIGURE 3.21 Coupling scheme energy level diagram and resultant NMR spectrum for an $I = \frac{1}{2}$ nucleus with a single neighbour nucleus with $I = 1$.

reflecting the equal probability of each m_I value on the $I = 1$ nucleus. Note how this differs from the triplet with 1:2:1 intensities derived from coupling to two equivalent $I = \frac{1}{2}$ nuclei. Coupling to more than one $I > \frac{1}{2}$ nucleus is also possible, provided they all lie in high symmetry environments, Section 3.16, and the resultant NMR spectrum can be derived by consideration of all the possible combinations of m_I values. The expression, given earlier, which states that a multiplet of $2nI + 1$, lines is obtained when a spin-$\frac{1}{2}$ nucleus is coupled to n equivalent nuclei of spin I can be used to derive the total number of resonances when coupling to quadrupolar nuclei.

EXAMPLE 3.11

Derive the expected form of the ^{19}F NMR spectra of the following species:

(i) $[AsF_6]^-$ (^{75}As 100% abundant, $I = \frac{3}{2}$);

(ii) $[^{121}Sb_2F_{11}]^-$ (Figure 3.22). For ^{121}Sb, $I = \frac{5}{2}$; ignore couplings beyond $^1J(Sb\text{-}F)$.

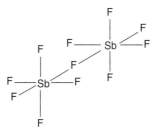

FIGURE 3.22 Structure of the molecular anion $[Sb_2F_{11}]^-$.

ANSWERS

(i) The octahedral $[AsF_6]^-$ anion has six symmetry equivalent fluorine atoms; these will couple with the $+\frac{3}{2}$, $+\frac{1}{2}$, $-\frac{1}{2}$, and $-\frac{3}{2}$ m_I levels of the ^{75}As nucleus giving a quartet with equal intensity resonances. The expression $2nI + 1$ with $n = 1$ and $I = \frac{3}{2}$ also gives four resonances.

(ii) There are three fluorine environments, namely the central bridging F^-, two equivalent terminal F^- (*trans* to the bridging F^-), and eight equivalent terminal F^- *cis* to the bridging F^-. Ignoring spin–spin coupling, the spectrum will consist of three resonances in the ratio 1:2:8. For each terminal fluorine atom, $^1J(Sb\text{-}F)$ coupling will give rise to a sextet (m_I on Sb taking values $\frac{5}{2}$, $\frac{3}{2}$, $\frac{1}{2}$, $-\frac{1}{2}$, $-\frac{3}{2}$, and $-\frac{5}{2}$) of equal intensities. The bridging fluorine can couple to $2 \times$ Sb with the $\sum m_I$ values 5 (= $\frac{10}{2}$), 4, 3, 2 . . . −5, producing an undecaplet (11-line multiplet); the intensities of these resonances will be in the ratio 1:2:3:4:5:6:5:4:3:2:1 representing the number of ways that each $\sum m_I$ value can be achieved. The full spectrum (with only $^1J(F\text{-}Sb)$ coupling considered) will, therefore, consist of two sextets (intensity ratio 2:8) and an undecaplet (eleven signal multiplet, relative intensity 1).

SELF TEST

Predict the ^{19}F NMR spectrum of the $[^{123}Sb_2F_{11}]^-$ anion; ^{123}Sb $I = \frac{7}{2}$.

3.8 **Magnitudes of coupling constants**

The sizes of $^2J(^1H\text{-}^1H)$ and the common $^3J(^1H\text{-}^1H)$ homonuclear coupling constants in organic molecules are typically 18 Hz or less. In contrast ^1H–X heteronuclear coupling constants can be several hundred hertz for ^1J; ^1J homonuclear and heteronuclear coupling between nuclei other than ^1H (such as ^{31}P and ^{19}F) can have coupling constants of many hundreds of hertz or kilohertz (Table 3.7). These large ^1J values reflect the fact that the

TABLE 3.7 Coupling constants for common inorganic NMR nuclei

Coupling constant	Magnitude/Hz
$^1J(^{31}P^{-1}H)$	150–200
$^1J(^{31}P^{-19}F)$	400–1500
$^2J(^{31}P^{-1}H)$	2–3
$^1J(^{11}B^{-19}F)$	65
$^1J(^{11}B^{-1}H)$ terminal	100–200
$^1J(^{11}B^{-1}H)$ bridging	50–100
$^1J(^{13}C^{-19}F)$	50–160
$^2J(^{19}F^{-19}F)$	40–370
$^1J(^{19}F^{-1}H)$	40–60
$^1J(^{29}Si^{-19}F)$	100–300
$^2J(^{195}Pt^{-1}H)$ and $^1J(^{195}Pt^{-13}C)$	15–25 and 60–200
$^1J(^{103}Rh^{-31}P)$	100–250
$^1J(^{103}Rh^{-13}C)$	5–15
$^1J(^{195}Pt^{-205}Tl)$	25 000–75 000
$^1J(^{19}F^{-77}Se)$	300–750

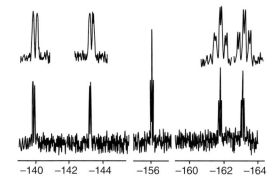

FIGURE 3.23 The ^{19}F NMR spectrum (282 MHz) of $C_5Ph_4(C_6F_5)OH$. (From H.K. Gupta, M. Stradiotto, D.W. Hughes, and M.J. McGlinchey, *J. Org. Chem.* **65** 3652 (2000).)

atoms are directly bonded, and in many cases the large gyromagnetic ratios of the nuclei involved (remembering that J depends on the product of the two γ values of the coupled nuclei).

The size of these coupling constants should be considered in relation to the chemical shift range for the nuclei involved and the typical linewidth. For proton NMR spectra recorded on a 500 MHz instrument the chemical shift range in Hz (with most environments in the range $\delta = 0$–10 ppm) corresponds to $\Delta\nu = (\Delta\delta$ (in ppm) $\times 500 \times 10^6)/10^6 = 5000$ Hz (from Eqn 3.2). The ^1H linewidth is normally a few hertz so ^3J homonuclear couplings, such as between hydrogen atoms on two adjacent carbon atoms, are clearly resolved in ^1H NMR spectra. For other nuclei, which usually have much larger chemical shift ranges than ^1H, the splitting patterns of individual resonances are often well spread out. For example, in Figure 3.23 the ^{19}F NMR spectrum of $C_5Ph_4(C_6F_5)OH$ shows coupling constants of around 20 Hz compared with the 300 000 Hz (~600 ppm) chemical shift range for ^{19}F on a 500 MHz spectrometer. Where even larger J values are found, for example, $^1J(^{31}P^{-19}F)$ ~1000 Hz, this again should be contrasted with the chemical shift range found for ^{31}P environments (−300 to +300 ppm, equivalent to 300 000 Hz on a 500 MHz NMR spectrometer) so that spin multiplets for different environments are normally well separated from each other despite the large coupling constant.

The sizes of coupling constants are often related to the geometry of a molecule, particularly the distance between the nuclei concerned, and to the magnitude of the gyromagnetic ratios of the two nuclei. As interactions between nuclei that occur through bonding electrons are the most important factor, then s-electrons, which have a finite probability of being at the nucleus, are most effective in transmitting spin–spin coupling throughout a molecule. As a result the degree of s-character in any bond can be used to predict the strength of the likely coupling constant between two adjacent nuclei (see Section 3.13).

Other trends or distinct values for certain structural relationships between nuclei can often be used to interpret NMR spectra. In square-planar Pt(II) phosphine complexes, $^1J(^{195}Pt^{-31}P)$ is sensitive to the group *trans* to a phosphine ligand and the value of $^1J(Pt-P)$ increases in the following order of *trans* ligands:

$$R^- < H^- < PR_3 < NH_3 < Br^- < Cl^-$$

For example, *cis*-[PtCl$_2$(PEt$_3$)$_2$], where Cl$^-$ is *trans* to P, has $J(Pt-P) = 3.5$ kHz, whereas *trans*-[PtCl$_2$(PEt$_3$)$_2$], with P *trans* to P, has $J(Pt-P) = 2.4$ kHz. These systematic variations allow *cis* and *trans* isomers to be distinguished quite readily from their NMR spectra. The rationalisation for the variation in the sizes of the coupling constants above stems from the fact that a ligand that exerts a large *trans* influence substantially weakens the bond *trans* to itself, causing a reduction in the NMR coupling between the nuclei.

3.9 Decoupling and decoupled spectra

Spin–spin coupling depends on the nucleus under observation (**A**) experiencing the local field generated by a nearby magnetic nucleus (**B**) during the period it undergoes resonance. This means that the spin–spin coupling

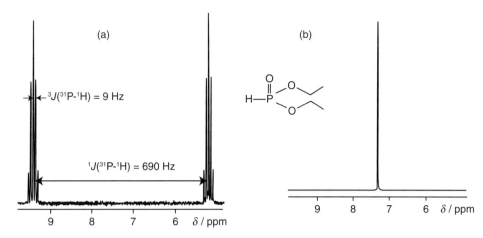

FIGURE 3.24 ^{31}P NMR spectra collected from diethylphosphite (inset): (a) ^{31}P spectrum showing ^{1}J and ^{3}J (P-H) coupling; (b) the ^{31}P{^{1}H} decoupled spectrum. Adapted from http://chem.ch.huji.ac.il/nmr/techniques/1d/row3/p.html.

interaction at **A** can be removed if nucleus **B** can be made to undergo rapid reorientation (transitions between its m_I states) such that the local field experienced at **A** is averaged. This can be achieved by irradiating the sample and, therefore, all **B** nuclei at their resonance frequencies while probing the different resonance frequency range and NMR spectrum of **A**. This is known as **broad band decoupling** and such spectra are denoted using the terminology A{B} as, for example, in the proton-decoupled fluorine NMR spectrum ^{19}F{^{1}H}. The collection of decoupled spectra, which are normally greatly simplified with respect to spectra that show all the spin–spin coupling interactions, can aid in the assignment of different resonances to specific sites in a molecule. The use of decoupling is particularly widespread for molecules that contain a lot of proton sites which otherwise show large numbers of complex multiplets through ^{n}J(A-H) coupling.

The two ^{31}P NMR spectra shown in Figure 3.24(a) and (b) were collected from diethylphosphite (Figure 3.24, inset) without (a) and with (b) proton decoupling. Without decoupling the spectrum shows complex multiplets due to ^{1}J(^{31}P-^{1}H) and ^{3}J(^{31}P-^{1}H) coupling; the ^{31}P{^{1}H} spectrum just shows a single resonance due to the one phosphorus environment in this compound.

EXAMPLE 3.12

Predict the form of the following spectra of the square planar complex [PdH(PPh$_3$)$_3$]$^+$. (Ignore coupling between phosphorus and the phenyl group protons, and assume Pd is NMR inactive)

(a) The ^{31}P NMR spectrum

(b) The ^{31}P{^{1}H} NMR spectrum

ANSWERS

(a) The complex [PdH(PPh$_3$)$_3$]$^+$ has two different phosphorus environments with two atoms *trans* to each other in the square planar molecule, labelled here as P(1), and the other *trans* to the proton, denoted here as P(2). These environments will be observed in the ^{31}P NMR spectrum as two resonances with different chemical shifts (intensity ratio 2P(1):1P(2)). Coupling with the hydrogen nucleus will split both resonances into doublets, with ^{2}J(^{31}P-^{1}H), and coupling with the inequivalent phosphorus site, ^{2}J(^{31}P(1)-^{31}P(2)), will produce a doublet of doublets for P(1) and a doublet of triplets for P(2); the total signal intensity in each multiplet will maintain the ratio 2P(1):1P(2).

(b) The ^{2}J(^{31}P-^{1}H) coupling will be removed in the ^{31}P{^{1}H} NMR spectrum but the homonuclear ^{2}J(^{31}P(1)-^{31}P(2)) coupling will be maintained and the spectrum will consist of a doublet (intensity 2) and a triplet (intensity 1).

SELF TEST

Predict the form of the ^{11}B{^{1}H} NMR spectrum of decaborane, ^{11}B$_{10}$H$_{14}$ (Figure 3.25). Comment on how this spectrum would compare with that obtained without decoupling; each boron atom has a terminal B–H bond and bridging hydrogen bonds exist along the B–B edges shown in Figure 3.25 (consider only ^{1}J (B-H) couplings).

In broad band decoupling, the whole range of frequencies for the decoupled nucleus is irradiated. It is also possible to undertake selective or narrow band decoupling (which when applied to protons is also known as **specific proton decoupling**) where only one coupled frequency, or a group of coupled frequencies, is irradiated and just

FIGURE 3.25 Decaborane, $B_{10}H_{14}$, showing boron framework (with B-B bonds < 1.8 Å) and the full molecular structure.

their associated coupling patterns are decoupled and removed. This technique is of particular use in assigning connectivities of nuclei in complex molecules with large numbers of resonances as, by sequentially and selectively decoupling at different frequencies, the assignment of individual resonances can more easily be undertaken.

3.10 Effects due to variations in natural abundance—isotopomers

Many elements occur naturally as isotopes with different numbers of neutrons in the nucleus and each isotope will have a different nuclear spin quantum number and gyromagnetic ratio. For example, silicon exists as the stable isotopes ^{28}Si (92.2%, $I = 0$), ^{29}Si (4.7%, $I = \frac{1}{2}$, $\gamma = -5.315 \times 10^7$ rad s^{-1} T^{-1}), and ^{30}Si (3.1%, $I = 0$). As well as being an important factor in determining the sensitivity of the NMR experiment for a particular experiment (silicon NMR spectra are much weaker than ^1H NMR spectra partly because only 4.7% of the molecules in a sample will contain the NMR active ^{29}Si nucleus) it will also determine the nature of any coupling pattern. This is because neighbouring nuclei may or may not have $I \neq 0$ and so be NMR active depending on which isotope is present in a particular molecule. Molecules that are identical except for differences in the isotopes of the atoms that are present are known as isotopomers.

Whether coupling is seen to a specific isotope will depend to a large degree on their natural abundance and the quality (signal-to-noise ratio) of the NMR spectrum collected. In carbon NMR spectra coupling between adjacent ^{13}C nuclei is not normally observed as with the low natural abundance of ^{13}C (1.1%, $I = \frac{1}{2}$; note ^{12}C $I = 0$, 98.9%) the likelihood of a probed ^{13}C nucleus having an adjacent ^{13}C is very small and, thus, would produce little intensity. Where different isotopes occur with moderate or higher abundances then the contribution of each to a coupling pattern needs to be considered. This may be illustrated by predicting the ^{19}F spectrum of the square planar molecule XeF_4 (Figure 3.26, inset), and considering ^{129}Xe (26.4%

abundant, $I = \frac{1}{2}$) and assuming all other Xe isotopes are NMR inactive. All the fluorine sites in XeF_4 are identical so without considering $^1J(^{19}F-^{129}Xe)$ the spectrum would consist of a single resonance ($\delta = 210$ ppm). 73.6% of the fluorine atoms in a sample will be bonded to non-NMR active xenon atoms so will be unperturbed by the adjacent xenon atom and, thus, resonate at 210 ppm. The remaining fluorine atoms in the sample will couple to their adjacent ^{129}Xe nucleus producing a doublet ($J(^{129}Xe-^{19}F)$ ~4000 Hz) so 26.4% of the intensity in the ^{19}F NMR spectrum will be apportioned in this doublet distributed either side of the main resonance. The summation of these resonances thus produces a central resonance (intensity ~74%) with two satellites each of intensity ~13% (Figure 3.26). Care should be taken in describing the form of this spectrum as it is not a 'triplet' but rather a singlet with an overlaid doublet. The resonances formed by coupling to a non-100% abundant nucleus are usually termed satellites and are often marked with asterisks (*).

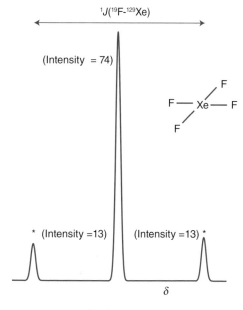

FIGURE 3.26 The ^{19}F NMR spectrum of XeF_4 (inset).

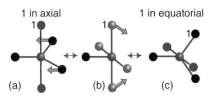

FIGURE 3.28 The Berry pseudo-rotation of an AL$_5$ molecule which interchanges the axial and equatorial positions.

EXAMPLE 3.13

Interpret the ^{19}F NMR spectrum (Figure 3.27) obtained from the [AgF$_4$]$^-$ anion. (^{107}Ag $I = \frac{1}{2}$, 52%; ^{109}Ag $I = \frac{1}{2}$, 48%).

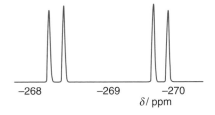

FIGURE 3.27 The ^{19}F NMR spectrum obtained from [AgF$_4$]$^-$.

ANSWER

Coupling from fluorine to an adjacent ^{107}Ag nucleus will occur in 52% of anions and to ^{109}Ag nuclei 48% of the time. Thus, each fluorine environment will be coupled to an adjacent silver nucleus to give two sets of doublets, with intensity ratio 52:48, these doublets will have different ^1J(Ag-F) values (as ^1J depends on the gyromagnetic ratio, note the values for ^{107}Ag and ^{109}Ag are in the ratio 1.089/1.252 (Table 3.2) consistent with the degree of coupling shown in the spectrum). The spectrum can, therefore, be explained as a single fluorine environment coupled to a single silver environment (consisting of 52% ^{107}Ag and 48% ^{109}Ag, note that the slightly more intense doublet has a smaller ^1J(Ag-F) value in agreement with a smaller γ for the more abundant ^{107}Ag nucleus). The spectrum is thus compatible with a single fluorine environment and either a tetrahedral or square planar shape for the [AgF$_4$]$^-$ anion.

SELF TEST

Predict the ^{19}F NMR spectrum of the [XeF$_3$]$^+$ cation which has the same shape as CℓF$_3$ shown in Figure 3.17 with Xe replacing Cℓ.

3.11 **Introduction to fluidity**

In many molecules the atomic positions are not fixed and different sites within a molecule can exchange with each other, chemical exchange. A simple case of this is the so-called Berry pseudo-rotation in a trigonal bipyramidal molecule A(L$_{eq}$)$_3$(L$_{ax}$)$_2$ (Figure 3.28), which exchanges equatorial ligands, L$_{eq}$, on a central A atom with the axial ligands (L$_{ax}$). If this process occurs faster than a technique timescale (how long the system is interrogated) then the method will see an averaged site.

As the energy level separation probed in NMR spectroscopy corresponds to ~10^8 Hz (radio waves) the timescale of NMR spectroscopy is 'slow' in comparison to other spectroscopic techniques used to characterise inorganic compounds (Chapter 1, Section 1.4). 'Slow' in this sense means that structures can be resolved provided their lifetime is not less than a few milliseconds. If a molecule undergoes some transformation, such as interconversion to a different geometry faster than a few times per microsecond, then the NMR experiment will sample more than one conformation and the NMR spectrum will represent an averaged structure. For example, Fe(CO)$_5$ shows just one ^{13}C resonance, indicating that, on the NMR timescale, all five CO groups are equivalent. However, the IR spectrum (of timescale about 1 ps) shows distinct axial and equatorial CO groups, and by implication has a trigonal-bipyramidal structure. The observed ^{13}C NMR spectrum of Fe(CO)$_5$ is the weighted average of the separate axial and equatorial site resonances.

Because the temperature at which an NMR spectrum is recorded can easily be changed, samples can often be cooled to a temperature at which the rate of interconversion becomes slow enough for separate resonances to be observed. Figure 3.29 shows the idealised ^{31}P-NMR

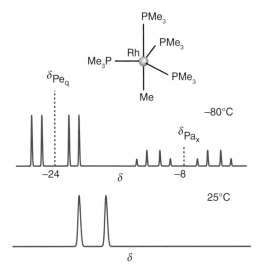

FIGURE 3.29 Idealised ^{31}P{^{1}H}-NMR spectra of [RhMe(PMe$_3$)$_4$] at room and low (−80°C) temperatures.

spectra of [RhMe(PMe$_3$)$_4$] at room temperature and at −80°C.

At low temperature the spectrum consists of a doublet of doublets, of relative intensity 3, near δ = −24 ppm, which arises from the equatorial phosphorus atoms (coupled to ^{103}Rh (I = ½), 100% abundant) and the single axial ^{31}P, and a doublet of quartets of intensity 1, near −8 ppm, derived from the axial phosphorus (coupled to ^{103}Rh and the three equatorial ^{31}P atoms). At room temperature, scrambling of the PMe$_3$ groups

makes them all equivalent and just a doublet is observed (from coupling to ^{103}Rh). The chemical shift observed is the weighted average arising from the two distinct axial and equatorial sites.

Variable temperature NMR (VT-NMR) allows determination of the temperature at which the spectrum changes from the high-temperature form to the low-temperature form ('the coalescence temperature') and, thence, to determine the barrier to interconversion (see Section 3.17).

EXAMPLE 3.14

Interpret the main changes seen in the variable temperature ^{1}H NMR spectra obtained from [Os$_3$(CO)$_8$(η^2-CH$_2$CH$_2$)(μ^6-C$_6$H$_6$)] (Figure 3.30).

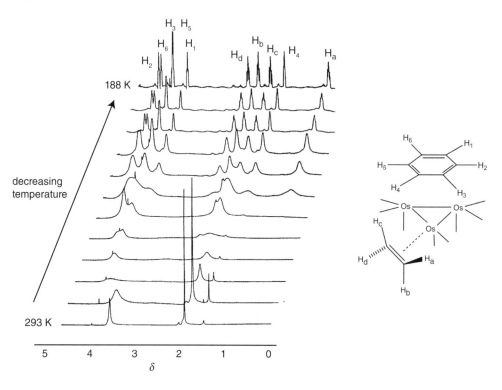

FIGURE 3.30 ^{1}H NMR spectra of [Os$_3$(CO)$_8$(η^2-CH$_2$CH$_2$)(η^6-C$_6$H$_6$)] (inset) as a function of temperature. (Data from M.A. Gallop, B.F.G. Johnson, J. Keeler, J. Lewis, S.J. Hayes, and C.M. Dobson, *J. Am. Chem. Soc.* **114** 2510 (1992).)

ANSWER

At the lowest temperature, 188 K, the C$_6$H$_6$ ring and ethene molecule have fixed orientations in the complex such that each proton has a distinct environment. Resonances in the range δ = 3–5 ppm and a resonance at ~1.5 ppm derive from protons in the C$_6$H$_6$ rings (labelled H$_1$–H$_6$ in Figure 3.30,

inset) and four resonances between δ = 0 and 3 ppm from the ethene molecule (H$_a$–H$_d$). On warming the sample both these ligands undertake rapid rotational motion around their Os–ligand axis. This averages the hydrogen positions and produces just two strong resonances one from the hydrogen sites in C$_2$H$_4$ at 2.1 ppm and one from the C$_6$H$_6$ group at 3.8 ppm.

SELF TEST

The ^{1}H NMR spectrum, as a function of temperature between 218 and 346 K, of Ge(η^1-C$_5$H$_5$)Me$_3$ (Figure 3.31a) is shown in Figure 3.31(b). Interpret the changes observed as a function of temperature.

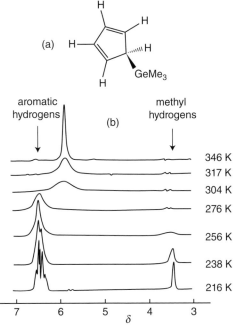

FIGURE 3.31 (a) Molecular structure of Ge(η_1-C$_5$H$_5$)Me$_3$ at low temperature and (b) its ^{1}H NMR spectra as a function of temperature. (Data from V.S. Shriro, Yu.A. Strelenko, Yu.A. Ustynyuk, N.N. Zemlyansky, and K.A. Kocheshkov, *J. Organomet. Chem.* **117** 321 (1976).)

Problems

These problems require knowledge of basic inorganic chemistry, such as common molecular geometries and VSEPR theory.

3.1 Basic NMR technique and resonance frequency problems

(a) Calculate the resonance frequency of a ^{103}Rh nucleus in (i) a 11.744 T magnetic field and (ii) a 11.740 T magnetic field using the data in Table 3.1. Using Eqn 3.1 and $E = h\nu$ calculate the difference in resonance frequency (in Hz) of the ^{103}Rh nucleus in the two fields.

(b) Using the data in Table 3.2 discuss why the preferred nucleus for studying tin compounds by NMR is ^{119}Sn.

3.2 NMR and the number of environments

(a) Use VSEPR theory to predict shapes for the following species and hence determine the number of distinct fluorine resonances expected in their ^{19}F NMR spectra.

(i) ClF$_5$; (ii) XeF$_6$ (assume a stereochemically inactive lone pair); (iii) XeOF$_4$; (iv) [ClOF$_3$]$^{2-}$

(b) The ^{31}P NMR spectrum of P$_4$O$_8$ shows two resonances of equal intensity. Describe a possible structure for this oxide based on the structure of P$_4$O$_9$ shown in Figure 3.7(b).

(c) Reaction of SF$_4$ with KF leads to a compound **A** whose solution ^{19}F NMR spectrum contains two resonances with intensity ratios 4:1. Describe a structure for the anion in **A** consistent with these data.

3.3 First-order coupling spectra

(a) Draw out the possible orientations of magnetic moments on four equivalent nuclei and hence show that the multiplet formed from coupling to these nuclei consists of five resonances in the ratio 1:4:6:4:1.

(b) Extend the prediction of the ^{19}F NMR spectra in Problem 3.2(a) to include first-order coupling and, hence, predict the high resolution ^{19}F NMR spectra of

(i) ClF$_5$; (ii) XeF$_6$; (iii) XeOF$_4$; (iv) [ClOF$_3$]$^{2-}$

(Ignore coupling to nuclei other than ^{19}F)

(c) Interpret the following NMR data

(i) $[IF_4]^-$ ^{19}F δ -120 (s); $[IF_4]^+$ ^{19}F δ -170 (t) δ -185 (t)

(ii) Square planar $[RhF_2Cl(CO)]^{2-}$

Isomer **A** ^{19}F δ -200 (d, $J = 420$ Hz);
^{103}Rh δ -9200 (t, $J = 420$ Hz)

Isomer **B** ^{19}F δ -190 (dd, $J = 420$ Hz, $J' = 30$ Hz),
δ -196 (dd, $J = 390$ Hz, $J' = 30$ Hz);
^{103}Rh δ -9750 (dd, $J = 420$ Hz, $J' = 390$ Hz)

(d) Predict the form of the ^{129}Xe NMR spectra of (i) $[XeOF_3]^+$ and (ii) $XeOF_2$.

3.4 Coupling to $I > \frac{1}{2}$ nuclei. Use Table 3.3.

(a) Predict the form of the ^{19}F NMR spectrum of the $[NbF_6]^-$ anion using the data in Table 3.3.

(b) The ^{19}F NMR spectrum of the $[AsF_4]^+$ cation consists of four lines of equal intensity. Explain this observation.

(c) The ^{19}F NMR spectrum of a sample containing the $[BF_4]^-$ anion consisted of a quartet of equal intensity peaks (total relative intensity 80%) and a septet of

equal intensity resonances (total relative intensity 20%). Explain this observation.

(d) Predict the form of the ^{19}F NMR spectrum of $^{81}BrF_3$.

3.5 Decoupled spectra

(a) Predict the form of the ^{13}C NMR and $^{13}C\{^1H\}$ spectra of the *cis* and *trans* forms of $W(CH_3)_2(CO)_4$ (take W as NMR inactive).

3.6 Isotopomers

(a) Extend the prediction of the ^{19}F NMR spectra in Problems 3.2(a) and 3.3(b) to include coupling to ^{129}Xe ($I = \frac{1}{2}$, 26% abundant) and, hence, predict the high resolution ^{19}F NMR spectra of

(i) XeF_6; (ii) $XeOF_4$

What would the ^{129}Xe NMR spectra of these compounds show?

(b) Explain why the ^{31}P NMR spectrum of *trans*-$[PtCl_2(PMe_3)_2]$ consists of three resonances with intensity ratios 1:4:1. Predict the expected features in the ^{31}P NMR spectrum of *cis*-$[PtCl_2(PMe_3)_2]$ (^{195}Pt, $I = \frac{1}{2}$, 33.8% abundant).

ADVANCED TOPICS

3.12 Factors determining chemical shift

Any parameter that affects the electron density around a nucleus will be likely to modify the level of shielding of the external magnetic field and thus cause changes to the chemical shift value. General trends in chemical shift values can be interpreted and provide useful information on the electronic environment of a nucleus studied using NMR. However, the interplay of numerous factors that modify electron distributions in molecules makes it difficult to predict exactly the relationship between any one parameter, such as oxidation state or bonding with electron withdrawing atoms, and chemical shift.

3.12.1 Oxidation state, molecular ion charge, and electronegativity

An increase in oxidation state corresponds to removing electrons from around a nucleus which would be expected to deshield it and produce more positive chemical shifts. In ^{51}V NMR (referenced to pure $VOCl_3$), V(V) compounds have resonances in the range 0 to -800 ppm, V(I) compounds

-1000 to -1500 ppm and V($-I$) -1700 to -1900 ppm, with the highest oxidation states at the most positive (deshielded) δ values. Generally, however, this is not always the case as increasing oxidation state often leads to high coordination numbers and shorter bonds which may act in an opposing direction and to a greater extent. Thus, the observed chemical shifts for ^{129}Xe in XeF_2, XeF_4, and XeF_6 are around -1800, 240, and -50 ppm, respectively, showing no direct trend with oxidation state, though higher oxidation states of Xe have, generally, more positive chemical shifts.

Increasing the negative charge on a species would be expected to provide additional shielding so complexes with the same oxidation states and ligands should show a trend to lower (more negative or less positive) chemical shifts in the order anionic (most positive δ) > neutral > cationic (most negative δ). A comparison of the ^{31}P NMR spectra chemical shift values of the P(V) chloro species, $[PCl_4]^+$ +86 ppm, PCl_5 -85 ppm, and $[PCl_6]^-$ -295 ppm, shows this trend.

Trends as a function of neighbouring atom electronegativity are easier to discern as direct comparison between similar species can be made. NMR data from the Group 14 halides are summarised in Table 3.8. The values show that increasing X^- anion electronegativity leads to more positive chemical shifts because the nucleus becomes increasingly deshielded as electrons are withdrawn from around it.

TABLE 3.8 Comparison of chemical shifts for Group IV tetrahalides

Halide	Halogen electronegativity (Pauling)	Nucleus/compound ^{29}Si SiX$_4$	Nucleus/compound ^{73}Ge ^{73}GeX$_4$	Nucleus/compound ^{119}Sn ^{119}SnX$_4$
Cl$^-$	3.16	−20	31	−50
Br$^-$	2.96	−94	−311	−1296
I$^-$	2.66	−346	−1081	−1705

3.12.2 Coordination number

Increasing the coordination number of an anion nucleus, such as $^{17}O^{2-}$ or $^{19}F^-$, by surrounding it by more cations, would be expected to change the shielding and, hence, the chemical shift. It is again difficult to predict with any certainty the exact direction of the shielding changes. For example, crystalline solids and glasses of the general formula MF_x, where M is a combination of Ca^{2+}, La^{3+}, or Al^{3+}, contain fluoride ions surrounded by those cations in varying numbers and mixtures of types: For F–La$_n$ and F–Ca$_n$ sites, where n represents the number of near neighbour cations of that type, increasing the value of n lowers the frequency of the chemical shift. However, for F–Al$_n$ increasing the value of n increases the chemical shift. Trends are also found for the ^{17}O nucleus in polyoxometallates where the O^{2-} anion may be a terminal group (V=O, designated μ^1), a bridge between two metal centres (W–O–W, μ^2), or capping over six metal atoms (μ^6). The chemical shifts then form a trend with μ^1-O at around 800 ppm, μ^2-O in the range 400–500 ppm, and μ^6-O at approximately 75 ppm.

3.12.3 Other influences on chemical shift

The electron distribution in a molecule, and, therefore, the shielding of an NMR active nucleus, will be subject to other parameters in addition to oxidation state and coordination number. The type of ligands on a metal centre and their electron donating or withdrawing nature is one such factor. In transition metal carbonyl phosphine complexes of the type $M(CO)_5(PR_3)$ the ^{13}C resonance frequency of the *cis* and *trans* carbonyl groups can be measured as a function of the phosphine R group. As R moves from electron donating (e.g. R = butyl) to electron withdrawing (e.g. R = OPh), the phosphine becomes less basic (overall less electron donating to the metal centre). In turn the metal electron back-donation through the π-bonding to the carbonyl group decreases, the carbon is less well shielded and the ^{13}C chemical shift decreases.

3.13 Trends in coupling constants

Trends may be found in the magnitudes of coupling constants as a function of the structure. Factors include changes in bond distances and angles, and of composition, but as with chemical shifts these are difficult to predict especially for systems where there may be subtle competing features.

One important parameter is the s-electron character in the bond between two nuclei that are coupled as the coupling is transmitted via electrons rather than through space. As s-electrons, in contrast to p and d electrons, have a finite probability of existing at the nucleus then they act as the strongest mechanism by which coupling can occur. Factors that affect the degree of s-electron character in a bond include hybridisation, electronegativity, oxidation state, and bond angles. A simple measure of the degree of s-character in a bond involving a main group element, such as boron or carbon, is through the concept of s–p hybridisation and its level in a bond. Thus, in polyhedral boranes such as $B_{10}H_{14}$ (Figure 3.32) the boron atoms can be described as sp hybridised (50% s-character) in their bonding to hydrogen (with the remaining p^2 electrons contributing to the bonding within

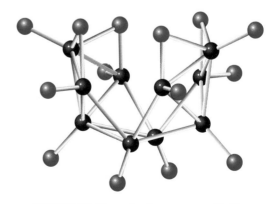

FIGURE 3.32 The molecular structure of $B_{10}H_{14}$.

the B_{10} cage) a ^1J (^{11}B-^{1}H) value of 120–170 Hz is found. This contrasts with that in the [BH_4]$^-$ anion, with formally sp^3 hybridisation and a 25% s-character in the B–H bond, where the coupling constant is much lower at 80 Hz. Measurement of coupling constants from NMR data can, therefore, provide a relative measure of the s-character in a bond.

EXAMPLE 3.15

Rationalise the experimental ^1J(^{13}C-^{13}C) coupling constants (Hz) for the following compounds: ethane, 36.8; benzene, 64.3; ethene, 70.6; ethyne, 140.8.

ANSWER

In hybridisation terms, the orbitals involved in forming the C–C bonds in each of the bonding in the compounds are, respectively, sp^3, sp^2, sp^2, and sp. So the s-character in the bond is 25%, 33%, 33%, and 50%, which follows the broad trends of increased coupling constant with increasing s-character in the bond.

SELF TEST

Explain the following ^1J(^{13}C-^{29}Si) values: R_2Si=C(OR)(R), 84 Hz, and R_3Si-COCMe$_3$, 34 Hz.

While there was no obvious trend in ^{129}Xe chemical shifts with xenon oxidation state (Section 3.12.1), ^1J (^{129}Xe-^{19}F) coupling constants do vary with oxidation state with values of −7600 to −5570 Hz for Xe(II), −3920 to −2384 for Xe(IV), −2700 to +1500 for Xe(VI), and +990 for Xe(VIII). This behaviour probably reflects the decreasing contribution of s orbitals to the Xe–F bonds as the xenon oxidation state increases.

3.14 Second-order coupling

In Section 3.7, the spin–spin coupling patterns in NMR spectra were interpreted in terms of coupling between non-equivalent sites within a molecule. Coupling constants were derived directly from the line separations and the intensity patterns could be predicted based on Pascal's triangle and the statistical nuclear spin distributions. Such coupling in the NMR spectrum is known as first order. However, in some molecules, especially where two coupled, non-equivalent nuclei have similar chemical shift values to their coupling constant, **second-order effects** occur which give rise to spectra where the line separations do not directly yield the coupling constants and the resonance multiplet

intensities differ from those expected on simple first-order coupling considerations.

First-order coupling of two $I = \frac{1}{2}$ nuclei **A** and **X** and the resultant NMR spectrum was described in Section 3.7. In summary, two transitions involving nucleus **A** are possible, αα→ βα (resonance frequency ($\nu_A - J/2$), denoted the A2 transition) and αβ → ββ; (resonance frequency ($\nu_A + J/2$), denoted A1) giving rise to the expected doublet for this nucleus (separation J) (Figure 3.33). When considering nucleus **X** two transitions associated with this nucleus are also possible, αα→ αβ (denoted B2) and βα→ ββ (B1), again separated by J. If **A** and **X** are homonuclear then their resonances will occur in the same NMR spectrum and the overall NMR spectrum of this case is two well-separated doublets (Figure 3.33).

If the resonance frequencies of the **A** and **X** sites become similar, that is, $\nu_A \sim \nu_X$, then the energies of the αβ and βα states must also become similar and they begin to 'mix'; the αβ state develops some βα character and vice versa (Figure 3.34). Under these conditions the **X** nucleus is re-labelled **B** to show the similarity in their chemical shifts. The energy of the βα state drops slightly and that of the αα state is raised. The energy of the transition αα→ βα now becomes slightly lower than ($\nu_A - J/2$) and that of the αβ → ββ transition ($\nu_A + J/2$) slightly higher. For nucleus **B** there are similar perturbations to the energies of the αα→ αβ and βα→ ββ transitions. In addition to these perturbations in energy levels, the probability of the transitions (i.e. line intensities) also varies—the A1 and B2 (previously labelled X2) transitions become weaker. The effect of this on the NMR spectrum is to produce an asymmetric pattern of four lines as shown in

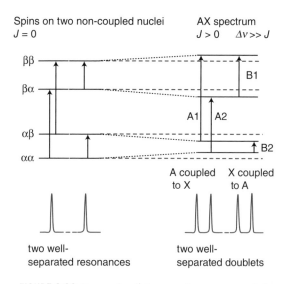

FIGURE 3.33 Energy level diagrams for a non-coupled homonuclear **AX** system and a homonuclear **AX** system with first order coupling, and the resulting NMR spectra.

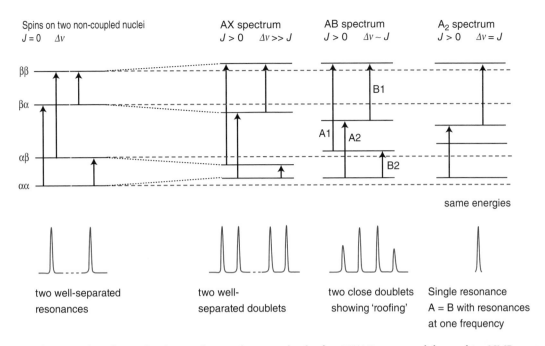

FIGURE 3.34 The effect of second-order coupling on the energy levels of an AX/AB system and the resulting NMR spectra.

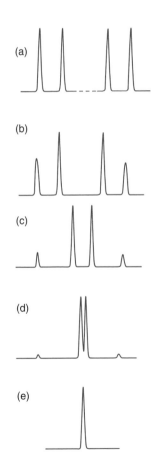

FIGURE 3.35 NMR spectra for **AX–AB–AA** systems with different $\Delta v/J$ values showing the evolution of 'roofing'. (a) **AX**, (b)–(d) **AB** with decreasing $\Delta v/J$, and (e) **AA**.

Figure 3.34. This evolution of the NMR spectrum continues as the chemical shift difference between **A** and **B** gets smaller and smaller tending towards the case where they become identical (i.e. chemically equivalent sites). At this point, two chemically identical sites in the spectrum (**A** and, now, **A**) revert to a single line (as the two sites are equivalent they do not show a coupled spectrum).

In summary, the spectra produced (Figure 3.35) show the behaviour as the two chemical shifts approach each other from being very different (denoted an '**AX**' spectrum as **A** and **X** are a long way apart in the alphabet) through similar chemical shifts (denoted '**AB**') to identical ('**AA**') where a single resonance is observed. These characteristic second-order spectra are often described as showing '**roofing**' where the inner two resonances approach the same frequency and gain intensity from the outer ones. It is possible through detailed analysis to extract the chemical shifts and coupling constants for second-order spectra.

3.15 Chemical and magnetic equivalence and inequivalence

Determining whether two nuclei are in chemically identical environments in a molecule is relatively straightforward and can be discerned from the symmetry of the system. For example, in the compound SiH_2F_2 with tetrahedral geometry the two fluorine

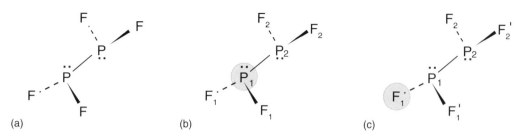

FIGURE 3.36 Molecular structure of *trans*-P_2F_4 showing (a) chemical equivalence, (b) magnetic inequivalence of fluorine nuclei from phosphorus P_1, and (c) magnetic inequivalence of phosphorus and fluorine nuclei from fluorine nucleus F_1.

atoms are chemically and, therefore, magnetically equivalent. Consider, however, the compound *trans*-diphosphorus tetrafluoride (P_2F_4, Figure 3.36a); chemically the two phosphorus atoms and the four fluorine atoms are equivalent. However, in terms of a specific phosphorus site and, therefore, the ^{31}P NMR spectra of the four fluorine atoms are not equivalent and are termed **magnetically inequivalent**. That is, from the viewpoint of P_1 in Figure 3.36(b), two fluorine atoms are directly attached to it (F_1) while the two others (F_2) are on the far side of the other ^{31}P nucleus. This means the coupling interactions from P_1 to F_1 and to F_2 will be different as they involve different distances through the bonds.

These nuclei are only magnetically distinguishable based on their different coupling to chemically equivalent nuclei. In labelling the molecule the sites can be designated as magnetically different by using prime symbols, as in A′, A″ and so on. So that the P_2F_4 molecule can be written strictly, treating each site as potentially magnetically distinct, as AA′XX′A″A‴ system, or using the labelling in Figure 3.36(c) $F_1F_1'P_1P_2F_2F_2'$, with different coupling constants between P_1 and F_1, and P_1 and F_2. From the viewpoint of a specific fluorine nucleus, say F_1, then F_2 and F_2' in Figure 3.36(c) lie at different distances away (one is *gauche* and the other *trans* to F_1) and are, therefore, magnetically inequivalent.

The chemical shifts of F_1, F_1', F_2, and F_2' in the ^{19}F NMR spectrum are identical (and also P_1 and P_2 in the ^{31}P NMR spectrum) as they are chemically equivalent. In the ^{19}F NMR spectrum first-order coupling would be expected as follows $^1J(F_1-P_1)$, $^2J(F_1-P_2)$, $^3J(F_1-F_2)$, and $^3J(F_1-F_2')$ with coupling constant values $^1J > ^2J > ^3J$. However, second-order coupling effects (Section 3.14) usually occur in systems with magnetically inequivalent nuclei giving rise to complex spectra. The ^{19}F NMR spectrum of *trans*-P_2F_4 (Figure 3.37) shows 26 lines, several different coupling constants, and second-order coupling. However, such spectra can be analysed by simulation and the coupling constants extracted (in this case $^1J(F_1-P_1) =$

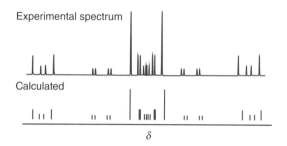

FIGURE 3.37 Experimental and calculated (using a $F_2PP'F'F''$ coupling scheme) ^{19}F NMR spectra of *trans*-P_2F_4. Adapted from F.A. Johnson and R.W. Rudolph, *J. Chem. Phys.* **47** 5449 (1967).

1198 Hz, $^2J(F_1-P_2') = 68$ Hz, $^3J(F_1-F_2) = 17$ Hz, and $^3J(F_1-F_2') \sim 0$ Hz), and so information on the molecular geometry and bonding extracted. In the case of *trans*-P_2F_4, this analysis, which included the need for two different $^3J(F-F_2')$ values and no $^2J(F_1-F_1')$ coupling due to chemical equivalence, allowed the molecular structure to be confirmed as the *trans* form at all the temperatures studied on cooling to $-140°C$.

EXAMPLE 3.16

Identify sites of chemical equivalence and magnetic inequivalence for the P and F atoms in the $[P_2O_5F_2]^{2-}$ anion (Figure 3.38), and hence indicate the likely form of the ^{19}F NMR spectrum.

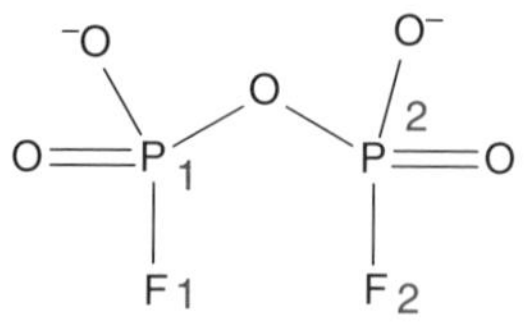

FIGURE 3.38 The structure of the $[P_2O_5F_2]^{2-}$ molecular anion.

ANSWER

Using the labelling in Figure 3.38 and considering initially phosphorus, P2 is chemically equivalent to P1 but has different magnetic relationships (coupling) with F1 and F2. Similarly F1 and F2 are chemically equivalent but magnetically inequivalent with respect to P1 and P2. This is termed as an XAA′X′ system. The main couplings for F1 will be $^1J(^{19}F1\text{-}^{31}P1)$ and $^3J(^{19}F1\text{-}^{31}P2)$. The phosphorus sites are chemically equivalent and, initially, we can consider just first-order coupling which predicts that the ^{19}F spectrum would consist of a doublet of doublets. In practice second-order effects will occur (see Section 3.14), that is, the system becomes more like a BAA′B′ system and a more complex form of the experimental spectrum with roofing would be predicted.

SELF TEST

Determine the expected form of the ^{31}P NMR spectrum of $[P_2O_5F_2]^{2-}$.

3.16 Relaxation and relaxation times: quadrupolar nuclei in NMR

The observation of an NMR resonance involves excitation of a nuclear spin with subsequent emission. Relaxation can also occur when an excited state returns to the ground state non-radiatively, that is, the energy is lost through some other process. This can occur via two processes, **spin–spin relaxation**, where the energy is transferred or exchanged with another nucleus, and **spin–lattice relaxation**, where the energy is lost to the surrounding structure or lattice as heat. Understanding how these processes occur and the factors that control their rates is important as they determine which nuclei and compounds can easily be studied using the NMR technique, controlling important factors such as data collection times. Spin–lattice relaxation is the route by which the Boltzmann equilibrium of nuclear magnetisation is restored and so it controls how often the NMR observation can be made for a particular nucleus (the faster the relaxation the more often the NMR spectrum can be sampled). Spin–spin relaxation controls the lifetime of the excited state and through the Heisenberg uncertainty principle the linewidth (the shorter the lifetime of the excited state and the broader the resonance). Both these relaxations occur through exponential processes and they are characterised by time

constants T_1 (spin–lattice relaxation) and T_2 (spin–spin relaxation).

These two relaxation mechanisms require an interaction with a magnetic field that oscillates at or near the resonance frequency of the nucleus under study. The dominant contributions to relaxation are dipole–dipole relaxation (involving transfer of energy to other nuclei or electron spins) and, for quadrupolar nuclei ($I \geq 1$), quadrupolar relaxation. Two further mechanisms, **chemical shift anisotropy (CSA)** and chemical exchange, may also contribute to the overall relaxation rate.

In systems with unpaired electrons, for example, many transition metal and lanthanide compounds, because of the large magnetic moment associated with an electron (around seven hundred times greater than a nuclear magnetic moment), the dipole-dipole relaxation route dominates. This relaxation route is in fact so fast that energy is quickly lost through such processes giving rise to broad signals, which also occur over a wide chemical shift range. While NMR spectroscopy can be undertaken on such paramagnetic systems the resolution is markedly reduced and coupling is rarely observed. Hence, high-resolution NMR spectroscopy is only generally possible for transition metal systems which are diamagnetic, for example, low-spin d^6 in an octahedral field (with the d_{xy}^2, d_{yz}^2, d_{xz}^2 configuration) or d^8 square planar complexes.

The rate of dipole–dipole relaxation involving other nuclei (rather than electrons) depends on $\gamma_a^2\gamma_\beta^2$, where γ_a and γ_β are the gyromagnetic ratios of the two nuclei involved so that relaxation by this method is fastest for nuclei with high γ values (see Table 3.2).

Quadrupolar relaxation depends on the size of the quadrupole moment of a nucleus (see Table 3.3) and the **electric field gradient (EFG)** experienced by a nucleus. In general, the larger the quadrupole moment (denoted Q, units fm^2, femtometres (10^{-15} m) squared) and also the larger the EFG the faster the quadrupolar relaxation. This manifests itself as broader resonances in the NMR spectrum and degrades the level of information that can be obtained in the NMR experiment. The EFG depends on the level of asymmetry in the environment of a nucleus so that a high symmetry site, such as one with locally cubic, octahedral or tetrahedral geometry, has a zero or small EFG and so quadrupolar relaxation will be slow. For quadrupolar nuclei in such high symmetry environments narrow lines and reasonably high resolution spectra can be obtained. Similarly a small quadrupole moment, Q, will produce reasonably sharp resonances in comparison with a nucleus with a larger value; thus ^9Be with $Q = 5.288$ fm^2 gives much sharper lines than ^{75}As with $Q = 31.4$ fm^2. A quadrupolar nucleus with a high quadrupole moment in a low symmetry environment will relax very quickly giving a very broad resonance which may be indistinguishable from the background. Generally quadrupolar moments increase as the size of the nucleus increases so that most light elements with quadrupolar nuclei, with atom numbers below calcium, give fairly sharp resonances in their NMR spectra especially in reasonably symmetric environments; this includes a number of important and widely studied species such as ^{7}Li, ^{11}B, ^{14}N, ^{17}O, ^{23}Na, ^{27}Al, ^{35}Cl, and ^{39}K.

CSA is the orientation-dependent variation in the chemical shift which derives from a non-spherical shielding around a nucleus. In solution rapid reorientation of a molecule normally averages this. In materials where the molecules do not re-orientate rapidly, such as solids, alternative methods of overcoming this rapid route to relaxation (and derived broad resonances) need to be used (see Section 3.19).

EXAMPLE 3.17

Explain the following observations:

(a) The ^{75}As NMR spectrum of $[AsF_6]^-$ anion is a sharp pattern consisting of a septet while the ^{75}As NMR spectrum of AsF_5 only shows a broad resonance feature.

(b) It is often possible to collect ^{59}Co NMR data from octahedral Co^{3+} complexes but not from Co^{2+} compounds.

ANSWERS

(a) The $[AsF_6]^-$ anion has octahedral geometry so the EFG around the arsenic nucleus is very small and quadrupolar relaxation also slow; the lifetime of the excited state is long giving rise to sharp lines and the multiplet is resolved. AsF_5 is trigonal bipyramidal, a low symmetry environment for arsenic, with fast quadrupolar relaxation and a broad resonance where $^{1}J(^{75}As-^{19}F)$ will not be resolved.

(b) Co^{3+} is a d^6 ion and in a strong octahedral field can adopt the low spin t_{2g}^6 configuration. This is diamagnetic and has no unpaired electron that would provide a pathway for extremely rapid relaxation. Hence the moderate relaxation times allow observation of good quality NMR spectra for this species despite having $I = \frac{1}{2}$.

SELF TEST

The ^{71}Ga ($I = \frac{3}{2}$) NMR spectrum of a solution of $GaCl_3$ in CH_3CN shows a broad resonance. Addition of HCl to this solution leads to an NMR spectrum showing a single sharp resonance, $\delta = -660$ ppm. Explain these observations.

3.17 The NMR timescale, site exchange, and fluxionality

Section 3.11 introduced the concept of fluxionality and its effect on the NMR spectrum. In this section a more detailed analysis of the NMR spectra of fluxional molecules is provided. In NMR spectroscopy, resonance frequencies, and the energy of the excited state, are normally between 10 and 1000 MHz yielding typical linewidths of around 0.1 Hz (Table 1.3). The excited state lifetime and relaxation mechanisms in NMR spectroscopy are such that the timescale of the technique (Section 1.4.2) is of the order of $10^{-7}-10^{-6}$ s.

If a nucleus observed in an NMR experiment exchanges between two environments with equal populations, the spectrum collapses to a single line, at the average resonance position, when the exchange rate, $1/\tau$, is large compared with the frequency difference between two resonances (δv). Two resonances separated by a chemical shift difference of δv (in Hz) will just merge to become a single resonance when

$$\tau = \frac{\sqrt{2}}{\pi \delta v}.$$

Eqn 3.4

Figure 3.39 shows the effect on the form of the spectrum for two resonances as τ changes relative to δv. Two lines will be resolved in the spectrum if the exchange rate is small compared with the chemical shift difference between the two lines and these will coalesce to give a single resonance when the exchange rate becomes large compared with Δv. To decide whether the timescale of the chemical exchange process will cause coalescence of the NMR resonances it is important to know the chemical shift difference (in hertz), which will vary between different nuclei, and to also recognise that chemical shifts depend on the applied field. Table 3.9 summarises some relevant data for several nuclei.

Because the chemical shift difference between two resonances increases linearly with the strength of the applied magnetic field, the exchange rate between two sites that leads to averaging, and thus a coalesced spectrum, must be five times larger for a 300 MHz spectrometer than in a 60 MHz instrument.

As we have seen in Section 3.11, the rate of a dynamic process, such as site exchange, can be altered by varying the temperature of a system. Thus, monitoring the NMR spectrum of a compound as a function of temperature can show the transition, provided the timescales are right. This transition occurs between a spectrum representative of a 'static' (sites exchanging slower than ~10^4 s^{-1}) arrangement, where different environments are distinguished through different chemical shifts, and a fully averaged arrangement and chemical shift value (sites exchanging faster than ~10^7 s^{-1}). Slow exchange NMR data are generally collected at low temperature with more rapid site exchange taking place on heating the sample.

Detailed studies of NMR spectra including analysis of line shapes and widths allow the determination of rate constants for exchange processes and their thermodynamic parameters such as the activation free energy, activation enthalpy and activation entropy for a molecular rearrangement. From the transition state theory of kinetics, the rate constant for a reaction is given by

$$k = \frac{k_B T}{h} \, e^{-\frac{\Delta G^{\ddagger}}{RT}} = \frac{1}{\tau} \qquad \textbf{Eqn 3.5}$$

where k_B is the Boltzmann constant, T the temperature, h is Planck's constant, and $\Delta G^{\ddagger}$ the free energy of activation for the reaction. Combining Eqns 3.4 and 3.5 gives for a system with exchange between two equally populated sites, at a coalescence temperature T_C, the following expression:

$$\frac{\Delta G^{\ddagger}}{RT_C} = \ln\left\{ \frac{\sqrt{2}k_B T_C}{\pi.\delta v.h} \right\} \qquad \textbf{Eqn 3.6}$$

Figure 3.40(a) shows the ^{13}C{^{1}H} NMR spectra as a function of temperature of the palladium complex [PdBr(o-C$_6$H$_4$CH$_2$^{13}CON(H)Bu){(S)-BINAP}], with the carbonyl carbon labelled with ^{13}C. At low temperature, $-35°$C, the spectrum consists of three sharp signals at δ = 172.8, 173.5, and 175.6 ppm, indicating that at this temperature the solution contains three distinct forms of the compound. On heating to $0°$C the signals at 173.5 and 175.6 ppm broaden and on further heating to $80°$C all the signals coalesce into a single resonance. The ^{31}P NMR spectrum of this compound shows similar signal averaging and coalescence at $80°$C. These variable temperature NMR spectra have been interpreted in terms of the three species shown in Figure 3.40(b) and their rapid interconversion above $80°$C. Using this coalescence temperature in Eqn 3.6 gives an approximate rate of interconversion of 10^2 s^{-1} and a free energy of activation ($\Delta G^{\ddagger}$) of 71 kJ mol^{-1}.

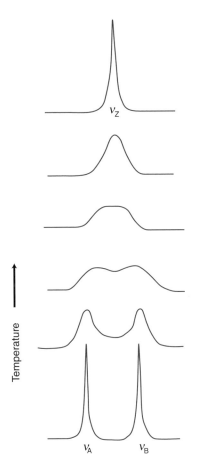

FIGURE 3.39 Coalescence NMR spectra showing the changes that occur as the rate of site exchange increases.

TABLE 3.9 The NMR timescales for different nuclei (adapted with permission from R.G. Bryant, *J. Chem. Educ.* **60** 933 (1983) copyright © 1983, American Chemical Society, and references therein).

Nucleus	Approximate chemical shift range, ppm	Typical difference in chemical shifts at 7 T field (300 MHz for ^{1}H)	Timescale in 7 T field
^{1}H	0–10	0–3	0.2 s–75 µs
^{13}C	0–200	0–15	0.2 s–15 µs
^{15}N	0–900	0–27	0.2 s–8 µs
^{19}F	0–300	0–85	0.2 s–3 µs
^{31}P	0–700	0–85	0.2 s–3 µs
^{59}Co	0–15000	0–1100	0.2 s–0.2 µs
^{199}Hg	0–3000	0–160	0.2 s–1.4 µs

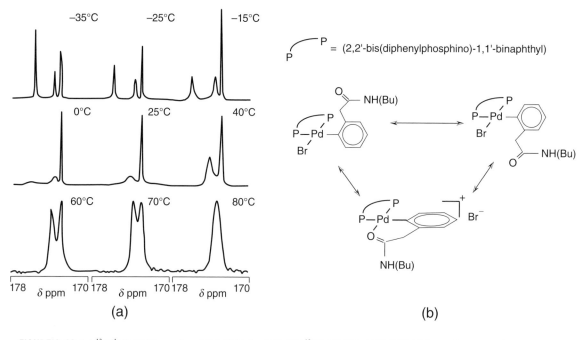

FIGURE 3.40 (a) $^{13}C\{^1H\}$ NMR spectra of (b) $[PdBr(o\text{-}C_6H_4CH_2{}^{13}CON(H)Bu)\{(S)\text{-}BINAP\}]$ as a function of temperature. (Data from A.L. Monteiro and W.M. Davis, *J. Braz. Chem. Soc.* **15** 83 (2004).)

3.18 Advanced NMR techniques and experiments: INEPT and COSY

The characterisation of inorganic compounds containing 100% abundant $I = \frac{1}{2}$ nuclei, such as ^1H, ^{31}P, and ^{19}F, can be undertaken rapidly and often provides, through analysis of chemical shifts and coupling schemes, definitive information on the structure of a molecule. For other nuclei, which have low abundance, sensitivities and receptivities (see Section 3.2.3) NMR spectra can often be difficult to obtain or the experiment requires very long data collection time to improve the signal-to-noise ratio to an acceptable level. Nuclei with $I = \frac{1}{2}$ in this class include ^{15}N, ^{119}Sn, ^{187}Os, ^{183}W, ^{99}Ru, and ^{57}Fe (see Table 3.2). Using more specialised NMR pulse techniques, such as INEPT discussed in Section 3.18.1, it is possible to enhance the NMR spectra of these nuclei and, thereby, collect spectra on more reasonable, shorter timescales. For complex molecules we have seen that spin decoupling (Section 3.9) can be used to simplify spectra and aid in the assignment of resonances; other sophisticated NMR techniques such as COSY (Section 3.18.2) allow more detailed information of coupling and, therefore, connectivity through bonds in a molecule to be extracted from NMR spectra.

3.18.1 INEPT

Insensitive nucleus enhancement by polarisation transfer (INEPT) uses the transfer of energy from a nucleus with high NMR sensitivity to one of low sensitivity to enhance the signal from the latter. The two nuclei must be coupled together. When a high sensitivity nucleus, typically ^1H, is excited at its resonance frequency using an appropriate radiofrequency pulse sequence, a portion of the energy is transferred through the coupling scheme to a neighbouring, low-sensitivity nucleus, whose NMR spectrum can then be obtained. The increase in signal intensity that results is proportional to the ratio of the gyromagnetic ratio of the two nuclei. In the case of ^{187}Os and ^1H this ratio is 26.75/0.619 (see Table 3.2) so the potential signal enhancement exceeds 40-fold. For ^{57}Fe–^1H INEPT this enhancement is approximately 30-fold. This is illustrated in Figure 3.41, where the ^{57}Fe–^1H INEPT NMR spectrum of ferrocene was collected in 100 s, instead of hours using standard NMR methods on the 2.1% abundant Fe nucleus. In this case collection of the INEPT NMR spectrum for 70 minutes improved the signal-to-noise ratio still further, such that the ^1J $(^{57}Fe\text{-}^{13}C)$ coupling constant could be determined as 4.8 Hz. Note that with the low natural abundance of ^{13}C and ^{57}Fe, the probability of a molecule containing both ^{57}Fe and a ^{13}C nuclei is approximately

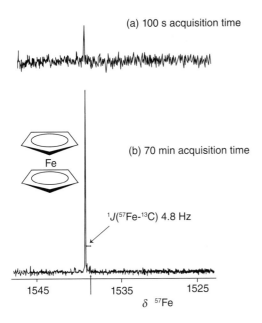

(a) 100 s acquisition time

(b) 70 min acquisition time

$^{1}J(^{57}Fe-^{13}C)$ 4.8 Hz

1545   1535   1525

δ ^{57}Fe

FIGURE 3.41 ^{57}Fe–^{1}H INEPT NMR spectrum of ferrocene (inset) with (a) short and (b) long acquisition times.

$0.021 \times 0.011 \times 10$ or two molecules in a thousand, so INEPT signal enhancement was essential to observe the coupled resonance (B. Wrackmeyer, O.L. Tok, and M. Herberhold *Organometallics* **20** 5774 (2001)).

3.18.2 **COSY NMR spectra**

For compounds that contain many different environments (including those that are magnetically as well as chemically inequivalent) and complex coupling patterns between nuclei, the derived NMR spectra can be extremely complicated. The complexity can reach the extent that it becomes difficult to identify which sites are bonded to each other or which nuclei are coupled in the molecule. Decoupled spectra discussed previously (Section 3.9) provide one way of simplifying spectra. Another way of probing the coupling between nuclei, and, thus, extract further structural and dynamic information from NMR spectra, is by undertaking **2-D NMR** (two-dimensional NMR). 2-D NMR overcomes the problems inherent in one-dimensional (1-D) NMR, where a single linear spectrum has to contain all the resonances and coupling patterns often leading to extensive peak overlap. In 2-D NMR multiple spectra are combined to produce a two-dimensional map showing which resonances, at different chemical shifts, couple together (seen as correlation peaks), and which do not (no correlation signal observed).

The **COSY (COrrelated SpectroscopY) NMR** spectrum allows the determination of the connectivity of a molecule by probing which nuclei are spin–spin coupled. The 2-D homonuclear NMR spectrum that results from the COSY experiment plots the chemical shifts or frequencies for a nucleus of interest along two perpendicular axes. It is also possible to obtain heteronuclear correlation spectra, in which the two axes correspond to different NMR active nuclei, such as ^{31}P and ^{1}H. A simple homonuclear COSY NMR spectrum for a system with two resonances is shown in Figure 3.42.

A diagonal cross section of this COSY NMR spectrum would be the simple NMR spectrum in the chemical shift domain of the compound, with resonances at chemical shifts marked A and B.

The off-diagonal peaks called cross peaks, marked X and Y, indicate **correlation** between pairs of nuclei and result from magnetisation transfer; the COSY experiment effectively involves excitation of one site at its resonance frequency and then sampling after a certain duration where that energy has been transferred through the coupling scheme. In the 2-D NMR spectrum each coupled pair of nuclei gives two symmetrical cross peaks above and below the diagonal. It is thus evident which atoms are connected or correlated to one another (within a small number of chemical bonds) by looking for cross peaks between various signals.

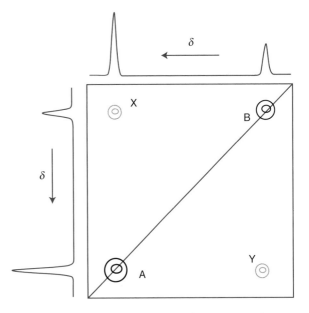

FIGURE 3.42 COSY NMR spectrum for a system with two resonances (at chemical shifts marked A and B) that transfer magnetisation between sites giving cross peaks at X and Y.

EXAMPLE 3.18

Interpret the ^{11}B–^{11}B COSY spectrum of decaborane shown in Figure 3.43.

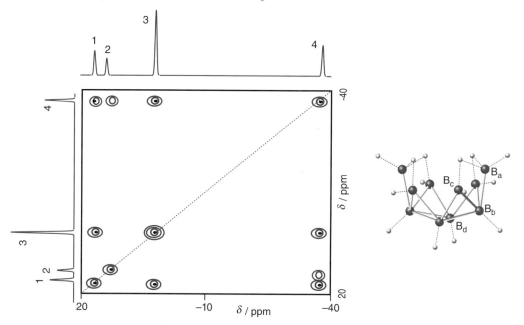

FIGURE 3.43 The ^{11}B–^{11}B COSY spectrum of decaborane, $B_{10}H_{14}$ (inset, boron framework). The resonances are labelled 1–4.

ANSWER

The decaborane cage contains four distinct boron site types as marked, B_a to B_d, giving rise to the four distinct resonances in the 1-D spectrum and on the diagonal of the COSY spectrum. Where cross terms are seen in the spectrum these boron nuclei must be adjacent to each other and where no cross term peak is seen these boron centres will not be directly bonded. Only the resonance marked 4 at around −38 ppm shows a link to all three other boron centres and so this allows assignment of this peak as B_b. Resonance 2 is only transferring energy to one other boron site so this can be assigned to B_a. Resonances 1 and 3 couple to two other boron sites so can be assigned to B_c or to B_d though based on the high intensity of peak 3 this can be assigned to B_c as there are twice as many of this site (4) in decaborane as there are of B_d (2).

SELF TEST

Predict the ^{11}B–^{11}B COSY NMR spectrum of closo-2-CB_6H_8 shown in Figure 3.44.

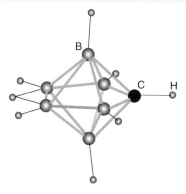

FIGURE 3.44 Molecular structure of closo-2-CB_6H_8.

3.19 Solid-state NMR

All the spectra presented so far in this chapter have been collected from molecules either as pure liquid or in solution. Under these conditions the individual molecules rotate rapidly, $>10^8\ s^{-1}$, and in the NMR timescale, of a few milliseconds, all orientations of the molecule are presented relative to the external magnetic field. It also means that interactions between spins within the molecule are orientationally averaged. These directionally averaged interactions are the dipolar interactions (met in the context of relaxation in Section 3.16) and any anisotropy of the shielding constants, where they have different values in different directions within a molecule. This orientational averaging produces the familiar sharp lines of solution NMR spectra. In a few solid materials, particularly at higher temperatures and near their melting points it is possible to obtain high resolution spectra using a normal NMR spectrometer; examples include $K[AsF_6]$ (^{19}F), P_4S_3 (^{31}P), and C_{60} (^{13}C) where the molecular ion, $[AsF_6]^-$, and molecules, P_4S_3 and C_{60}, rotate on their lattice sites, thereby producing the same orientational averaging as occurs in liquids.

In a solid the nuclei are normally on fixed sites either as part of a molecule (such as P_4 in white phosphorus) or in an extended lattice (as with carbon in graphite). Consideration of a nucleus in a molecule that is part of a powdered solid placed in an external magnetic field shows that it can experience a range of local fields (Figure 3.45) determined by the orientation of the molecule with respect to the external field and the orientations of other nearby magnetic nuclei.

A range of different fixed orientations of the molecule will be present across the whole sample causing variations in the directions associated with shielding constants (giving rise to CSA) and dipolar interactions. This static arrangement gives rise to the characteristic solid powder NMR pattern shown in Figure 3.46 for

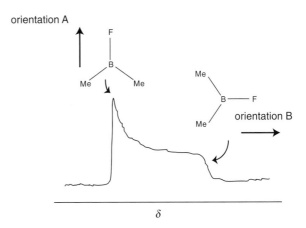

FIGURE 3.46 Characteristic powder NMR spectrum showing a range of chemical shifts as a result of different molecular orientations and the varying magnetic fields it experiences.

an asymmetric molecule. The overall resonance is very broad with the extreme frequencies representing orientation parallel and perpendicular to the external field, with their differing shielding constants.

Resonances between these extremes represent intermediate molecular orientations with respect to the external field and the overall shape of the spectrum represents their combined probabilities. Obviously while there is some useful information in such NMR spectra the fine detail corresponding to different environments, available from solution NMR, is not resolved. To obtain high resolution NMR spectra from solids a method of averaging the dipolar interactions and CSA is needed.

The strength of the dipolar interactions between two nuclear spins in a solid, **A** and **B**, is given by an expression with the form:

$$dipolar\ interaction = C\frac{\gamma_A\gamma_B}{r_{AB}^2}(3\cos^2\theta_{AB}-1) \qquad \textbf{Eqn 3.7}$$

The dipolar interaction can be seen to depend on the gyromagnetic ratios of **A** and **B** (γ_A, γ_B), their separation (r_{AB}), and the angle between the r_{AB} vector and the applied magnetic field (θ_{AB}) (Figure 3.47a). In common with the dipolar interactions, the CSA also depends on the angle dependent factor ($3\cos^2\theta - 1$). In liquids the tumbling of the molecules averages the value of ($3\cos^2\theta - 1$) over all values of θ, which gives the value zero—hence the ability to undertake solution NMR and obtain high resolution spectra.

The form of the expression $|(3\cos^2\theta - 1)|$ as a function of θ is shown in Figure 3.48 and, as can be seen, it has zero value when $\theta = 54.74°$, the so-called **magic angle**. To remove the orientation-dependent effects in the NMR spectrum from a solid, a method of exploiting the

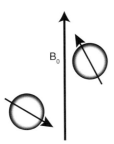

FIGURE 3.45 Nuclei fixed in a solid experience different external magnetic fields relative to their own nuclear orientation and those of neighbouring magnetic nuclei.

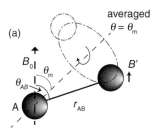

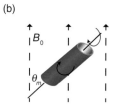

FIGURE 3.47 (a) Dipolar interactions depend upon γ_A, γ_B, r_{AB}, and θ_{AB} but are averaged by rotation around an angle relative to the external magnetic field, B_0. (b) Averaging to zero can be achieved in a solid materials by rotating a sample holder at the magic angle, θ_m, with respect to B_0.

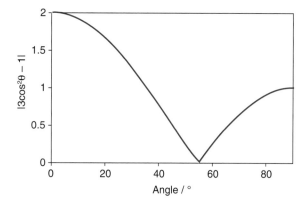

FIGURE 3.48 Variation of the expression $|3\cos^2\theta - 1|$ with θ; the expression equals zero when $\theta = 54.74°$, the magic angle.

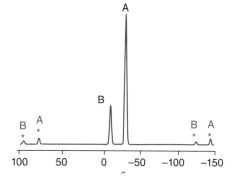

FIGURE 3.49 ^{31}P MASNMR spectrum of a mixture of $Cs_4P_2Se_9$ and $Cs_4P_2Se_{10}$ showing two distinct phosphorus environments, **A** (from $Cs_4P_2Se_9$) and **B** (from $Cs_4P_2Se_{10}$); spinning side bands are marked with*. (Data from M.A. Cave, C.G. Canlas, I. Chung, R.G. Iyer, M.G. Kanatzidis, and D.P. Weliky, *J. Solid State Chem.* **180** 2877 (2007).)

the orientation-dependent effects. Under more typical experimental conditions, a powdered sample spinning at 2–70 kHz, magic-angle spinning NMR (MASNMR) spectra can be obtained from solids for most $I = \frac{1}{2}$ nuclei. These spectra, while not normally as highly resolved as solution spectra, show the key feature of different chemical shifts for each unique environment in a compound. For example, in Figure 3.49 two different phosphorus environments in the compound mixture under investigation show resonances at approximately −5 and −35 ppm, marked A and B. Note that spin–spin coupling is not observed in MASNMR spectra as it is necessarily averaged by the magic-angle spinning technique.

One additional feature often observed in MASNMR spectra are **spinning side bands**, marked as A* and B* in Figure 3.49. These are additional resonances which appear symmetrically either side of the main resonances at a chemical shift that corresponds to the rotation rate of the sample.

For some nuclei the residual broadening of signals is comparable to the chemical shift range meaning that MASNMR spectra are diagnostically relatively unhelpful. This is a particular problem for ^{1}H, with its narrow chemical shift range of $\Delta\delta \sim 10$ and problems in fully narrowing the resonance due to the strength of the dipolar interactions. As a result, the MASNMR method is not applicable to this important nucleus except for very simple systems with few proton environments and access to expensive MASNMR instruments with the highest sample spinning speeds. Figure 3.50 shows the considerable improvement in the resolution in ^{1}H MASNMR spectra of glycine at very fast MAS spinning speeds, up to 70 kHz, but even so resonances remain at around 1 ppm in width.

averaging of angle dependent interactions at the angle $\theta = 54.74°$ is needed. This can be achieved by rotating the sample rapidly at the magic angle with respect to the external magnetic field (Figure 3.47b). Spinning rates are typically in the range 2–70 kHz and at these speeds most orientation-dependent effects are averaged. The exceptions are in systems where $\gamma_A\gamma_B$ is large and r_{AB} is small when the dipolar coupling is very strong (Eqn 3.7). This is the case for ^{1}H where spinning rates need to be exceptionally high to even start to average

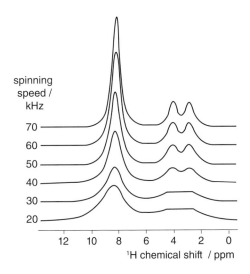

FIGURE 3.50 Improvement in the resolution of ¹H MASNMR spectra of glycine on using very fast spinning. (Data from http://nmr900.ca/images/Glycine-FastMAS.jpg.)

3.20 Applications of MASNMR

Many important functional materials are insoluble so conventional NMR techniques are not appropriate to their study. These include many silicates, aluminosilicates, such as zeolites, aluminophosphates (ALPOs), battery materials, carbon nanotubes and fullerides, and materials used in fuel cells (such as complex metal oxides). Many of these solid compounds contain nuclei amenable to study by NMR, for example, ^{29}Si and ^{31}P in zeolites and ALPOs, respectively, and ^{7}Li in rechargeable battery systems. Thus MASNMR has been widely employed in characterising and understanding the properties of these material types. The interpretation of these MASNMR spectra typically follows the methods used for simple solution NMR spectra. Example applications include identifying the different environments present from the number of resonances observed in the spectrum, and the interpretation of chemical shift values in terms of bonding and resultant shielding of the different environments of the nuclei in the material.

EXAMPLE 3.19

Interpreting simple MASNMR spectra

(a) The ^{31}P MASNMR spectrum obtained from solid $AlH_2(P_3O_{10}).2H_2O$, containing the anion $[P_3O_{10}]^{5-}$, shows two resonances close to −21 ppm and one at

−30 ppm. Describe a likely structure for the $[P_3O_{10}]^{5-}$ anion in this compound.

ANSWER

The MASNMR spectrum implies two similar sites and one further distinct environment; this is consistent with the linear triphosphate anion (Figure 3.51).

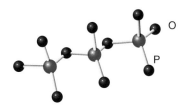

FIGURE 3.51 Linear triphosphate anion $[P_3O_{10}]^{5-}$.

(b) Predict the form of the ^{11}B MASNMR spectrum at high magnetic field of borax, $Na_2B_4O_5(OH)_4.8H_2O$ (sodium tetraborate), which contains the polyborate anion shown in Figure 3.52.

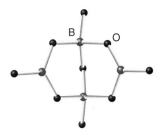

FIGURE 3.52 The polyborate anion, $[B_4O_5(OH)_4]^{2-}$ (hydrogen atoms not shown), present in borax, $Na_2B_4O_5(OH)_4.8H_2O$.

ANSWER

While ^{11}B is a quadrupolar nucleus, at high magnetic field the signals are narrowed sufficiently to resolve different environments for boron nuclei in the tetraborate anion. Inspection of the molecular anion shows two distinct boron environments—one four-coordinate, tetrahedral site and a trigonal planar site. Therefore, the NMR spectrum would be expected to show two resonances of equal intensity. The experimental spectrum collected in a 21.1 T field shows resonances at 64 ppm (four-coordinate, tetrahedral boron) and 47 ppm (trigonal planar boron).

SELF TEST

Describe the main features expected in the ^{11}B and ^{31}P MASNMR spectra of $Zn_5[BP_3O_{14}]$, which contains the $[BPO_6]^{4-}$ anion and two different $[BO_3]^{3-}$ anion sites in its structure.

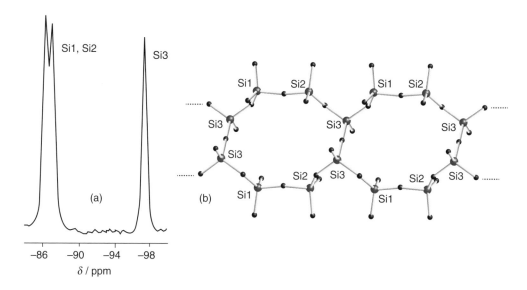

(a)

(b)

FIGURE 3.53 (a) The ^{29}Si MASNMR spectrum of xonotlite, $Ca_6Si_6O_{17}(OH)_2$ and (b) the repeating unit of the silicate chains in xonotlite. (Data from E. Lippmaa, M. Maegi, A. Somoson, G. Englehardt, and A.R. Grimmer, *J. Am. Chem. Soc.* **102** 4889 (1980).)

TABLE 3.10 Characteristic chemical shift ranges for $Si(O)_{4-n}(OSi)_n$

Species and designation	SiO$_4$ Q0	SiO$_3$(OSi) Q1	SiO$_2$(OSi)$_2$ Q2	SiO(OSi)$_3$ Q3	Si(OSi)$_4$ Q4
Chemical shift range/ppm	−65 to −74	−75 to −84	−85 to −94	−95 to −104	−105 to −120

Another application of MASNMR is illustrated by consideration of the ^{29}Si MASNMR spectrum of the silicate mineral xonotlite, $Ca_6Si_6O_{17}(OH)_2$, shown in Figure 3.53(a). Three peaks are visible, two with similar chemical shifts. This is in agreement with the repeat unit present in the silicate chains of xonotlite (Figure 3.53b), where Si1 and Si2 have similar environments with two bridging and two terminal oxygen atoms each, while Si3 has three bridging and one terminal oxygen atom. The resonance at around −97.5 ppm can be assigned to Si3.

^{29}Si MASNMR studies of numerous silicates have shown that the five different types of silicon environment have characteristic ^{29}Si chemical shift ranges as shown in Table 3.10. These range from discrete $[SiO_4]^{4-}$ ions (denoted Q0) with four terminal oxygen atoms through to $Si(OSi)_4$ with all oxygen atoms bridging (denoted Q4); the Q-notation defines the number of bridging oxygen atoms around silicon. These characteristic chemical shift data values can be used to characterise silicate compounds.

EXAMPLE 3.20

The ^{29}Si MASNMR spectrum of the mineral wollastonite ($CaSiO_3$) is shown in Figure 3.54. Show that this is consistent with the compound containing $Si_3O_9^{6-}$ rings formed from three SiO_4 tetrahedra as shown. Are the three silicon sites equivalent in the crystal structure?

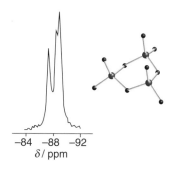

FIGURE 3.54 The ^{29}Si MASNMR spectrum of the mineral wollastonite ($CaSiO_3$) and the $Si_3O_9^{6-}$ ring structure. (Data from the same source as Figure 3.53.)

ANSWER

The chemical shift values lie in the range −85 to −91 ppm. According to Table 3.10 these values all correspond to Q2 environments for silicon (two terminal and two bridging oxygen on silicon, $Si(-OSi)_2(-O^-)_2$). This is consistent with the tricyclic ring structure shown. The fact that three different resonances are observed in the ^{29}Si MASNMR spectrum of the solid implies that three slightly different environments exist for silicon. This is likely to be due to the rings not being symmetric (with a threefold rotation axis) when they are packed in the solid making the silicon sites inequivalent.

SELF TEST

Predict the ^{29}Si MASNMR spectrum of a linear trisilicate ion of the formula $[Si_3O_{10}]^{8-}$, $[O_3Si-O-SiO_2-O-SiO_3]^{8-}$.

Knowledge of the amount of silicon present, and its distribution, is of great use in understanding the important catalytic applications of the aluminosilicate zeolites. The general formula of a zeolite $M_x[Si_{1-x}Al_xO_2].nH_2O$, $0 \leq x \leq 0.5$, represents a fully connected three-dimensional structure which is constructed of AlO_4 and SiO_4 tetrahedra linked through an oxygen atom, with M^+ cations and water molecules occupying extra framework sites (see Figure 3.55). When $x = 0$ the framework is formed solely of silicon-centred tetrahedra, so that each Si centre is connected through four oxygen atoms to four other silicon-centred tetrahedra. For $x > 0$ the situation is equivalent to the simple silicate chemistry discussed above (and uses the similar Q-notation) but with the potential for five different environments for silicon ranging from $Si(OSi)_4$ (Q4) to $Si(OAl)_4$ (Q0) (Table 3.11).

In a zeolite the silicon to aluminium ratio can vary over a large range from purely silicon on the tetrahedral sites to a 50:50 mixture of Si and Al. The distribution over the tetrahedral sites in the structure is statistical, with the constraint that there is no Al–O–Al link present in the structure. For a purely siliceous zeolite, only Q4 resonances will be observed but increasing aluminium content means increasing intensity in resonances corresponding to Q3 and Q2 frequencies will start to appear (see Figure 3.56), followed at higher aluminium levels by resonances in the Q1, and, eventually, Q0 ranges. When the Si:Al ratio reaches 1:1, that is, $M[SiAlO_4].nH_2O$, each

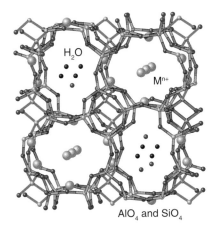

FIGURE 3.55 A typical zeolite structure consisting of three-dimensionally connected AlO_4 and SiO_4 tetrahedra, each linked through an oxygen atom (shown as lines), with M^{n+} cations and water molecules in the pores.

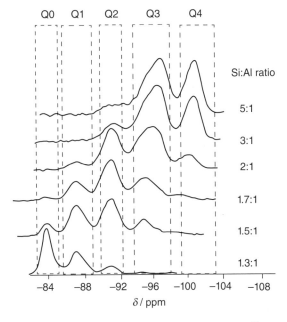

FIGURE 3.56 Effect of increasing Si:Al ratio on the ^{29}Si MASNMR spectrum of a zeolite. The increasing silicon results in increasing intensity for resonances corresponding to Q3 and Q4 environments.

TABLE 3.11 Chemical shift ranges for $Si(OAl)_{4-n}(OSi)_n$ environments

Species and designation	$Si(OAl)_4$ Q0	$Si(OAl)_3(OSi)$ Q1	$Si(OAl)_2(OSi)_2$ Q2	$Si(OAl)(OSi)_3$ Q3	$Si(OSi)_4$ Q4
Approximate chemical shift range/ppm	−78 to −85	−86 to −91	−92 to −97	−98 to −103	−104 to −110

SiO_4 tetrahedron must be linked to four AlO_4 tetrahedra, that is, this composition will show only Q0 resonances (all silicon is present as $Si(-OAl)_4$). The ^{29}Si MASNMR spectrum can, therefore, be modelled for all silicon to aluminium ratios in a zeolite and comparison with an experimental spectrum from a zeolite of unknown composition then allows the Si:Al ratio to be determined.

EXAMPLE 3.21

Assuming a random distribution of silicon and aluminium in a zeolite of framework composition $[Si_{0.75}Al_{0.25}O_2]$, calculate the form of the expected ^{29}Si MASNMR spectrum.

ANSWER

We can work out the probability of the five different environments for a silicon atom. The probability of a $Si(OAl)_4$ being present is $(\frac{1}{4})^4$ as each of the surrounding four tetrahedral sites has a $\frac{1}{4}$ chance of being centred on aluminium; the probability of $Si(OSi)(OAl)_3$ is $4 \times (\frac{1}{4})^3(\frac{3}{4})$; note there are four ways of choosing on to which site to place silicon. Carrying this calculation on gives peak ratios of 0.004(Q0):0.047(Q1):0.211(Q2):0.422(Q3):0.316(Q4). These resonances should occur in the typical chemical shift ranges given in Table 3.11. The form of this spectrum is close to that seen in Figure 3.56 for the Si:Al ratio of 3:1 (~0.75:0.25).

SELF TEST

Calculate the form of the ^{29}Si MASNMR spectrum expected for a zeolite with a Si:Al ratio of 2:1.

3.21 Other advanced techniques in MASNMR

The techniques of homonuclear and heteronuclear decoupling described in Section 3.9 can be used to simplify solution NMR spectra, and the use of multiple pulse sequences such as INEPT (Section 3.18.1) allows the collection of spectra with difficult to observe nuclei. In a similar vein advanced NMR techniques can be applied to solids to enhance the quality and usefulness of spectra. The high-resolution technique **CP-MASNMR** is a combination of MASNMR with **cross polarisation (CP)**, which has been used to study many

compounds containing ^{13}C, ^{31}P, and ^{29}Si. Cross polarisation in MASNMR is similar to the INEPT experiment in that an abundant high receptivity nucleus is excited, and its energy is then transferred to the observed nucleus. This is achieved in CP-MASNMR by using a long, low power radiofrequency pulse and at a frequency, known as the **nutation frequency**, that effectively transfers the magnetisation. For systems that have strongly dipolar-coupled nuclei, such as ^1H and ^{13}C in many organometallic compounds, high power decoupling can be simultaneously applied improving spectral resolution.

The technique is also used to study molecular compounds in the solid state. For example, the ^{13}C-CP-MASNMR spectrum of $[Fe_2(C_8H_8)(CO)_5]$ at $-160°C$ indicates that all C atoms in the cyclooctatetraene (C_8H_8) ring are equivalent on the timescale of the experiment. The interpretation of this observation is that the molecule is fluxional even in the solid state.

3.22 Quantitative NMR (qNMR) and analysis

The area under a resonance peak in a solution NMR spectrum is directly proportional to the amount of material in the solution and, thus, its concentration. **Quantitative NMR**, also known as **qNMR**, can be used to measure quantitatively, and to a high degree of accuracy, the amount of material in a sample. qNMR applications are mainly in quality and process control, standardisation, and forensics. While the majority of analytical applications are for organic species and the common NMR nuclei, ^1H and ^{13}C, other nuclei such as ^{31}P, ^{11}B, and ^{19}F can be routinely used in the analysis of compounds such as pharmaceuticals and agrochemicals. One example of the use of qNMR is in the analysis of fluoridisers used in toothpaste. The two common fluoridisers used in toothpaste are sodium fluoride, NaF, and sodium monofluorophosphate, Na_2PO_3F. On dissolution in D_2O the former gives a single resonance in the NMR spectrum at -100 ppm and the latter two resonances (from coupling to ^{31}P) at -52.5 ppm and -54.8 ppm. By comparing the intensities of these resonances, and the use of standard solutions of known fluoride ion concentration, the fluorine concentration of each fluoridiser in a sample of unknown composition can be determined. This can be achieved to a high level of accuracy, within a few percent of the actual value and at a reasonable precision, demonstrating the quantitative nature of analysis.

Bibliography

J.K.M. Saunders and B.K. Hunter. (1993) *Modern NMR Spectroscopy, a Guide for Chemists*. Oxford: Oxford University Press.

J.A. Iggo. (1999) *NMR Spectroscopy in Inorganic Chemistry*. Oxford: Oxford University Press.

J.W. Akitt and B.E. Mann. (2000) *NMR and Chemistry*. Cheltenham, UK: Stanley Thornes.

K.J.D. MacKenzie and M.E. Smith. (2002) *Multinuclear Solid-State Nuclear Magnetic Resonance of Inorganic Materials*. Amsterdam Pergamon.

R.K. Harris. (1986) *Nuclear Magnetic Resonance Spectroscopy*. London: Longman.

D.C. Apperley, R.K. Harris, and P. Hodgkinson. (2012) *Solid State NMR: Basic Principles & Practice*. New York Momentum Press.

J. Mason (Editor). (1987) *Multinuclear NMR*. New York: Plenum Press.

J. Keeler. (2010) *Understanding NMR spectroscopy*. Chichester: Wiley.

P. Hore (2015) *Nuclear Magnetic Resonance*, 2nd ed. Oxford: Oxford University Press.

4 Vibrational spectroscopy

FUNDAMENTALS

4.1 Introduction to vibrational spectroscopy

Vibrational spectroscopy involves the excitation of the vibrational motion of atoms in molecules by electromagnetic radiation. This is used to characterise inorganic compounds as vibrational frequencies are characteristic of the functional groups present and the molecular shape of the material under study. These transitions are usually observed either by direct absorption of **infrared radiation** (100–$4000 \, \text{cm}^{-1}$, 100–$2.5 \, \mu\text{m}$, 3×10^{12}–$1 \times 10^{14} \, \text{Hz}$, 0.012–0.5 eV) or by **Raman scattering** of higher energy radiation (near infrared (NIR), visible, or ultraviolet (UV)). The information obtained from infrared (IR) absorption or Raman scattering spectroscopy can be used in both a qualitative or quantitative fashion. Qualitative vibrational spectroscopy is used to determine quickly whether a molecule contains a specific functional group, for example, a carbonyl group or ligand. Quantitative analysis of the spectra can provide information on the shape and symmetry of a molecule. For example, whether an octahedral transition metal complex has its three carbonyl ligands arranged in the facial or meridional isomeric forms.

4.1.1 Underlying theory

Vibrational spectroscopy is essentially about balls and springs, and the frequencies of the vibrations are dependent on both the masses of the balls (atoms) and the strength of the spring (chemical bond) connecting them. Heavier masses will vibrate at lower frequencies than lighter masses, and stronger chemical bonds (springs) will vibrate at a higher frequency than weaker bonds. It is this relationship between atomic structure and vibrational frequency which gives the technique its chemical specificity.

Quantum theory does not allow continuous variation of vibrational energy as predicted by classical mechanics, instead the **allowed vibrational energies, E_{vib},** are given by

$$E_{vib} = h\omega\left(\text{v} + \tfrac{1}{2}\right) \qquad \text{Eqn 4.1}$$

where $\text{v} = 0, 1, 2, \ldots$ is the **vibrational quantum number**.

NOTE ON GOOD PRACTICE

Although v (italic v) is often used for the vibrational quantum number, this can lead to confusion with ν (Greek nu); therefore to avoid this, v is used in this text. Other authors use n to avoid the confusion completely.

ω is the **harmonic vibrational frequency** of a diatomic molecule about the equilibrium bond length, r_e, in a **quadratic potential**, and is given by

$$\omega = \frac{1}{2\pi}\sqrt{\frac{k}{\mu}} \qquad \mu = \frac{m_1 m_2}{m_1 + m_2} \qquad \text{Eqn 4.2}$$

where k is the bond force constant, in N m^{-1}, and μ is the **reduced mass**. In a diatomic molecule μ depends on the masses of the two atoms at the end of the bond, m_1 and m_2.

ω (omega) is used for harmonic frequencies, and ν (nu) is used to denote the experimentally observed, **anharmonic frequencies**. Almost all of the chemical vibrational spectroscopic data is presented in terms of wavenumber with units of cm^{-1}, so Eqn. 4.2 needs to be modified as in Eqn. 4.3. A tilde, ~, is sometimes used to denote units of wavenumbers.

$$\tilde{\omega} = \frac{1}{2\pi c}\sqrt{\frac{k}{\mu}} \qquad \mu = \frac{m_1 m_2}{m_1 + m_2} \qquad \text{Eqn 4.3}$$

NOTE ON GOOD PRACTICE

Although vibrational data is usually presented in cm⁻¹ and called 'frequency', this is incorrect and should be avoided, and wavenumber used instead.

EXAMPLE 4.1

Calculate the reduced masses of $H^{35}Cl$, $H^{37}Cl$, and $D^{35}Cl$ using Eqn 4.2. What does this information tell us about the relevant motion of the atoms involved?

ANSWER

Using Eqn 4.2 the reduced masses are: $H^{35}Cl$, 0.972 g mol⁻¹; $H^{37}Cl$, 0.974 g mol⁻¹; $D^{35}Cl$, 1.89 g mol⁻¹. As these reduced masses are very close to those of H (or D) it means that the vibrational movement is largely confined to the hydrogen/deuterium atom moving back and forth against an almost static chlorine atom. Therefore, changing from ^{35}Cl to ^{37}Cl will not make very much difference to the vibrational frequency, but substituting H for D will.

SELF TEST

Use Eqn 4.3 and the reduced masses to calculate the ratio of the harmonic vibrational wavenumbers, $\tilde{\omega}(H^{35}Cl)/\tilde{\omega}(H^{37}Cl)$ and $\tilde{\omega}(H^{35}Cl)/\tilde{\omega}(D^{35}Cl)$.

If $\tilde{\omega}(H^{35}Cl)$ is at 2991.0 cm⁻¹, use your ratios to predict the positions of $\tilde{\omega}(H^{37}Cl)$ and $\tilde{\omega}(D^{35}Cl)$.

EXAMPLE 4.2

The experimental stretching mode, $\tilde{\nu}$, in carbon monoxide ($^{12}C^{16}O$) is at 2143 cm⁻¹. Calculate the force constant, k, for this vibrational mode.

ANSWER

$$\tilde{\nu} = \frac{1}{2\pi c}\sqrt{\frac{k}{\mu}} \implies k = \mu(2\pi c\tilde{\nu})^2$$

$$\mu = \frac{12.000 \times 15.995}{12.000 + 15.995} \text{ g mol}^{-1} = 6.8562 \text{ g mol}^{-1}$$
$$= \frac{6.8562}{6.0221 \times 10^{23} \times 1000} = 1.1385 \times 10^{-26} \text{ kg}$$

$$k = \mu(\tilde{\nu}2\pi c)^2 = 1.1385 \times 10^{-26}$$
$$\times (2143 \times 2 \times \pi \times 2.9979 \times 10^{10})^2$$
$$= 1855 \text{ kg s}^{-2} = 1855 \text{ N m}^{-1}$$

Note that the masses are for ^{12}C and ^{16}O, and that c is in cm s⁻¹ as wavenumber units are cm⁻¹.

SELF TEST

On isotopic substitution it is assumed that k remains constant. Making use of this, predict the position of the CO stretching mode in (a) $^{13}C^{16}O$ and (b) $^{12}C^{18}O$.

4.1.2 Selection rules

In the **harmonic approximation** for a diatomic molecule the vibrational **selection rule** is $\Delta v = \pm 1$, and if anharmonicity is included, this can be relaxed to include $\Delta v = \pm 2$ etc. However, not all vibrational modes are IR and/or Raman active, and some in fact are neither. More specific selection rules for IR absorption and Raman scattering spectroscopy need to be considered.

4.1.2.1 IR selection rule

The IR selection rule states that for a vibrational mode to be IR active there must be a change in the dipole moment of the molecule during the vibration. The dipole moment can be resolved into its x, y, and z components. There does not need to be a permanent dipole in the molecule, except in diatomics. The stretching modes in homonuclear diatomics such as O_2, N_2, etc. will be IR inactive, but the stretching modes in heteronuclear diatomics such as CO, HCl, HF, etc. will be IR active. In the case of CO_2, where there is no permanent dipole moment, there are two stretching modes: a symmetric one where both the C=O bonds extend and compress in phase, and an antisymmetric (or asymmetric) one where one C=O bond lengthens while the other shortens (or alternatively the carbon atom is moving backwards and forwards between the two oxygen atoms) (Figure 4.1). During the symmetric stretch the dipole moment of CO_2 is zero at all times, and so there is no dipole moment change during this vibrational mode. In contrast, during the antisymmetric stretching mode there is a zero dipole moment at equilibrium, but a non-zero dipole moment at all other positions, indicating a change in the dipole moment during the vibration. Therefore, the symmetric stretching mode is IR inactive, but the antisymmetric stretching mode is IR active (Figure 4.1).

4.1.2.2 Raman selection rule

The selection rule for vibrational Raman scattering states that a vibrational mode will only be Raman active if there is a change in the **polarisability** of the molecule during the vibration. The polarisability is a measure of the ease

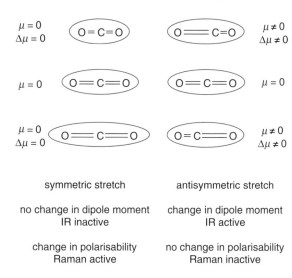

$\mu = 0$ $\quad$ O=C=O $\qquad$ O==C=O $\quad$ $\mu \neq 0$
$\Delta\mu = 0$ $\qquad\qquad\qquad\qquad\qquad$ $\Delta\mu \neq 0$

$\mu = 0$ $\quad$ O=C=O $\qquad$ O=C=O $\quad$ $\mu = 0$

$\mu = 0$ $\quad$ O==C=O $\qquad$ O=C==O $\quad$ $\mu \neq 0$
$\Delta\mu = 0$ $\qquad\qquad\qquad\qquad\qquad$ $\Delta\mu \neq 0$

symmetric stretch $\qquad\qquad$ antisymmetric stretch

no change in dipole moment $\qquad$ change in dipole moment
IR inactive $\qquad\qquad\qquad$ IR active

change in polarisability $\qquad$ no change in polarisability
Raman active $\qquad\qquad\qquad$ Raman inactive

FIGURE 4.1 IR and Raman activity of the symmetric and antisymmetric stretching modes in CO_2.

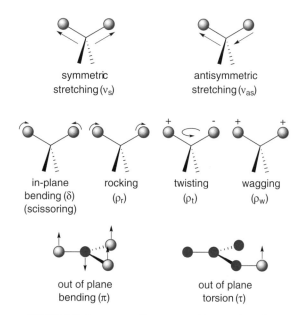

symmetric stretching (ν_s) $\qquad$ antisymmetric stretching (ν_{as})

in-plane bending (δ) (scissoring) $\quad$ rocking (ρ_r) $\quad$ twisting (ρ_t) $\quad$ wagging (ρ_w)

out of plane bending (π) $\qquad$ out of plane torsion (τ)

FIGURE 4.2 Summary of important vibrational modes.

of deforming the electron density and hence inducing a dipole moment in a molecule when it is placed in an electric field. While the dipole moment and its change can be fairly easily spotted, the polarisability and any change in it is harder to visualise. However, it can be imagined to be like a rubber sheet or balloon surrounding the molecule. In the symmetric stretch in CO_2 (Figure 4.1), this surface is extending and contracting with the vibrational mode, so it is Raman active. In contrast, in the antisymmetric stretching mode it remains unchanged, and therefore this mode is Raman inactive. In later sections (Section 4.4.5) a more rigorous way of determining IR and Raman activity will be developed.

4.1.3 **Number of vibrational modes**

A molecule with N atoms has $3N$ degrees of freedom as each atom can move in the x, y, and z directions. Three of these degrees of freedom describe the translational motion of the molecule as a whole. If the molecule is non-linear, another three describe the rotational motion of the whole molecule. Therefore, there are $3N - 6$ remaining degrees of freedom which describe the vibrations of the atoms within the molecule. If the molecule is linear, there are $3N - 5$ vibrational degrees of freedom because there is no **moment of inertia** about the internuclear axis, so there are only two rotational degrees of freedom.

4.1.4 **Normal modes of vibration**

The independent vibrational degrees of freedom are called the vibrational (or normal) modes of a molecule and involve displacements of the atoms

with respect to each other, all at the same frequency. These **normal modes** can be divided into whether they principally involve bond stretching (ν) or angle deformation modes (δ, ρ, π, τ). A summary of the important modes is shown in Figure 4.2. For example, a bent XY_2 triatomic has $3N - 6 = 3$ vibrational degrees of freedom, which can be thought of as two stretching modes and one bending mode. A linear XY_2 triatomic has $3N - 5 = 4$ vibrational degrees of freedom, which correspond to two stretching modes and two bending modes.

EXAMPLE 4.3

Determine the IR and Raman activity of the vibrational modes in water shown pictorially in Figure 4.3.

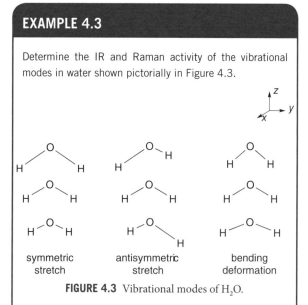

symmetric stretch $\qquad$ antisymmetric stretch $\qquad$ bending deformation

FIGURE 4.3 Vibrational modes of H_2O.

ANSWER

In the symmetric stretching mode there is a change in the dipole moment in the z direction and a change in the polarisability so it is both IR and Raman active. In the antisymmetric stretching mode, there is a change in the dipole moment in the y direction, and a change in the polarisability, so it is both IR and Raman active. In the bending mode there is a change in the dipole moment in the z direction and a change in the polarisability so it is both IR and Raman active.

SELF TEST

Determine the IR and Raman activity of the stretching modes of BF_3 shown in Figure 4.4.

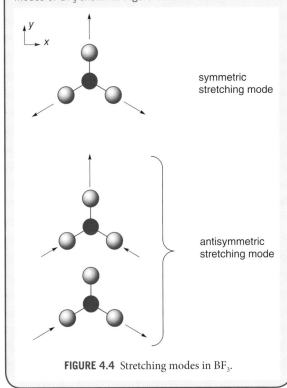

FIGURE 4.4 Stretching modes in BF_3.

4.2 **Experimental considerations**

The values of the characteristic vibrational frequencies are obtained in two main ways, IR absorption spectroscopy and Raman scattering spectroscopy. These are complementary and often both are applied to obtain the maximum information.

IR absorption is a direct measurement of the characteristic vibrational frequency. The sample is irradiated with IR light and those frequencies which correspond to the energy of the vibrational modes are absorbed and measured. This is shown in the left-hand side of Figure 4.5. Note that the lower, ground, state is denoted by v'' whereas the upper state is given by v'.

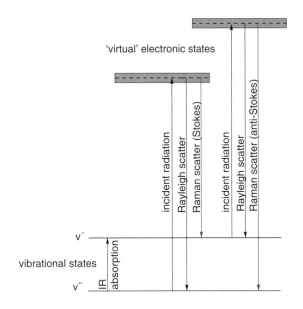

FIGURE 4.5 Processes occurring during IR absorption and Raman scattering.

For Raman scattering (Figure 4.5), the sample is irradiated with high energy light (usually UV, visible, or NIR, wavelength 200–1200 nm), most of which is scattered elastically (Rayleigh scatter), but a very small proportion (10^{-4}–10^{-5}) is scattered inelastically. In this case the incident photon either gives up a small part of its energy to a vibrational mode (denoted Stokes), or gains a small part of energy from a vibrational mode (termed anti-Stokes). The difference between the incident and inelastically scattered Stokes and anti-Stokes Raman radiation is called the Raman shift and provides the information about the energies of the vibrational modes. In conventional Raman scattering the incident radiation results in transitions to 'virtual' electronic states, but if the incident laser energy closely matches an allowed electronic transition then either fluorescence or resonance Raman scattering can be observed (see Section 4.5.2).

4.2.1 **IR spectrometers**

Traditionally IR spectra were collected using instruments equipped with prisms, diffraction gratings and thermal detectors connected to chart recorders. Spectra would take minutes to acquire, even for modest resolution, as each energy point was collected sequentially. With the increase in computing power, IR spectrometers now almost exclusively use interferometers and Fourier transform data processing techniques. Such instruments are known as Fourier transform IR (FTIR) spectrometers.

A schematic diagram of an FTIR spectrometer is shown in Figure 4.6. Light from the IR source (heated

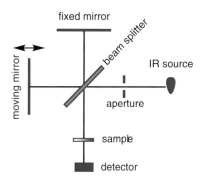

fixed mirror

beam splitter

IR source

moving mirror

aperture

sample

detector

FIGURE 4.6 Schematic diagram of principal components of a Fourier transform IR (FTIR) spectrometer.

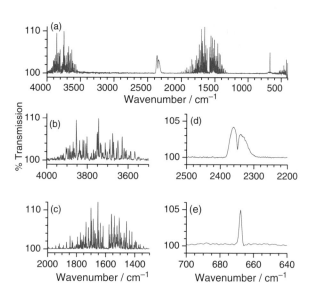

FIGURE 4.7 IR spectra (2 cm^{-1} resolution) of (a) atmospheric gases, (b) H_2O stretching modes, (c) H_2O bending mode, (d) CO_2 antisymmetric stretching mode, (e) CO_2 bending mode.

element) is defined by an aperture and impinges on a beam splitter which partially transmits and partially reflects the light to the fixed and moving mirrors. As the moving mirror moves backwards and forwards a complex interference pattern known as an interferogram is generated, which is converted to a spectrum by Fourier transform techniques. A red laser is used to calibrate the instrument, as well as for sample alignment. As each spectrum only takes a matter of seconds to acquire, there is a very significant increase in signal-to-noise statistics over scanning instruments with grating monochromators. As these instruments have few moving parts they can be miniaturised for portability and made very robust.

FTIR spectrometers are single-beam instruments, so it is necessary to record both a background and a sample spectrum, and then ratio these to obtain a traditional transmission spectrum. As the amount of atmospheric gases may vary between the background and sample spectrum it is quite common to observe features due to atmospheric water and carbon dioxide in the spectra. In Figure 4.7, the concentration of the atmospheric gases was slightly higher when the background spectrum was collected, compared to when the sample spectrum (which only contained atmospheric gases) was acquired.

These features can be subtracted either manually or automatically by the data acquisition software. As they will be present in all spectra (except those recorded on research instruments with a vacuum bench) it is important to be aware of their effect on the spectra obtained. For water the rotational fine structure on the antisymmetric and symmetric stretching modes overlap in the 4000–3500 cm^{-1} region, (Figure 4.7a), and the rotational fine structure on the bending mode is in the 2000–1350 cm^{-1} region (Figure 4.7b). These features due to water often appear as 'noise' in IR spectra and can be readily identified. However, it is easy to misinterpret the CO_2 features between 2400 and 2300 cm^{-1} and at 667 cm^{-1}. The anti-symmetric stretching mode of CO_2 centred at 2349 cm^{-1} (Figure 4.7c) has a characteristic rotational fine structure (at higher resolution the individual components can be resolved). The

bending mode of CO_2 at 667 cm^{-1} (Figure 4.7d) is characterised by an intense central absorption (Q-branch) with much weaker rotational fine structure (P- and R-branches) on either side. These unwanted spectral features are one of the few disadvantages of single channel FTIR instruments compared to double beam grating instruments.

4.2.2 Raman spectrometers

Raman scattering (Figure 4.5) is typically 10^6 times weaker than Rayleigh scattering (where the energy of the photon is unchanged). Modern vibrational Raman instruments conventionally use lasers as intense light sources in the UV, visible, or NIR part of the spectrum, and the intensity of Raman scattering is proportional to $(1/\lambda)^4$. Therefore, with 266 nm excitation the Raman intensity is 16 times that at 532 nm and 64 times that at 1064 nm. Although the use of UV and visible-light lasers as the excitation source yields much more intense Raman scattering than a NIR laser, Raman spectroscopy is plagued by fluorescence. Collecting data using longer wavelength NIR sources can reduce fluorescence but even with NIR excitation, sample burning and IR emission can be observed for materials with low energy visible absorptions close to the 1064 nm excitation. By deliberately selecting a transition that coincides with an allowed electronic transition, resonance Raman scattering can be observed (Section 4.5.2).

A number of possible geometries are used to collect the scattered radiation, with the most common being the 90° or 180° scattering of the incident radiation. Raman spectrometers use either dispersive monochromator optics, where the wavelengths of the scattered radiation are scanned sequentially onto a photomultiplier detector, or the wavelengths can be dispersed onto a position sensitive detector. These configurations are most commonly employed with visible or UV laser sources. With the ready availability of FTIR spectrometers, it is common to find a NIR excitation source (e.g. Nd-YAG laser at 1064 nm) and Raman spectrometer employing the same interferometer as the FTIR bench, but with modified beam splitter and detector. The key requirement in all Raman spectrometers is the need to remove the Rayleigh scatter to avoid saturating or damaging the detector. This can be achieved with edge filters acting as high- and low-pass filters, or notch filters. The former allow spectra to be obtained much closer to the excitation wavelength, but a notch filter enables the collection of both the Stokes and anti-Stokes lines.

EXAMPLE 4.4

What is the position in nm for a 200 cm^{-1} Stokes shift for 532.0 nm excitation?

ANSWER

532.0 nm is equivalent to 18 797 cm^{-1}, (Chapter 1, Section 1.2.4) and, as the Stokes line will be at lower wavenumber, this corresponds to 18 597 cm^{-1}, equivalent to 537.7 nm.

SELF TEST

What is the position for a 200 cm^{-1} Stokes shift (in nm) when using 1064 nm excitation?

4.2.3 IR and Raman sampling

IR and Raman spectra can be obtained from gas, liquid, and solid-phase materials, as well as more exotic sample environments, such as surfaces or cryogenic matrices. In principle, Raman experiments are easier because the requirement to use IR transparent materials is removed, and glass or quartz sample holders can be employed. Aqueous solutions are also much easier to study with Raman spectroscopy as water is a strong IR absorber, but a weak Raman scatterer. However, as noted above, Raman scattering can be a weak effect, and can easily be swamped by fluorescence, including that from the sample holder.

4.2.3.1 IR sampling

For IR absorption experiments, the path length or penetration depth of the radiation into the sample is a crucial factor to avoid distortion of the absorption data. If the penetration depth, which will vary as a function of wavelength, is shorter than the particle size, the accuracy of the ordinate (y-axis) data will be compromised. For gas phase samples the concentration may be so low that long path length cells are required, whereas in solid and liquid/solution samples the reverse is more of a problem and the effective path length needs to be kept to a minimum. Historically, solid samples have been measured either using 'mulls' or thin KBr discs, although diffuse reflectance can also be used. For mulls, the material of interest is diluted in a mulling agent (such as Nujol®, a paraffin oil) and then presented to the instrument compressed between two IR transparent plates (e.g. NaCl, KBr, CsI, or polyethylene depending on the wavenumber range required). KBr discs are made by grinding a small quantity of the material of interest with KBr, and then compressing the mixture to form a self-supporting transparent disc. This has the advantage of not having the spectral features of the mulling agent, but KBr under moderate pressure is not completely inert.

Liquids can be studied as thin films on halide plates (e.g. NaCl, KBr) by allowing the majority of the material to evaporate before the addition of a second plate. Solutions can be studied in special cells which are usually constructed from halide windows with variable thickness shims made of PTFE to control the path length.

There are many other specialist sampling devices, especially for solid samples, such as **diffuse reflectance** and **attenuated total reflectance** (ATR). ATR devices originally used ZnSe crystal optics which limited their use. However, the advent of diamond ATR has revolutionised IR sampling, so that for most liquid and solid samples it is now the technique of choice for mid-IR (MIR) spectroscopy as there is no or minimal sample preparation required. In ATR, the IR beam is passed through an internally reflecting element, such as ZnSe or diamond, but a small fraction escapes from the surface as an **evanescent wave** (Figure 4.8). If there is a sample in good contact with the ATR optical element, this can absorb

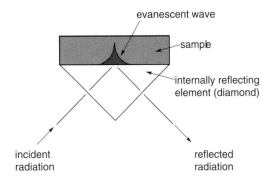

FIGURE 4.8 Schematic diagram of a diamond attenuated total reflection (ATR) accessory for IR spectroscopy.

radiation from the evanescent wave and an absorption spectrum is obtained.

The striking difference between solid-state and gas-phase IR spectra is shown in Figure 4.9 for GeH_4. In the solid state spectrum (Figure 4.9a) there is one peak for each IR active vibrational mode, ν_{Ge-H} stretching mode at 2095 cm^{-1} and a δ_{Ge-H} bending mode at 798 cm^{-1}. In contrast, in the gas phase (Figure 4.9b), the molecule is free to rotate so rotational fine structure is observed on each of these bands. This rotational fine structure contains important structural information which can yield moments of inertia and hence structural parameters. In Figure 4.9(c) the effect of increasing the resolution from 1.0 cm^{-1} to 0.25 cm^{-1} is shown, and this allows for the observation of individual features from the Ge isotopes in GeH_4 to be resolved. In addition, weak bands that are formally IR forbidden can gain intensity in the gas phase due to an interaction between the rotational and vibrational modes known as the Coriolis effect, and an example of this is a second δ_{Ge-H} bending mode observed at 935 cm^{-1}.

The majority of IR spectra are presented in percentage transmission mode, almost certainly because of the ease of recording such data on early instruments. However, this results in a nonlinear ordinate (y-axis) scale, and if quantitative ordinate data are required, the transmission data must be converted to absorbance via the Beer–Lambert law (Chapter 5, Section 5.2). The difference between the percent (%) transmission and absorbance data is shown in Figure 4.9(b) and (c). One of the benefits of the nonlinearity of the transmission data is that it is much easier to identify weaker features.

4.2.3.2 Raman sampling

As Raman spectroscopy involves scattering rather than absorption, sample preparation is more straightforward, especially as glass containers can be used for solids or liquids (including aqueous solutions) and neat solids can also be sampled.

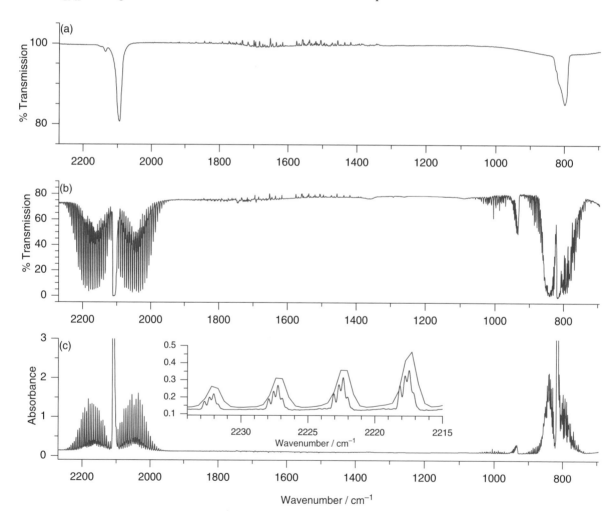

FIGURE 4.9 Transmission IR spectra of (a) solid GeH_4 at 50 K, (b) gas-phase GeH_4 at 298 K, and (c) absorbance IR spectrum of gas-phase GeH_4 at 298 K. The insert shows a portion of the gas-phase absorbance spectrum at 1.0 and 0.25 cm^{-1} resolution.

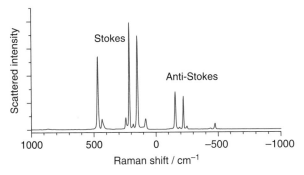

FIGURE 4.10 Raman spectrum of sulfur (recorded with 1064 nm excitation).

Both Stokes and anti-Stokes transitions can be observed, as demonstrated in the 1064 nm Raman spectrum of sulfur at room temperature (Figure 4.10). Normally the Stokes lines are collected due to their higher intensity.

4.2.4 **IR and Raman microscopy**

It is possible to interface both IR and Raman spectrometers to optical microscopes in order to improve the spatial resolution down to the micron scale and to correlate the vibrational data with the structural motifs of samples such as biological cells and mixtures of minerals.

4.3 **Qualitative considerations**

In the analysis of vibrational spectroscopic data the whole molecule needs to be considered as, in principle, all the atoms are involved in each vibrational mode. However, it is often possible or convenient to assign vibrational modes as being characteristic of the presence of specific functional groups. It is usually the higher

wavenumber features that are easiest to assign to specific groups, rather than peaks in the fingerprint region (<1500 cm^{-1}), but this region is characteristic for each compound.

4.3.1 **X–H groups and hydrogen bonds**

Some of the shortest and, therefore, strongest bonds are those involving hydrogen, especially those to the elements of the second row of the periodic table: B–H, C–H, N–H, O–H, F–H. The stretching vibrations (ν_{X-H}) have characteristic IR wavenumber ranges, as shown in Table 4.1, together with those of the heavier hydrides for comparison.

These values are highly diagnostic, particularly for the common species O–H, N–H, and C–H. Thus, inorganic compounds containing coordinated water or hydroxyl groups show absorptions at around 3600 cm^{-1} and organometallic compounds containing C–H, bands near 3000 cm^{-1}. Trends within these values reflect the strength of the X–H bond both across a period and down a group in the periodic table. Molecules containing two or more X–H bonds, for example, water and ammonia, also show bending vibrations. For water molecules, this mode, also sometimes referred to as a scissor mode, is observed at ~1600 cm^{-1} and in the ammonium cation at 1400 cm^{-1}.

Coordination to a metal can shift these values; for example, in $[Me_2N(C_2H_4)_2NMe_2]^{2+}[InBr_5(H_2O)]^{2-}$. H_2O the ν_{O-H} modes for water coordinated to the indium are at 3500 and 3430 cm^{-1} together with a broader band at 3280 cm^{-1} for uncoordinated (lattice) water involved in hydrogen bonding. In $Cs_2[InBr_5(H_2O)]$, there is no band at 3280 cm^{-1} confirming this assignment. In addition to these shifts a number of very diagnostic rocking, twisting, and wagging modes of the coordinated water molecules can also be observed at lower wavenumbers. The ν_{N-H} symmetric and antisymmetric stretching modes

TABLE 4.1 IR data for common X–H units and XH$_n$ compounds

Bond, X–H	B–H	C–H	N–H	O–H	F–H
ν_{X-H}/cm^{-1}	2650–2350 (terminal) 2100–1600 (bridge)	3060–2850	3500–3200	3700–3500	3958
Molecule, XH$_n$		SiH$_4$	PH$_3$	SH$_2$	ClH
ν_{X-H}/cm^{-1}		2183	2421, 2327	2627, 2615	2885
Molecule, XH$_n$		GeH$_4$	AsH$_3$	SeH$_2$	BrH
ν_{X-H}/cm^{-1}		2114	2185, 2122	2358, 2345	2559
Molecule, XH$_n$		SnH$_4$	SbH$_3$	TeH$_2$	IH
ν_{X-H}/cm^{-1}		1901	1894, 1891	2092, 2085	2230

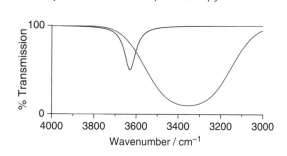

FIGURE 4.11 Effect of hydrogen bonding on the ν_{O-H} mode, no hydrogen bonding (black) and with extensive hydrogen bonding (blue).

in ammine complexes (e.g. $[Cr(NH_3)_6]^{3+}$ (3185 and 3257 cm^{-1}) and $[Co(NH_3)_6]^{3+}$ (3160 and 3240 cm^{-1})) are at lower wavenumber than those of 'free' ammonia (3336 and 3414 cm^{-1}). These shifts reflect the strength of the N–H bond, which weakens upon coordination while the M–N bond gets stronger.

Hydrogen bonding plays a very significant role in mineralogy, bioinorganic chemistry, and biochemistry (e.g. DNA), and is also important in directing the structures of simple inorganic compounds and materials. The ν_{O-H} modes are strongly affected by hydrogen bonding, so that the sharp intense bands observed for isolated/free O–H groups around 3650–3590 cm^{-1} shift to lower wavenumber, increase in intensity, and usually become very broad (Figure 4.11).

Water itself and waters of crystallisation (e.g. in hydrated salts) also exhibit features in this region (3550–3200 cm^{-1}), and a broad band centred at 3450 cm^{-1} is often observed in IR spectra of KBr discs due to its hygroscopic nature. If there is intramolecular hydrogen bonding, the shift is even larger (3200–2500 cm^{-1}) and values as low as 1750 cm^{-1} have been observed for symmetrical

hydrogen bonds in compounds such as CrOOH in the solid state. While the frequencies of the O–H stretching modes decrease when involved in hydrogen bonding, the bending mode at ca. 1620 cm^{-1} shifts to a higher wavenumber. The features due to N–H stretches are also shifted to lower wavenumber and broadened by hydrogen bonding, but to a lesser extent than for the O-H bands. Table 4.2 summarises the properties of different hydrogen bonds, including the effect on the ν_{X-H} modes.

Because of the large change in reduced mass in X–H bonds when hydrogen, ^{1}H, is replaced by deuterium, D (^{2}H) (Section 4.1.1), there is a large change in stretching frequency of X–H bonds. The ratio of the wavenumbers is close to that given theoretically, that is,

$$\frac{\tilde{\nu}_H}{\tilde{\nu}_D} \sim \sqrt{\frac{\mu_D}{\mu_H}} \sim \sqrt{2} \qquad \textbf{Eqn 4.4}$$

so that the O–H stretching mode near 3600 cm^{-1} shifts to 2550 cm^{-1} on deuteration and, for example, the N–H stretch in $(NH_4)[ReO_4]$ moves from 3480–3390 cm^{-1} to 2340 cm^{-1} in $(ND_4)[ReO_4]$. This provides a highly diagnostic tool for the presence of exchangeable hydrogen or protons in a compound. For example, the acidic forms of zeolites have surface OH groups that absorb in the IR spectrum at ~3600 cm^{-1}; on exposure to D_2O vapour these are replaced by peaks at ~2500 cm^{-1} showing rapid deuterium–hydrogen exchange.

4.3.2 Functional groups in organometallic complexes

A summary of the important values found for 'free' organic molecules is given in Table 4.3 and these are useful in the characterisation of organometallic complexes. It should be noted that the fingerprint region

TABLE 4.2 Summary of hydrogen bond characteristics (D is the hydrogen bond donor atom, A is the acceptor atom)

D–H···A hydrogen bond	Weak H-bond	Moderate H-bond	Strong H-bond
Bond lengths	D–H << H···A	D–H < H···A	D–H ≈ H···A
H···A (Å)	3.2–2.2	2.2–1.5	1.5–1.2
Bond enthalpy (H···A) E_{HB} (kJ mol^{-1})	<16	16–60	60–190
Decrease in ν_{D-H} (%)	<10	10–25	25–80
Typical hydrogen bonds	C–H···O C–H···N	–O–H···O= –O–H···N≡ –N–H···O= –N–H···N≡	[F···H···F]$^-$

TABLE 4.3 IR data for common organic functional groups

Functional group	IR band/cm^{-1}	Functional group	IR band/cm^{-1}	Functional group	IR band/cm^{-1}
ν_{C-H} (aliphatic)	2960–2850	$\nu_{C\equiv C}$	2260–2100	ν_{C-F}	1400–1000 (s)
ν_{C-H} (aromatic)	3040–3010 (weak)	$\nu_{C=C}$	1660–1640	ν_{C-Cl}	800–600 (s)
ν_{C-H} (C=C–H)	3100–3000	ν_{C-C}	1150–950	ν_{C-Br}	750–500 (s)
ν_{C-H} (C≡C–H)	3300	$\nu_{C=O}$	1850–1650	ν_{C-I}	500 (s)

(1500–500 cm^{-1}) is usually very complex containing features due to a large variety of stretching and deformation modes that are interacting with each other, so while the spectrum as a whole is characteristic of a given compound, it is not usually possible to assign individual features to particular vibrational modes. Therefore, the table concentrates on the higher wavenumber bands which are more easily assigned to specific functional groups. Cyclopentadienyl complexes such as [FeCp$_2$] display ν_{CH} at 3077 cm^{-1} and ν_{CC} at 1110 and 1410 cm^{-1}, and [Cr(C$_6$H$_6$)$_2$] has ν_{CH} at 3037 cm^{-1} and ν_{CC} at 1426 cm^{-1}. The $\nu_{C=O}$ modes in organic molecules are very characteristic of the specific functional group (e.g. aldehydes vs. ketones, and saturated vs. unsaturated substituents), and tables of these data are widely available.

4.3.3 IR data for common anions

Many of the common inorganic oxoanions or fluoroanions are used as counterions for complexes and these have characteristic and intense IR absorptions, and a summary is given in Table 4.4.

These data can be used as a quick diagnostic characterisation to check that the anion is present in the compound. For example, an intense peak at ca. 1070 cm^{-1} will indicate the presence of [BF$_4$]$^-$, whereas peaks around 850 cm^{-1} are indicative of [PF$_6$]$^-$.

4.3.4 Compounds and complexes containing oxo and nitrido groups

The $\nu_{M=O}$ modes in compounds containing oxo groups (M=O) usually occur around 1100–900 cm^{-1} (Table 4.5). The position is sensitive to coordination environment: for example, it is highest for the fluorides in both OVX$_3$ and OPX$_3$ compounds; dependent on the oxidation state (e.g. in manganates), and the metal (in Group 6 compounds). The corresponding $\nu_{M=S}$ modes are much lower in energy.

Oxo groups are also found in coordination complexes, and some of the most common examples are vanadyl (VO^{2+}) where the $\nu_{V=O}$ mode is sensitive to coordination environment. Dioxo complexes (O=M=O) also exhibit

TABLE 4.4 IR data for common anions

Ion	Vibrational mode	IR band cm^{-1}	Ion	Vibrational mode	IR band cm^{-1}
[CO$_3$]$^{2-}$	ν_{C-O}	1450–1410	[BF$_4$]$^-$	ν_{B-F}	1070
[SO$_4$]$^{2-}$	ν_{S-O}	1130–1080	[SiF$_6$]$^{2-}$	ν_{Si-F}	740
[SO$_3$]$^{2-}$	ν_{S-O}	970–930	[PF$_6$]$^-$	ν_{P-F}	865, 835
[NO$_3$]$^-$	ν_{N-O}	1380–1350	[AsF$_6$]$^-$	ν_{As-F}	700
[NO$_2$]$^-$	ν_{N-O}	1250–1230	[SbF$_6$]$^-$	ν_{Sb-F}	670
[PO$_4$]$^{3-}$	ν_{P-O}	1100–1000			
[SiO$_4$]$^{4-}$	ν_{Si-O}	1100–900			
[ClO$_4$]$^-$	ν_{Cl-O}	1120			
[IO$_4$]$^-$	ν_{I-O}	850			

TABLE 4.5 Representative $\nu_{M=O}$ and $\nu_{M=S}$ modes

	$\nu_{M=O}$/cm^{-1}		$\nu_{M=O}$/cm^{-1}		$\nu_{M=O}$/cm^{-1}		$\nu_{M=O}$/cm^{-1}		$\nu_{M=O/S}$/cm^{-1}
OVF$_3$	1058	OPF$_3$	1415	[MnO$_4$]$^-$	902	[CrO$_4$]$^{2-}$	890	[ReO$_4$]$^-$	920
OVCl$_3$	1035	OPCl$_3$	1290	[MnO$_4$]$^{2-}$	820	[MoO$_4$]$^{2-}$	837	[ReS$_4$]$^-$	486
OVBr$_3$	1025	OPBr$_3$	1261	[MnO$_4$]$^{3-}$	778	[WO$_4$]$^{2-}$	838		

TABLE 4.6 Metal–halogen stretching modes (cm^{-1}) for some nickel complexes (dppe = 1,2-bis(diphenylphosphino)ethane, py = pyridine)

	linear	*trans*-planar	*cis*-planar	tetrahedral	*trans*-octahedral
	NiX$_2$	[Ni(PEt$_3$)$_2$X$_2$]	[Ni(dppe)X$_2$]	[Ni(PPh$_3$)$_2$X$_2$]	[Ni(py)$_4$X$_2$]
ν_{Ni-Cl}	521	403	341, 328	341, 305	207
ν_{Ni-Br}	414	338	290, 266	265, 232	140

IR bands in the 1100–850 cm^{-1} region. The N^{3-} ion is iso-electronic with O^{2-}, and due its strong π donating character, a M≡N triple bond is formed with $\nu_{M≡N}$ in the range 1100–1000 cm^{-1}.

SELF TEST

Explain why *trans*-[Os(bipy)$_2$(O)$_2$] has one $\nu_{M=O}$ IR band at 872 cm^{-1}, whilst *cis*-[Os(bipy)$_2$(O)$_2$] (bipy = 2,2'-bipyridine) has two $\nu_{M=O}$ IR bands at 863 and 833 cm^{-1}.

EXAMPLE 4.5

Explain the positions of the $\nu_{V=O}$ modes in [VO(acac)$_2$] and [VO(acac)$_2$(pyridine)] in Figure 4.12.

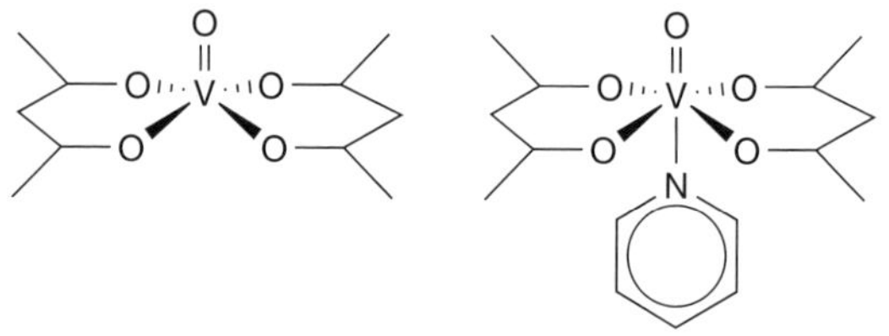

$\nu_{V=O}$ 999 cm^{-1} $\nu_{V=O}$ 973 cm^{-1}

FIGURE 4.12 The structures and $\nu_{V=O}$ modes of [VO(acac)$_2$] and [VO(acac)$_2$(pyridine)].

ANSWER

The oxygen in the oxo group in the vanadyl complexes is acting as a π base (as well as a σ base) towards the vanadium. As the metal orbitals that are involved in the σ donation/bonding from the pyridine are the same as those involved with the V=O unit, less donation of electron density from the oxygen to the vanadium can take place, resulting in a weaker bond, and lower stretching frequency.

4.3.5 Complexes with halide ligands

The low-energy ν_{M-X} modes, where M is a metal and X is a halogen, often lie outside the range of conventional (KBr beam splitter) FTIR instruments. With more advanced instrumentation these modes can be observed and give valuable information as they are dependent on both the halogen and molecular geometry, as shown in Table 4.6 for some nickel complexes.

In the related square planar palladium(II) and platinum(II) complexes, the position of the ν_{M-X} mode is sensitive to the ligand *trans* to the halogen (because of the *trans*-influence). For example, for [PtCl$_3$CO]$^-$ $\nu_{Pt-Cl(trans)}$ is 322 cm^{-1}, but this drops to ca. 275 cm^{-1} for the PPh$_3$, PMe$_3$, and PEt$_3$ analogues, indicating that there is a stronger *trans*-influence in the phosphine complexes than the carbonyl one. The $\nu_{Pt-Cl(cis)}$ modes are not affected by the change in ligand.

4.3.6 Complexes with carbonyl ligands

Carbon monoxide coordinates to the metal atom via the carbon atom, but can act either as a terminal or bridging ligand, and these two bonding modes are often readily identifiable from the IR spectra, as shown in Figure 4.13.

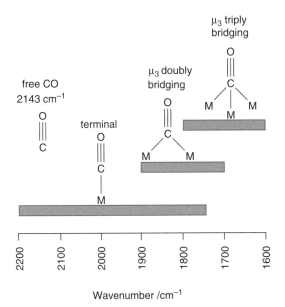

FIGURE 4.13 Representative ranges of ν_{CO} modes in different complexes.

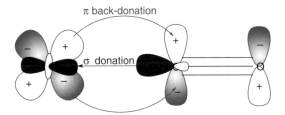

FIGURE 4.15 Summary of bonding in carbonyl complexes.

The CO gas phase ν_{CO} stretching wavenumber is 2143 cm^{-1}. The vast majority of transition metal carbonyl complexes have the ν_{CO} modes at lower wavenumber because the CO ligand acts as a σ donor and π acceptor using its π* antibonding orbital to accept electron density from the metal orbitals with the correct symmetry (t_{2g} in octahedral and e in tetrahedral) (Figure 4.15).

This back-bonding has the effect of increasing the strength of the M–C bond, but reducing the strength of the C≡O bond. As the amount of back-bonding increases, ν_{M-C} increases and ν_{CO} reduces. In general, this effect is greater than the increase in ν_{CO} due to σ bonding (from a slightly antibonding orbital on CO), and ν_{CO} is usually lower than that for 'free' CO at 2143 cm^{-1}. These are known as classical carbonyl complexes. While this is the case for the vast majority of carbonyl complexes there are some, known as 'non-classical' where the ν_{CO} modes are at a higher wavenumber than free CO, indicating very little back-bonding is taking place. These are often associated with carbonyl complexes containing high oxidation state metals.

The position of the ν_{CO} mode can be used to indicate the extent of back-bonding arising from different metals, ligands, and oxidation states. As the electron density on the metal increases, more back-bonding can take place resulting in higher wavenumber ν_{M-C} modes and lower wavenumber ν_{CO} modes. If CO is replaced by a stronger σ base and/or poorer π acid, then there is greater back-bonding to the remaining CO ligands. As can be seen from Figure 4.13 there is some overlap between terminal and bridging ν_{CO} modes so the data needs interpreting carefully and thoughtfully, especially for complexes containing electron-rich metals.

EXAMPLE 4.6

Use the ν_{CO} IR data to comment on the presence of terminal and bridging carbonyl ligands in [Fe(CO)$_5$] (2002 and 1979 cm^{-1}) and [Fe$_2$(CO)$_9$] (2066, 2038, 1855, 1851 cm^{-1}).

ANSWER

For [Fe(CO)$_5$] both CO stretching modes are characteristic of terminal CO ligands as in Figure 4.14. In [Fe$_2$(CO)$_9$] there are two sets of CO stretching modes, one clearly in the terminal CO region and one that is in the overlap region of terminal and bridging CO. As these two sets of bands are separated by ca. 200 cm^{-1} it is reasonable to assume that they come from CO in two different environments, therefore there are both terminal and bridging CO ligands present in [Fe$_2$(CO)$_9$] (Figure 4.14).

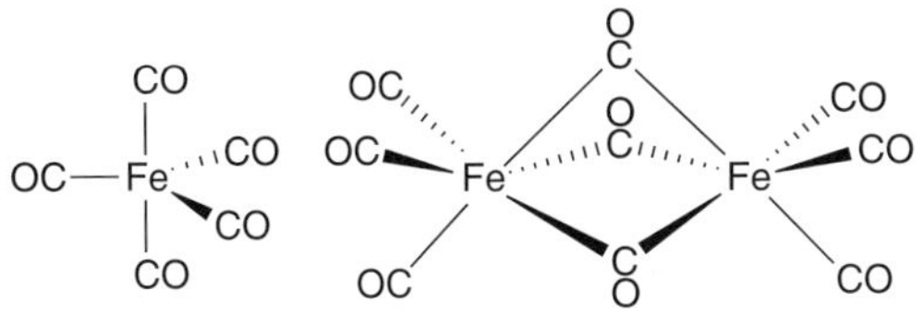

FIGURE 4.14 Structures of [Fe(CO)$_5$] and [Fe$_2$(CO)$_9$].

SELF TEST

The solid-state IR spectrum of [Co$_2$(CO)$_8$] displays bands at 2071, 2044, 2042, 1866, and 1857 cm^{-1}, but in a non-coordinating solvent there are only bands at 2069, 2055, and 2032 cm^{-1}. Account for these observations.

EXAMPLE 4.7

Explain why the positions of the IR active ν_{CO} and ν_{M-C} mode vary in the following complexes: [Mn(CO)$_6$]$^+$, ν_{CO} 2095 cm^{-1}, ν_{M-C} 412 cm^{-1}; [Cr(CO)$_6$], ν_{CO} 2000 cm^{-1} ν_{M-C} 441 cm^{-1}; [V(CO)$_6$]$^-$, ν_{CO} 1858 cm^{-1} ν_{M-C} 460 cm^{-1}.

ANSWER

The electron density increases on the metal as the complexes become more anionic, therefore there is more back-bonding so the v_{CO} mode reduces and the v_{M-C} mode increases in wavenumber.

SELF TEST

(a) Account for the changes in v_{CO} in the following complexes: [Ni(CO)$_3$(PMe$_3$)], 2064.1 cm^{-1}; [Ni(CO)$_3$(PPh$_3$)], 2068.9 cm^{-1}; [Ni(CO)$_3$(PF$_3$)], 2110.8 cm^{-1}. (Cy = cyclohexyl)

(b) Account for the changes in the IR active v_{CO} modes in the following tetrahedral complexes: [Cu(CO)$_4$]$^+$, 2184 cm^{-1}; [Ni(CO)$_4$], 2058 cm^{-1}; [Co(CO)$_4$]$^-$, v_{CO} 1883 cm^{-1}; [Fe(CO)$_4$]$^{2-}$, v_{CO} 1729 cm^{-1}; [Mn(CO)$_4$]$^{3-}$, 1670 cm^{-1}.

4.3.7 Complexes with dinitrogen as a ligand

N$_2$ is isoelectronic with CO, but as it is both a weaker σ donor and a weaker π acceptor than CO it is a relatively poor ligand, and complexes are usually limited to good metal π donors. The N$_2$ stretching mode for free N$_2$ is at 2331 cm^{-1}, and as for the majority of CO complexes, this shifts to lower wavenumber (2220–1850 cm^{-1}) on coordination, due to back-bonding from the metal. N$_2$ usually

binds to the metal end-on η^1, analogous to CO, but both side-on (η^2) and linear bridging geometries between two metals are known.

4.3.8 Complexes with dioxygen as a ligand

The O–O bond length and O–O stretching wavenumbers of compounds containing the O$_2$ unit are given in Table 4.7 and these can be related to the differing population of the antibonding orbitals in O$_2$.

Complexes of O$_2$ are important in biology (haemoglobin, myoglobin, haemerythrin, and haemocyanin) and in catalytic oxidation, and as shown in Table 4.8 these are usually separated into superoxo and peroxo complexes. The superoxo complexes have v_{O-O} in the 1200–1075 cm^{-1} range, similar to that observed for potassium superoxide (KO$_2$) (1108 cm^{-1}). The [O$_2$]$^-$ unit can either be bonded end on or bridging, but in both cases the M–O–O units are bent. The peroxo complexes have v_{O-O} in the 935–750 cm^{-1} range, similar to that observed for sodium peroxide (Na$_2$O$_2$) (760 cm^{-1}), and can be either side-on, or bridging. For example, the v_{O-O} mode in cytochrome P-450 is at 1140 cm^{-1}, indicating the presence of a superoxo complex.

As the O–O stretching mode is essentially a symmetric mode in each case, these are easier to observe by Raman rather than IR spectroscopy. In addition to the position of the v_{O-O} modes, their behaviour under

TABLE 4.7 Geometric and spectroscopic data for compounds containing an O$_2$ unit

	[O$_2$]$^+$[AsF$_6$]$^-$	O$_2$	K$^+$[O$_2$]$^-$	(Na$^+$)$_2$[O$_2$]$^{2-}$
	dioxygenyl	dioxygen	superoxo	peroxo
O–O bond length/Å	1.123	1.207	1.28	1.49
v_{O-O}/cm^{-1}	1858	1555	1108	760
Bond order	2.5	2.0	1.5	1.0

TABLE 4.8 Spectroscopic properties of complexes containing different O$_2$ units

	free O=O	superoxo (end-on, asymmetric) M—O	superoxo (doubly bridging μ$_2$)	peroxo (side-on, symmetric)	peroxo (doubly bridging μ$_2$)
v_{O-O}/cm^{-1}	1555 (IR inactive)	1195–1130	1122–1075	932–800	884–790
O–O bond length / Å	1.207	1.25–1.35	1.26–1.36	1.30–1.55	1.44–1.49
Bond order	2	1.5	1.5	1	1

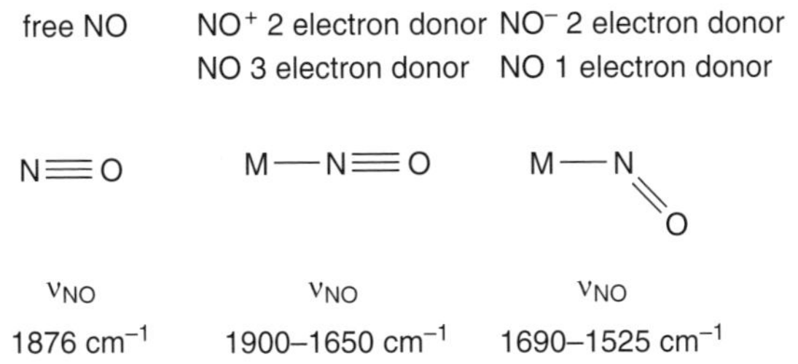

free NO | NO⁺ 2 electron donor NO⁻ 2 electron donor
         | NO 3 electron donor   NO 1 electron donor

N≡O   M—N≡O   M—N
                                   \
                                    O

ν_{NO} ν_{NO} ν_{NO}

1876 cm⁻¹   1900–1650 cm⁻¹   1690–1525 cm⁻¹

FIGURE 4.16 Representative values of ν_{NO} modes in different coordination environments.

isotopic substitution can also differentiate between the end-on and side-on modes. For the $^{16}O^{18}O$ species, the ν_{O-O} mode will be split into two bands for the end-on configuration as the two oxygen atoms are not equivalent, but in the side-on configuration the ν_{O-O} mode in the $^{16}O^{18}O$ species will be one band as the two oxygen atoms are equivalent. (See Problem 4.3 for an example with N_2 complexes.)

4.3.9 Complexes with nitrosyl ligands

Nitric oxide (NO) can act as a σ donor and π acceptor much like CO, except that it is a radical species with 11 valence electrons. When bound as a ligand it can be thought of as NO⁺ (linear, isoelectronic with CO) or as NO⁻ (bent, isoelectronic with O_2) both acting as two electron donors. NO⁺ is a worse σ donor, but better π acceptor than CO. Linear NO⁺ is a better π acid than bent NO⁻. (NO can also be regarded as neutral but as a three electron donor when linear (i.e. donates the lone pair plus single electron), or as a one electron donor when bent.) As the formal N–O bond order, and extent of back-bonding, is different for the linear and bent forms, IR spectroscopy can be used to distinguish between them, as shown in Figure 4.16.

EXAMPLE 4.8

The IR spectrum of solid $[Ir(\eta^3-C_3H_5)(NO)(PPh_3)_2]^+[PF_6]^-$ displays ν_{NO} at 1763 cm⁻¹, while the $[BF_4]^-$ salt shows ν_{NO} at 1631 cm⁻¹. Room temperature IR spectra of solutions of both complexes have bands at 1763 and 1631 cm⁻¹ of equal intensity. On cooling the solutions, the band at 1763 cm⁻¹ increases in intensity, while the 1631 cm⁻¹ band decreases. This process is reversible. Explain this behaviour.

ANSWER

ν_{NO} at 1763 cm⁻¹ indicates linear M–N≡O, whereas ν_{NO} at 1631 cm⁻¹ indicates bent M–N=O; therefore there is an equilibrium between an 18- and a 16-electron complex (Figure 4.17).

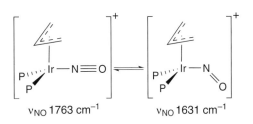

FIGURE 4.17 Isomerisation of $[Ir(\eta^3-C_3H_5)(NO)(PPh_3)_2]^+[PF_6]^-$.

SELF TEST

The ν_{NO} mode in $[RhCl_2(NO)(PPh_3)_2]$ is at 1630 cm⁻¹, in $[CoCl_2(NO)(PPh_3)_2]$ there are two ν_{NO} modes at 1725 and 1640 cm⁻¹. Predict whether the NO ligand is bent or linear in these complexes.

4.3.10 Complexes with cyanide, cyanate, thiocyanate, and selenocyanate ligands

The ν_{CN} mode in cyanide (CN⁻), cyanate (NCO⁻), thiocyanate (NCS⁻), and selenocyanate (NCSe⁻) ligands is between 2200 and 2000 cm⁻¹. CN⁻ behaves analogously to CO in terms of synergistic bonding, but it is a better σ donor, and poorer π acceptor than CO. As for CO, the value of ν_{CN} is dependent on metal oxidation state (e.g. $K_3[Fe(CN)_6]$, 2118 cm⁻¹; $K_4[Fe(CN)_6].3H_2O$, 2044 cm⁻¹).

In addition to identifying the type of functional groups present, IR spectroscopy can also be used to distinguish between different types of bonding in complexes containing **ambidentate ligands** such as SCN⁻ leading to **linkage isomers**. SCN⁻ can bind to the metal via either the S or N atom, and the different coordination modes give rise to different characteristic IR wavenumbers (Table 4.9). For example, in $[Zn(\underline{N}CS)_4]^{2-}$ the ν_{CN} mode is at 2074 cm⁻¹ and the ν_{CS} mode is at 832 cm⁻¹, but in $[Hg(\underline{S}CN)_4]^{2-}$ the ν_{CN} mode is at 2134 cm⁻¹ and the ν_{CS} mode is at 716 cm⁻¹.

4.3.11 Complexes with nitrite ligands

Nitrite, NO_2^-, ligands can also display linkage isomerisation and can bind to a metal either via an oxygen atom (nitrito) or the nitrogen atom (nitro). A classic

TABLE 4.9 Table of representative ν_{CN} and ν_{CS} modes in complexes containing NCS ligands

	M–SCN		M–NCS
	thiocyanato		isothiocyanato
ν_{CN}	ca. 2100 cm⁻¹	>	ca. 2050 cm⁻¹
ν_{CS}	720–690 cm⁻¹	<	860–780 cm⁻¹

example of how IR spectroscopy can be used to discriminate between the nitrito and nitro complexes of cobalt is shown in Figure 4.18. In the nitrito complex, the NO_2^- ligand is bound via one of the oxygen atoms, with the result that the two N-O bonds are different and give rise to two well separated ν_{NO} modes. In contrast, in the nitro complex, the NO_2^- ligand is bound via the central N, so that in this case the two ν_{NO} modes are much more similar as they can be thought

of as symmetric and antisymmetric N–O stretching modes. On standing (or gentle warming) the salmon pink nitrito complex isomerises to the yellow-brown nitro complex, and this can be followed by IR spectroscopy as shown in Figure 4.19. By following the rate of increase or decrease of characteristic peaks it is possible to derive the enthalpy of isomerisation. (The UV-vis spectra associated with this process are shown in Chapter 5, Figure 5.23.)

4.4 Applications of group theory to vibrational spectroscopy

The previous sections indicate that vibrational spectroscopy is a very powerful way of identifying functional groups, and in some cases the mode of bonding. However, with the application of symmetry and group theory it is possible to extract much more information from the spectra, such as the shape of a molecule.

4.4.1 Symmetry operations and symmetry elements

Before considering the symmetry operations found in isolated molecules or ions, it is necessary to differentiate between two similar but different entities, **symmetry operations** and **symmetry elements**. Symmetry operations are motions of an object (e.g. rotations, mirror planes) which carry it into an equivalent configuration, that is, it is not possible to tell the final and initial states apart. Symmetry elements are the axes and planes around which the symmetry operations are carried out. Therefore, the symmetry elements are geometrical constructions such as lines, planes, and points, whereas symmetry operations are the movements of the various parts of the molecule which carry it from one configuration into another indistinguishable configuration. It is the symmetry operations that are most important for molecular symmetry not the elements.

4.4.1.1 Indistinguishable, equivalent, and identical configurations

The essential concept about symmetry operations is that the molecule must have an **indistinguishable, equivalent,** or **identical** configuration before and after a symmetry operation is carried out. This means that the molecular configuration after the symmetry operation cannot be distinguished from the original one. The configuration does not have to be identical

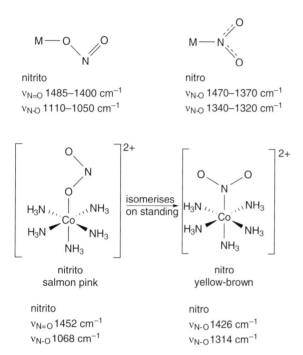

nitrito
$\nu_{N=O}$ 1485–1400 cm^{-1}
ν_{N-O} 1110–1050 cm^{-1}

nitro
ν_{N-O} 1470–1370 cm^{-1}
ν_{N-O} 1340–1320 cm^{-1}

nitrito
salmon pink

nitro
yellow-brown

nitrito
$\nu_{N=O}$ 1452 cm^{-1}
ν_{N-O} 1068 cm^{-1}

nitro
ν_{N-O} 1426 cm^{-1}
ν_{N-O} 1314 cm^{-1}

FIGURE 4.18 Chemical and spectroscopic properties of cobalt nitrito and nitro complexes.

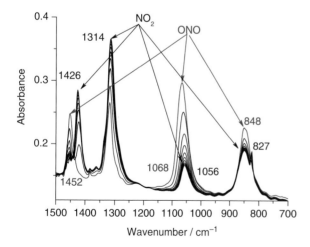

FIGURE 4.19 In-situ IR spectra of isomerisation of [Co(NH$_3$)$_5$(ONO)]Cl$_2$ to [Co(NH$_3$)$_5$(NO$_2$)]Cl$_2$ over a 90 min. period.

as some equivalent parts may have been interchanged. Indistinguishable or equivalent means it looks the same, identical means it is the same. It is now possible to systematically list all the symmetry operations and elements found in molecules and ions. It is important to identify all possible ways of interchanging the atoms. The symmetry operation is given first: then the element. The use of a molecular model kit will significantly enhance the understanding of this section.

4.4.1.2 Proper rotations: axes of symmetry C_n

Proper rotations involve the rotation (normally taken to be clockwise) of the molecule about an axis which results in an indistinguishable or identical configuration. The **rotation axis** may be directed along a bond or between bonds. The proper rotation in H_2O shown in Figure 4.20 involves a 180° rotation about an axis that bisects the two O–H bonds, and both of the hydrogen atoms are interchanged indicated by a change in colour from black to blue. The atom labels are for our convenience, if H_a and H_b were different, the symmetry operation would not exist. Proper rotation axes are given the symbol C_n, where n is an integer and is the number of symmetry operations required to return the molecule to an identical configuration. Thus $360/n$ is the minimum rotation necessary to give an equivalent/indistinguishable configuration. The rotation axis in water is labelled C_2. In BF_3 (Figure 4.20) there are both 120° (C_3) and 180° (C_2) proper rotation symmetry operations, which are also known as threefold and twofold rotation axes, respectively.

Each C_n symmetry element generates nC_n symmetry operations. For H_2O, two consecutive C_2 symmetry operations return the molecule to its starting point. In BF_3 two consecutive C_3 symmetry operations do not return it to its starting point. A rotation such as this is given the symbol C_n^m, where both n and m are integers.

The second 240° rotation in BF_3 is labelled C_3^2. (C_3^2 can also be thought of as a rotation of −120°.) A C_2 axis generates C_2^1, C_2^2 ($\equiv E$; see Section 4.4.1.3), and a C_3 axis generates C_3^1, C_3^2, and C_3^3 ($\equiv E$).

In XeF_4 (Figure 4.20) there is a C_4 symmetry operation where all the fluorine atoms are interchanged. In addition there are three distinct C_2 rotation axes. One labelled C_2 is the C_4^2 symmetry operation, the other two are perpendicular to this axis, one is directed along the Xe–F bonds (C_2'), the other bisects the Xe–F bonds (C_2'').

A special case of rotation is found in linear molecules such as XeF_2 (Figure 4.20), where a rotation of any kind about the internuclear axis results in an equivalent configuration; this is called a C_∞ rotation axis, and the symmetry operation is C_∞.

The highest order rotation axis is called the **principal rotation axis** and this defines the z direction.

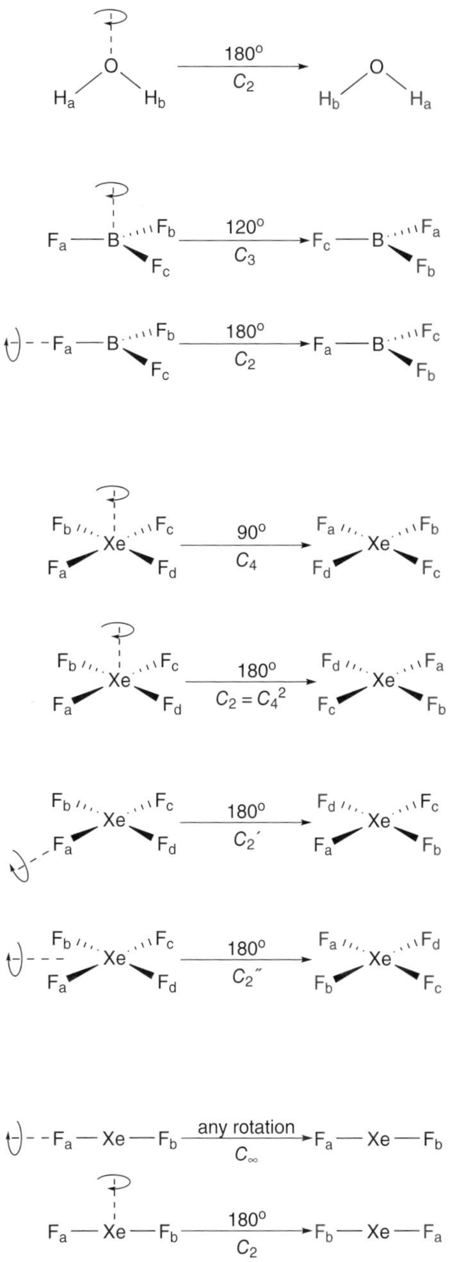

FIGURE 4.20 Examples of proper rotations in H_2O, BF_3, XeF_4, and XeF_2. (Blue atoms are the ones interchanging position.)

EXAMPLE 4.9

Identify and sketch the C_2 axis in cis-$[PtCl_2(NH_3)_2]$.

ANSWER

Although square planar cis-$[PtCl_2(NH_3)_2]$ would normally be drawn in perspective as in the left-hand side of Figure 4.21, it is easier to identify the C_2 axis in the right-hand diagram, which also helps to define the z-axis, which is required when identifying horizontal and vertical directions.

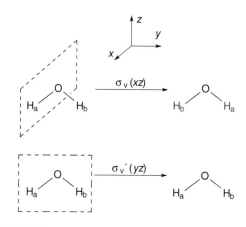

FIGURE 4.21 C_2 axis in cis-$[PtCl_2(NH_3)_2]$.

SELF TEST

Identify and sketch the C_3 and C_2 symmetry operations in tetrahedral $SnCl_4$.

(Hint: look for axes that are along bonds, bisect bonds, or are perpendicular to planes.)

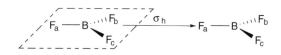

FIGURE 4.22 Vertical mirror planes, σ_v, in H_2O (blue labels indicate interchanged atoms).

4.4.1.3 Identity operation, E

The **identity symmetry operation** is equivalent to either a rotation of 0° or a rotation of 360° about an arbitrary axis (i.e. C_1). For the subsequent analysis it is much better to think of this as a 'do nothing' operation, rather than a 360° rotation. This important operation is called the **identity operation** and has the symbol E.

4.4.1.4 Reflections: symmetry planes, σ

A **plane of symmetry** (element) is often called a **mirror plane**, or a **reflection plane**. This, and the reflection in it (the operation), are both given the symbol σ. If a molecule possesses a mirror plane, it will be bisected by that plane. There are three different, but related, reflection symmetry operations.

4.4.1.5 Vertical mirror planes, σ_v

Water possesses two mirror planes (Figure 4.22), both of which contain the principal (C_2) rotation axis, z. One is in the plane of the molecule (yz), the other perpendicular to it (xz). In this case it is very important to define the coordinate or axis system being used. As the plane of symmetry contains the highest order rotation axis of the molecule it is described as σ_v (v = vertical); the perpendicular plane, which also contains the principal rotation axis, is labelled $\sigma_v{'}$. The ' (prime) designation is not that useful and in practice they are labelled $\sigma_v(xz)$ and $\sigma_v{'}(yz)$.

4.4.1.6. Horizontal mirror planes, σ_h

If a molecule possesses a plane of symmetry that is perpendicular to the highest rotation axis, such as the C_3

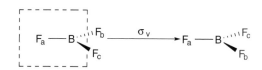

FIGURE 4.23 Horizontal, σ_h, and vertical, σ_v, mirror planes in BF_3 (blue labels indicate interchanged atoms).

axis in BF_3, then it is said to have a horizontal plane of symmetry, σ_h (Figure 4.23). BF_3 also has three σ_v mirror planes directed along each of the B–F bonds and bisecting the remaining F–B–F bond angle.

It is important to note that the 'v' and 'h' labelling refers to the relationship between the planes and the principal rotation axis z, and not necessarily to the plane of the molecule.

4.4.1.7 Dihedral mirror planes, σ_d

The final type of symmetry plane contains the main rotation axis, but it is different from the σ_v planes. These are known as dihedral planes, and are labelled σ_d. In general, they bisect the σ_v planes, but to be completely accurate they have to bisect two C_2 axes (except in the C_{4v} point group). Square planar molecules such as XeF_4 contain σ_v, σ_h, and σ_d planes, as shown in Figure 4.24.

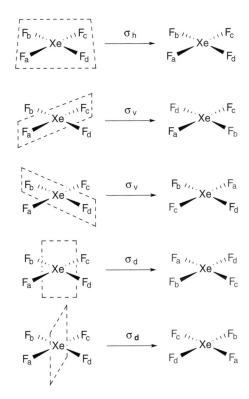

FIGURE 4.24 Horizontal, σ_h, vertical, σ_v, and dihedral, σ_d, mirror planes in XeF_4 (blue labels indicate interchanged atoms).

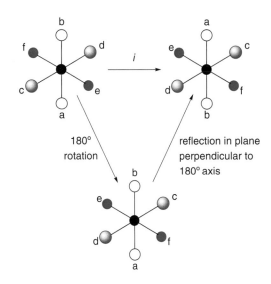

FIGURE 4.25 Inversion centre in octahedral complexes.

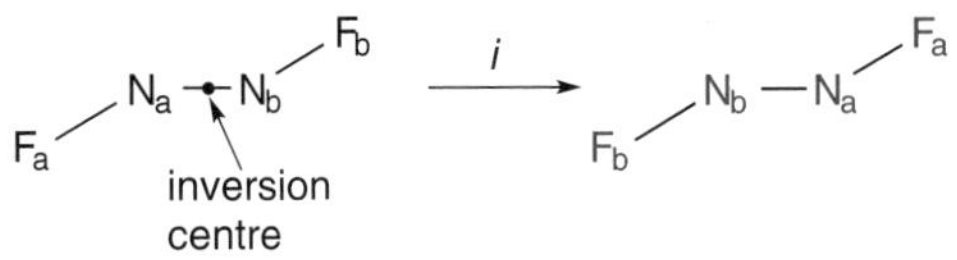

FIGURE 4.26 Inversion centre in *trans*-N_2F_2 (blue labels indicate interchanged atoms).

If a molecule possesses a plane of symmetry, but does not have a rotation axis (other than $C_1 \equiv E$) then the symbol for this plane is σ_h. (This would be the case in a partially deuterated water molecule, HOD). It should also be noted that $\sigma^2 \equiv E$, that is, a second reflection returns the molecule to its starting point.

Rotations and reflections are single symmetry operations. Two slightly more complex symmetry elements and their operations now need to be considered: inversion and improper rotations.

4.4.1.8 Inversion: centre of symmetry, *i*

An inversion symmetry operation converts the *x*, *y*, *z* coordinates for each atom to −*x*, −*y*, −*z*, as shown in Figure 4.25.

Each pair of atoms is related to each other through one point in space, called the **inversion centre** or **centre of symmetry**. The symmetry operation is the inversion. Both the operation and element are given the symbol *i*. If a molecule possesses an inversion centre it is said to be **centrosymmetric**. In Figure 4.25, the inversion centre is located at the black atom at the centre of the octahedron. Carrying out two inversions returns us to an identical configuration (i.e. $i^2 \equiv E$).

The inversion centre, *i*, does not have to be situated at an atom, as in octahedral systems as above, but can be in space (e.g. the centre of a benzene ring), or in the middle of a bond as in *trans*-N_2F_2 (Figure 4.26).

4.4.1.9 Improper rotations: rotation–reflection axes

Figure 4.25 also shows that inversion can be thought of as a 180° rotation, followed by reflection in the plane perpendicular to this rotation axis. This is an example of the final symmetry operation that is required to describe the symmetry properties of molecules, and is known as an **improper rotation**. Improper rotations consist of a rotation about an axis, followed by reflection in a plane perpendicular to it. The symmetry operation is the combined action; the individual rotations and reflections may or may not be also present. Improper rotations are given the symbol S_n.

In tetrahedral molecules there is an S_4 improper rotation symmetry operation (Figure 4.27).

In this case there are no separate C_4 or σ_h symmetry operations. S_4 symmetry operations need to be included

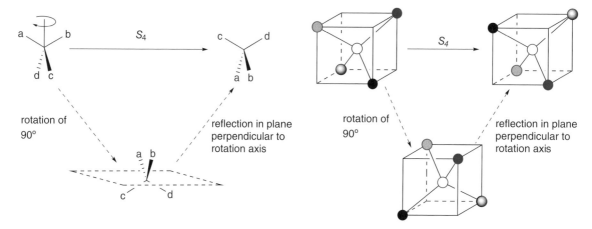

FIGURE 4.27 S_4 symmetry operation in tetrahedral molecules.

for tetrahedral molecules as they are the only symmetry operations that interconvert all of the ligand atoms. In BF_3 the S_3 and in XeF_4 the S_4 improper rotation symmetry operations are often 'masked' by the C_3 and C_4 axes and the σ_h mirror planes that are also present in these molecules. S_2 is equivalent to an inversion. S_1 is a 360° rotation followed by σ and is therefore equivalent to σ.

Improper rotations are important symmetry operations to identify as a molecule cannot be chiral if it possesses an improper rotation axis, S_n.

As for the proper rotation axes, C_n, the S_n improper rotation axes give rise to more than one distinct symmetry operation, and the number is dependent on whether S_n is odd or even. If n is even, there are n symmetry operations S_n^1, S_n^2, ... S_n^n. For S_4 there are S_4^1, S_4^2 ($\equiv C_2^1$), S_4^3, and S_4^4 ($\equiv E$). Therefore, the presence of an S_n axis with even n means that a $C_{n/2}$ axis must also be present. If n is odd there are $2n$ symmetry operations because S_n^n is equivalent to σ_h and not E, which is equivalent to S_n^{2n}. For example, for S_5 there are S_5^1, S_5^2 ($\equiv C_5^2$), S_5^3, S_5^4($\equiv C_5^4$), S_5^5 ($\equiv \sigma_h$), S_5^6 ($\equiv C_5^1$), S_5^7, S_5^8 ($\equiv C_5^3$), S_5^9, and S_5^{10} ($\equiv E$). Therefore, the presence of an S_n axis with odd n means that a C_n axis and a σ_h plane must also be present.

EXAMPLE 4.10

Sketch the effect of the E, C_3, and σ_v symmetry operations on a trigonal pyramidal molecule such as NH_3.

ANSWER

Figure 4.28 shows the effect of the symmetry operations. It needs to be remembered that the a, b, and c labels are just for convenience.

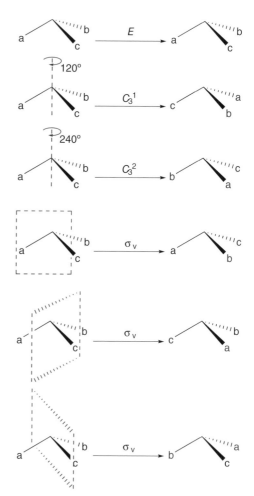

FIGURE 4.28 Effect of the C_{3v} symmetry operations on a trigonal pyramidal molecule (blue labels indicate interchanged atoms).

SELF TEST

Sketch the effect of the E, C_3, C_2, S_4, and σ_d symmetry operations for tetrahedral molecules such as $SnCl_4$ and the E, C_3, C_2, S_3, σ_h, and σ_v symmetry operations for trigonal bipyramidal molecules such as PF_5.

4.4.2 Point groups

The particular symmetry operations that a molecule possesses can have profound consequences on its spectroscopic properties. While the molecule can be completely specified by listing all of the symmetry elements it possesses (e.g. E, C_n, σ, i, and S_n) it is much more convenient to summarise these in what is known as the **point group** of the molecule.

Water and other apparently diverse molecules, such as ClF_3, SO_2Cl_2, and *cis*-N_2F_2, all possess the same set of symmetry elements. They are said to belong to the same C_{2v} point group. The point group of *trans*-N_2F_2 is C_{2h}. Every molecule can be assigned to a point group according to its symmetry elements. They are called point groups because there is always one point that remains unmoved by all of the symmetry operations; this point may be at an atom, in the middle of a bond, or in free space. Point group labels are the shorthand used by chemists to describe molecular shape. For example, saying that BF_3 has D_{3h} point group is the same as saying that it is trigonal planar.

Table 4.10 below shows the most important classes of point groups and their characteristic symmetry elements.

TABLE 4.10 Classes of point groups and characteristic symmetry elements

Point group	Characteristic symmetry elements
C_s	E and one σ only
C_i	E and a centre of inversion i only
C_n	E and one C_n axis
C_{nv}	E, one C_n axis, and n σ_v planes
C_{nh}	E, one C_n axis, one σ_h plane, and an inversion centre i
D_{nh}	E, one C_n axis, C_2 axis $\perp$ to C_n, and one σ_h plane
D_{nd}	E, one C_n axis, C_2 axis $\perp$ to C_n, and n σ_v planes
'Special' point groups	
$D_{\infty h}$	Linear molecule with a centre of inversion
$C_{\infty v}$	Linear molecule without a centre of inversion
T_d	Tetrahedron
O_h	Octahedron
I_h	Icosahedron

Note that, except where stated, only the most easily identified elements are listed.

The point group of a molecule can always be identified by listing all of the symmetry elements present in a molecule and using a table such as that above. However, a more convenient way of assigning point groups is to use a flowchart such as in Figure 4.29.

The flowchart is followed from the top by answering 'yes' or 'no' to the questions at each branch. The n in the point group label is the value of the principal rotation axis in the second question.

4.4.3 Determining point groups

4.4.3.1 'Special groups'

The first query in the flow diagram asks whether the molecule belongs to a 'special' group. There are two types of special groups, linear molecules and high symmetry molecules. These are dealt with at the start as they are easy to identify.

4.4.3.1.1 Linear point groups

There are two linear point groups, $D_{\infty h}$ and $C_{\infty v}$. $D_{\infty h}$ is for molecules with a centre of symmetry, such as H_2, N_2, and CO_2, and $C_{\infty v}$ is for those without a centre of symmetry, including HF, CO, and OCS (Figure 4.30).

EXAMPLE 4.11

Assign the following linear molecules to their point groups: XeF_2 and HBr.

ANSWER

XeF_2 is linear with a centre of symmetry so belongs to the $D_{\infty h}$ point group, but HBr has no centre of symmetry so has the $C_{\infty v}$ point group.

SELF TEST

Assign the following linear molecules to their point groups: N_2O, NO, HC≡CH, C_3O_2 (O=C=C=C=O).

4.4.3.1.2 High symmetry or cubic point groups

There are three high symmetry point groups; two of these, the tetrahedron, T_d, and the octahedron, O_h, are very common, but the icosahedron, I_h, is much rarer. These are also known as **cubic groups**, as the x, y, and z directions are equivalent.

Tetrahedral coordination of a central atom can be constructed by placing four ligand atoms on opposing corners of a cube, as shown in Figure 4.31 for GeH_4 and $[MnO_4]^-$. A T_d molecule does not need to have an atom at the centre, for example, P_4 (white phosphorus).

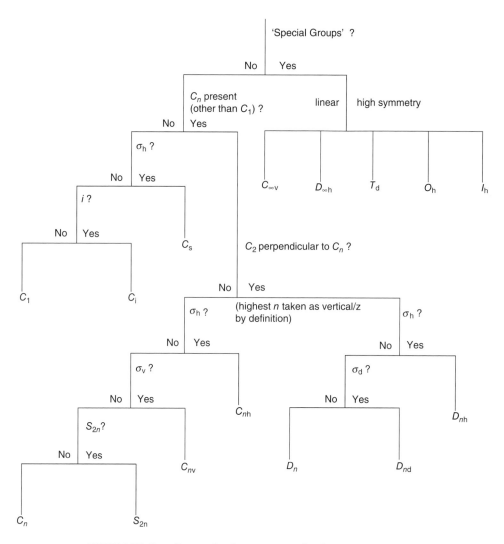

FIGURE 4.29 Flow diagram for determining molecular point groups.

H——H   H——F

N≡≡≡N   C≡≡≡O

O══C══O   O══C══S

$D_{\infty h}$ $C_{\infty v}$

FIGURE 4.30 Examples of molecules having $D_{\infty h}$ and $C_{\infty v}$ point groups.

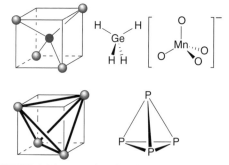

FIGURE 4.31 Examples of molecules and ions with T_d point group.

The tetrahedron is so called because it has four faces. If there is an atom at the centre this is also four-coordinate. An octahedron has eight faces, but the central atom, such as in SF_6 and $[Ni(H_2O)_6]^{2+}$, is six-coordinate and the molecule or ion belongs to the O_h point group (Figure 4.32). $[Ni(H_2O)_6]^{2+}$ is an example where the ligand is taken as a single object, so the orientation effect of the hydrogen atoms is assumed to be small (and is hence usually ignored). This is also the case for NH_3, CH_3 units, etc. As in the tetrahedral case, there is no requirement

for an atom at the centre of the polyhedron, for example, $[B_6H_6]^{2-}$.

When dealing with tetrahedral and octahedral complexes, it is very important to distinguish between whether a compound is strictly octahedral (or

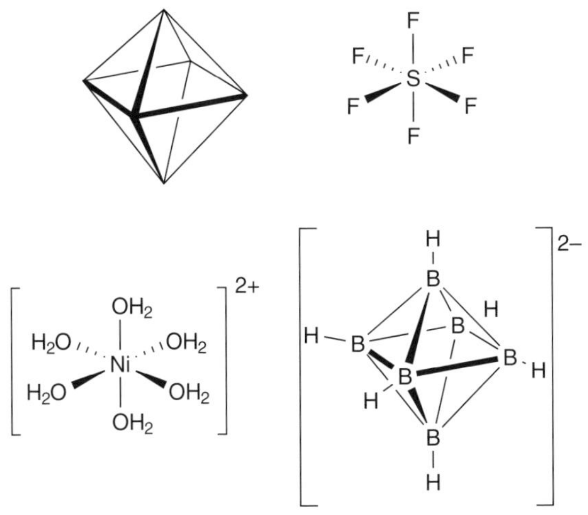

FIGURE 4.32 Examples of molecules and ions with O_h point group.

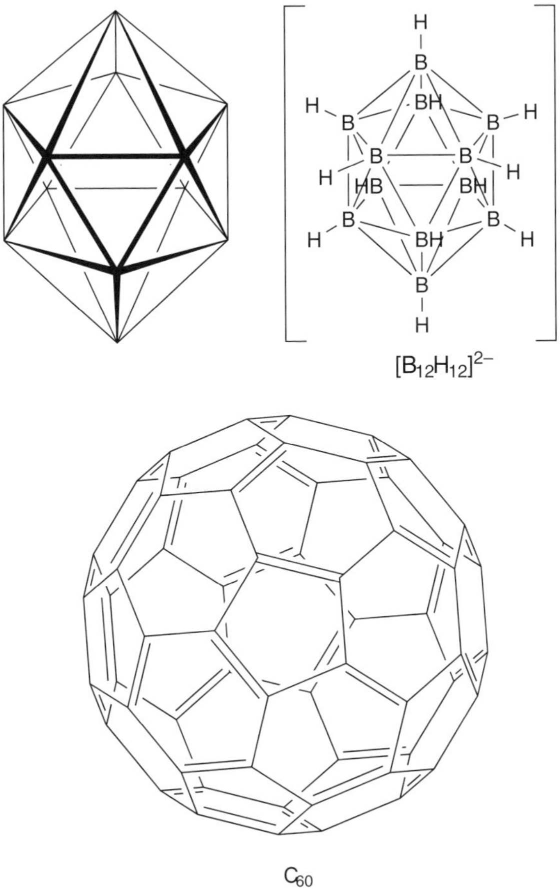

FIGURE 4.33 Examples of molecules and ions with I_h point group.

tetrahedral) or has an overall octahedral shape, but with lower symmetry (e.g. $ML_4L'_2$ or $ML_2L'_2$, where L and L′ are different ligands).

The final high symmetry or cubic point group is I_h, which is exemplified by an icosahedron. This is a fairly rare point group, but examples include $[B_{12}H_{12}]^{2-}$ and C_{60} (Figure 4.33).

4.4.3.2 Other common point groups

The use of a molecular model kit is beneficial for this section where two commonly occurring point groups are identified through examples.

Representative examples of the common point groups are given in Figure 4.34.

EXAMPLE 4.12

Determine the point group of water.

ANSWER

Starting at the top of the flow diagram in Figure 4.29, H_2O does not belong to a special group as it is neither linear nor cubic. The next question is whether there is a C_n rotation axis, and in this case water can be rotated 180° to give an indistinguishable configuration, therefore there is a C_2 axis. The next question asks whether there are any perpendicular C_2 axes, and as there is none in H_2O the left-hand branch is selected. As the C_2 axis defines the *z* direction, there is no horizontal (σ_h) mirror plane, but there are two vertical (σ_v) mirror planes. Therefore, the point group of water is C_{2v}.

SELF TEST

Determine the point groups of (a) NH_3, (b) SO_2Cl_2, (c) $XeOF_4$, and (d) *cis*-$[PtCl_2(NH_3)_2]$.

EXAMPLE 4.13

Determine the point group of trigonal planar BF_3.

ANSWER

Using the flow diagram (Figure 4.29), the point group of BF_3 can be determined as D_{3h} because it does not belong to a special group, it has a C_3 axis, there are C_2 axes perpendicular to the C_3 principal rotation axis, and there is a horizontal (σ_h) mirror plane. The key difference between *C* and *D* point groups is that the presence of at least one C_2 axis perpendicular to the principal rotation axis results in *D* point groups.

SELF TEST

Determine the point groups of (a) $[PtCl_4]^{2-}$, (b) *trans*-$[PtCl_2(NH_3)_2]$, (c) XeF_4, and (d) $[XeF_5]^-$.

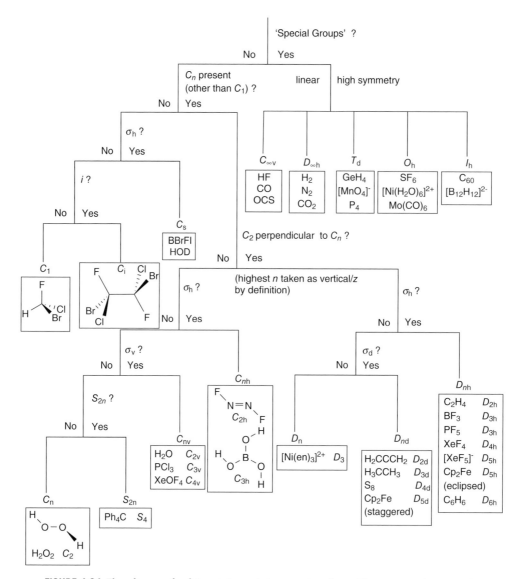

FIGURE 4.34 Flow diagram for determining point groups, together with representative examples.

4.4.4 Character tables

Associated with each point group is a '**character table**' which contains all of the symmetry operations as well as the information needed to work out the properties that depend on the molecular symmetry. Once the point group has been determined it can be confirmed by checking that all of the symmetry operations present in the table are in the molecule and vice versa. The character tables for D_{3h} and D_{4h} are shown in Tables 4.11 and 4.12; others are available in Appendix 1.

The point group symbols (D_{3h} and D_{4h}) in the top left-hand corner are known as **Schönflies symbols**, and these are used in the majority of group theory applications, except crystallography, where the **international system** is used. The symmetry operations for the point group (E, C_3, etc.) are given in the top row. The numbers below the symmetry operations are called 'characters' and have the symbol χ (chi). These represent the effect of the symmetry operations on the **basis set** chosen, for example, for a vibrational mode +1 means that it remains the same and −1 that it is reversed. The penultimate column at the right gives important information about symmetry properties with x, y, z vectors (e.g. dipole moments, p orbitals) as well as rotations (R_x, R_y, R_z). The final column provides information about binary combinations (polarisability, d orbitals). Some character tables include ternary combinations, such as xyz etc. used to describe f orbitals.

The A, B, E, T labels on the left-hand side are **Mulliken symbols** and these are called the **irreducible**

TABLE 4.11 D_{3h} character table

D_{3h}	E	$2C_3$	$3C_2$	σ_h	$2S_3$	$3\sigma_v$		
A_1'	1	1	1	1	1	1		x^2+y^2, z^2
A_2'	1	1	−1	1	1	−1	R_z	
E'	2	−1	0	2	−1	0	(x, y)	(x^2-y^2, xy)
A_1''	1	1	1	−1	−1	−1		
A_2''	1	1	−1	−1	−1	1	z	
E''	2	−1	0	−2	1	0	(R_x, R_y)	(xz, yz)

TABLE 4.12 D_{4h} character table

D_{4h}	E	$2C_4$	C_2	$2C_2'$	$2C_2''$	i	$2S_4$	σ_h	$2\sigma_v$	$2\sigma_d$		
A_{1g}	1	1	1	1	1	1	1	1	1	1		x^2+y^2, z^2
A_{2g}	1	1	1	−1	−1	1	1	1	−1	−1	R_z	
B_{1g}	1	−1	1	1	−1	1	−1	1	1	−1		x^2-y^2
B_{2g}	1	−1	1	−1	1	1	−1	1	−1	1		xy
E_g	2	0	−2	0	0	2	0	−2	0	0	(R_x, R_y)	(xz, yz)
A_{1u}	1	1	1	1	1	−1	−1	−1	−1	−1		
A_{2u}	1	1	1	−1	−1	−1	−1	−1	1	1	z	
B_{1u}	1	−1	1	1	−1	−1	1	−1	−1	1		
B_{2u}	1	−1	1	−1	1	−1	1	−1	1	−1		
E_u	2	0	−2	0	0	−2	0	2	0	0	(x, y)	

representations for each point group. The vibrational modes can be described in terms of the A, B, E, and T representations, as they contain information in a very concise manner. All one-dimensional (**non-degenerate**) irreducible representations are given the symbol A or B. Two-dimensional (**doubly degenerate**) irreducible representations are labelled E (not to be confused with the identity operation, *E*). Three-dimensional (**triply degenerate**) irreducible representations are labelled T (F is used in some old tables, books, and articles). Four and five dimensions are labelled G and H, respectively. These labels are sometimes referred to as **symmetry species** or **symmetry labels**. In general, at least a threefold rotation axis must be present for degenerate representations to exist.

One-dimensional irreducible representations are labelled A if they are symmetric with respect to rotation about the highest order C_n axis, that is, the character for the C_n symmetry operation is +1. If it is antisymmetric with respect to rotation about C_n, that is, $\chi = -1$ for C_n, then the irreducible representation is B. A subscript 1, as in A_1, means that the irreducible representation is symmetric

with respect to rotation about a C_2 axis perpendicular to C_n, or in the absence of such an axis, to reflection in a σ_v plane perpendicular to the molecule. An antisymmetric representation is, therefore, given a subscript 2, for example, A_2. If the molecule is centrosymmetric, a subscript g is added if the irreducible representation is symmetric with respect to inversion, and a u subscript is used if it is antisymmetric with respect to inversion. If the molecule has a horizontal mirror plane σ_h, ′ (prime) and ″ (double prime) are used if the irreducible representation is symmetric and antisymmetric, respectively, with respect to reflection in the σ_h plane.

4.4.5 A more rigorous approach to selection rules

4.4.5.1 IR selection rule

In Section 4.1.2.1 it was stated that for a vibrational mode to be IR active there must be a change in the dipole moment (μ) of the molecule during the vibration. (The molecule does not need a permanent dipole moment.) With the use of group theory this statement can

TABLE 4.13 C_{2v} character table

C_{2v}	E	C_2	$\sigma_v\,(xz)$	$\sigma_v'(yz)$		
A_1	1	1	1	1	z	x^2, y^2, z^2
A_2	1	1	−1	−1	R_z	xy
B_1	1	−1	1	−1	x, R_y	xz
B_2	1	−1	−1	1	y, R_x	yz

be applied more rigorously. As the change in the dipole moment can be resolved into its x, y, and z components, these will transform in exactly the same way as a vector in the x, y, or z direction (or as a p_x, p_y, or p_z orbital). This information is easily extracted from character tables by simply looking for the x, y, or z entries (sometimes given as T_x, T_y, T_z) in the penultimate column.

In C_{2v} molecules such as water, *cis*-N_2F_2, and CH_2Cl_2, Table 4.13 shows that vibrational modes having A_1, B_1, or B_2 symmetry will be IR active as these irreducible representations have a z, an x, or a y in the penultimate column. As there is only x, y, and z, it follows that there can never be more than three irreducible representations which give rise to IR active vibrational modes. This does not mean that there cannot be more than three IR active modes, as there may be several modes with the same symmetry label.

4.4.5.2 Raman selection rule

Section 4.1.2.2 stated that for a vibrational mode to be active in the Raman effect there must be a change in the polarisability (α) of the molecule during the vibration. Just as with the IR selection rule, group theory readily allows for the identification of changes in polarisability. The polarisability is related to the ease with which the electron density in the molecule can be deformed. As the polarisability can be thought of as a sphere or ellipsoid, it is given by a polarisability tensor (this is just a 3×3 matrix), which is made up of terms of the form α_{xx}, α_{xy}, α_{xz}, α_{yx}, α_{yy}, α_{yz}, α_{zx}, α_{zy}, and α_{zz}:

$$\begin{bmatrix} \alpha_{xx} & \alpha_{xy} & \alpha_{xz} \\ \alpha_{yx} & \alpha_{yy} & \alpha_{yz} \\ \alpha_{zx} & \alpha_{zy} & \alpha_{zz} \end{bmatrix}$$

These correspond to the binary products such as xy or z^2 in the last column of the character tables. For C_{2v} molecules, vibrational modes having A_1, A_2, B_1, or B_2 symmetry will be Raman active, as they all have entries in the penultimate column. Because there are more binary combinations, there are more possible Raman active irreducible representations than IR active ones. If a polarised excitation source (e.g. a laser) is used for the Raman experiment, the only mode that retains this polarisation information is the **totally symmetric representation**, that is, the top one in the character table. All the rest will be **depolarised**. This is easy to measure experimentally by placing a piece of Polaroid in front of the detector, and noting any change in intensity when it is rotated from parallel ($I_{\parallel}$) to perpendicular ($I_{\perp}$) with respect to the incident laser polarisation (see Section 4.5.1). A large change indicates a polarised totally symmetric mode. The other depolarised modes only show a modest reduction of three-quarters in intensity on going from $I_{\parallel}$ to $I_{\perp}$ (see Section 4.5.1 for further details). As an example for the Raman active vibrations of a molecule having C_{2v} symmetry, the A_1 modes will be **polarised** (p), the others (A_2, B_1, B_2) will be depolarised (dp).

4.4.6 Intensities

While group theory can tell us whether a vibrational mode is IR or Raman active, it is unable to tell us anything about the relative intensities directly. However, in general, the more symmetric modes are the most intense in Raman spectra, while the more asymmetric modes are more intense in the IR spectra.

4.4.7 Mutual exclusion principle

The **mutual exclusion principle**, which is the extreme of the case in Section 4.4.6, states that if a molecule has a centre of symmetry (inversion centre), i, then no vibrational mode can be both IR and Raman active. This is because IR active modes will be drawn from the u representations, and the Raman active modes from the g representations. This can be useful in assigning point groups, and hence molecular shapes. For example, if there is an IR and a Raman band at essentially the same wavenumber, then there is almost certainly no centre of symmetry present. While the absence of evidence is not the evidence of absence, the lack of any coincident IR and Raman features is a good indication of the presence of an inversion centre.

EXAMPLE 4.14

The IR spectrum of CO_2 has bands at 2349 cm^{-1} and 667 cm^{-1} due to $\nu_{C=O}$ and $\delta_{O=C=O}$ modes, respectively. The Raman spectrum only has a symmetric stretching mode at 1337 cm^{-1}. Use the mutual exclusion principle to identify whether CO_2 is linear or bent.

ANSWER

As there are no bands due to $\nu_{C=O}$ stretching modes that are coincident in the IR and Raman spectrum, the mutual exclusion principle indicates that CO_2 is linear.

SELF TEST

Use the data in Table 4.14 and the mutual exclusion principle to identify whether the molecules are linear or bent.

TABLE 4.14

Molecule	Vibrational mode	IR/cm^{-1}	Raman/cm^{-1}
XeF_2	ν_{Xe-F}	555	497
O_3	ν_{O-O}	1135, 1089	1135, 1089
BrOBr	ν_{Br-O}	587, 504	587, 504

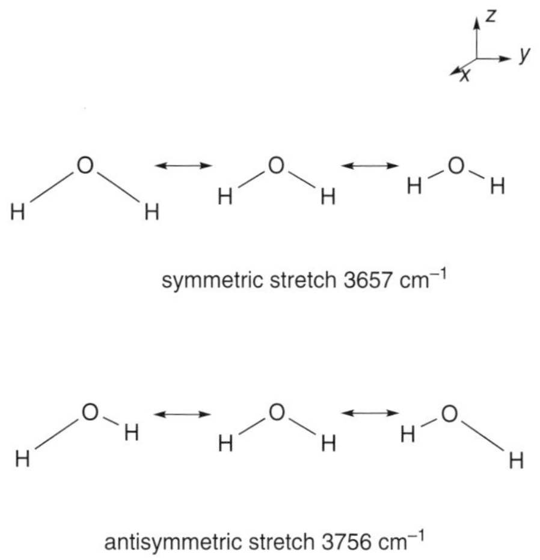

4.4.8 Reducible and irreducible representations

In order to understand what the A_1 and B_2 labels mean it is necessary to identify how atomic displacements can be associated with them. When the two O–H bonds in water

symmetric stretch 3657 cm^{-1}

antisymmetric stretch 3756 cm^{-1}

FIGURE 4.35 Representation of the O–H stretching modes (ν_{O-H}) in H_2O.

vibrate, their motions are coupled together in two ways so that they either both extend and compress in phase to give a symmetric stretching mode, or while one is extending, the other is contracting to give an antisymmetric stretching mode (sometimes also known as an asymmetric stretching mode). These are shown in Figure 4.35.

The effect of each of the C_{2v} symmetry operations on these vibrations when the molecule is frozen at some point in the vibration (but not at the mid-point) is now considered. Each of the symmetry operations is performed on this frozen molecule, as shown in Figure 4.36. The effect of the operation is represented by +1 if the molecule is completely unchanged and by −1 if the molecule is reversed.

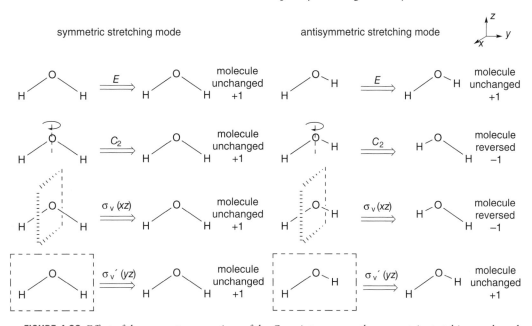

FIGURE 4.36 Effect of the symmetry operations of the C_{2v} point group on the symmetric stretching mode and the antisymmetric stretching mode of H_2O.

TABLE 4.15 Identification of irreducible representations for the O–H stretching modes in water

	E	C_2	$\sigma_v(xz)$	$\sigma_v'(yz)$	Irreducible representation (symmetry label)
Symmetric stretch	1	1	1	1	A_1
Asymmetric stretch	1	−1	−1	1	B_2

This collection of numbers or '**characters**' is then compared to those in the C_{2v} character table (Table 4.13) to identify the symmetry label (irreducible representation) for the vibration (Table 4.15).

The symmetric O–H stretching mode is said to 'span' or 'transform' as A_1 and the antisymmetric stretch as B_2. Alternatively, the symmetric stretch has A_1 symmetry and the antisymmetric stretch has B_2 symmetry. (If the molecule is put in the xz plane, instead of the yz plane, the irreducible representation becomes $A_1 + B_1$.)

> **NOTE ON GOOD PRACTICE**
>
> By convention the symmetry labels for vibrational modes are always written in either all upper case, as $A_1 + B_2$ or all lower case as in $a_1 + b_2$ (but not a mixture of cases).

As individual stretching modes behave exactly the same as the bonds, the number of 'unshifted' O–H bonds labelled 1 and 2 can be used. The effect of the symmetry operations on the two individual O–H stretches/bonds is shown in Figure 4.37. If an O–H bond is unshifted it contributes +1, but if it moves it contributes 0.

For the two stretches, the characters are given in Table 4.16. There is no row of the C_{2v} character table (Table 4.13) with these numbers, but they are those obtained from the sum of the symmetric (A_1) and antisymmetric (B_2) values (Table 4.17).

This is a unique combination as addition of any other rows does not lead to the same set of numbers. Therefore, the symmetry of the vibrational modes can be worked out by calculating the number of stretching modes (or bonds) that are unshifted by the symmetry operations and identifying which rows of the character table add together to give these numbers.

When the characters generated by a set of objects correspond to a row in the character table, the objects are said to span an **irreducible representation** of the point group, and this has the symbol, Γ, and in this case Γ_{O-H}. The symmetric stretch spans the A_1 irreducible representation. The antisymmetric stretch spans the B_2 irreducible representation. When the characters generated by a set of objects correspond to a sum of rows in the character table, the objects are said to span

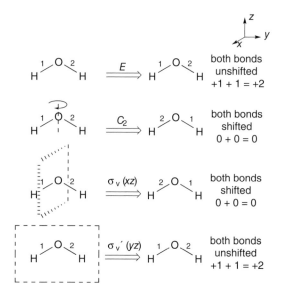

FIGURE 4.37 Effect of the C_{2v} symmetry operations on the two O–H stretches in water.

TABLE 4.16 Number of unshifted O–H stretching modes in H_2O

	E	C_2	$\sigma_v(xz)$	$\sigma_v'(yz)$
Number of unshifted O–H stretching modes	2	0	0	2

TABLE 4.17 Summation of characters from symmetric (A_1) and antisymmetric (B_2) stretching modes

	E	C_2	$\sigma_v(xz)$	$\sigma_v'(yz)$
A_1 symmetric	1	1	1	1
B_2 antisymmetric	1	−1	−1	1
Two O–H bonds	2	0	0	2

a **reducible representation** which consists of two or more irreducible representations. Reducible representations also have the symbol, Γ, which in this case is Γ_{O-H} as well.

Combining this information with Section 4.4.5, it can be seen that once group theory has identified the symmetry characteristics of the vibrational modes, it is then possible to identify which modes will be IR active, and which will be Raman active simply by consulting the character tables. From the C_{2v} character table (Table 4.13), the A_1 symmetric stretch in H_2O is associated with a z in the penultimate column so is IR active, and x^2, y^2, z^2 in the final column so it is also Raman active. The B_2 antisymmetric stretching mode has a y in the penultimate column and yz in the final column, so it is also both IR and Raman active.

4.4.9 Determining reducible and irreducible representations for stretching modes, Γ_{str}

Section 4.4.8 demonstrated what the terminology A_1 and B_2 represents and how they are generated for a simple case of an H_2O molecule. This procedure can now be extended to more complex molecules and ions.

The basis of the method is to take something that can be visualised and has some chemical significance such as stretching modes, atoms or orbitals and see how these are affected by the symmetry operations of the point group to get a reducible representation. This information can be converted into very concise, accurate irreducible representations consisting of the Mulliken symbols (A_1, E', etc.) and these can be used to yield important information about IR and Raman vibrational activity.

4.4.9.1 Pyramidal XY_3 molecules

As an example the symmetry properties of the X–Y stretching modes in a pyramidal XY_3 molecule (e.g. NH_3, PF_3, $[SO_3]^{2-}$, $[ClO_3]^-$) will be considered. The first stage in the process is to identify how many of the X–Y stretching modes remain unshifted under the symmetry operations of the C_{3v} point group shown in Table 4.18.

For C_{3v} the symmetry operations are E, $2C_3$, and $3\sigma_v$. Under the identity, E, all three X–Y stretching modes are unshifted (Figure 4.38); under the threefold rotation axis, C_3, all three move, so zero are unshifted (Figure 4.38). For the σ_v mirror planes directed along each X–Y bond, one of

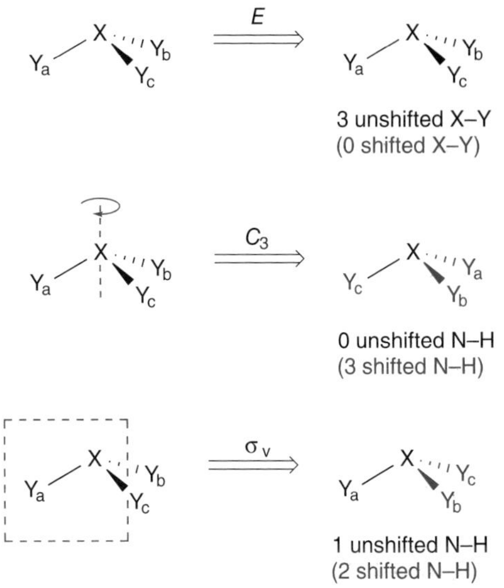

FIGURE 4.38 Determining the reducible representation, Γ_{X-Y}, for an XY_3 molecule.

the X–Y stretching modes is unshifted, while two interchange, as shown in Figure 4.38. (It is the number of X–Y stretching modes that are unshifted per symmetry operation (i.e. 1 for σ_v) that is required; the fact that there is more than one of them (i.e. $3\sigma_v$) is taken into account later.) This generates the reducible representation, Γ_{str} or Γ_{X-Y}.

At this stage, a reducible representation, Γ_{X-Y}, has been generated and this now needs to be broken down (reduced) into its irreducible representations.

4.4.9.1.1 Pyramidal XY_3 molecules—by reduction formula

The fail-safe method of obtaining an irreducible representation from the reducible representation is to use the reduction formula:

$$n_i = \frac{1}{h}\sum_c g_c \chi_r \chi_i \qquad \textbf{Eqn 4.5}$$

where

n_i is number of times the irreducible representation i occurs in the reducible representation;

h is the order of the point group, that is, total number of symmetry operations present;

TABLE 4.18 C_{3v} character table

C_{3v}	E	$2C_3$	$3\sigma_v$		
A_1	1	1	1	z	$x^2 + y^2, z^2$
A_2	1	1	−1	R_z	
E	2	−1	0	$(x, y), (R_x, R_y)$	$(x^2 - y^2, xy), (xz, yz)$

g_c is the order of the symmetry class (the number in front of the symmetry operation in the top row of the character table);

χ_r are the characters of the reducible representation;

χ_i are the characters of the irreducible representation.

In this case, $h = 6$ as there are six symmetry operations in the C_{3v} point group ($E + 2C_3 + 3\sigma_v$). g_c is the number of symmetry operations in each symmetry class, which for the C_{3v} point group is 1 for E, 2 for C_3, and 3 for σ_v. χ_r are the 3, 0, 1 values determined by considering the number of unshifted X–Y stretching modes in Figure 4.38. χ_i are the characters in the centre of the character tables, that is, 1, –1, 2, 0. Table 4.19 shows the reducible representation, Γ_{X-Y}, followed by the analysis using the reduction formula using the C_{3v} character table.

Therefore, Γ_{str}, $\Gamma_{X-Y} = A_1 + E$.

4.4.9.1.2 Pyramidal XY$_3$ molecules—tabular method

The only values that changed between the rows in the calculation above were those for χ_i. Based on this observation, there is an easier way of carrying out this exercise known as the tabular method, where the structure of the character table is used to help keep track of the arithmetic. This can be achieved either by working across each row, or down each column. The former is more physically correct, but the latter is usually easier. This is summarised in Table 4.20 and is the recommended method.

Therefore, Γ_{str}, $\Gamma_{X-Y} = A_1 + E$.

4.4.9.1.3 Pyramidal XY$_3$ molecules—'by eye'

For small, high symmetry molecules such as H_2O, NH_3, BF_3, and SiF_4 there is a quick and easy way to determine the irreducible representation using the 'by-eye' method. The key initial step is to assume that there will be at least one totally symmetric representation present. That is, every molecule or ion has as a **breathing mode**, where all the same bonds are getting longer (or shorter) in phase. This will always be the top entry in the character table, that is, A_1 in the case of C_{3v} point group, and will have characters of 1 for each symmetry operation, that is, whichever symmetry operation is applied it looks the same before and after. These values of 1 are then subtracted from the reducible representation to give another representation, which may or may not correspond to one of the other irreducible representations in the character table. It is simply a case of looking up in the character table what the residual representation is. This is shown for XY$_3$, C_{3v} in Table 4.21. In this case the residual values correspond to E. Therefore, as before, $\Gamma_{str} = A_1 + E$ for pyramidal XY$_3$ molecules.

TABLE 4.19 Determination of Γ_{X-Y} for XY$_3$ using the formula method

C_{3v}	$1E$	$2C_3$	$3\sigma_v$		
Γ_{str}, Γ_{X-Y} (no. of unshifted X–Y stretches)	3	0	1		
A_1	1	1	1	z	$x^2 + y^2$, z^2
A_2	1	1	–1	R_z	
E	2	–1	0	(x, y), (R_x, R_y)	$(x^2 - y^2, xy)$, (xz, yz)

$n(A_1) = (1/6)\sum\{(1 \times 3 \times 1) + (2 \times 0 \times 1) + (3 \times 1 \times 1)\} = 1A_1$

$n(A_2) = (1/6)\sum\{(1 \times 3 \times 1) + (2 \times 0 \times 1) + (3 \times 1 \times -1)\} = 0A_2$

$n(E) = (1/6)\sum\{(1 \times 3 \times 2) + (2 \times 0 \times -1) + (3 \times 1 \times 0)\} = 1E$

$\therefore \Gamma_{str}$, $\Gamma_{X-Y} = A_1 + E$.

TABLE 4.20 Determination of Γ_{X-Y} for XY$_3$ using the tabular method.

C_{3v}	$1E$	$2C_3$	$3\sigma_v$	$h = 6$	
Γ_{str}, Γ_{X-Y}	3	0	1	$\sum$	$\sum/h$
$n(A_1)$	3	0	3	6	1
$n(A_2)$	3	0	–3	0	0
$n(E)$	6	0	0	6	1
$\therefore \Gamma_{str}$, $\Gamma_{X-Y} = A_1 + E$.					

TABLE 4.21 Determination of Γ_{X-Y} for XY$_3$ using the 'by-eye' method

C_{3v}	E	$2C_3$	$3\sigma_v$	
Γ_{str}, Γ_{X-Y}	3	0	1	
Assume $1A_1$	1	1	1	
residual	2	–1	0	$\Rightarrow$ E
E	2	–1	0	
$\therefore \Gamma_{str}$, $\Gamma_{X-Y} = A_1 + E$				

If after the first subtraction the remaining reducible representation does not correspond to a single row in the character table, it is worth assuming that there is another totally symmetric representation present; if this does not work, the tabular method should be used.

4.4.9.1.4 Pyramidal XY$_3$ molecules—identifying IR and Raman activity

From the C_{3v} character table (Table 4.18) it can be seen that for the A_1 irreducible representation there is a z in the penultimate column and $x^2 + y^2$, z^2 in the final column so the A_1 mode is both IR and Raman active. Likewise for the E mode there is (x, y) in the penultimate column and $(x^2 - y^2, xy)$, (xz, yz) in the final column; therefore the E mode is also both IR and Raman active. The E mode is doubly degenerate, that is, there are two vibrational modes at exactly the same energy/frequency/wavenumber, so only one band is observed in the spectrum.

Therefore, in the X–Y stretching region ($\nu_{X–Y}$) two bands (A_1 + E) are expected in the IR spectrum and two bands (A_1 + E) in the Raman spectrum, with the A_1 being polarised and the E being depolarised.

EXAMPLE 4.15

The Raman and IR spectra of [ClO$_3$]$^-$ in the Cl–O stretching region ($\nu_{Cl–O}$) are shown in Figure 4.39. Assign the spectral features to their symmetry species.

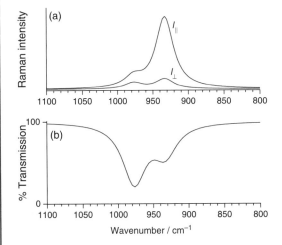

FIGURE 4.39 Raman (a) and IR (b) spectra of aqueous solutions of [ClO$_3$]$^-$. (Data from D.J. Gardiner, R.B. Girling, and R.E Hester, *J. Mol. Struct.* **13** 105 (1972).)

ANSWER

[ClO$_3$]$^-$ is pyramidal and belongs to the C_{3v} point group, so the analysis used for XY$_3$ molecules can be used. Both the Raman and IR spectra contain peaks at 977 and

---

933 cm^{-1}, but with different relative intensities. In addition, there are Raman polarisation data available. In the IR spectrum it is expected that the less symmetric modes will be more intense than the highly symmetric ones. Therefore, the 977 cm^{-1} peak can be assigned to the E mode, with the 933 cm^{-1} band belonging to the A_1 mode. This is confirmed in the Raman spectrum as the 933 cm^{-1} is the most intense feature, and when the polarisation is changed from $I_{\parallel}$ to $I_{\perp}$, its intensity reduces dramatically, but that of the 977 cm^{-1} peak is much less affected. As the totally symmetric modes are the only ones that are polarised, this confirms the assignment of the 933 cm^{-1} band to A_1.

SELF TEST

Sketch the IR and Raman spectra in the $\nu_{N–H}$ region for solid NH$_3$ (A_1, 3300 cm^{-1}; E, 3375 cm^{-1}).

4.4.9.2 Γ_{str} for linear molecules

All three of the methodologies outlined for XY$_3$ molecules will work for all point groups, except the linear ones. In the case of linear molecules the formula and tabular methods do not work because in this case $h = \infty$. However, for small linear molecules the 'by-eye' method works.

EXAMPLE 4.16

Determine Γ_{str} for OCS.

ANSWER

The procedure is summarised in Table 4.22.

TABLE 4.22 Determination of Γ_{str} for OCS

$C_{\infty v}$	E	$2C_{\infty}^{\Phi}$	...	$\infty\sigma_v$		
Γ_{str}	2	2		2		
Assume 1 Σ^+	1	1		1		
residual	1	1		1	$\Rightarrow \Sigma^+$	
$\Sigma^+ \equiv A_1$	1	1	...	1	z	$x^2 + y^2$, z^2
$\Sigma^- \equiv A_2$	1	1	...	-1	R_z	
$\Pi \equiv E_1$	2	2cos Φ	...	0	(x, y), (R_x, R_y)	(xz, yz)
$\Delta \equiv E_2$	2	2cos 2Φ	...	0		$(x^2 - y^2, xy)$
$\Phi \equiv E_3$	2	2cos 3Φ	...	0		

In this case the totally symmetric representation is Σ^+, and subtracting the characters for this from the reducible representation gives another Σ^+. Therefore, $\Gamma_{str} = 2\Sigma^+$, both of which are IR and Raman active. The IR spectrum of OCS has bands at 2062 cm^{-1} and 859 cm^{-1}. As these are well separated, the former can be thought of as predominantly the C=O stretching mode ($\nu_{C=O}$), and the latter the C=S stretching mode ($\nu_{C=S}$), rather than a symmetric and antisymmetric combination.

SELF TEST

Use the $D_{\infty h}$ character table in Appendix 1 to show that for XeF$_2$ $\Gamma_{Xe-F} = \Sigma_g^+ + \Sigma_u^+$ and assign the bands in Table 4.14 to their symmetry species.

4.4.10 Checking calculations

There are a number of checks that can be carried out to ensure that the analysis has been applied correctly: (i) the values of the irreducible representations (A$_1$ etc.) must be integers; (ii) the order of Γ_{str} must be the same as the number of stretching modes in the basis that was used to generate it (remember that E and T modes are doubly and triply degenerate, respectively). If the number of totally symmetric representations is non-integer, this usually indicates an error in calculating the reducible representation, χ_r. Non-integer values for the other irreducible representations often indicates a mixing up of +1 and −1 in the calculation.

EXAMPLE 4.17

The IR spectrum of TiCl$_4$ has one band in the Ti–Cl stretching region (ν_{Ti-Cl}) at 498 cm^{-1}. The Raman spectrum of TiCl$_4$ has two ν_{Ti-Cl} bands at 498 and 389 cm^{-1}, with depolarisation ratios of approximately 0.7 and 0.05, respectively. Use these data to confirm that TiCl$_4$ has T_d point group.

TABLE 4.23 T_d character table

T_d	E	$8C_3$	$3C_2$	$6S_4$	$6\sigma_d$		
A$_1$	1	1	1	1	1		$x^2 + y^2 + z^2$
A$_2$	1	1	1	−1	−1		
E	2	−1	2	0	0		$(2z^2 - x^2 - y^2, x^2 - y^2)$
T$_1$	3	0	−1	1	−1	(R_x, R_y, R_z)	
T$_2$	3	0	−1	−1	1	(x, y, z)	(xy, xz, yz)

ANSWER

If it is assumed that TiCl$_4$ has T_d point group, then calculation of Γ_{Ti-Cl} will allow the determination of the number of IR and Raman active Ti–Cl stretching modes using the T_d character table (Table 4.23).

Table 4.24 shows Γ_{Ti-Cl} and the 'by-eye' analysis to calculate $\Gamma_{Ti-Cl} = A_1 + T_2$. Of these, A$_1$ is Raman active (and polarised) and T$_2$ is both IR active and Raman active (and depolarised). Therefore, the 498 cm^{-1} band can be assigned to the T$_2$ mode and the 389 cm^{-1} band to the A$_1$. Hence the vibrational spectra are consistent with TiCl$_4$ being tetrahedral.

TABLE 4.24 Determination of Γ_{Ti-Cl} for TiCl$_4$

T_d	E	$8C_3$	$3C_2$	$6S_4$	$6\sigma_d$	
Γ_{Ti-Cl}	4	1	0	0	2	
Assume 1 A$_1$	1	1	1	1	1	
Residual	3	0	−1	−1	1	$\Rightarrow$ T$_2$

SELF TEST

(a) The IR spectrum of ClF$_3$ has Cl–F stretching modes at 752, 702, and 529 cm^{-1}, with the 702 cm^{-1} band being the most intense. The Raman spectrum also contained three bands at the same wavenumbers; those at 752 and 529 cm^{-1} were the most intense and the band at 702 cm^{-1} was barely visible. Use these data to identify whether ClF$_3$ is planar, pyramidal, or 'T-shaped', and assign the vibrational modes to their symmetry species.

(b) The Raman spectrum of [XeF$_5$]$^+$ has Xe-F stretching modes at 671 (29) cm^{-1}, 664 (5) cm^{-1}, 629 (100) cm^{-1}, and 623 (5) cm^{-1} (the relative intensities are given in parentheses). Use these data to determine whether [XeF$_5$]$^+$ has trigonal bipyramidal (D_{3h}) or square based pyramidal (C_{4v}) geometry. Explain how IR data could be used to assign the vibrational modes to their symmetry species.

(c) The Raman spectrum of (n–Bu$_4$N)$_2$[PtCl$_4$]$^{2-}$ in chloroform has ν_{Pt-Cl} modes at 328 cm^{-1} (polarised) and 305 cm^{-1} (depolarised), whereas the IR spectrum has one band at 313 cm^{-1}. Use these data to confirm a square planar geometry for [PtCl$_4$]$^{2-}$.

4.4.11 Carbonyl and dinitrogen complexes

The above treatment has considered all of the stretching modes in the molecules. Just as the stretching modes can be separated from the bending modes because they are energetically well separated, in some circumstances it is possible to factor out the stretching modes for different functional groups if they are also reasonably well separated in wavenumber, as there will be little or no interaction between the various stretching modes. Although this approach can be used for the stretching modes involving O and Cl, for example, in $OPCl_3$ (ν_{PO}, 1290 cm^{-1}, ν_{PCl}, 581, 486 cm^{-1}), probably the best examples are the carbonyl and dinitrogen complexes where the ν_{CO} and ν_{NN} stretching modes are well separated in energy from all the other possible modes in the complex. By considering the number of unshifted CO (or NN) stretching modes it is possible to determine the number of IR and Raman active modes, and, hence, identify the shape of the complex present. This is a very powerful way of characterising metal carbonyl complexes.

The IR and Raman active CO stretching modes for the common octahedral carbonyl complexes are given in Table 4.25. The approximate form of these vibrational modes is shown schematically in Figures 4.40 and 4.41. The arrows indicate the significant changes in relative atom positions that occur. Note that they will not always be of the same size, and there will be other displacements to ensure that the centre of gravity of the molecule remains fixed.

4.4.12 Analysis of metal carbonyl IR spectra

The IR spectra shown in Figure 4.42 follow the reaction of $[Mo(CO)_6]$ with pyridine (py) under reflux in toluene.

The analysis of these IR spectra requires a fundamentally different approach compared with the explanation of NMR spectra in Chapter 3. The interpretation of NMR spectra is based on the number of equivalent nuclei, so for $[Mo(CO)_6]$, one ^{13}C NMR peak would be expected as all the carbon atoms are equivalent. In the case of $[Mo(CO)_5(py)]$ there would be two ^{13}C NMR features, one from the axial CO and the other from the four equatorial CO ligands. In IR spectroscopy, the molecule as a whole normally needs to be considered, but for carbonyl complexes the ν_{CO} modes can be dealt with separately. For $[Mo(CO)_6]$, $\Gamma_{CO} = A_{1g} + E_g + T_{1u}$ (Table 4.25), but only the T_{1u} mode at 1981 cm^{-1} is IR active. The schematic representations in Figure 4.40 show that this involves the simultaneous compression and extension of *trans* C≡O bonds, and that the three of these are degenerate. Therefore, while the IR and NMR spectra of $[Mo(CO)_6]$ both contain one feature, this is for very different reasons. For $[Mo(CO)_5(py)]$, $\Gamma_{CO} = 2A_1 + B_1 + E$, and of these $2A_1$ and E are IR active, therefore the IR spectrum (Figure 4.42) contains three bands. A sharp weak one at 2074 cm^{-1} is due to the totally symmetric (left-hand) A_1 mode in Figure 4.40, together with two more intense peaks at 1938 cm^{-1} (E) and 1897 cm^{-1} (A_1). The 1938 cm^{-1} feature comes from the doubly degenerate E mode, where for each pair of CO bonds *trans* to each other, one extends while the other compresses. These assignments are based on calculations, but it would be expected that the E mode would generate the most intense peak.

For *cis*-tetracarbonyl complexes with a C_{2v} point group, $\Gamma_{CO} = 2A_1 + B_1 + B_2$, and as all of these are IR active four IR active ν_{CO} modes are expected. The spectrum of *cis*-$[Mo(CO)_4(py)_2]$ (Figure 4.42) is very characteristic for this type of complex, with a high wavenumber peak (at 2013 cm^{-1}) and three bands at lower wavenumber (1892, 1873, and 1831 cm^{-1}) that are often overlapping. The 2013 cm^{-1} band is due to the A_1 mode involving the symmetric in-phase stretch of all four of the CO bonds, as shown in

TABLE 4.25 Summary of the IR and Raman active ν_{CO} modes in octahedral metal carbonyl complexes

	Point group	Γ_{CO}	IR active	Raman active
$[M(CO)_6]$	O_h	$A_{1g} + E_g + T_{1u}$	T_{1u}	$A_{1g} + E_g$
$[M(CO)_5L]$	C_{4v}	$2A_1 + B_1 + E$	$2A_1 + E$	$2A_1 + B_1 + E$
cis-$[M(CO)_4L_2]$	C_{2v}	$2A_1 + B_1 + B_2$	$2A_1 + B_1 + B_2$	$2A_1 + B_1 + B_2$
trans-$[M(CO)_4L_2]$	D_{4h}	$A_{1g} + B_{1g} + E_u$	E_u	$A_{1g} + B_{1g}$
fac-$[M(CO)_3L_3]$	C_{3v}	$A_1 + E$	$A_1 + E$	$A_1 + E$
mer-$[M(CO)_3L_3]$	C_{2v}	$2A_1 + B_2$	$2A_1 + B_2$	$2A_1 + B_2$
cis-$[M(CO)_2L_4]$	C_{2v}	$A_1 + B_2$	$A_1 + B_2$	$A_1 + B_2$
trans-$[M(CO)_2L_4]$	D_{4h}	$A_{1g} + A_{2u}$	A_{2u}	A_{1g}
$[M(CO)L_5]$	C_{4v}	A_1	A_1	A_1

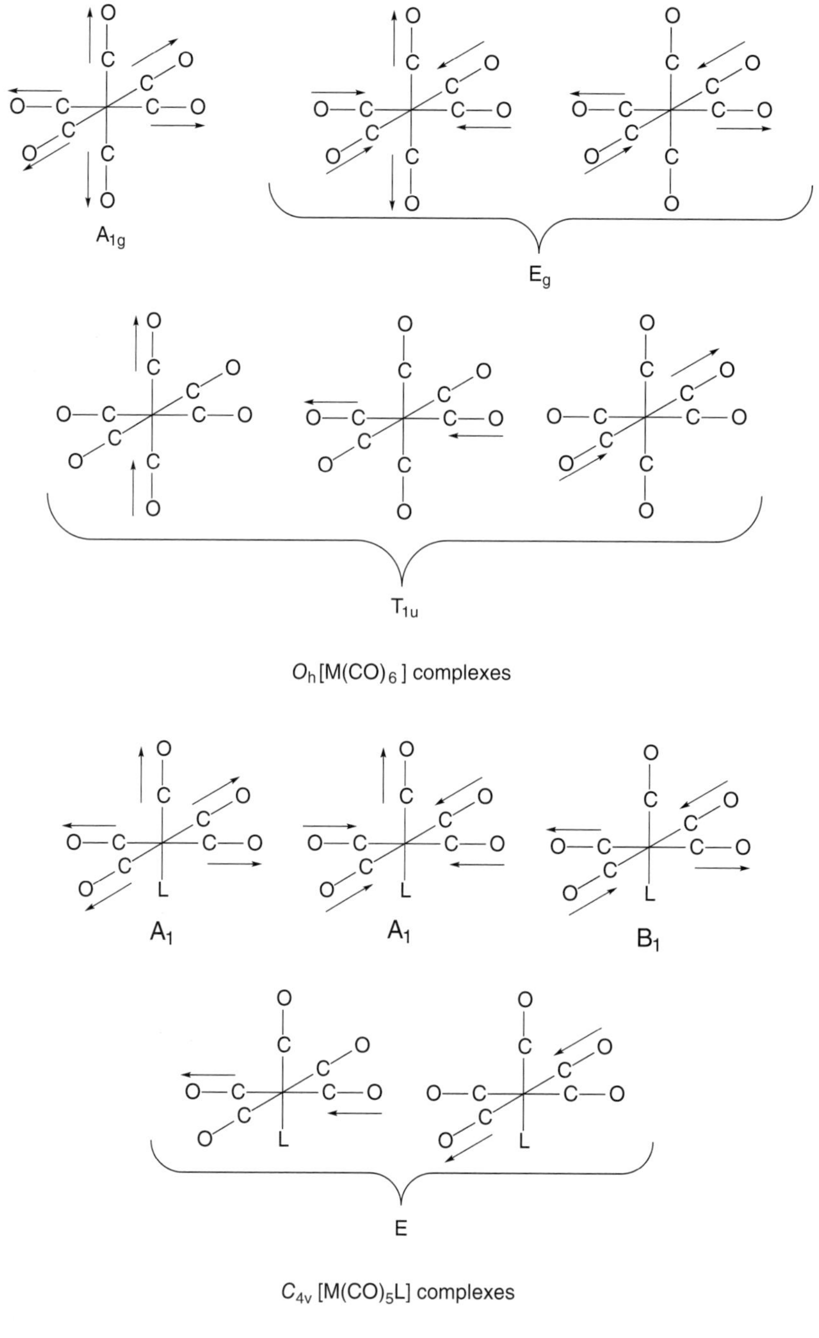

FIGURE 4.40 Representations of the CO stretching modes in $[M(CO)_6]$ and $[M(CO)_5L]$ complexes.

Figure 4.41. The three overlapping bands between 1900 and 1800 cm^{-1} are due to the other A$_1$ stretching mode and the B$_1$ and B$_2$ asymmetric stretching modes. For *fac*-$[Mo(CO)_3(py)_3]$ $\Gamma_{CO} = A_1 + E$, both of which are IR active and these are observed at 1905 cm^{-1} (A$_1$) and 1775 cm^{-1} (E) (Figure 4.41, 4.42). These values are at the extreme for terminal CO, indicating the strong σ donor and weak π acceptor capability of pyridine.

This set of spectra also shows how the position of the ν$_{CO}$ modes is sensitive to the number of CO ligands present. Pyridine is a better σ donor and worse π acid than CO, therefore as each CO is replaced by a pyridine, there is more back-bonding per CO, and hence the ν$_{CO}$ modes move to lower wavenumber (Figure 4.15).

If *cis*-$[Mo(CO)_4(py)_2]$ is allowed to react with a bulky phosphine ligand such as triphenylphosphine (PPh$_3$)

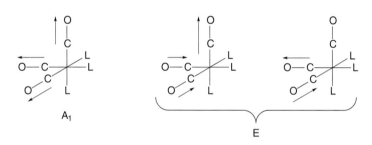

D_{4h} *trans*-[M(CO)$_4$L$_2$] complexes

C_{2v} *cis*-[M(CO)$_4$L$_2$] complexes (Note the *z*-axis bisects the M–L axes, and L are in the *yz* plane)

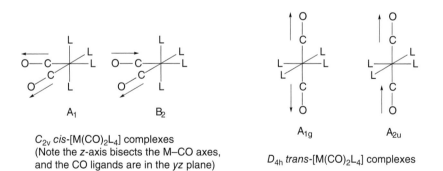

C_{3v} [M(CO)$_3$L$_3$] complexes (Note the *z*-axis trisects the three M–L axes)

C_{2v} *cis*-[M(CO)$_2$L$_4$] complexes
(Note the *z*-axis bisects the M–CO axes,
and the CO ligands are in the *yz* plane)

D_{4h} *trans*-[M(CO)$_2$L$_4$] complexes

FIGURE 4.41 Representations of the CO stretching modes in [M(CO)$_4$L$_2$], [M(CO)$_3$L$_3$], and [M(CO)$_2$L$_4$] complexes.

under mild conditions (short reflux in CH$_2$Cl$_2$), ligand exchange takes place to yield *cis*-[Mo(CO)$_4$(PPh$_3$)$_2$] (Figure 4.43). The shift of the ν_{CO} modes to slightly higher wavenumber indicates that there is less back-bonding per CO in the phosphine complex, implying that the phosphine is a better π acid than pyridine. If more forcing conditions are employed (i.e. reflux in toluene)

the *cis* complex isomerises to *trans*-[Mo(CO)$_4$(PPh$_3$)$_2$] (Figure 4.43).

For *trans*-[M(CO)$_4$(PPh$_3$)$_2$], Γ_{CO} = A$_{1g}$ + B$_{1g}$ + E$_u$ of which the E$_u$ mode is IR active. While there is one intense peak in the IR spectrum at 1896 cm^{-1} (Figure 4.43), there are also two much weaker bands at 1954 and 2022 cm^{-1}. These are at the same position as peaks observed in the

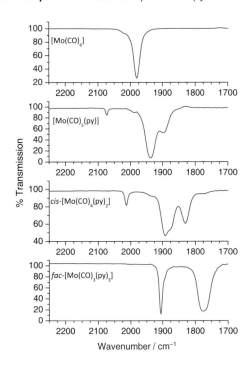

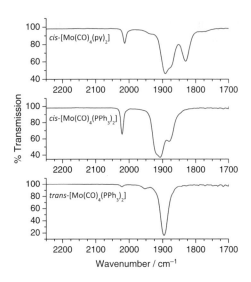

FIGURE 4.42 IR solution (CH$_2$Cl$_2$) spectra of reaction products of [Mo(CO)$_6$] and pyridine (py).

FIGURE 4.43 IR solution (CH$_2$Cl$_2$) spectra of *cis*-[Mo(CO)$_4$(py)$_2$] and its reaction products with PPh$_3$.

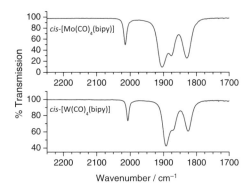

FIGURE 4.44 IR solution spectra (CH$_2$Cl$_2$) of *cis*-[Mo(CO)$_4$(bipy)] and *cis*-[W(CO)$_4$(bipy)] (bipy = 2,2′-bipyridine).

Raman spectrum and indicate that the molecule does not have perfect D_{4h} symmetry.

Figure 4.44 shows the IR solution spectra of *cis*-[Mo(CO)$_4$bipy] and *cis*-[W(CO)$_4$bipy] (bipy = 2,2′-bipyridine) which also display the pattern characteristic of *cis*-tetracarbonyl complexes. While the CO stretching modes are at similar wavenumbers for the pyridine, PPh$_3$ and bipy complexes, on closer inspection the order is PPh$_3$ > py ≈ bipy, showing that there is less back-bonding per CO in the PPh$_3$ complex compared to those of py and bipy due to the different σ and π bonding characteristics of the ligands. For *cis*-[Mo(CO)$_4$bipy] and *cis*-[W(CO)$_4$bipy] complexes, the CO stretching modes are at lower wavenumber in the tungsten complex, indicating greater back-bonding, which is usually associated with the more diffuse 5d rather than 4d orbitals.

4.4.13 Analysis of all vibrational modes, Γ_{vib}, Γ_{str} and Γ_{def}

So far group theory has been used to predict the IR and Raman activity of stretching modes. For all but diatomics there are also bending/deformation modes (Figure 4.2) which can also be used to characterise inorganic compounds. Molecules have $3N - 6$ vibrational modes ($3N - 5$ if linear), and in general for every X–Y bond, there will be one X–Y stretching frequency, with the remainder of

the $3N - 6$ (or $3N - 5$) modes being deformation or bending modes, as shown in Table 4.26.

Unfortunately, it is not usually possible to employ the same methodology as used in the analysis of stretching modes, by considering the number of unshifted bends, as these are often quite hard to visualise. Therefore, an alternative approach is used, where the total number of vibrational modes (Γ_{vib}) is made up of the sum of the stretching (Γ_{str}) and deformation modes (Γ_{def}) thus:

$$\Gamma_{vib} = \Gamma_{str} + \Gamma_{def} \qquad \textbf{Eqn 4.6}$$

However, there is still the problem of visualising all of the vibrational modes. The $3N - 6$ vibrational modes (Γ_{vib}) are part of the total degrees of freedom Γ_{3N} as are the three translational degrees of freedom

TABLE 4.26 Summary of number of stretching and bending/deformation modes in XY$_n$ molecules

	Total number of vibrational modes	No. of stretching modes	No. of bending/deformation modes
XY$_2$ (Linear)	4 (3N – 5)	2	2
XY$_2$ (Bent)	3 (3N – 6)	2	1
XY$_3$	6	3	3
XY$_4$	9	4	5

(Γ_{trans}) and the two or three rotational degrees of freedom (Γ_{rot}). Therefore, the following relationship can be used:

$$\Gamma_{3N} = \Gamma_{vib} + \Gamma_{rot} + \Gamma_{trans} \qquad \textbf{Eqn 4.7}$$

and the irreducible representations of Γ_{trans} and Γ_{rot} are easily obtained from the character tables.

In order to calculate Γ_{3N} it is necessary to consider how the atoms move in three dimensions. To do this three Cartesian vectors (x, y, z) are placed on each atom (Figure 4.45), and how many of these remain unshifted by the symmetry operations of the point group is determined. This gives the reducible representation Γ_{3N} (also known as Γ_{mol}).

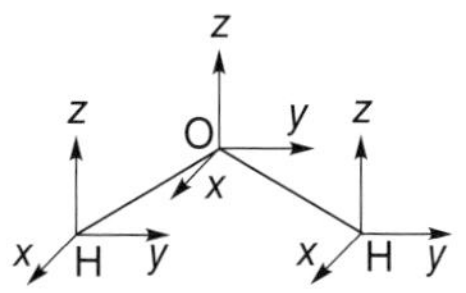

If an atom moves through space during a symmetry operation, then all the Cartesian vectors have also shifted and they contribute zero to Γ_{3N}. If the atom remains unshifted, as in the case of the O atom in H$_2$O during the C_2 symmetry operation, the Cartesian vectors will contribute to Γ_{3N}, but the direction is now important. If the vector such as the oxygen z vector is pointing in the same direction before and after the C_2

FIGURE 4.45 Cartesian vectors in H$_2$O.

symmetry operation (Figure 4.46) it contributes +1 to Γ_{3N}, but if as in the case of the oxygen x and y vectors the direction is reversed, these each contribute −1 to Γ_{3N}. Therefore, overall the C_2 symmetry operation contributes −1 to Γ_{3N} (0 from each H, and +1 + −1 + −1 from the O). For $\sigma_v(xz)$ only the oxygen is unshifted, and the x and z vectors both contribute +1 as they are in the same direction, while the y vector contributes −1 as it is reversed (Figure 4.46), therefore overall Γ_{3N} is 1. For $\sigma_v'(yz)$ all three of the atoms are unshifted, and each atom contributes +1, so overall Γ_{3N} is 3. For E all the atoms remain unshifted, and all the vectors are in the original direction, therefore this contributes 3 × 3 = 9 to Γ_{3N}.

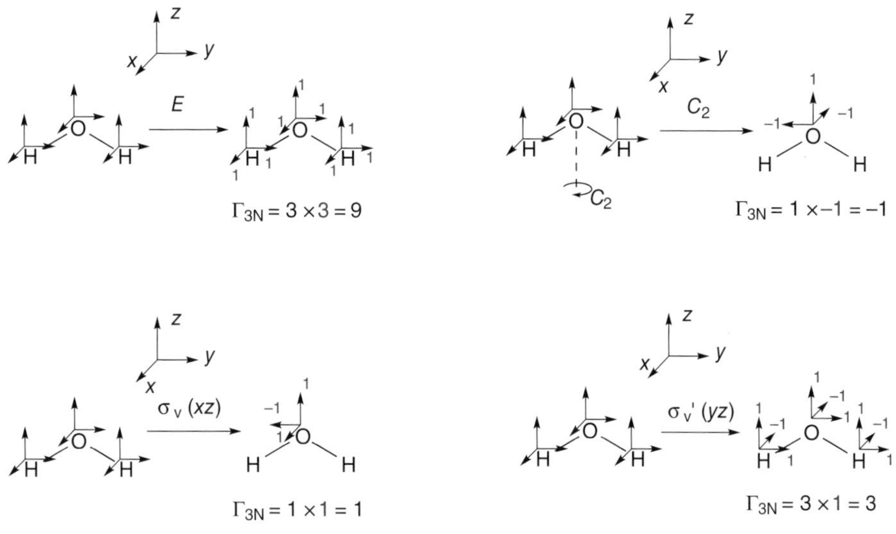

FIGURE 4.46 Determination of Γ_{3N} for H$_2$O.

While it is possible to carry out this process for large molecules it soon becomes very cumbersome, and it is far easier to identify the number of unshifted atoms (N_{us}), and then work out the contribution or character per unshifted atom (χ_{us}) and multiply these together to give Γ_{3N} ($\Gamma_{3N} = N_{us}\chi_{us}$). This is the calculation shown in Figure 4.46 where the first number (in black) is N_{us} and the second number (in blue) is χ_{us}.

It has already been seen for the C_{2v} point group that χ_{us} for E is 3, for σ_v it is +1 and for C_2 it is −1. These are the values of χ_{us} for these symmetry operations in all point groups. The values of χ_{us} for the other symmetry operations are in Table 4.27.

It is also always possible to extract the values of χ_{us} directly from the character tables by simply adding up the characters (1, 0, etc.) in the table corresponding to the x, y, and z vectors for each symmetry operation. If two or more vectors transform as degenerate (e.g. E), then the value in the table is simply used. If two or

more vectors are non-degenerate (e.g. B_u, in C_{2h}), then the values should be added together. A good check is whether a value of three for χ_{us} is obtained for the identity element, E.

The method of extracting χ_{us} for the symmetry operations in the C_{2v} and C_{3v} point groups is shown in Tables 4.28 and 4.29.

Having worked out the contribution per unshifted atom, χ_{us}, this is multiplied by the number of unshifted atoms, N_{us} to give Γ_{3N}, which is then reduced in the normal way. However, even for small and simple molecules such as H_2O or NH_3, it is better to use the tabular method.

4.4.13.1 Determination of Γ_{vib}, Γ_{str}, and Γ_{def} for pyramidal XY_3

The complete process for a pyramidal XY_3 molecule is shown in Table 4.30. The most common error when carrying out this calculation is to forget the atom at

TABLE 4.27 Values of character per unshifted atom (χ_{us}) for different symmetry operations

Symmetry operation	χ_{us}	Symmetry operation	χ_{us}	Symmetry operation	χ_{us}
E	3	C_2	−1	S_3^1, S_3^5	−2
i	−3	C_3^1, C_3^2	0	S_4^1, S_4^3	−1
σ	+1	C_4^1, C_4^3	+1	S_6^1, S_6^5	0
		C_6^1, C_6^5	+2		
		C_n	$1 + 2\cos(360/n)$	S_n	$-1 + 2\cos(360/n)$

TABLE 4.28 Extracting χ_{us} from the C_{2v} character table

C_{2v}	E	C_2	$\sigma_v(xz)$	$\sigma_v'(yz)$		
A_1	1	1	1	1	z	x^2, y^2, z^2
A_2	1	1	−1	−1	R_z	xy
B_1	1	−1	1	−1	x, R_y	xz
B_2	1	−1	−1	1	y, R_x	yz
χ_{us}	3	−1	1	1		
	$(1 + 1 +1)$	$(1 − 1 − 1)$	$(1 + 1 − 1)$	$(1 − 1 + 1)$		

TABLE 4.29 Extracting χ_{us} from the C_{3v} character table

C_{3v}	$1E$	$2C_3$	$3\sigma_v$		
A_1	1	1	1	z	$x^2 + y^2, z^2$
A_2	1	1	−1	R_z	
E	2	−1	0	$(x, y), (R_x, R_y)$	$(x^2 − y^2, xy), (xz, yz)$
χ_{us}	3	0	1		

TABLE 4.30 Calculation of Γ_{3N} for pyramidal XY_3

C_{3v}	E	$2C_3$	$3\sigma_v$	$h = 6$	
N_{us} (no. of unshifted atoms)	4	1	2		
χ_{us} (contribution per unshifted atom)	3	0	1		
Γ_{3N} ($N_{us} \times \chi_{us}$)	12	0	2	Σ	Σ/h
$n(A_1)$	12	0	6	18	3
$n(A_2)$	12	0	-6	6	1
$n(E)$	24	0	0	24	4

the centre when determining the number of unshifted atoms, N_{us}. Therefore, $\Gamma_{3N} = 3A_1 + A_2 + 4E$.

As the starting point was four atoms, there should be $3N$ vectors, that is, 12 representations at the end of the calculation. Although it only looks like there are eight, the E representations are doubly degenerate, and when this is taken into account there are indeed twelve. This irreducible representation is for all the degrees of freedom within the molecule, three of which are taken up by translational (Γ_{trans}) and three by rotational (Γ_{rot}) degrees of freedom to leave $3N - 6$ vibrational degrees of freedom (Γ_{vib}) (Eqn 4.7).

The values of Γ_{trans} and Γ_{rot} need to be identified, and these can be looked up in the C_{3v} character table (Table 4.18), as their irreducible representations are indicated by x, y, z for Γ_{trans} and R_x, R_y, and R_z for Γ_{rot}. Therefore, for C_{3v} $\Gamma_{trans} = A_1 + E$ and $\Gamma_{rot} = A_2 + E$ and these are subtracted from Γ_{3N} to give the required Γ_{vib} (Table 4.31). Therefore, $\Gamma_{vib} = 2A_1 + 2E$.

As $\Gamma_{str} = A_1 + E$ (Section 4.4.9.1), this means that Γ_{def} ($\Gamma_{vib} - \Gamma_{str}$) $= A_1 + E$, both of which are IR and Raman active.

TABLE 4.31 Calculation of Γ_{vib} from Γ_{3N}, Γ_{trans}, and Γ_{rot}

Γ_{3N}	$3A_1$	$1A_2$	$4E$
Γ_{trans}	$1A_1$	$0A_2$	$1E$
Γ_{rot}	$0A_1$	$1A_2$	$1E$
Γ_{vib}	$2A_1$	$0A_2$	$2E$

EXAMPLE 4.18

The Raman spectrum of a 2.0 M solution of XeO_3 in H_2O is shown in Figure 4.47. Using the information in Section 4.4.13, explain how the assignments have been made.

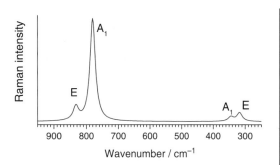

FIGURE 4.47 Raman spectrum of 2.0 M XeO_3 in H_2O. (Data from H.H. Claassen, and G. Knapp, *J. Am. Chem. Soc.* **86** 2341 (1964).)

ANSWER

XeO_3 is a C_{3v} pyramidal molecule as it is formally xenon(VI) with one lone pair of electrons. Therefore, the analysis in Section 4.4.13 can be used to assign the spectra. This predicted two Raman active Xe=O stretching modes ($A_1 + E$) and two Raman active bending modes ($A_1 + E$). The peaks at 833 and 780 cm^{-1} can be assigned as stretching modes as they are at highest energy and to the E and A_1 stretching modes in particular on the basis of their relative intensity. The 780 cm^{-1} band is polarised in the Raman experiment, but the 833 cm^{-1} peak is depolarised, confirming their assignment as A_1 and E, respectively. In the IR spectrum, the relative intensity of these two peaks was reversed. The relative intensity of the two lower wavenumber deformation modes 344 and 317 cm^{-1} would suggest an assignment of E and A_1, respectively, but polarisation data indicate a reverse assignment.

SELF TEST

Calculate Γ_{vib}, Γ_{str}, and Γ_{def} and determine the number of IR and Raman active modes in: (a) BF_3, (b) GeF_4, (c) XeF_4, (d) SeF_4.

4.4.14 Labelling and assigning of vibrational modes

The final stage in the assignment of spectra and the characterisation of molecules is the labelling of the vibrational modes, such as ν_1 and ν_3 for the two stretching modes in H_2O. In order to do this the vibrational modes are ranked in order of the irreducible representations in the character table. If there are two or more vibrational modes with the same symmetry (irreducible representation), they are ranked within the symmetry species by energy/frequency/wavenumber with the highest first.

So for H_2O it is:

A_1	ν_1	symmetric stretching mode	3657 cm^{-1}
A_1	ν_2	bending mode	1595 cm^{-1}
B_2	ν_3	antisymmetric stretching mode	3756 cm^{-1}

and for XeO_3 it is:

A_1	ν_1	symmetric stretching mode	780 cm^{-1}
A_1	ν_2	'umbrella' bending mode	344 cm^{-1}
E	ν_3	asymmetric stretching mode	833 cm^{-1}
E	ν_4	bending mode	317 cm^{-1}

EXAMPLE 4.19

Assign the main spectral features below 1750 cm^{-1} in the IR and Raman spectra of calcite (calcium carbonate) shown in Figure 4.48.

ANSWER

The spectral region below 1750 cm^{-1} contains the fundamental vibrational modes associated with the CO_3^{2-} unit. In the IR spectrum the fundamentals are observed at 1440 cm^{-1} (vs), 876 cm^{-1} (s), and 712 cm^{-1} (m), and in the Raman spectrum at 1436 cm^{-1} (vw), 1086 cm^{-1} (vs), and 712 cm^{-1} (m). (The labels vs, s, m, w, and vw, refer to the relative intensity of the features: vs (very strong), s (strong), m (medium), w (weak), vw (very weak), with sh referring to a shoulder and br to broad.)

For a D_{3h} CO_3^{2-} ion, the following irreducible representations can be determined:

$$\Gamma_{vib} = A_1' + A_2'' + 2E'$$
$$\Gamma_{str} = A_1' + E'$$
$$\Gamma_{def} = E' + A_2''$$

The A_2'' and E' modes are IR active, and the A_1' and E' modes are Raman active. Therefore, one ν_{C-O} mode (E') and two bending modes ($E' + A_2''$) are expected in the IR spectrum. In the Raman spectrum, there will be two Raman active stretching modes ($A_1' + E'$) and one bending mode (E'). From the Raman spectrum it is clear that the intense band at 1086 cm^{-1} which is absent in the IR spectrum can be assigned to the A_1' stretching mode. Likewise, the band at 876 cm^{-1} that is present in the IR spectrum, but absent from the Raman spectrum, can be assigned to the A_2'' bending/deformation mode. Therefore, the bands at 1440/1436 cm^{-1} and 712 cm^{-1} must belong to the E' modes. (Although IR and Raman bands are expected at the same value, small differences are not unusual). As stretching modes usually occur at higher wavenumber than bending modes, the 1440/1436 cm^{-1} bands can be

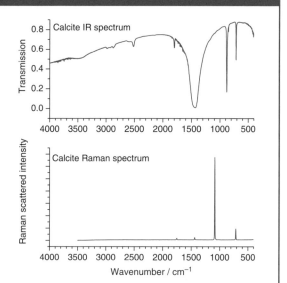

FIGURE 4.48 IR and Raman (1064 nm) spectra of calcite $(CaCO_3)$.

assigned to the E' stretching mode and the 712 cm^{-1} bands to the E' bending mode. The form of these vibrational modes is shown in Figure 4.49, together with the traditional labelling of the modes in terms of ν_1, ν_2, etc. The A_2'' mode can be identified as an out-of-plane mode, as it is associated with a dipole change in the z direction.

In modern D_{3h} character tables the ordering of the irreducible representations is given as A_1', E', and A_2'', which is not consistent with the above labelling. The reason for this is that the original assignments and labelling used what are known as the 'Herzberg' character tables where the ordering was in terms of A, B, E, T, whereas the modern character tables separate out the $'$ and $''$ irreducible representations. (This is also found in other point groups with $'$ and $''$ or g and u labels such as D_{nh}, O_h, etc.)

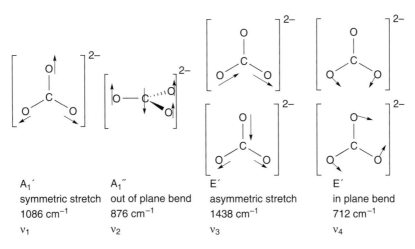

A_1'	A_1''	E'	E'
symmetric stretch	out of plane bend	asymmetric stretch	in plane bend
1086 cm^{-1}	876 cm^{-1}	1438 cm^{-1}	712 cm^{-1}
ν_1	ν_2	ν_3	ν_4

FIGURE 4.49 Schematic representation of the vibrational modes of the carbonate anion, together with the traditional labelling of the modes.

SELF TEST

The solid phase Raman spectrum of GeH$_4$ has bands at 2114, 2106, 931, and 819 cm^{-1}, with the 2106 cm^{-1} band being the most intense and also polarised. The solid phase IR spectrum has bands at 2114 and 819 cm^{-1} (Figure 4.9). Assign these to the correct symmetry species and label them in terms of ν_1, ν_2, etc.

Problems

4.1 Describe how IR spectroscopy could be used to determine the following.

(a) Whether the water molecules in CrCl$_3$(H$_2$O)$_6$ are water of crystallisation or coordinated to the metal centre?

(b) Whether the hydrogen bond in solid β-CrOOH was stronger or weaker than that in a different phase, α-CrOOH?

(c) The mode of coordination of SeCN$^-$ ligands in transition metal complexes.

(d) The rate of an isomerisation reaction of the [Ru(NH$_3$)$_4$(H$_2$O)(SO$_2$)]$^{2+}$ cation, which involves a change in the coordination by the SO$_2$ ligand, from sulfur to oxygen, to the ruthenium centre.

4.2

(a) Explain the ordering of the ν_{CO} modes in the following *fac*-isomers: [Mo(CO)$_3$(PF$_3$)$_3$], 2090, 2055 cm^{-1}; [Mo(CO)$_3$(PPh$_3$)$_3$], 1934, 1835 cm^{-1}; [Mo(CO)$_3$(py)$_3$], 1905, 1775 cm^{-1}.

(b) Explain the position of the ν_{CO} and ν_{NO} modes in [RuCl(CO)(NO)(P^{t}Bu$_2$Me)$_2$] (ν_{CO} 1914 cm^{-1}, ν_{NO} 1570 cm^{-1}) and [Ru(CO)(NO)(P^{t}Bu$_2$Me)$_2$]$^+$ (ν_{CO} 1966 cm^{-1}, ν_{NO} 1709 cm^{-1}).

4.3 Use the IR spectra of isotopically enriched [Co(N$_2$)] and [Ni(N$_2$)] in Figure 4.50 to identify whether the N$_2$ is side-on or end-on bonded in these compounds. The isotopic enrichment was ^{14}N$_2$:^{14}N^{15}N:^{15}N$_2$ 3:2:1.

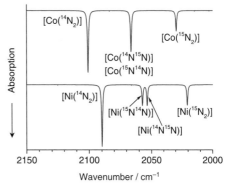

FIGURE 4.50 IR spectra of isotopically enriched [Co(N$_2$)] and [Ni(N$_2$)] in solid argon. (Data from G.A. Ozin and A. van der Voet, *Can. J. Chem.* **51** 637 (1973).)

4.4 The carbonylation of RuCl$_3$ using a mixture of HCl and HCOOH followed by precipitation with CsCl gave complexes Cs$_2$[Ru(CO)$_2$Cl$_4$] **A** after four hours and Cs[Ru(CO)$_3$Cl$_3$] **B** after 20 hours. The IR spectrum of complex **A** contained two ν_{CO} modes at 2074 (vs) and 2006 (vs) cm^{-1}, and IR spectrum of complex for **B** had ν_{CO} modes 2141 (s) and 2077 (vs) cm^{-1}. Use these data to identify the isomers formed and comment on the position of the ν_{CO} modes.

4.5 The IR spectrum from [Rh(CO)$_3$] is shown in Figure 4.51. Determine whether [Rh(CO)$_3$] is trigonal planar, pyramidal, or T-shaped.

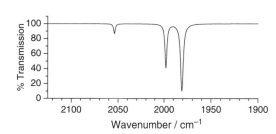

FIGURE 4.51 IR spectrum of [Rh(CO)$_3$] in an argon matrix at 10 K.

4.6 The IR spectrum of VF$_5$ has bands in the V–F stretching region at 810 and 784 cm^{-1}, and at 719 and 608 cm^{-1} in the Raman spectrum, together with a very weak feature at 810 cm^{-1}. Use these data to determine the shape of VF$_5$ and assign the vibrational modes to their symmetry species.

4.7 The planar molecule N$_2$F$_2$ can be isolated as both *cis* (C_{2v}) and *trans* (C_{2h}) isomers. One of these isomers yielded the following IR and Raman data.

IR: 1525, 950, 890, 728, 341 cm^{-1}
Raman: 1525(p), 950, 890(p), 728, 550, 341(p) cm^{-1}

Use the mutual exclusion principle to identify which isomer has a structure consistent with these data. Calculate Γ_{vib}, Γ_{str}, and Γ_{def} for both isomers and use this to confirm the assignment. Assign all the bands to their symmetry species.

4.8 The Raman spectrum of white phosphorus (P$_4$) has three bands at 614, 467, and 372 cm^{-1}, the first of which is polarised. In the IR spectrum there is a very intense peak at 467 cm^{-1}. Determine Γ_{str} and Γ_{vib} for white phosphorus and assign the observed vibrational modes to their symmetry species. Comment on the apparent breakdown of separating the stretching and deformation modes in clusters such as P$_4$.

ADVANCED TOPICS IN VIBRATIONAL SPECTROSCOPY

The following parts build on the material in the preceding sections, provide the detailed background for Raman polarisation studies, and discuss some advanced Raman experiments, before applying more advanced aspects of group theory to vibrational spectra.

4.5 Advanced topics in Raman spectroscopy

This section provides the necessary background to understanding the application of polarisation measurements before considering some advanced aspects of Raman spectroscopy that can be applied to characterising inorganic compounds.

4.5.1 Applications of polarisation in Raman spectroscopy

In the preceding sections polarisation information in the Raman spectra was used to help with assignments. The key features are: (i) the totally symmetric vibrational modes are the only ones that remain polarised (p), all the others become depolarised (dp); (ii) the depolarisation ratio, ρ, is <0.75 for the polarised modes, and is 0.75 for the depolarised modes.

4.5.1.1 Experimental considerations

As a laser is usually used for Raman experiments, the polarisation of the exciting source can be exploited to obtain extra information from the spectra. Raman polarisation measurements do not require oriented samples, but either rapidly tumbling molecules in the gas or liquid phase, or randomly oriented molecules in the solid state.

A schematic diagram of how polarisation measurements can be undertaken for Raman spectra is shown in Figure 4.52. The plane polarised laser is configured so that the electric vector is in the z direction. When this interacts with the sample, scattered light is given off in all directions, but the light scattered at 90° is collected and analysed (many commercial instruments use 180° scattering geometries, but the principle is the same). In front of the detector is an analyser, which can be thought of as like a letter-box, so that it only lets the scattered light through in one polarisation direction at a time. This is used to measure the intensity of the light perpendicular to the incident laser polarisation (I_y ($I_\perp$)) and parallel to the incident laser polarisation (I_z ($I_\parallel$)) The ratio of these, $I_\perp/I_\parallel$, is called the depolarisation ratio, ρ.

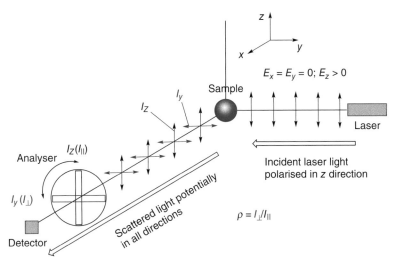

FIGURE 4.52 Schematic diagram of Raman polarisation experiment.

4.5.1.2 Depolarisation ratio, ρ

The totally symmetric vibrational mode for any point group is the only one that retains the polarisation information inherent in the exciting source, that is, z^2 on its own; the others scramble the polarisation, that is, mix z with x and/or y. While this helps to explain why there is different polarisation behaviour, it does not tell us what the extent of this variation is. The depolarisation ratio, ρ, is measured experimentally as:

$$\rho = \frac{I_y(I_\perp)}{I_z(I_\|)}$$ **Eqn 4.8**

Theoretically for plane polarised light it can be shown that for a totally symmetric mode $\rho < ¾$, and for all other modes $\rho = ¾$. (See Online Resource Centre for further details.) Therefore, for all vibrational modes except the totally symmetric one, $\rho = ¾$, and these are known as depolarised (dp) modes. For the totally symmetric modes $\rho < ¾$ and these are called polarised (p) modes. This analysis only states that $\rho < ¾$, not what the value actually is. For highly symmetric molecules such as CCl_4 ρ can approach 0, but in other cases the difference may be less marked.

In the spectra of CCl_4 shown in Figure 4.53 there is only one band that is very strongly affected by changing the analyser orientation and, therefore, this is the totally symmetric A_1 mode. The others are reduced slightly in intensity, so these are depolarised and belong to the other symmetry species. The double peak at 775 cm^{-1} is due to Fermi resonance (see Section 4.6.5).

4.5.2 Advanced Raman techniques

In addition to the Raman scattering experiments described in previous sections, there are a number of related techniques such as **resonance Raman** and **surface-enhanced Raman scattering** (**SERS**). These are useful techniques for some specific characterisation applications. Hyper Raman and coherent anti-Stokes Raman spectroscopy (CARS) are discussed in the Online Resource Centre.

4.5.2.1 Resonance Raman scattering

Figure 4.5 shows that Raman scattering involving transitions to 'virtual' electronic states resulting in Rayleigh, Stokes, and anti-Stokes scattering. In resonance Raman spectroscopy, the excitation source coincides (or nearly coincides) with an allowed electronic transition within the sample, and this results in an increase in the scattering intensities by 10^2–10^6. Because specific vibrational states (usually the totally symmetric ones) in the excited

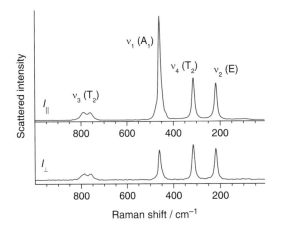

FIGURE 4.53 Raman spectra (1064 nm excitation) of CCl_4 showing effect of polarisation on peak intensity.

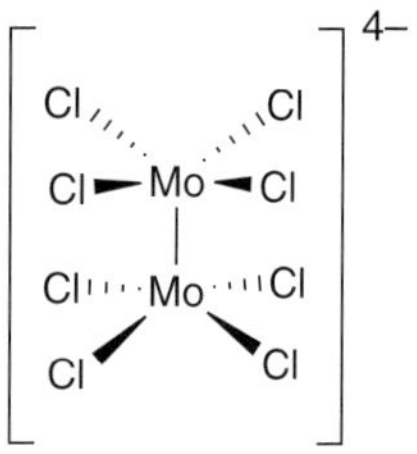

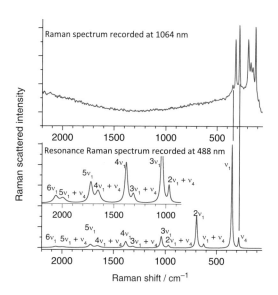

FIGURE 4.54 Structure of the $[Mo_2Cl_8]^{4-}$ anion.

electronic state are involved, the spectrum is often simplified. In addition to resonance Raman, resonance fluorescence is also possible, which can mask the Raman bands. Resonance fluorescence tends to dominate in the gas phase, but because of the broader vibronic bands in the solid or solution, resonance Raman is observed in these phases.

The $K_4[Mo_2Cl_8]$ complex (Figure 4.54) provides a good example of the differences between conventional and resonance Raman spectra. $K_4[Mo_2Cl_8]$ is one of a number of dinuclear molybdenum complexes which have very short metal–metal bonds of around 2.15 Å, compared to 2.73 Å in metallic molybdenum. The short bond-length results from a metal–metal quadruple bond made up of one σ bond, two π bonds, and one δ bond formed between the d orbitals of the two molybdenum atoms.

The top spectrum in Figure 4.55 is a conventional Raman spectrum collected with 1064 nm excitation and shows the fundamental vibrational modes below 400 cm^{-1} associated with the D_{4h} point group of the $[Mo_2Cl_8]^{4-}$ ion. The resonance Raman spectrum recorded with 488 nm excitation is very different. In the fundamental region (<400 cm^{-1}) there are only two bands observed, but there are many at higher wavenumber. The 488 nm laser excitation wavelength is within the electronic transition centred at 19 000 cm^{-1}, which gives the complex its characteristic pink-purple colour. The resonance Raman spectrum is dominated by the totally symmetric modes, v_1 and v_4, which correspond to the A_{1g} Mo–Mo stretch and A_{1g} symmetric Mo–Cl stretching modes, respectively. While these are both present in the conventional Raman spectrum, their relative intensity has changed, and the intensity of the v_1 band is strongly enhanced. In addition there is a long progression of bands due to the overtones of v_1, as well as a weaker set of combination bands involving both v_1 and v_4 (see Section 4.6.1). This is very characteristic of resonance Raman spectra.

The huge intensity enhancement of the totally symmetric bands in resonance Raman spectra means that it finds application in bioinorganic chemistry as a metal centred spectroscopic technique. For example, if a metalloenzyme has a suitable metal-based electronic transition, the vibrational properties of the metal centre can be probed by resonance Raman spectroscopy ignoring

FIGURE 4.55 Raman and resonance Raman spectra of $K_4[Mo_2Cl_8]$. (Resonance Raman data adapted from R.J.H. Clark and M.L. Franks, *J. Am. Chem. Soc.* **97** 2691 (1975).)

all the vibrational modes associated with the organic backbone which would dominate conventional IR or Raman spectroscopy. Most of the data on O_2 complexes in Table 4.8 were obtained using resonance Raman spectroscopy.

4.5.2.2 Surface-enhanced Raman spectroscopy (SERS)

Resonance Raman experiments require the presence of a suitable chromophore and laser to achieve very significant intensity enhancements. Enhancement of 10^{10} in signal intensity can also be achieved with SERS. This was first observed in the early 1970s for pyridine adsorbed onto a silver surface roughened electrochemically. The key part in the SERS experiment is the use of **plasmon resonances** in the surface to increase the intensity of the Raman scattering. Although colloidal solutions can be used, the most common SERS method is to deposit a liquid sample onto a glass or silicon surface that has a nanostructured metal surface, commonly silver, although other noble metals are used.

As a result of the sensitivity arising from the intensity enhancement SERS has been used to detect the presence of low abundance proteins in biomolecules, and has potential for cancer screening and detection. The selection rules can be different because of the interaction of the molecule with the surface, and the orientation of the molecule with respect to the surface. The combination of resonance Raman with SERS improves the detection sensitivity even further, especially as the fluorescence decay pathway is nearly completely quenched.

4.6 **Advanced applications of group theory in vibrational spectroscopy**

Symmetry and group theory are able to provide more information than just the number of IR and Raman active vibrational modes. This section summarises the use of direct products, descent of symmetry and projection operators to enable a deeper understanding of vibrational spectroscopy and its application to the characterisation of inorganic compounds.

4.6.1 **Direct products—application to overtones and combinations in vibrational spectroscopy**

In the earlier sections the IR and Raman activities of the fundamental vibrational modes were determined. In the harmonic oscillator approximation the vibrational levels are all equidistant, and the only allowed transitions are those between adjacent levels (i.e. $\Delta v = \pm 1$). However, all real vibrations are anharmonic with the result that the vibrational selection rule is relaxed so that $\Delta v = \pm 2, \pm 3$, etc. are now no longer totally forbidden. In addition, as the vibrational quantum number, v, increases the energy levels get (slightly) closer together. Features due to these anharmonic effects are often seen in the spectra usually at higher wavenumber and are often at approximately twice the wavenumber of the fundamentals, or close to the sum of two vibrational modes. These transitions are also usually weak compared to the fundamentals. The two most common examples are **overtone** and **combination bands**. Liquid water has a very pale blue colour if a very long path length is viewed, and this is due to very weak overtones and combinations of the v_{O-H} modes absorbing in the red/orange part of the spectrum (see Chapter 5, Figure 5.1). As v_{O-D} is lower than v_{O-H}, D_2O appears colourless as these bands are shifted to lower energy and out of the visible light region. The overtone and combination bands may be a contributing effect, together with the scattering of light, as to why the sea and icebergs can appear blue.

Overtones arise from vibrational transitions in which the molecule gains two or more quanta in the same vibrational mode. Therefore, if the fundamental band is found at v_i the first overtone will be at approximately $2v_i$, the second overtone at $3v_i$ etc. The difference between the overtone frequency and an integral multiple of the fundamental is a measure of the anharmonicity of the vibration. The observed overtones are always at a lower wavenumber than the integral multiples of the fundamentals. Combination bands arise from transitions to an excited state where the molecule has acquired two or more vibrational quanta, distributed between two or more vibrational modes. If two fundamentals are found at v_i and v_j, then a binary

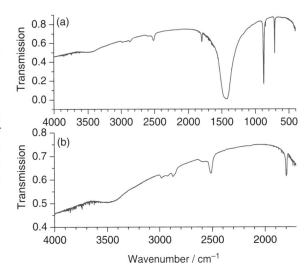

FIGURE 4.56 IR spectrum of calcite ($CaCO_3$); (a) full range, (b) expansion to show overtone and combination bands.

combination will be found at $\approx (v_i + v_j)$. Difference bands such as $v_i - v_j$ can also sometimes be observed.

The IR spectrum of the carbonate ion shows features below 1750 cm^{-1} that can be assigned to the fundamental stretching and deformation modes (Section 4.4.14). The IR spectrum of calcite is shown in Figure 4.56 and in addition to these fundamentals (v_3, 1438 cm^{-1}; v_2, 876 cm^{-1}, v_4, 712 cm^{-1} (v_1 at 1086 cm^{-1} is IR inactive)) there are weak features at 1796 and 2514 cm^{-1} together with a weaker cluster of peaks at 2854, 2875, 2925, and 2984 cm^{-1}, as shown in the expansion.

In order to identify if these bands are due to overtones or combinations, the first stage is to calculate the approximate positions of the possible overtones and combinations, and these are shown in the second column of Table 4.32.

It is then necessary to identify the IR activity of these modes. This is done by multiplying the characters of the appropriate irreducible representations to generate a new reducible representation for the overtone or combination, and then reducing this to its irreducible representation(s) and looking up their IR and Raman activities in the character tables in the normal way. Table 4.33 shows the analysis for some of the overtones and combinations.

For the $2v_1$ overtone the result of multiplying A_1' by A_1' is clearly A_1', and this will therefore only be Raman active. In the case of the $2v_2$ overtone multiplying A_2'' by A_2'' is A_1' and this will also only be Raman active. This is an example of the general principle that when an irreducible representation is multiplied by itself, the answer will always contain the totally symmetric representation. Therefore, all first overtones will be Raman active, but the IR activity is dependent on the point group. For the $v_1 + v_2$ combination, the result of multiplying A_1' by A_2'' is A_2'', and therefore this band will only be IR active.

TABLE 4.32 Potential overtone and combination bands for calcite

Overtone/combination	Approximate wavenumber/cm^{-1}	Symmetry labels	Direct product	IR activity
$2\nu_1$	2172	$A_1' \times A_1'$	A_1'	none
$2\nu_2$	1752	$A_2' \times A_2'$	A_1'	none
$2\nu_3$	2876	$E' \times E'$	$A_1' + A_2' + E'$	E'
$2\nu_4$	1424	$E' \times E'$	$A_1' + A_2' + E'$	E'
$\nu_1 + \nu_2$	1962	$A_1' \times A_2''$	A_2''	A_2''
$\nu_1 + \nu_3$	2524	$A_1' \times E'$	E'	E'
$\nu_1 + \nu_4$	1798	$A_1' \times E'$	E'	E'
$\nu_2 + \nu_3$	2314	$A_2'' \times E'$	E''	none
$\nu_2 + \nu_4$	1588	$A_2'' \times E'$	E''	none
$\nu_3 + \nu_4$	2150	$E' \times E'$	$A_1' + A_2' + E'$	E'

TABLE 4.33 Calculation of irreducible representations for D_{3h} calcite

D_{3h}	E	$2C_3$	$3C_2$	σ_h	$2S_3$	$3\sigma_v$		
$2\nu_1 \approx 2172$ cm^{-1}								
$A_1' \times A_1'$								
A_1'	1	1	1	1	1	1		$x^2 + y^2, z^2$
A_1'	1	1	1	1	1	1		$x^2 + y^2, z^2$
$\Gamma(A_1' \times A_1') = A_1'$	1	1	1	1	1	1		$x^2 + y^2, z^2$
$2\nu_2 \approx 1752$ cm^{-1}								
$A_2'' \times A_2''$								
A_2''	1	1	−1	−1	−1	1	z	
A_2''	1	1	−1	−1	−1	1	z	
$\Gamma(A_2'' \times A_2'') = A_1'$	1	1	1	1	1	1		$x^2 + y^2, z^2$
$\nu_1 + \nu_2 \approx 1962$ cm^{-1}								
$A_1' \times A_2''$								
A_1'	1	1	1	1	1	1		$x^2 + y^2, z^2$
A_2''	1	1	−1	−1	−1	1	z	
$\Gamma(A_1' \times A_2'') = A_2''$	1	1	−1	−1	−1	1	z	
$\nu_1 + \nu_3 \approx 2524$ cm^{-1}								
$\nu_1 + \nu_4 \approx 1798$ cm^{-1}								
$A_1' \times E'$								
A_1'	1	1	1	1	1	1		$x^2 + y^2, z^2$
E'	2	−1	0	2	−1	0	(x, y)	$(x^2 - y^2, xy)$
$\Gamma(A_1' \times E') = E'$	2	−1	0	2	−1	0	(x, y)	$(x^2 - y^2, xy)$
$2\nu_3 \approx 2876$ cm^{-1}								
$\nu_3 + \nu_4 \approx 2150$ cm^{-1}								
$E' \times E'$								
E'	2	−1	0	2	−1	0	(x, y)	$(x^2 - y^2, xy)$
E'	2	−1	0	2	−1	0	(x, y)	$(x^2 - y^2, xy)$
$\Gamma(E' \times E') = A_1' + A_2' + E'$	4	1	0	4	1	0		

This is an example of a second general principle that when any irreducible representation is multiplied by the totally symmetric representation, the answer always contains the original irreducible representation. On this basis the $v_1 + v_3$ and $v_1 + v_4$ combinations will both have E′ symmetry and will be both IR and Raman active. For the $2v_3$, $2v_4$ overtones and the $v_3 + v_4$ combination band the result of multiplying E′ by E′ is now no longer a single irreducible representation, and this is to be expected as the effect of multiplying two doubly degenerate representations together results in an answer that must have order 4, and none of these have been encountered. This can be reduced using the normal techniques (Section 4.4.9) and the answer is $A_1′ + A_2′ + E′$, but it is much quicker and easier to use 'tables of direct products' to work out these out (and all the others), and a table is provided in Appendix 2. $A_1′$ is Raman active, $A_2′$ is neither IR nor Raman active, and E′ is both IR and Raman active. Multiplying a representation by itself again results in the totally symmetric representation (amongst others in this case).

Using this analysis it is possible to assign the band at 1796 cm^{-1} to be due to the $v_1 + v_4$ combination band, the 2514 cm^{-1} feature to a $v_1 + v_3$ combination, and the weak cluster of bands at 2875 and 2854 cm^{-1} are probably due to the $2v_3$ overtone.

As the bands are usually very weak, unambiguous assignment is not always or often as clear cut as in the calcite case, especially if several lie at similar energies. There can also be complications when degenerate modes are involved, as the band may be split into the individual sub-bands in the direct product.

Overtones and combinations can also be used to identify the energy of the fundamentals that are both IR and Raman inactive, or hard to identify for other reasons. A good example of the former application is in D_{4h} species such as XeF$_4$ where the B_{2u} (v_5) mode is silent in both the IR and Raman, but the first overtone will be totally symmetric (A_{1g}) and, therefore, be Raman active. In the Raman spectrum a weak band at 442 cm^{-1} has been assigned to $2v_5$ implying that v_5 is approximately 221 cm^{-1}.

4.6.2 Direct products—application to selection rules

Direct products are at the heart of selection rules, as a spectroscopic transition will involve an initial state, a final state, and a dipole operator. A spectroscopic transition is only allowed if the transition moment is greater than zero, which can only happen if the product of the initial state, final state, and dipole operator spans the totally symmetric representation. The symmetry labels for the dipole operators can be easily obtained from the character tables by looking for the irreducible representations

that span x, y, z. (In the case of Raman spectroscopy it is the binary products that are required.) There is a general and very important observation that wavefunctions for normal vibrations in their ground states transform as the totally symmetric irreducible representation of the point group of the molecule (NO is a rare exceptions to this). For the vast majority of vibrational examples, this simply reduces the problem from a triple product calculation to a double product calculation because of the totally symmetric nature of the ground state, and multiplying by the totally symmetric representation leaves things unchanged. Multiplying two identical irreducible representations always results in the totally symmetric representation. This explains why a vibrational mode must have the same symmetry properties as the dipole moment operator to be IR active.

4.6.3 Descent of symmetry and correlation diagrams—application to calcite and aragonite IR spectra

The IR spectrum of calcite (Figure 4.57) has been used previously to demonstrate how spectral features are assigned to fundamentals (Section 4.4.14) and overtone and combination bands (Section 4.6.1). Aragonite is a different morphological form of calcium carbonate, and its IR spectrum is also shown in Figure 4.57. The broad intense C–O stretching mode shifts to higher wavenumber (1480 cm^{-1}) than in calcite (1438 cm^{-1}). However, the most significant change is in the 1200–500 cm^{-1} region, where a new band at 1083 cm^{-1} appears in the aragonite spectrum, and the peak at 712 cm^{-1} in the calcite spectrum is now replaced by a doublet at 713 and 700 cm^{-1} in the aragonite spectrum.

The Raman spectrum of calcite (Figure 4.48) has an intense band at 1086 cm^{-1} assigned to the symmetric v_{C-O} mode, and while this is IR inactive in calcite it gains some IR intensity in aragonite. The origin of these observations is that while 'free' CO$_3^{2-}$ (in solution or the gas phase) has D_{3h} point group, when it is in either calcite or aragonite this is no longer the case. Embedding an isolated, free ion or molecule in a solid lattice usually decreases the local symmetry, and, consequently, the number of symmetry operations because of the effect of the neighbouring atoms and ions. The splitting of the band at 712 cm^{-1} in aragonite occurs because there is no symmetry operation such as a C_3 axis that is required for the presence of a degenerate E mode. This is known as 'lifting the degeneracy'.

The symmetry operations in 'free' carbonate from the D_{3h} character table (Table 4.11) are E, $2C_3$, $3C_2$, σ_h, $2S_3$, $3\sigma_v$. In calcite, only the E, the C_3, and the three C_2 rotation axes of the isolated carbonate ion apply to the

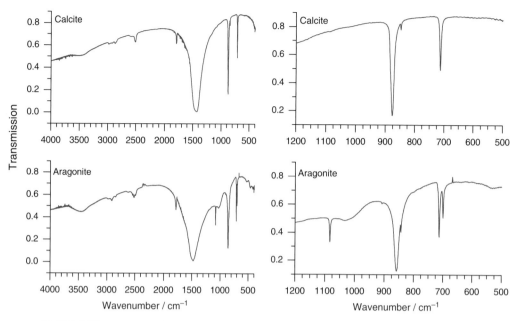

FIGURE 4.57 IR spectra of calcite (top) and aragonite (bottom) together with expansion (right).

environment of the carbonate ion in the crystal. The σ_v, σ_h, and the S_3 symmetry operations are no longer present. With this collection of symmetry operations the site symmetry for carbonate in calcite is D_3, as confirmed by the D_3 character table shown in Table 4.34.

In contrast, the carbonate site symmetry in aragonite is C_s as the only symmetry operation that applies to the ion in the solid state is one mirror plane (labelled σ_h). Therefore, the symmetry of the carbonate ion has been reduced from D_{3h} in free carbonate to D_3 in calcite and C_s in aragonite. In general, when the symmetry properties of any species are lowered the number of IR and Raman active bands changes, either because previously inactive modes become active or because the degeneracy is lifted. Although it is possible to calculate the new Γ_{vib} and Γ_{str} from scratch using the processes described in Section 4.4.13, it is easier to use **correlation tables** to deal with this **descent of symmetry**, and these are available in Appendix 3. These correlation tables show how the irreducible representations of a high symmetry point group 'map' onto those in the lower symmetry one. The D_{3h}

example of the carbonate ion is shown in Table 4.35, with the D_{3h} irreducible representations in the centre, and the D_3 and C_s derived ones to the left and right. The IR and Raman activity is obtained from the character tables in the normal way.

The descent of symmetry has no effect on the IR and Raman activity in calcite. In contrast, in aragonite the formerly IR inactive A_1' symmetric stretching mode becomes A' which is both IR and Raman active. This analysis also shows that the doubly degenerate E' mode in D_{3h} remains doubly degenerate (E) in D_3, but in aragonite it is split into two components (A' and A'').

The presence of the mode around 1085 cm^{-1} and the splitting of the 712 cm^{-1} band can be used to identify the presence of aragonite, rather than calcite, in naturally occurring carbonates found in bird's eggs and crustacean shells as shown in the spectra in Figure 4.58.

Descent of symmetry/correlation tables can also be useful when carrying out ligand substitution reactions, or isotopic substitution. For example, once Γ_{str} has been worked out for [M(CO)$_6$] or [M(CO)$_4$] complexes, the correlation

TABLE 4.34 D_3 character table

D_3	E	$2C_3$	$3C_2$		
A_1	1	1	1		$x^2 + y^2$, z^2
A_2	1	1	−1	z, R_z	
E	2	−1	0	(x, y), (R_x, R_y)	$(x^2 - y^2, xy)$, (xz, yz)

TABLE 4.35 Correlation table for descent of symmetry from D_{3h} to D_3 and C_s

IR and Raman activity	D_3 CO_3^{2-} in calcite	D_{3h} Free CO_3^{2-}	C_s CO_3^{2-} in aragonite	IR and Raman activity
Raman	A_1	A_1'	A'	IR Raman
IR	A_2	A_2'	A''	IR Raman
IR Raman	E	E'	$A' + A''$	IR Raman
Raman	A_1	A_1''	A''	IR Raman
IR	A_2	A_2''	A'	IR Raman
IR Raman	E	E''	$A' + A''$	IR Raman

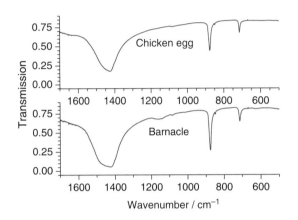

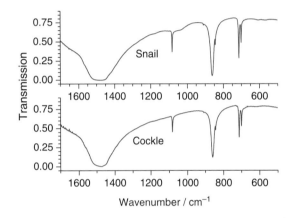

FIGURE 4.58 IR spectra of shells showing major carbonate component is calcite (left) and aragonite (right).

tables in Appendix 3 can be used to determine Γ_{str} for all the $[M(CO)_{6-n}L_n]$ and $[M(CO)_{4-n}L_n]$ reaction products.

4.6.4 Γ_{vib} for linear molecules

In Section 4.4.9.2, Γ_{str} was determined for linear molecules using the 'by-eye' method as the tabular method was not possible because of the C_∞ axis. The way to solve the C_∞ problem is to calculate Γ_{str} and Γ_{vib} for a non-infinite group using D_{2h} for $D_{\infty h}$ and C_{2v} for $C_{\infty v}$ with the z direction along the internuclear axis and then use the correlation tables in Appendix 3 to map back to the infinite groups.

4.6.5 Fermi resonance

In some spectra, such as the Raman spectrum of CCl_4 in Figure 4.59, there are unexpected doublets observed, such as those at 787 and 759 cm^{-1}. The three peaks at 458, 313, and 217 cm^{-1} can be readily assigned to the ν_1 (A_1), ν_4 (T_2), and ν_2 (E) modes, respectively. However, where the ν_4 (T_2) fundamental is expected around 770 cm^{-1} a doublet of approximately equal intensity is observed.

This is due to Fermi resonance, which occurs when a fundamental vibrational mode and a combination (or overtone) mode of the same symmetry are in close proximity. In this case the fundamental has T_2 symmetry, and by considering the possible overtones and combinations it can be seen that none of the overtones are involved. But the ($\nu_1 + \nu_4$) combination is predicted to occur at ca. 770 cm^{-1} and this is the most likely mode, as it has T_2 symmetry ($A_1 \times T_2 = T_2$). When two modes of the same symmetry interact they push each other apart energetically and can 'steal' intensity from each other. While it might be expected that the fundamental would have much greater intensity than the combination mode (the $2\nu_3$ overtone at 1535 cm^{-1} is much weaker than either of the two components of the Fermi doublet), it is the intensity stealing that explains why the doublet is of almost equal intensity. As a result of the mixing it is not possible to say that one of the features is due to the fundamental and the other to the combination, rather that the doublet overall is due to the Fermi resonance of the ν_3 and ($\nu_1 + \nu_4$) modes.

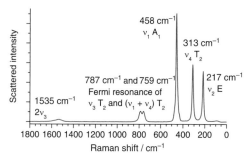

FIGURE 4.59 Raman spectrum (1064 nm excitation) of CCl_4 showing Fermi resonance.

Fermi resonance is pernicious and present in many vibrational spectra. Other classic examples where Fermi resonance is encountered include the interaction of the overtone of the CO_2 bending mode (667.4 cm^{-1}) with the symmetric stretch to give bands at 1285.4 and 1388.2 cm^{-1} in the Raman spectrum. In KNCO, bands are observed at 1282 and 1202 cm^{-1} due to Fermi resonance of the overtone of the bending mode at 630 cm^{-1} and the CO stretching mode expected at 1240 cm^{-1}. Therefore, it is important to be aware of the potential impact Fermi resonance can have on spectra. However, to mis-quote Samuel Johnson it may be noted that "Fermi resonance is the last refuge of the scoundrel" in trying to assign "interesting" features in the spectrum.

4.6.6 Projection operators and symmetry coordinates to visualise vibrational modes

It is now relatively routine to use computational chemistry programs to calculate the frequencies/wavenumbers of the vibrational modes and also to visualise them. These are becoming ever more accurate, but a discrepancy between calculated and observed values is expected as the calculated values are normally harmonic while the experimental ones are anharmonic. However, it is also possible to visualise the various vibrational modes by simple application of symmetry. The schematic diagrams in this chapter have used both of these approaches. While it is possible to employ this approach for bending modes, it is most commonly applied to stretching modes as they are much easier to visualise. The process involves identifying internal coordinates such as bond stretching modes, and then seeing how these behave, or are projected, under the various symmetry operations of the point group. For XY_3, the bond stretching modes are labelled Δr_1, Δr_2, and Δr_3. The symmetry coordinate (S) allows the visualisation of the vibrational mode and this is expressed for stretching modes (Δr) as:

$$S = N \sum_K \chi_i(K).K(\Delta r)$$ **Eqn 4.9**

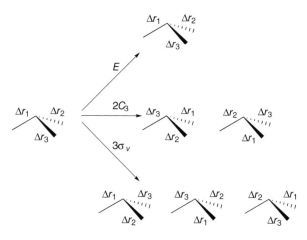

FIGURE 4.60 Effect of symmetry operations on Δr_1, Δr_2, and Δr_3 internal coordinates in XY_3.

where $\chi_i(K)$ is the character of the irreducible representation K and $K(\Delta r)$ is the transformation of the basis vector (i.e. internal coordinate Δr) under that symmetry operation. N is a normalisation constant, and is the inverse of the square root of the sum of the squares of the coefficients in the expression.

The effect of using the three changes in the X–Y bond length (Δr_1, Δr_2, and Δr_3) as the internal coordinates and how they are projected by the symmetry operations is shown in Figure 4.60 and Table 4.36. It is important to note that all of the symmetry operations are included, that is, both C_3^1 (120°) and C_3^2 (240°) and all three of the σ_v reflections.

Application of Eqn 4.9 leads to:

$$S_{A_1} = \frac{1}{\sqrt{3}}(\Delta r_1 + \Delta r_2 + \Delta r_3)$$
$$S_{A_2} = 0$$
$$S_E = \frac{1}{\sqrt{6}}(2\Delta r_1 - \Delta r_2 - \Delta r_3)$$

S_{A_2} being equal to zero should come as no surprise as $\Gamma_{str} = A_1 + E$ for pyramidal XY_3 C_{3v} species (Section 4.4.9.1). Therefore, it is only normally necessary to generate the

TABLE 4.36 Effect of symmetry operations on Δr_1, Δr_2, and Δr_3 internal coordinates in XY_3 molecules

C_{3v}	E	$2C_3$	$3\sigma_v$
Δr_1	Δr_1	$\Delta r_3 + \Delta r_2$	$\Delta r_1 + \Delta r_2 + \Delta r_3$
A_1	1	1	1
A_2	1	1	−1
E	2	−1	0

TABLE 4.37 Use of combination and difference internal coordinates to identify second E symmetry coordinate

C_{3v}	E	$2C_3$	$3\sigma_v$
Δr_1	Δr_1	$\Delta r_3 + \Delta r_2$	$\Delta r_1 + \Delta r_2 + \Delta r_3$
Δr_2	Δr_2	$\Delta r_1 + \Delta r_3$	$\Delta r_3 + \Delta r_2 + \Delta r_1$
Δr_3	Δr_3	$\Delta r_2 + \Delta r_1$	$\Delta r_2 + \Delta r_1 + \Delta r_3$
$\Delta r_2 + \Delta r_3$	$\Delta r_2 + \Delta r_3$	$\Delta r_1 + \Delta r_3 + \Delta r_2 + \Delta r_1$ $= 2\Delta r_1 + \Delta r_3 + \Delta r_2$	$\Delta r_3 + \Delta r_2 + \Delta r_1 + \Delta r_2 + \Delta r_1 + \Delta r_3$
$\Delta r_2 - \Delta r_3$	$\Delta r_2 - \Delta r_3$	$\Delta r_1 + \Delta r_3 - \Delta r_2 - \Delta r_1$ $= \Delta r_3 - \Delta r_2$	$\Delta r_3 + \Delta r_2 + \Delta r_1 - \Delta r_2 - \Delta r_1 - \Delta r_3$ $= 0$
A_1	1	1	1
E	2	−1	0

symmetry coordinates for the components of Γ_{str}. For molecules with at least one threefold axis, degenerate vibrational modes are found. As the E mode is doubly degenerate, with a dipole moment change in both the x and y directions, there must be a second symmetry coordinate orthogonal (perpendicular) to the first one. (To help visualise this it is useful to recall that the p orbitals also transform with e symmetry (orbitals use lower case symbols) in the C_{3v} point group.) If Δr_2 or Δr_3 are used as the basis, these generate symmetry coordinates of the form $(1/\sqrt{6})(2\Delta r_2 - \Delta r_1 - \Delta r_3)$ and $(1/\sqrt{6})(2\Delta r_3 - \Delta r_1 - \Delta r_2)$, which although appear different are really just the same as the original and are not orthogonal, and will not result in a dipole moment change perpendicular to the one when Δr_1 is used. What is required is to take the combination $\Delta r_2 + \Delta r_3$ and difference $\Delta r_2 - \Delta r_3$, and this is shown in Table 4.37.

Using $\Delta r_2 + \Delta r_3$,

$$S_E = \frac{1}{\sqrt{6}}(-2\Delta r_1 + \Delta r_2 + \Delta r_3)$$

which is just the same form as the previous one.

Using $\Delta r_2 - \Delta r_3$,

$$S_E = \frac{1}{\sqrt{2}}(\Delta r_2 - \Delta r_3)$$

Therefore, the symmetry coordinates for the X–Y stretching modes in a XY_3, C_{3v} molecule are

$$S_{A_1} = \frac{1}{\sqrt{3}}(\Delta r_1 + \Delta r_2 + \Delta r_3)$$

$$S_E = \begin{cases} \dfrac{1}{\sqrt{6}}(2\Delta r_1 - \Delta r_2 - \Delta r_3) \\[2ex] \dfrac{1}{\sqrt{2}}(\Delta r_2 - \Delta r_3) \end{cases}$$

and these are summarised pictorially in Figure 4.61.

The consequence of this is that the relative motion of the atoms is different in the two contributions to the E mode. In the left-hand vibrational E mode, all the Y atoms are in motion, and the dipole moment change is directed along Δr_1. In contrast, in the right-hand mode only two of the Y atoms are involved, and now the dipole moment change is directed perpendicular to Δr_1. This has important consequences on isotopic substitution in degenerate modes, where non-binomial patterns are observed, and is discussed in detail in the Online Resource Centre. The arrows are schematic, and it must be remembered that the central atom will also be involved to maintain the centre of gravity at the same point.

This approach can be extended to other simple molecules, and can also be used to identify the form of the symmetry adapted linear combination of atomic orbitals used to construct molecular orbitals, where the + and − represent the sign of the wavefunction. A similar analysis could be carried out for D_{3h} trigonal planar molecules and the symmetry coordinates are very similar.

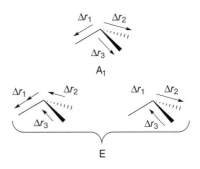

FIGURE 4.61 Representation of vibrational modes in pyramidal XY_3 molecule derived using symmetry coordinates.

Bibliography

R.L. Carter. (1997) *Molecular Symmetry and Group Theory.* New York: Wiley.

J.S. Ogden. (2001) *Introduction to Molecular Symmetry.* OUP Primer.

P.H. Walton. (1998) *Beginning Group Theory for Chemistry.* Oxford: Oxford University Press.

A. Vincent. (2001) *Molecular Symmetry and Group Theory.* 2nd ed. Chichester: Wiley.

S.F.A. Kettle. (1995) *Symmetry and Structure.* 2nd ed. Chichester: Wiley.

G. Davidson. (1991) *Group Theory for Chemists.* Basingstoke, UK: Macmillan.

K.C. Malloy. (2011) *Group Theory for Chemists.* 2nd ed. Cambridge, UK: Woodhead Publishing.

K. Nakamoto. (2008) *Infrared and Raman Spectra of Inorganic and Coordination Compounds.* Oxford: Wiley.

E.B. Wilson, J.C. Decius, and P.C. Cross. (1955) *Molecular Vibrations.* New York: Dover Publications.

E. Smith and G. Dent. (2005) *Modern Raman Spectroscopy: A Practical Approach.* Chichester: Wiley.

M.T. Weller, T.L. Overton, J.A. Rourke, and F.A. Armstrong. (2014) *Inorganic Chemistry*, 6th ed. Oxford: Oxford University Press.

C.E. Housecroft and A.G. Sharpe. (2012) *Inorganic Chemistry*, 4th ed. Harlow: Pearson.

P.W. Atkins and J. de Paula. (2014) *Physical Chemistry* 10th ed. Oxford: Oxford University Press.

Electronic absorption and emission spectroscopy

5

FUNDAMENTALS

5.1 Introduction

The electronic absorption and emission spectra of atoms and molecules involves the rearrangement of electrons and provides a wealth of information on the structure and properties of such compounds, as well as their identification. While a variety of electronic transitions are possible in inorganic compounds, the most important are those involving electrons in the d orbitals of transition metal compounds and complexes. Many of these transitions have energies in the visible region of the electromagnetic spectrum, or just outside it in the ultraviolet (UV) or near infrared (NIR), and the characteristic colour of many transition metal complexes is the result of electronic transitions involving the d orbitals.

The approximate wavelength (nm), wavenumber (cm^{-1}), and energy (eV) values for the visible part of the electromagnetic spectrum are shown in the **colour wheel** in Figure 5.1. Blue light has an approximate wavelength range 490–430 nm (20400–23300 cm^{-1}, 2.5–2.9 eV), while red light is associated with 800–620 nm (12500–16100 cm^{-1}, 1.5–2.0 eV) radiation.

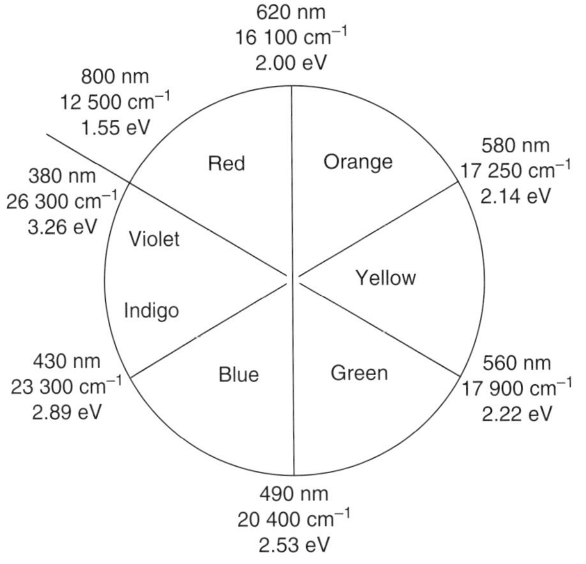

In visible light absorption, which is the more common experimental technique, it is the component of the incident radiation that is not absorbed that gives rise to the colour seen by the human eye. The colour observed is opposite that of the colour being absorbed in the colour wheel shown in Figure 5.1, and a representative spectrum is shown in Figure 5.2. If a compound has an absorption of the indigo/violet component of visible light 430–380 nm (23300–26300 cm^{-1}, 2.9–3.3 eV), this results in a yellow material. This is why compounds with intense UV absorption bands that have tails into the visible region often appear yellow. Green colouration can be caused by either absorption of violet light close to 400 nm, as well as red light at 700 nm. When there are multiple absorption features it gets harder to use this approach, but it is still illustrative.

FIGURE 5.1 Colour wheel and complementary colours.

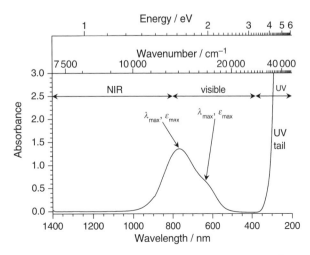

FIGURE 5.2 UV-vis-NIR spectrum of 0.1 M [VO(H$_2$O)$_5$]$^{2+}$.

EXAMPLE 5.1

If a transition metal compound absorbs green light, what colour will the human eye observe?

ANSWER

From the colour wheel in Figure 5.1, the absorption of green light at ca. 500 nm (20 000 cm^{-1}) produces red coloured compounds.

SELF TEST

What is the colour of the solution giving rise to the spectrum shown in Figure 5.2?

Electronic absorption in transition metal compounds may also occur to higher or lower energies than the visible region. These correspond to the short-wavelength UV between 190 nm and 380 nm (the onset of the visible region with violet light) and long-wavelength NIR radiation between 800 nm and 2000 nm (Figure 5.2). As most of the electronic transitions are in the UV and visible regions of the electromagnetic spectrum, electronic spectroscopy is also often known as UV–visible spectroscopy, shortened to **UV-vis spectroscopy**. If the spectral range is extended to include the NIR, then it is known as **UV-vis-NIR spectroscopy**.

5.2 Experimental UV-vis spectroscopy

UV-vis spectrophotometers are among the cheapest of analytical instruments and usually consist simply of a light source, a monochromator, and a detector. For NIR and visible measurements, quartz halogen lamps are used that produce electromagnetic radiation smoothly across the wavelength range 360–2000 nm. For the UV region, a deuterium discharge lamp is used producing light in the wavelength range 190–400 nm. The monochromator is usually a diffraction grating, and this can be used in either scanning or dispersive mode.

Figure 5.3 shows a schematic optical layout for a double beam scanning spectrophotometer. The diffraction grating acts as the monochromator and is rotated so that one wavelength at a time is passed through the sample and reference to the detector. The slits either side of the diffraction grating define the beam shape and resolution (narrower slits lead to higher resolution, but longer experimental times for the same signal-to-noise) and reduce stray light. In the optical layout shown in Figure 5.3 there is only one detector. The chopper (or beam splitter) is half-mirrored and half open, and as it rotates the light path alternates between the sample channel and the reference channel. In contrast to grating IR spectrophotometers (which are very rare these days) the monochromator is placed before the sample to reduce the risks of photochemistry and the sample undergoing a reaction. A photomultiplier detector is conventionally used in scanning instruments for visible and UV regions, with a diode detector for the NIR. Single beam instruments are also found with just one position in the sample compartment. Spectra take the order of a few minutes to acquire.

In the dispersive set-up shown in Figure 5.4, the static diffraction grating disperses a range of wavelengths onto a position sensitive detector such as photodiode array (PDA) or charged coupled device (CCD). This can result in a very short acquisition time of seconds or less, and is only limited by the readout time of the detector, but there may be reduced spectral range or resolution compared to the scanning instruments. As there is no moving part within the system it can be modularised and miniaturised with the source, sample, and detector separated from each other by optical fibres, and the box containing the detector components reduced to the size of a large matchbox.

Fourier transform instruments similar to those used for IR spectroscopy (see Chapter 4, Figure 4.6) are available, but because of the shorter wavelengths involved, the tolerances on the optical

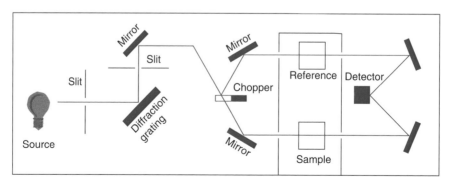

FIGURE 5.3 Schematic diagram of double beam scanning UV-vis spectrophotometer.

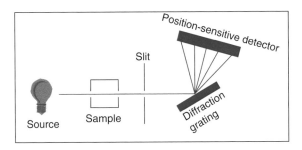

FIGURE 5.4 Schematic diagram of dispersive UV-vis spectrophotometer.

components are much tighter, adding considerably to the cost, and they are only used as research tools.

The majority of measurements using electronic absorption spectroscopy of inorganic compounds, and especially transition metal complexes, use solutions, which are contained in glass cuvettes for visible wavelengths, or quartz cuvettes for UV-vis studies. These normally have a path length of 1 cm, but a large selection of shapes and sizes are available, as well as cuvettes with taps for work with air sensitive materials. For high precision work, 'matched pairs' of cuvettes should be used, which have identifying marks or codes. In recent years, 'plastic' cuvettes, which are much more robust and cheaper than quartz cuvettes, have been introduced and some are transparent down to about 300 nm.

The important parameters reported from the spectra are the peak position, λ_{max}, given either in nm or cm^{-1} units, and the molar absorptivity, ε_{max}, and their identification is shown in Figure 5.2. The molar absorptivity, ε_{max} (in older literature it is often called the extinction coefficient), is determined from the experimental spectra using the Beer–Lambert law,

$$A = \varepsilon c l = -\log_{10} T = -\log_{10} \frac{I_t}{I_0} = \log_{10} \frac{I_0}{I_t} \quad \textbf{Eqn 5.1}$$

where c is the concentration of the solution and l is the path length of the cuvette. As 1 cm path length cuvettes are normally used, the conventional units of ε are dm^3 mol^{-1} cm^{-1} (or M^{-1} cm^{-1}). The absorbance A is defined as the absorbance at λ_{max}, and is calculated by the instrument software as the negative $\log_{10}$ of the transmission, T (I_t/I_0), of light through the sample. In optical spectroscopy, $\log_{10}$ is used to calculate the absorption, but in X-ray techniques, the natural $\log_e$ (ln) is used to calculate the mass absorption coefficient. If the Beer–Lambert law is not obeyed at different concentrations, this implies that some chemistry is taking place in the sample.

EXAMPLE 5.2

Calculate ε_{max} for a 0.1 M solution of [Cr(H$_2$O)$_6$]$^{3+}$ in a 1.0 cm cuvette that has d–d electronic transitions with λ_{max} at 17 100 cm^{-1}, 24 000 cm^{-1}, and 37 800 cm^{-1} with absorbance of 1.3, 1.5, and 0.40, respectively.

ANSWER

As there are three d–d transitions each will be expected to have a different ε_{max} value. $\varepsilon_{max} = A/cl$. So for the band at 17 100 cm^{-1}, $\varepsilon_{max} = 1.3/(0.1 \times 1) = 13$ dm^3 mol^{-1} cm^{-1}. For the 24 000 cm^{-1} peak, $\varepsilon_{max} = 1.5/(0.1 \times 1) = 15$ dm^3 mol^{-1} cm^{-1} and for the 37 800 cm^{-1} band, $\varepsilon_{max} = 0.4/(0.1 \times 1) = 4$ dm^3 mol^{-1} cm^{-1}.

SELF TEST

(a) Calculate the absorbance of peaks that have 50%, 10%, and 1% transmission.

(b) Calculate the percentage transmission for peaks with an absorbance of 2, 3, and 4.

Figures 5.2 and 5.5 show typical UV-vis-NIR spectra for 0.1 M [VO(H$_2$O)$_5$]$^{2+}$ and 0.2 M [Ni(H$_2$O)$_6$]$^{2+}$, respectively, and these cover both the UV-vis and NIR spectral ranges. Most commercial and teaching laboratory **UV-vis spectrophotometers** scan and collect the data in nanometre units, the majority of which have an upper

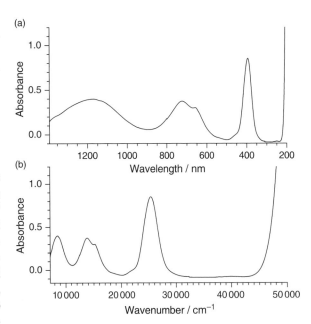

FIGURE 5.5 Electronic absorption spectra of 0.2 M [Ni(H$_2$O)$_6$]$^{2+}$ solution as a function of (a) wavelength (nm) and (b) wavenumber (cm^{-1}).

wavelength limit of 900 nm, or sometimes 1100 nm, though more expensive research instruments extend into the NIR region as far as 3000 nm. Therefore, the UV-vis spectra in Figure 5.2 and Figure 5.5(a) are those normally encountered with teaching laboratory nm based instruments.

The nm, cm^{-1}, and eV scales in Figure 5.2 clearly show that the wavelength scale is not linear in energy, and Figure 5.5 shows that one consequence of spectra presented in wavelength (nm) units is that they display very broad peaks at long wavelengths and sharper ones at shorter wavelengths. Spectroscopists prefer spectra that are linear in energy, and for electronic absorption spectra, wavenumber units (cm^{-1}) are normally used. Spectra obtained from materials in the solid state may use eV as the energy unit (1 eV = 8065.5 cm^{-1}) as electronvolts are normally used to describe band gaps in semiconductors. The lower spectrum in Figure 5.5 is plotted in wavenumber units of cm^{-1}, and the bandwidths are now much more equal. Although it is easy to convert between nm and cm^{-1}, the data point interval will not be constant. Subsequent spectra are plotted in wavenumber, but with a nanometre scale present to aid identification and comparison with spectra obtained in teaching laboratories. Figures 5.2 and 5.5 also show that UV-vis spectra often contain features of very different absorbance values. The peaks at short wavelength/high wavenumber are off the top of the scale used to show the λ_{max} and ε_{max} values of the other features. It is often necessary to collect UV-vis data at different sample concentrations to be able to identify the peak positions of all of the features.

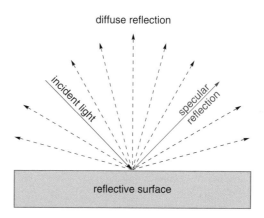

FIGURE 5.6 Specular (solid line) and diffuse reflectance (dotted lines).

Although UV-vis spectra can be obtained from solid-state materials, the sample preparation is crucial because of the short penetration depth of visible and especially UV light. While very thin homogeneous films or KBr discs can be studied, **diffuse reflectance** techniques are a popular alternative for inorganic solid-state materials. The scattering from solids can be divided into two parts (illustrated in Figure 5.6): the **specular component**, where the scattered radiation is at the same angle as the incident beam, and the diffuse scatter where the light is scattered in all directions.

If the sample has no absorption in the visible part of the spectrum, it appears white as all the colours are diffusely reflected at equal amounts. If the sample has absorption bands in the visible part of the spectrum, these are removed from the diffusely scattered component, and the material appears coloured. It is this diffuse scatter that gives colours to everyday objects. As the diffusely scattered light is spread over a large number of angles, it is important to be able to collect as much of this as possible, while excluding the specular component. The majority of diffuse reflectance accessories for UV-vis spectrometers use what is known as an **integrating sphere** which is coated with highly reflective $BaSO_4$ or PTFE to collect the diffuse reflectance. An alternative optical arrangement known as a 'praying mantis', which can also be used for IR measurements, allows for smaller quantities of samples to be used. In both of these cases the sample under investigation is diluted in a highly reflective material, such as $BaSO_4$. A spectrum of the neat $BaSO_4$ is collected first and acts as a reference or background spectrum to compare the reflectivity of the compound of interest diluted in $BaSO_4$. The parameter that is measured is the reflectivity (or percent reflectivity) of the sample at any given wavelength, and a representative spectrum of the pigment indigo (used to make blue denim) is shown in Figure 5.7(a).

EXAMPLE 5.3

Estimate the λ_{max} and ε_{max} values for the three peaks shown in the spectrum of $[Ni(H_2O)_6]^{2+}$ in Figure 5.5.

ANSWER

The λ_{max} (A) values are 8500 cm^{-1} (0.4), 14500 cm^{-1} (0.35), and 25500 cm^{-1} (0.85); therefore, as the concentration is 0.2 M, and assuming the path length is 1.0 cm, the λ_{max} (ε_{max}) values are 8500 cm^{-1} (8 dm^3 mol^{-1} cm^{-1}), 14500 cm^{-1} (7 dm^3 mol^{-1} cm^{-1}), and 25500 cm^{-1} (17 dm^3 mol^{-1} cm^{-1}). (The central peak is made up of one at 13800 cm^{-1} (0.37) and 15200 cm^{-1} (0.30), see Example 5.26 for its explanation.)

SELF TEST

Convert the following wavelength values to their corresponding wavenumber (cm^{-1}) units: 1000 nm, 600 nm, 400 nm, 250 nm. (A useful conversion is the value in nm = 10^7/cm^{-1}.)

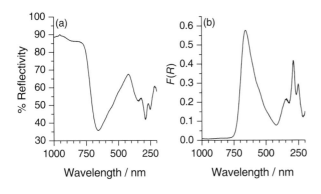

FIGURE 5.7 Diffuse reflectance spectra of indigo diluted in BaSO$_4$: (a) % reflectivity; (b) after conversion to Kubelka–Munk function, $F(R)$.

While it is possible to determine the peak positions from the reflectivity, these are inverted compared to the conventional absorption spectra. To obtain absorption-like spectra from diffuse reflectance data it is necessary to use the **Kubelka–Munk function**,

$$F(R) = \frac{(1-R)^2}{2R}$$ **Eqn 5.2**

where R is the reflectivity. The application of the Kubelka–Munk function to the diffuse reflectance data of indigo is shown in Figure 5.7(b). The Kubelka–Munk function can be used quantitatively under the correct conditions, but for most inorganic characterisation examples it is the peak position (λ_{max}) that is most important. The form of the diffuse scattering spectrum can also be affected by particle size especially when the crystallite size is of the order of the wavelength of UV-vis light (0.2–1 μm), a fairly typical particle size for powder materials. Further corrections for the scattered intensity due to this effect, known as Mie scattering, may also be necessary if quantitative absorption data are required.

5.3 Types of electronic transitions in inorganic compounds and complexes

A number of different electronic transitions are possible in inorganic compounds and complexes. These may involve: electrons moving from one d orbital to another which are known as d–d transitions; electrons moving from ligand based p orbitals to metal based d orbitals (or vice versa) known as ligand to metal charge transfer (LMCT) or metal to ligand charge transfer (MLCT)

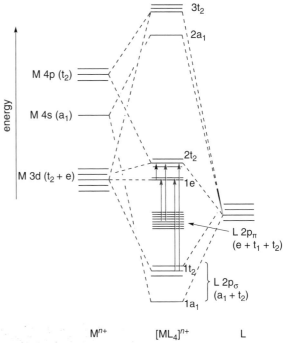

FIGURE 5.8 Schematic molecular orbital diagram for a tetrahedral complex showing d–d transitions (in black) and ligand to metal charge transfer (LMCT) transitions (in blue).

transitions; electrons moving between metals of different oxidation states, called intervalence charge transfer (IVCT) transitions; and electrons moving between orbitals within the ligands, known as intra-ligand transitions. Figure 5.8 shows an energy level diagram (molecular orbital (MO) diagram) for a tetrahedral transition metal complex and shows example electronic excitations that can occur in this system corresponding to d–d transitions and LMCT transitions. Figure 5.8 also indicates that the d–d transitions between the 1e and 2t$_2$ MOs (shown in black) occur at a lower energy than the charge transfer ones from the non-bonding and bonding ligand based orbitals (shown in blue). Before discussing these d–d transitions (Section 5.5) and charge transfer transitions (Section 5.6) in detail it is necessary to consider which of the possible electronic transitions will be observed in different circumstances and their intensities. This is controlled by the **selection rules**.

5.4 Selection rules and absorption intensities in UV-vis spectroscopy

The electronic selection rules for molecules are derived from those for atomic spectroscopy, and are broken down into two parts, a spin selection rule and a dipole selection rule. To a good first approximation they can be dealt with separately.

5.4.1 Spin selection rule

As photons cannot give up or accept any spin during the interaction with the sample, electronic transitions are only allowed if the number of unpaired electrons (including zero) is the same in the ground state and the excited state, that is, $\Delta S = 0$.

5.4.2 Dipole selection rules

Photons have one unit of angular momentum; therefore one unit of angular momentum must either be absorbed or emitted by the species under investigation during UV-vis spectroscopy. This is the origin of the dipole selection rule, as electromagnetic radiation consists of alternating electric and magnetic dipoles. In the case of electronic absorption spectroscopy it is the electric dipole that couples with the sample. In principle magnetic dipole and electric quadrupole transitions are also possible. For the d-block elements these are found to be much weaker than the electric dipole transitions and are usually not considered, but magnetic dipole transitions are more common in the spectra of lanthanides and actinides.

5.4.2.1 Orbital selection rule

In atomic spectroscopy the dipole selection rule is expressed as the **orbital selection rule**, $\Delta l = \pm 1$, as this explicitly involves one unit of angular momentum. As a result, s $\leftrightarrow$ p, p $\leftrightarrow$ d, and d $\leftrightarrow$ f transitions are **allowed** but d $\leftrightarrow$ d, d $\leftrightarrow$ s, etc. are **forbidden**. For molecules, one unit of angular momentum is still involved. As d orbitals can be considered to have a limited role in bonding in coordination complexes they retain considerable 'atomic' character, and it is useful to continue the link with atomic spectroscopy and express the dipole selection rule in terms of the orbital selection rule, $\Delta l = \pm 1$. Therefore, all d–d transitions in molecules are forbidden by this selection rule, and are expected to be weak. Those that are effectively p $\leftrightarrow$ d transitions, such as LMCT, are allowed.

5.4.2.2 Parity/Laporte selection rule

A second variant on the dipole selection rule is known as the **parity selection rule**. If the molecule/ion has a centre of symmetry (inversion centre; see Section 4.4.1.8) transitions are only allowed between orbitals of different parity. This gives rise to the parity selection rule, which states that g $\leftrightarrow$ u are allowed but g $\leftrightarrow$ g and u $\leftrightarrow$ u are forbidden (g stands for gerade—German for even—and

u stands for ungerade—German for odd). The parity selection rule is also known as the **Laporte selection rule**. The Laporte/parity selection rule is very important for octahedral and square planar complexes as it also makes the d $\leftrightarrow$ d transitions forbidden in these geometries. But for complexes with other geometries where there is no inversion centre such as tetrahedral, the application of this selection rule is relaxed, and the observed transitions are more intense. The lack of a centre of symmetry, as in tetrahedral complexes, is related to the nuclear coordinates, and even if it is absent, the d orbitals retain their 'even' atomic character, so d–d transitions will still be weak.

For high-symmetry compounds with a centre of symmetry, the orbital and Laporte/parity selection rules are two ways of expressing the same thing. For lower symmetry species it is important to remember that the d–d transitions will still be forbidden by the orbital selection rule, even though the Laporte selection rule appears to be relaxed.

EXAMPLE 5.4

Identify which of the following complexes have a centre of symmetry: $[Ni(H_2O)_6]^{2+}$, *cis*-$[CoCl_2(NH_3)_4]^+$, *trans*-$[CoCl_2(NH_3)_4]^+$, $[NiCl_4]^{2-}$, $[Ni(CN)_4]^{2-}$ and, therefore, have more weakly absorbing d–d transitions.

ANSWER

See Figure 5.9.

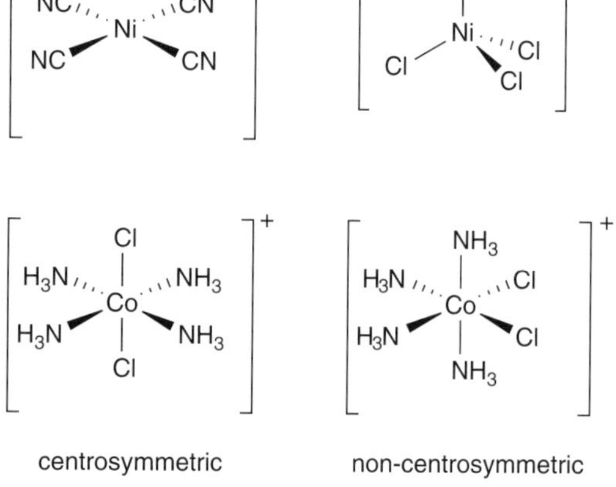

centrosymmetric   non-centrosymmetric

FIGURE 5.9 Examples of centrosymmetric and non-centrosymmetric complexes.

5.4.3 Relaxation of selection rules

The discussion above implies that all d–d transitions are forbidden by the dipole selection rule, and that none will be observed. However, as many transition metal compounds are coloured due to d–d transitions, there must be mechanisms which 'relax' the selection rules. Such transitions are often described as 'less forbidden' or 'more allowed' when discussing UV-vis spectra.

5.4.3.1 Relaxation of spin selection rule

The spin selection rule ($\Delta S = 0$) can be relaxed when the electronic spin states acquire some orbital character due to spin–orbit coupling so that they are no longer treated as having pure spin character. As spin–orbit coupling increases with atomic number, the degree of breaking of this rule, sometimes termed 'allowedness' of spin-forbidden transitions, increases from 3d to 4d to 5d complexes.

5.4.3.2 Relaxation of dipole selection rule

The orbital selection rule ($\Delta l = \pm1$) is relaxed if there is mixing of the orbitals so that they are no longer pure d orbitals. If the metal d orbitals mix with p orbitals then the d–d transitions are no longer pure d to pure d, but have some p character, and become more intense. This can happen to a small degree in octahedral complexes where the metal e_g orbitals acquire some p character as they are involved in the σ-bonding framework with the ligand p orbitals. This mixing of d and p character is much greater when the metal p and d orbitals have the same symmetry properties or labels, such as for tetrahedral molecules (Figure 5.8) where the p$_x$, p$_y$, and p$_z$ orbitals and the d$_{xy}$, d$_{xz}$, and d$_{yz}$ orbitals both have t$_2$ symmetry.

There are two significant ways that the Laporte/parity selection rule can be relaxed in centrosymmetric complexes. The first is a static effect where the complex is not perfectly centrosymmetric due to packing effects or other small distortions. The second is a dynamic effect and is due to the vibrations of the molecule. If one or more of the vibrational modes of the complex causes the loss of the inversion centre, the Laporte selection rule is relaxed. For example, in the T$_{1u}$ stretching mode in an octahedral complex (see Chapter 4, Figure 4.40), one of the ligands approaches the metal, whilst the other goes further away, thus removing the centre of symmetry. This is known as **vibronic coupling**. The vibrational motion of the complex is also the reason for relatively broad bands in electronic absorption spectra of complexes. The electronic transition occurs over a much shorter timescale than the vibrational motion, so that every electronic transition samples a molecule in a slightly different geometry, so resulting in a broad spread of energies.

In the case of tetrahedral complexes, both the orbital mixing and the lack of an inversion centre can be used to explain why the intensity of the d–d transitions can be up to two orders of magnitude greater than in the corresponding octahedral complexes.

5.4.4 Typical UV-vis intensities in transition metal compounds

The typical molar absorptivities (ε) of bands observed in electronic absorption spectra of transition metal complexes based on the selection rules are summarised in Table 5.1.

Therefore, it is usually possible to assign the bands observed to one of the above types using ε_{max}, but often it is necessary to record the spectra at different concentrations in order to get information about both d–d and charge transfer transitions.

TABLE 5.1 Representative molar absorptivities (ε_{max}) for electronic transitions in transition metal complexes

Band type	Selection rules		ε_{max}/(dm^3 mol^{-1} cm^{-1})
	Spin	Dipole	
LMCT or MLCT	Spin allowed	Orbital allowed	Laporte allowed	3000–50 000
d–d T$_d$	Spin allowed	Orbital forbidden	Laporte not relevant	250–1000
d–d O$_h$	Spin allowed	Orbital forbidden	Laporte forbidden	5–100
d–d O$_h$	Spin forbidden	Orbital forbidden	Laporte forbidden	<1

When concentrated HCl is added to a pale pink solution of $[Co(H_2O)_6]^{2+}$, the solution turns an intense blue colour. Explain the difference in the intensity in colour of these two solutions.

ANSWER

The addition of excess Cl^- to octahedral $[Co(H_2O)_6]^{2+}$ (present in aqueous solutions of Co(II)) results in the formation of the tetrahedral complex $[CoCl_4]^{2-}$. While the d–d transitions in octahedral $[Co(H_2O)_6]^{2+}$ are spin allowed, they are both orbitally and Laporte forbidden, resulting in ε_{max} values of ca. 5 dm^3 mol^{-1} cm^{-1}. In tetrahedral $[CoCl_4]^{2-}$ the spin selection rule has not changed, but now there is no longer a centre of symmetry, so the Laporte selection is not operating, and there is also considerable p–d orbital mixing and the ε_{max} values are now ca. 750 dm^3 mol^{-1} cm^{-1}. Note that the change in colour is mainly because Δ_{tet} is less than Δ_{oct}, but because there are transitions in the NIR for both of these complexes it is not straightforward to relate the colour change, from pink to blue issue, to a single energy separation, Δ.

SELF TEST

What sort of coordination environments would be best for a commercial pigment based on d–d transitions in a transition metal complex?

5.5 d–d electronic transitions in transition metal complexes

One of the main applications of UV-vis spectroscopy is in the characterisation of transition metal complexes and in particular their d–d transitions, where the technique can give very detailed information on the species present.

5.5.1 Common colours of transition metal complexes

The majority of first row transition metal complexes are coloured due to d–d transitions and some have colours that are characteristic of the metal ion in specific coordination environments. While such colours are not diagnostic they can provide a quick insight into the most likely metal ion present and its coordination geometry from visual inspection of the metal salt or solution. Octahedral hexaaqua $[M(H_2O)_6]^{n+}$ ions will generally form when a metal salt is dissolved in water and the addition of concentrated HCl gives tetrahedral $[MCl_4]^{n-}$. Table 5.2 summarises the colours of some of the common first row transition metal ions.

The colours of these species are usually the result of d–d electronic transitions in the visible region, each of different intensity as a result of the selection rules described in Section 5.4. For example, green $[Ni(H_2O)_6]^{2+}$

TABLE 5.2 Summary of colours associated with different transition metal ions

Ion	Typical octahedral example	Colour	Typical tetrahedral example	Colour
Ti^{3+}	$[Ti(H_2O)_6]^{3+}$	purple, blue		
V^{2+}	$[V(H_2O)_6]^{2+}$	purple/lilac		
V^{3+}	$[V(H_2O)_6]^{3+}$	green		
V^{4+}	$[VO(H_2O)_5]^{2+}$	blue		
Cr^{2+}	$[Cr(H_2O)_6]^{2+}$	blue		
Cr^{3+}	$[Cr(H_2O)_6]^{3+}$	green		
Mn^{2+}	$[Mn(H_2O)_6]^{2+}$	very pale pink	$[MnCl_4]^{2-}$	yellow-brown
Mn^{3+}	$[MnF_6]^{3-}$	green, purple		
Fe^{2+}	$[Fe(H_2O)_6]^{2+}$	pale green-blue	$[FeCl_4]^{2-}$	yellow-brown
Fe^{3+}	$[FeF_6]^{3-}$	pale violet		
Co^{2+}	$[Co(H_2O)_6]^{2+}$	pink-red	$[CoCl_4]^{2-}$	intense blue
Co^{3+}	$[Co(NH_3)_6]^{3+}$	golden-brown		
Ni^{2+}	$[Ni(H_2O)_6]^{2+}$	green	$[NiCl_4]^{2-}$	yellow-green
Ni^{2+}	$[Ni(en)_3]^{2+}$	violet		
Cu^{2+}	$[Cu(H_2O)_6]^{2+}$	blue	$[CuCl_4]^{2-}$	yellow

whose visible region spectrum is shown in Figure 5.5 has three spin-allowed d–d transitions between 7000 and 35 000 cm^{-1}, where the electrons are promoted from the t$_{2g}$ to the e$_g$ orbitals in an octahedral environment. These are relatively weak because they are forbidden by both the orbital and Laporte selection rules. The green colour of the $[Ni(H_2O)_6]^{2+}$ arises as the ion absorbs red light (11 150–17 000 cm^{-1}, 900–590 nm) and blue-violet light (23 100–28 000 cm^{-1}, 430–360 nm), but there is very little absorption at 20 000 cm^{-1} (500 nm), leaving the green portion to be observed. While the colour of $[Fe(H_2O)_6]^{3+}$ is pale violet, this is only observed in strong perchloric acid, as most iron(III) salts in water give brown solutions/suspensions due to hydrolysis and the formation of species with Fe–O–Fe units.

5.5.2 d–d spectra of 3d element complexes

In the set of spectra of the d^1 to d^9 hexaaqua complexes of the first row transition elements shown in Figure 5.10, there appears at first sight to be a random number of bands observed. However, on closer inspection and analysis it is found that there are only ever zero, one, or three spin-allowed transitions for the 3d high-spin octahedral complexes. For the spectra with seemingly only two bands, the third one is either too weak, is obscured by a higher energy charge transfer band, or occurs at a lower energy than the spectrometer limit. For Mn(II), the intensities, and hence the molar absorptivities indicate that these are spin-forbidden bands. The electronic absorption spectra of aqueous solutions of the first row transition metals in Figure 5.10 are experimental spectra, and the low energy cut-off is at 7200 cm^{-1} (1390 nm) as below this the overtone and combination bands of water make aqueous solutions opaque. The spectrum of Cr(II) also contains some Cr(III) due to its extreme air sensitivity. A detailed analysis is given in Section 5.13, but using reasonably simple concepts a lot of information can still be extracted from these spectra.

EXAMPLE 5.6

Calculate ε_{max} for the most intense band in the Ti(III) spectrum in Figure 5.10.

ANSWER

For λ_{max} of 20 000 cm^{-1}, A is 0.4. Using the Beer–Lambert law and assuming a path length of 1 cm, this means that $\varepsilon_{max} = 0.4/(0.08 \times 1) = 5$ dm^3 mol^{-1} cm^{-1}.

SELF TEST

Calculate ε_{max} for the most intense bands in the other spectra in Figure 5.10.

5.5.2.1 d^1 octahedral complexes

For transition metal complexes with d^1 electron configurations the interpretation of the d–d electronic transitions observed in their UV-vis spectra is relatively straightforward and can provide very useful information on the bonding within the complex. The most common coordination geometry for first row transition metal ions is octahedral and under the influence of a crystal field the five d orbitals, which are degenerate in a free ion or in a spherical field, become non-degenerate (Figure 5.11).

The lower energy t$_{2g}$ set consists of the d$_{xy}$, d$_{xz}$, and d$_{yz}$ orbitals directed between the ligands, and the higher energy e$_g$ set of d$_{x^2-y^2}$ and d$_{z^2}$, directed towards the point charges of an octahedral crystal field. The separation of the t$_{2g}$ and e$_g$ energy levels is denoted as Δ_{oct}. (In the older literature it is often given as 10Dq.) In principle the value of Δ_{oct} can be obtained directly from the UV-vis spectrum for d^1 configurations as the transition energy between the t$_{2g}$ and e$_g$ states is always Δ_{oct}, as shown in Figure 5.12. The transition energy, seen as a single peak in the UV-vis spectrum, therefore characterises the interaction between the ligands and the central metal.

EXAMPLE 5.7

Calculate Δ_{oct} for $[Ti(H_2O)_6]^{3+}$ from the spectrum in Figure 5.10.

ANSWER

The spectrum of $[Ti(H_2O)_6]^{3+}$ in Figure 5.10 has a broad asymmetric peak at 20 000 cm^{-1} (500 nm) which corresponds to the transition of an electron from the t$_{2g}$ orbital to the e$_g$ orbital. The energy of this transition provides a direct measure of the energy separation between the two orbitals and thus Δ_{oct}. So Δ_{oct} for $[Ti(H_2O)_6]^{3+}$ is 20 000 cm^{-1}. The values of Δ_{oct} are normally tabulated in cm^{-1}.

SELF TEST

The spectrum of $[TiCl_6]^{3-}$ has an absorption at 769 nm, that of $[TiF_6]^{3-}$ at 518 nm and $[TiBr_6]^{3-}$ at 877 nm. Calculate Δ_{oct} (in cm^{-1}) for each of these, and comment on the values.

5.5.2.2 High-spin d^5 octahedral complexes

A saturated solution of $[Mn(H_2O)_6]^{2+}$ is very pale pink and the spectrum in Figure 5.10 shows a large number of very weak peaks. The molar absorptivity values derived from this spectrum are very small (0.03–0.07 dm^3 mol^{-1} cm^{-1}) and by use of Table 5.1 they can be assigned to spin, orbital, and Laporte forbidden d–d transitions and this is consistent with there being no spin-allowed transitions

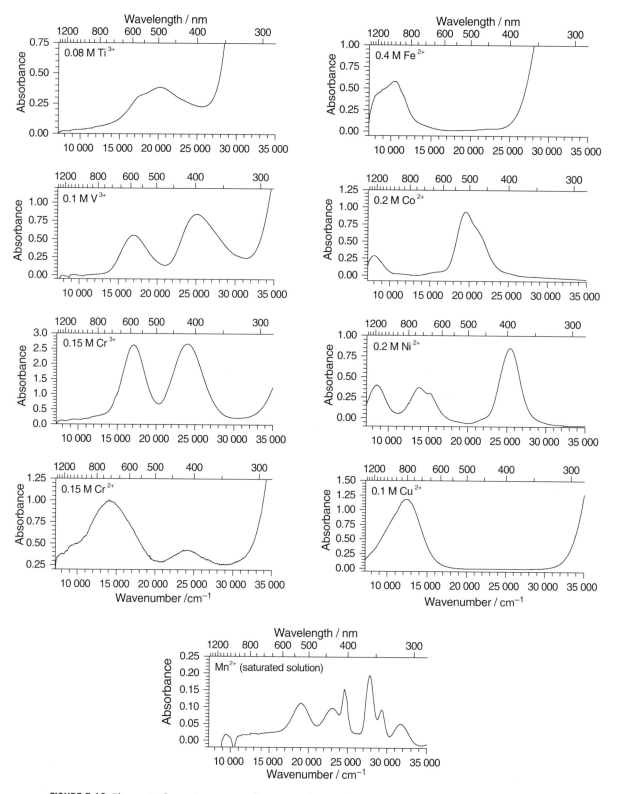

FIGURE 5.10 Electronic absorption spectra of aqueous solutions of the first row transition metals, $[M(H_2O)_6]^{n+}$.

in high-spin d^5 octahedral (and tetrahedral) complexes. These weak features are easily masked by other spectral transitions. Section 5.13.5.6 illustrates how a value of Δ_{oct} can be determined from such data.

5.5.2.3 d^9 octahedral complexes

The UV-vis spectra of compounds containing octahedral d^9 metal centres can be interpreted in a similar fashion

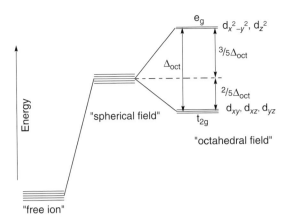

FIGURE 5.11 Schematic crystal field splitting diagram for octahedral complexes.

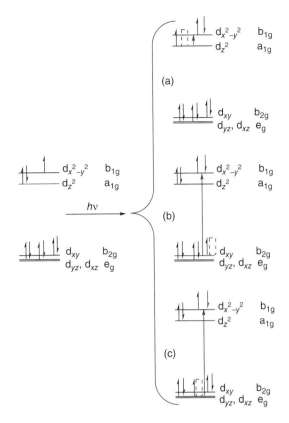

FIGURE 5.13 Schematic diagram of spin-allowed transitions in a d^9 Cu(II) complex with a tetragonally elongated Jahn–Teller distortion. The dashed blue boxes indicate the location of the blue electron before the transition. The blue arrows indicate the transition energies.

to those of d^1 ions as the electron transition from a t_{2g} orbital to an e_g orbital in d^1 is equivalent to a hole moving from an e_g orbital to a t_{2g} orbital in d^9. Strictly octahedral d^9 complexes are extremely rare as a Jahn–Teller distortion removes the degeneracy of both the t_{2g} and the e_g orbitals. The Jahn–Teller theorem states that orbitally degenerate electronic states in nonlinear molecules will distort to remove the degeneracy. In Cu(II) it is the unpaired electron that can be in either of the e_g orbitals that causes the Jahn–Teller distortion. In the vast majority of Cu(II) complexes this Jahn–Teller distortion manifests itself as a tetragonal elongation so that the d_{z^2} and $d_{x^2-y^2}$ orbitals are no longer degenerate (Figure 5.13). The symmetry labels for the d orbitals given in Figure 5.13 can be obtained by identifying xy etc. in the O_h character table. While there are three possible transitions in Figure 5.13, only one peak is often observed in the spectrum, as the transition within the e_g orbitals (d_{z^2} (a_{1g}) to $d_{x^2-y^2}$ (b_{1g})) may be at low energy (Figure 5.13a), and the relatively small splitting of the t_{2g} orbitals (b_{2g} and e_g) results in unresolved broadening of the band associated with a transition from the t_{2g}

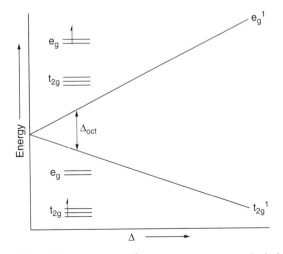

FIGURE 5.12 Splitting of a d^1 configuration in an octahedral crystal field.

set to the e_g set (Figure 5.13b and c). This can be seen for the spectrum of $[Cu(H_2O)_6]^{2+}$ in Figure 5.10 where only one d–d peak is observed at 12 500 cm^{-1} (800 nm), which is broad and asymmetric and is assigned to the transitions from the d_{xy} and (d_{xz}, d_{yz}) orbitals to the $d_{x^2-y^2}$ orbital. The lower energy d_{z^2} to $d_{x^2-y^2}$ transition is observed at ca. 6500 cm^{-1} in the solid state. In spectra of copper complexes containing nitrogen donor ligands such as $[Cu(NH_3)_6]^{2+}$, one broad band at 16 000 cm^{-1} is assigned to the d_{xy} (d_{xz}, d_{yz}) to $d_{x^2-y^2}$ transition, with a shoulder at 11 000 cm^{-1} due to the d_{z^2} to $d_{x^2-y^2}$ transition.

The asymmetry of the peak in the spectrum of $[Ti(H_2O)_6]^{3+}$ in Figure 5.10 is also due to a Jahn–Teller distortion, but this time in the excited state, where the two e_g orbitals no longer have the same energy (Figure 5.14a). As the splitting of the t_{2g} orbitals is ca. 400–800 cm^{-1}, and that of the e_g orbitals is 1000–4500 cm^{-1} two overlapping transitions are expected to be observed in the spectrum from the e_g orbitals to the a_{1g} and b_{1g} orbitals. These are observed at 17 720 and 20 240 cm^{-1} in the Ti(III) spectrum in Figure 5.10, and the higher energy transition from e_g to b_{1g} is usually taken to be Δ in this case.

A similar situation is observed in the spectrum of d^1 $[VO(H_2O)_5]^{2+}$ (Figure 5.2), but this time it is a distortion of

(a)

(b)

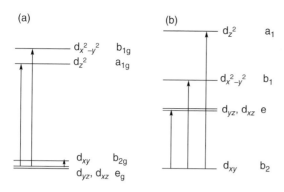

$d_{x^2-y^2}$ b_{1g}
d_{z^2} a_{1g}

d_{xy} b_{2g}
d_{yz}, d_{xz} e_g

d_{z^2} a_1

$d_{x^2-y^2}$ b_1

d_{yz}, d_{xz} e

d_{xy} b_2

FIGURE 5.14 Schematic diagram of orbital energy levels and transitions in (a) a d^1 complex with a Jahn–Teller distorted excited state and (b) a d^1 [VO(H$_2$O)$_5$]$^{2+}$ complex.

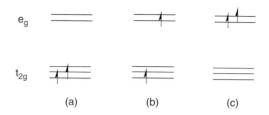

e_g

t_{2g}

(a)   (b)   (c)

FIGURE 5.16 Energy levels in (a) ground and (b, c) spin-allowed excited states of a d^2 octahedral complex (ignoring electron repulsion).

the ground state not because of a Jahn–Teller effect but because the V=O (vanadyl) unit has a very short bond length (ca. 1.7 Å) and so the complex has C_{4v} symmetry. The energy level diagram in Figure 5.14(b) can be used; the symmetry labels can be obtained from the C_{4v} character table, and as there is no longer a centre of symmetry there are no g subscripts. As a result of the very short V=O bond, all the orbitals with a z component increase in energy relative to those that just involve x and y. This results in three d–d transitions, the two lowest energy ones correspond to the band at 13 000 and shoulder at 16 000 cm^{-1} in Figure 5.2, the highest energy transition from the d_{xy} orbital to the d_{z^2} orbital is hidden by the tail of a strong absorption in the UV. An estimate of Δ can be made from the position of the shoulder at 16 000 cm^{-1}, which corresponds to the electron moving from the d_{xy} to the $d_{x^2-y^2}$ orbital.

5.5.2.4 High-spin d^4 and d^6 octahedral complexes

In Figure 5.10, the spectrum of the sky blue solution of [Cr(H$_2$O)$_6$]$^{2+}$, which contains high-spin d^4 Cr(II), has one broad, asymmetric feature at 14 000 cm^{-1} (714 nm) and is similar to that of the Cu(II) d^9 spectrum (the second peak at

24 000 cm^{-1} (420 nm) is due to some Cr(III) due to the extreme air sensitivity of Cr(II)). High-spin Cr(II) has a d^4 configuration with one electron in the e_g orbitals (Figure 5.15) just as d^9, and is also another classic example of a Jahn–Teller distorted ground state. The energy level diagram in Figure 5.13 can also be modified for high-spin d^4. The spectrum of the largely colourless solution of [Fe(H$_2$O)$_6$]$^{2+}$ containing high-spin d^6 Fe(II) (Figure 5.10) is similar to that of d^1 with evidence of a double peak due to a Jahn–Teller distorted excited state, and Figure 5.14(a) can be used. The similarity is because both d^1 and high-spin d^6 complexes have one unpaired electron in the t_{2g} orbitals, as shown in Figure 5.15.

For all of these configurations it is possible to obtain an estimate of Δ_{oct} directly from the spectra, but because of the Jahn–Teller distortions in either the ground or excited states, accurate determinations need to take this properly into account.

5.5.2.5 d^2, d^3, high-spin d^7, and d^8 octahedral complexes

The spectra of d^2, d^3, d^7, and d^8 complexes shown in Figure 5.10 contain two or three bands. For d^2, for example, V(III), the ground state, $t_{2g}^2 e_g^0$, is shown in Figure 5.16(a), and the two spin-allowed excited states of $t_{2g}^1 e_g^1$ and $t_{2g}^0 e_g^2$ are shown in Figure 5.16(b) and (c), respectively. As the e_g orbitals are higher in energy than t_{2g}, the energy ordering is $t_{2g}^2 < t_{2g}^1 e_g^1 < e_g^2$.

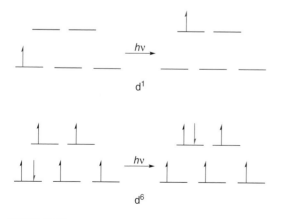

d^1

$\xrightarrow{h\nu}$

d^6

$\xrightarrow{h\nu}$

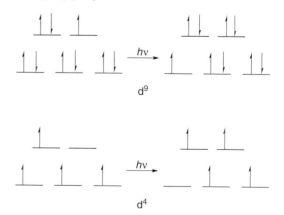

d^9

$\xrightarrow{h\nu}$

d^4

$\xrightarrow{h\nu}$

FIGURE 5.15 Summary of spin-allowed transitions in octahedral d^1, d^9, high-spin d^6, and high-spin d^4 complexes (ignoring effects of Jahn–Teller distortions).

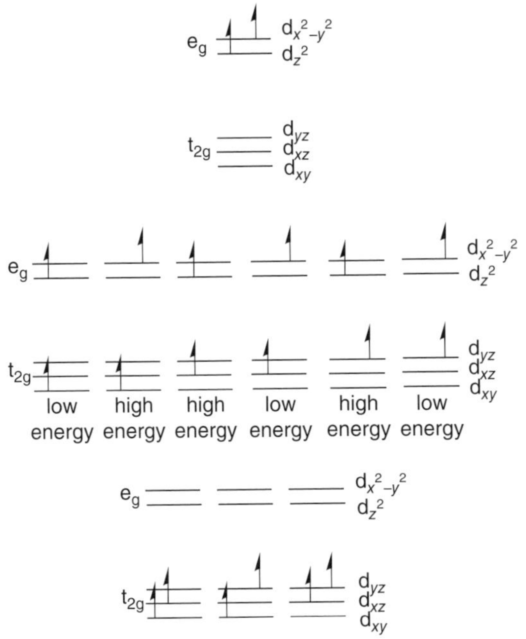

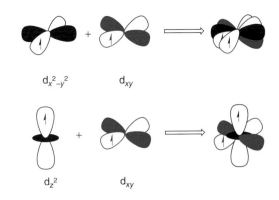

FIGURE 5.17 Possible arrangements of the $t_{2g}^2e_g^0$, $t_{2g}^1e_g^1$, and $t_{2g}^0e_g^2$ configurations of a d^2 octahedral complex.

While this appears to account for the spectra with two d–d bands, it does not explain why there are three clear bands in $[Ni(H_2O)_6]^{2+}$ (Figure 5.10). This is because electron–electron repulsion has not been taken into account, particularly in the $t_{2g}^1e_g^1$ configuration. The full analysis of the effects of electron repulsion in the ground and excited states is covered in the advanced section of this chapter (Section 5.13). However, a simple picture can be used to explain the UV-vis spectra of d^2, d^3, high-spin d^7, and d^8 octahedral complexes. Figure 5.17 indicates that there are three ways of representing Figure 5.16(a), six for Figure 5.16(b), and one for Figure 5.16(c).

The three representations for t_{2g}^2 all have the same energy, and are therefore degenerate. However, no representation in the O_h point group has sixfold degeneracy, so the six arrangements for $t_{2g}^1e_g^1$ cannot all have the same energy, and must split into groups of one, two or three configurations. In fact they separate into a pair of triply degenerate terms and this can be rationalised by considering the relationship between the electrons and the orbitals that are occupied. The arrangements $(d_{z^2})^1(d_{xy})^1$ and $(d_{x^2-y^2})^1(d_{xy})^1$ are shown in Figure 5.18.

The top arrangement leads to the two electrons both being in the xy plane and therefore there is greater electron–electron repulsion as they are much closer together than in the bottom one, where they are spread out in the x, y, and z directions. The higher repulsion that results from the top arrangement results in it having a higher energy than the bottom arrangement. Therefore, the six configurations are split into a pair of triply degenerate configurations labelled as high and low (in energy) in Figure 5.17.

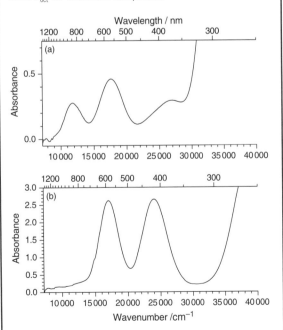

FIGURE 5.18 Different electron–electron repulsion in the $t_{2g}^1e_g^1$ excited state.

d^3, high-spin d^7, and d^8 can all be dealt with in a similar manner, and there are three spin-allowed d–d transitions in d^2, d^3, high-spin d^7, and d^8 octahedral complexes. This is clearly the case in Figure 5.10 for $[Ni(H_2O)_6]^{2+}$, but for $[V(H_2O)_6]^{3+}$ and $[Cr(H_2O)_6]^{3+}$ there appear to only be two peaks because the third peak is lost under the high-energy charge transfer peak at high wavenumber.

Δ_{oct} can in principle be read directly from the spectra for d^3 and d^8 configurations as the lowest energy d–d transition is equivalent to Δ_{oct}, but this is not the case for d^2 and d^7.

EXAMPLE 5.8

The electronic absorption spectra of $[V(H_2O)_6]^{2+}$ and $[Cr(H_2O)_6]^{3+}$ are shown in Figure 5.19. Use these to determine Δ_{oct} for these two complexes.

FIGURE 5.19 Electronic absorption spectra of 0.1 M $[V(H_2O)_6]^{2+}$ and 0.15 M $[Cr(H_2O)_6]^{3+}$.

ANSWER

Both of these complexes are d^3. In $[V(H_2O)_6]^{2+}$ there are three d–d bands at 11 800, 17 650, and 26 900 cm^{-1}. In contrast, in $[Cr(H_2O)_6]^{3+}$ there only appear to be two bands at 17 080 and 24 030 cm^{-1}, but the third appears as a weak shoulder at 37 800 cm^{-1} on the intense charge transfer band. Therefore, as Δ_{oct} corresponds to the lowest energy d–d transition in d^3 complexes it can be determined as 11 800 cm^{-1} for $[V(H_2O)_6]^{2+}$ and 17 080 cm^{-1} for $[Cr(H_2O)_6]^{3+}$. Δ_{oct} is larger for higher oxidation state metals of the same d^n configuration as the increased positive charge will reduce the metal–ligand bond length and hence increase the difference in energy of the t_{2g} and e_g orbitals. The intense charge transfer band is also at lower energy in the spectrum of $[V(H_2O)_6]^{2+}$ compared to $[Cr(H_2O)_6]^{3+}$ indicating the relatively stability of the two oxidation states.

SELF TEST

Use the spectra of $[Ni(H_2O)_6]^{2+}$ and $[Ni(en)_3]^{2+}$ (en = ethylenediamine, or 1,2-diaminoethane) in Figure 5.20 to determine Δ_{oct} for these two complexes. Comment on the values obtained.

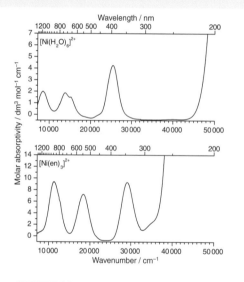

FIGURE 5.20 Electronic absorption spectra of $[Ni(H_2O)_6]^{2+}$ and $[Ni(en)_3]^{2+}$.

5.5.2.6 Low-spin d^6 octahedral complexes

Determining the number of spin-allowed transitions in low-spin complexes is more difficult, but for low-spin d^6 octahedral complexes such as Co(III) and Fe(II) two bands are usually observed. The value of Δ_{oct} cannot be read directly from the spectra, but an estimate can be obtained by taking a quarter of the difference between the two values and adding this to the lower value. (Section 5.13.5.4 outlines a more rigorous analysis.)

EXAMPLE 5.9

Estimate Δ_{oct} for $[Co(NH_3)_6]Cl_3$ and $[CoCl(NH_3)_5]Cl_2$ using the spectra in Figure 5.21.

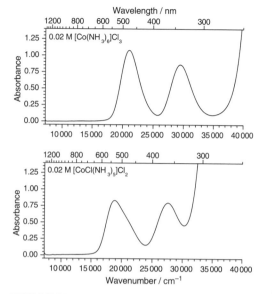

FIGURE 5.21 Electronic absorption spectra of 0.02 M $[Co(NH_3)_6]Cl_3$ and 0.02 M $[CoCl(NH_3)_5]Cl_2$.

ANSWER

Both of the complexes contain low-spin Co(III), d^6. The ε_{max} value of ca. 50 $dm^3\ mol^{-1}\ cm^{-1}$ for both spectra indicates that these features are due to d–d transitions. Although there is a centre of symmetry in $[Co(NH_3)_6]^{3+}$, which is formally absent in $[Co(NH_3)_5Cl]^{2+}$, there is no appreciable difference in the molar absorptivity between these two complexes, indicating that the effect of NH_3 and Cl^- are similar. In $[Co(NH_3)_6]Cl_3$ the peaks are at ca. 21 000 and 29 500 cm^{-1}. Adding a quarter of the difference to the lower value gives an estimate of 23 000 cm^{-1} for Δ_{oct} for this complex. In $[CoCl(NH_3)_5]Cl_2$ the two peaks are at ca. 19 000 and 27 500 cm^{-1}, and using the same approach an estimate of 21 000 cm^{-1} for Δ_{oct} is obtained for this complex. The broadening of the 19 000 cm^{-1} peak and shoulder at 20 000 cm^{-1} in the spectrum of $[CoCl(NH_3)_5]Cl_2$ are because the local symmetry is no longer O_h, but C_{4v}. $[Co(NH_3)_6]Cl_3$ is an orange-brown colour, while $[Co(NH_3)_5Cl]Cl_2$ is purple, indicating that relatively subtle shifts in the absorption bands can have a profound effect on the perceived colour.

SELF TEST

The spectrum of $[Rh(NH_3)_6]^{3+}$ has two d–d transitions at 32 800 and 39 200 cm^{-1}, and in the spectrum of $[Ir(NH_3)_6]^{3+}$ they are at 39 800 and 46 800 cm^{-1}. Estimate Δ_{oct} for these complexes, and compare the values to that of $[Co(NH_3)_6]^{3+}$.

5.5.2.7 Tetrahedral complexes

Spectra of tetrahedral complexes can be analysed in a similar manner, because a d^n tetrahedral configuration is analogous to octahedral d^{10-n}, but they will all be high spin for first row transition metal complexes. For complexes with one spin-allowed transition (d^1, d^4, d^6, d^9), Δ_{tet} can be determined directly from the spectra. For those with three spin-allowed transitions (d^2, d^3, d^7, and d^8) it is now the d^2 and d^7 examples where Δ_{tet} can in principle be obtained directly from the spectra. However, the transitions will be at lower energy as $\Delta_{tet} \approx {}^4/_9\ \Delta_{oct}$ (Figure 5.22) because the d orbitals now no longer point either directly at the ligands or directly between them.

Care needs to be exercised as the transitions may lie outside the range of conventional instruments. The other key change is that because there is no centre of symmetry, the d orbitals are labelled e and t_2, and the parity/Laporte selection rule is relaxed. There is also considerable d–p orbital mixing as the $3d_{xy}$, $3d_{xz}$, and $3d_{yz}$ orbitals and $4p_{x,y,z}$ orbitals have the same symmetry (t_2). The combined effect is that the transitions may be up to two orders of magnitude more intense than for octahedral complexes.

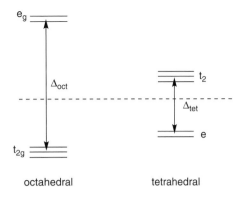

FIGURE 5.22 Crystal field splitting in octahedral and tetrahedral environments.

5.5.2.8 Other geometries

For other geometries, such as square planar, trigonal planar, and trigonal bipyramidal, the simple crystal field approach will give a general indication of the ordering of the d orbitals, but in these cases detailed calculations are required to determine the relative order and energies accurately as the reduction in symmetry of the complex means more transitions will be observed.

5.5.3 The spectrochemical series

The preceding sections have demonstrated that the value of Δ_{oct} can be obtained from the UV-vis spectra of transition metal complexes, and that its value is dependent on the ligands as well as the metal, and its oxidation state.

5.5.3.1 Ligand effects

When the ligands are placed in increasing value of their Δ_{oct} values for the same metal in the same oxidation state the following sequence is found:

weak field
high spin

$I^- < Br^- < S^{2-} < \underline{S}CN^- < Cl^- < \underline{N}O_2^- < N^{3-} < F^- < OH^- < C_2O_4^{\ 2-} < O^{2-} < H_2O < \underline{N}CS^- < MeC\underline{N} < py < NH_3 < en < bipy < phen < \underline{N}O_2^- < PPh_3 < \underline{C}N^- < \underline{C}O$

strong field
low spin

This ordering of Δ_{oct} is known as the spectrochemical series, because the values are determined from the UV-vis spectra. The series has the weak-field ligands favouring high-spin complexes at the left-hand side, and the strong-field ligands, favouring low-spin complexes are at the right-hand side, with the intermediate field ligands in the centre. For (potentially) ambidentate ligands, the

---

EXAMPLE 5.10

$K_2[VF_6]$ is a pale pink solid, mainly as a result of the d–d absorption band centred on $20\,120\ cm^{-1}$ (497 nm). Predict a colour for tetrahedral V(IV) fluoro compounds.

ANSWER

Octahedral V(IV) is a d^1 ion and $20\,120\ cm^{-1}$ will correspond to Δ_{oct}. For tetrahedral V(IV) $\Delta_{tet} = {}^4/_9\ \Delta_{oct}$, so $\Delta_{tet} = 8940\ cm^{-1}$ equivalent to 1120 nm. This absorption will be in the NIR but is likely to be broad and have a tail into the red visible region making tetrahedral vanadium fluoride compounds blue. VF_4 is lime green, but this is because in the solid state at room temperature it consists of fluorine bridged tetragonally compressed octahedron, which give rise to an absorption band at $15\,000\ cm^{-1}$ (667 nm) with a weaker shoulder at $23\,000\ cm^{-1}$ (435 nm) on an intense UV tail starting at $26\,500\ cm^{-1}$ (380 nm).

SELF TEST

When an excess of NCS^- is added to a solution of $[Co(H_2O)_6]^{2+}$, the pale pink solution turns a very deep blue. What does this indicate about the structure of cobalt in the two complexes? Propose a structure for the NCS^- complex.

donor atom is underlined in bold. From these data it can be seen that halide ligands tend to give smaller values of Δ_{oct} than O-donor ligands, or N-donor ligands, with cyano complexes giving large values of Δ_{oct}.

Crystal field theory alone cannot explain this ordering as it is related to the extent of π bonding in the complex. Ligands which act as π bases, donating electrons to the metal through a π bonding interaction, such as the halides and oxygen donors, tend to be weak-field ligands, giving rise to high-spin complexes. Ligands such as CO, CN^-, and phosphines, which can act as π acids, accepting electrons from the metal through a π bonding interaction, are strong-field ligands. Those ligands that only have weak or no π behaviour (e.g. NH_3) tend to be the intermediate field ligands.

EXAMPLE 5.11

Figure 5.23 shows UV-vis spectra of $[Co(NH_3)_5(ONO)]Cl_2$ and $[Co(NH_3)_5(NO_2)]Cl_2$. Explain why their linkage isomerisation affects Δ_{oct} and the positions of the absorption maxima in the spectra.

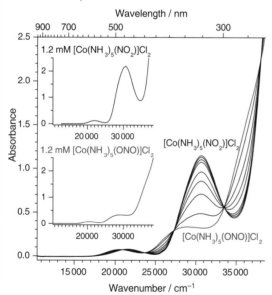

FIGURE 5.23 Electronic absorption spectra of isomerisation of 1.2 mM solution of $[Co(NH_3)_5(ONO)]Cl_2$ to $[Co(NH_3)_5(NO_2)]Cl_2$ in 0.1 M $HClO_4$ at 40°C for 90 min. Inserts show spectra of initial and final complexes.

ANSWER

These are both low-spin d^6 complexes, and two spin-allowed d–d transitions are usually observed. While the detailed analysis in Section 5.13.5.4 is required to determine Δ_{oct} accurately, an estimate of Δ_{oct} can be obtained by adding a quarter of the difference between the two transition energies to the lower one. For $[Co(NH_3)_5(ONO)]Cl_2$ with transitions at 20500 and 29000 cm^{-1}, this gives $\Delta_{oct} =$

22600 cm^{-1} and for $[Co(NH_3)_5(NO_2)]Cl_2$ with transitions at 22000 and 30500 cm^{-1} $\Delta_{oct} = 24100$ cm^{-1}. In nitrite, NO_2^-, the negative charge is located in a non-bonding π orbital localised on the oxygen atoms. Therefore, when it binds via oxygen atoms to give nitrito ligands (M–ONO) it acts as a strong π donor, and as expected this results in a relatively weak-field ligand, between Cl^- and F^-, in the spectrochemical series. When it bonds via the nitrogen atom in the nitro ligands, the π^* antibonding orbitals can overlap with the metal t_{2g} orbitals and in nitro coordination (M–NO_2) the ligand can act as a π acceptor. In this case it lies higher in the spectrochemical series between bipy/phen and PPh_3. This change is clearly shown in Figure 5.23 when $[Co(NH_3)_5(ONO)]Cl_2$ isomerises to $[Co(NH_3)_5(NO_2)]Cl_2$ as the peaks move to higher wavenumber. The change in intensity is due to differential effects of the nitrito and nitro ligands on the orbital and Laporte/parity selection rules.

SELF TEST

Calculate ε_{max} for the d–d peaks in Figure 5.23, assuming a cell pathlength of 1 cm.

5.5.3.2 Metal effects

The spectrochemical series for metals with the same ligand, which is shown below, shows that Δ_{oct} is dependent on both the metal and its oxidation state.

$$Mn^{2+} < Ni^{2+} < Co^{2+} < Fe^{2+} < V^{2+} < Fe^{3+} < Cr^{3+}$$
$$< V^{3+} < Co^{3+} < Mn^{4+} < Mo^{3+} < Rh^{3+} < Ru^{3+} < Pd^{4+}$$
$$< Ir^{3+} < Pt^{4+}$$

This series shows that Δ_{oct} increases with oxidation state as the ligands are pulled closer to the metal resulting in greater interaction/repulsion. There is also a substantial increase in Δ_{oct} on going from 3d to 4d to 5d complexes of the same group, because the d orbitals get larger and there is greater interaction/repulsion between the metal and ligand orbitals. For the divalent 3d metals, the order is related to the extent of covalent bonding.

A selection of values of the ligand field splitting parameter, Δ, are given in Table 5.3.

EXAMPLE 5.12

Interpreting UV-vis spectra using the spectrochemical series

Explain the following observations for compounds which all contain Cu(II) surrounded by six ligands. CuF_2 is almost colourless, $[Cu(H_2O)_6]^{2+}$ is blue, and $[Cu(NH_3)_6]^{2+}$ is deep dark blue.

ANSWER

Cu(II) has one d–d transition which occurs in the range 9000–14500 cm^{-1} (1100–700 nm) in the UV-vis spectrum. CuF_2 has the weakest field ligand, F^-, in the spectrochemical series, so its absorption will be at the lowest energy and highest wavelength and this absorption lies in the NIR region and the material is colourless. With the next strongest field ligand, H_2O, the absorption band moves to a shorter wavelength and the edge of this lies in the red part of the spectrum imparting a blue colour to the hexaaqua cation. With ammonia as a ligand the absorption moves fully into the red region and the cation is observed as the complementary deep blue colour.

SELF TEST

Some iron(II) complexes can be changed reversibly between high spin and low spin by changing the temperature because Δ and the electron pairing energy, P, are very delicately balanced. These are known as spin-crossover complexes and the iron(II) examples tend to contain a FeN_6 coordination environment. By consideration of the spectrochemical series suggest ligand donor atoms which might be used to produce spin-crossover complexes of Fe(III).

5.6 Charge transfer transitions in transition metal complexes and compounds

Intense charge transfer transitions are common in the spectra of transition metal complexes and compounds, and can often obscure the weaker d–d transitions. Figure 5.8 shows that transitions from both the ligand σ and π MOs to the metal based orbitals can be involved in charge transfer transitions.

5.6.1 Ligand to metal charge transfer (LMCT) transitions

In metal complexes it is also possible to promote electrons from the ligand based orbitals, which are mainly p orbital in character, to the metal based d orbitals (see Figure 5.8). As $\Delta l = \pm 1$, these are orbitally allowed, and therefore, very

TABLE 5.3 Table of ligand field splitting parameters, Δ (in cm^{-1}), obtained from electronic absorption spectra. (Data taken in part from A.B.P. Lever. (1984) *Inorganic Electronic Spectroscopy*, 2nd Ed. Elsevier and B.N. Figgis and M.A. Hitchman. (2000) *Ligand Field Theory and its Applications*. Wiley-VCH.)

| No of d electrons | Metal ion | Coordination environment | | | | | | | | |
		$6F^-$	$6Cl^-$	$6Br^-$	$6H_2O$	$6NH_3$	3en	$6CN^-$	$4Cl^-$	$4Br^-$
d^1	Ti^{3+}	19300	13000	11400	20100			22300		
	V^{4+}	20250	15300						9000	
d^2	V^{3+}	16100	12000		19100			23600		
d^3	V^{2+}		9000	8600	11800	14800	15500	22300		
	Cr^{3+}	14900	13200		17100	21550	22000	26600		
d^4	Cr^{2+}	14700	13000		14000		18000			
	Mn^{3+}	22000	20000		21000			*34000*		
d^5	Mn^{2+}	9400	8000	7000	8500		11000	*30000*	3300	3100
	Fe^{3+}	14000	11000		14000			*35000*	5000	
d^6	Fe^{2+}				10400			*33800*	4050	3000
	Co^{3+}	13000			*18300*	*22900*	*23350*	*34500*		
	Rh^{3+}	*22800*	20500	18900	*27100*	*33800*	*34600*	*44000*		
	Ir^{3+}		*25000*	*23000*		*41500*				
d^7	Co^{2+}		7500	6500	9200	10200	11200		3120	
d^8	Ni^{2+}	7250	7700	7500	8500	10750	11700		4090	3790
d^9	Cu^{2+}				13000	15000	16000			

Values in *italics* are for low-spin compounds.

intense. In octahedral complexes the metal d orbitals have g character while the ligand p orbitals have u character so they are also allowed under Laporte/parity selection rule. These transitions are often referred to as dipole allowed or fully allowed transitions. As shown in Table 5.1, they have a molar absorptivity, ε, in the thousands and often tens of thousands of dm^3 mol^{-1} cm^{-1}. These electronic transitions can be considered as a transient move of negative charge from the ligand p orbitals to the metal d orbitals and are called ligand to metal charge transfer (LMCT) transitions. Formally, the ligand has been transiently oxidised and the metal transiently reduced. This is much more likely to happen for ligands with lone pairs of electrons and metals in high oxidation states. The energy of these transitions characterises the energy separation between the ligand based orbitals and the metal d orbitals. In most transition metal complexes this energy separation is large, and these electronic transitions occur in the UV part of the electromagnetic spectrum below 400 nm (above 25 000 cm^{-1}). However, when the metal is in a high oxidation state the energy of the metal d orbital to which the electron is excited becomes lower, reducing the energy separation and bringing the transitions into the visible part of the spectrum. Ligands with higher energy p orbital electrons, such as those containing the heavier p block elements, for example, iodide, also reduce the LMCT energy towards, and into, the visible region.

A classic example of a charge transfer transition is in permanganate and the spectrum of a 0.50 mM solution of potassium permanganate, K[MnO$_4$], is shown in Figure 5.24. The features at ca. 19 000 cm^{-1} (525 nm), with a molar absorptivity of 3000 dm^3 mol^{-1} cm^{-1}, are a LMCT transition (from filled O^{2-} p orbitals to empty Mn^{7+} d orbitals) and give rise to the very intense purple colour characteristic of permanganate solutions. The spectrum and its

expansion shows evidence of regular fine structure due to vibrational progressions on the 19 000 cm^{-1} band, as well as less resolved structure on the higher energy bands at ca. 32 000 cm^{-1}. (Further information on the origin of the vibrational fine structure is available in the Online Resource Centre.) As these are absorption spectra, the progressions are related to the vibrational intervals in the excited state. The separation of the peaks in the well-resolved vibrational progression at 19 000 cm^{-1} is ca. 680 cm^{-1}. The totally symmetric $v_{\text{Mn-O}}$ mode has been observed at 839 cm^{-1} in the Raman spectrum. Therefore, as expected the vibrational frequency is lower in the excited state, indicating a lower force constant, and hence a weaker bond.

Similar transitions occur in other tetrahedral oxo anions, such as [CrO$_4$]$^{2-}$ and [VO$_4$]$^{3-}$, and the UV-vis spectroscopic data can be used to quantify the energy level separation between the oxygen and transition metal orbitals. As the oxidation state of the metal decreases the LMCT band shifts to higher energies and the absorption maximum moves towards and into the UV giving the characteristic colours of the anions: yellow, [CrO$_4$]$^{2-}$ (27 000 cm^{-1}, 370 nm), and pale yellow, [VO$_4$]$^{3-}$ (36 200 cm^{-1}, 276 nm). This analysis of LMCT transitions can be extended in terms of optical electronegativities (Section 5.6.4).

EXAMPLE 5.13

K[ReO$_4$] is a colourless compound and its UV-vis spectrum has an absorption maximum at 230 nm (43 700 cm^{-1}). What does this spectrum tell you about the oxidising properties of [ReO$_4$]$^-$ compared with [MnO$_4$]$^-$? The LMCT band in K[ReS$_4$] is at 505 nm (19 800 cm^{-1}); explain why this absorption occurs at a much lower energy than found for the perrhenate anion. What colour would you predict for K[ReS$_4$]?

ANSWER

The relatively low energy of the LMCT transitions in [MnO$_4$]$^-$ (19 000 cm^{-1}, 525 nm, 2.35 eV) indicates that Mn^{7+} readily gains electrons and, is therefore, a very strong oxidising agent ([MnO$_4$]$^-$/MnO$_2$, $E^{\ominus}$ = +1.69 V). The equivalent transition in [ReO$_4$]$^-$ is at much higher energy (43 500 cm^{-1}, 230 nm, 5.39 eV) meaning that Re^{7+} has a much lower tendency to gain an electron and so will be a much weaker oxidising agent ([ReO$_4$]$^-$/ReO$_2$, $E^{\ominus}$ = +0.51 V).

Replacing oxygen by sulfur increases the energy of the ligand electron energy levels (3p orbitals lie at a higher energy than 2p). This shifts the LMCT transition from the UV (230 nm) into the visible region (532 nm). The LMCT band at 505 nm in K[ReS$_4$] corresponds to green; therefore blue and red are not absorbed, so the colour of the powder is purple.

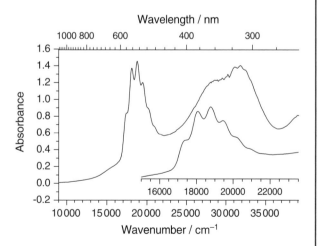

FIGURE 5.24 Electronic absorption spectrum of 0.5 mM K[MnO$_4$] solution.

SELF TEST

In general, transition metals in high oxidation states become weaker oxidising agents as the Group is descended, as we have seen with $[ReO_4]^-$ and $[MnO_4]^-$. Predict a colour and position of the LMCT band in the UV-vis spectrum for the $[MoO_4]^{2-}$ and the $[MoS_4]^{2-}$ ions.

5.6.2 **Characterisation of pigments, semiconductors, and band gaps**

LMCT absorptions observed in metal complexes in solution have analogous transitions in many solids where an electron can be promoted from the valence band (filled and usually formed from the overlap of anion (X^{n-}) p orbitals) to the conduction band (usually empty or partially empty and formed from overlap of orbitals centred on the metal (M^{n+})) (Figure 5.25).

Transitions between the valence and conduction bands are seen as **absorption edges** in the UV-vis spectrum, as there is a continuity of excited states above the bottom of the conduction band into which excitation can take place. The onset of the absorption determines the band gap, ΔE, in the material under study. Many solid pigments are characterised using solid state UV-vis spectroscopy and their absorption edge and band gap extracted from the experimental data. The white pigment TiO_2 has a large separation of the valence and conduction bands, and the absorption edge falls in the UV part of the spectrum making this material colourless. The rutile and anatase polymorphs of TiO_2 are both used as photocatalysts and there is considerable effort to chemically modify them so that the band gap is accessible by sunlight.

Materials used in solar cells, photovoltaics, need to absorb sunlight across the whole of the visible region in order to have high efficiencies. Therefore, the best photovoltaics

have a band gap of 1.1–1.6 eV (9000–13000 cm^{-1}, 1100–770 nm) which means they absorb right across the visible region. As a consequence of absorbing all visible wavelengths they are grey-black in colour. Silicon, widely used in solar cells, has a band gap of 1.1 eV and absorbs light with wavelengths out to 1100 nm. Alternative materials such as CdTe and Cu_2ZnSnS_4 have band gaps of 1.5 eV. UV-vis spectroscopy is widely used to characterise new potential photovoltaic materials such as the recently discovered lead halide perovskites with band gaps of 1.7 eV.

EXAMPLE 5.14

Figure 5.26 shows the solid state UV-vis absorption spectra of three common inorganic pigments used by artists. Determine the colour of these pigments and explain the origin of the trend in absorption edge position for this series of compounds.

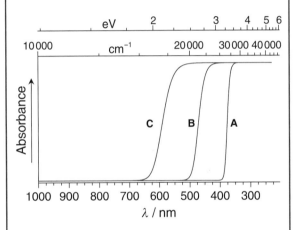

FIGURE 5.26 Solid state UV-vis absorption spectra of some inorganic pigments..

ANSWER

The colours of these pigments can be determined from their absorption edges (band gaps) and the origin of colour discussed in Section 5.1. Pigment **A** only absorbs UV radiation so will be white (or colourless if a single crystal). Pigment **B** absorbs blue and violet light below 490 nm and the colour observed is therefore bright yellow. Pigment **C** absorbs light below 630 nm, leaving just the red part of the spectrum to be observed. **A** is the 'zinc white' pigment which is zinc oxide, ZnO. Pigment **B** is 'cadmium yellow', which is cadmium sulfide, CdS, and **C** is 'cadmium red', which is a cadmium selonosulfide, Cd(SeS), solid solution. The absorption edge (band gap) for CdSe is ca. 710 nm resulting in a red-black solid. By controlling the proportion of selenium in the Cd(SeS) solid solution the band gap can

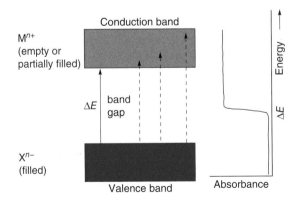

FIGURE 5.25 Schematic diagram of band gap and associated transitions.

be tuned to give 'cadmium orange' as well as 'cadmium red' pigments. The band gaps, and hence colours, of CdS and CdSe are also dependent on particle size, and the band gap shifts to higher energy as the particle size is reduced from about 100 Å. The emission is also very sensitive to the size of the nanoparticles. The reduction in the energy of the absorption edge (band gap) is due to the increasing energy of the chalcogenide originated band as the orbitals that form this band change from O^{2-}(2p), to S^{2-}(3p) and Se^{2-}(4p). This behaviour mirrors that seen for the $[ReO_4]^-$ and $[ReS_4]^-$ ions discussed in Example 5.13.

SELF TEST

Predict the form of the UV-vis spectrum of CdTe.

What colour would you expect of rutile (TiO_2) that had been chemically modified so that its band gap absorption had a tail that was just in the visible region?

5.6.3 Metal to ligand charge transfer (MLCT) transitions

If the metal is in a low oxidation state with the t_{2g} (and possibly e_g) orbitals populated then a metal to ligand charge transfer (MLCT) band can be observed if there are low-lying empty orbitals on the ligands, which are often of a π^* nature. In this case the metal is transiently oxidised and the ligand transiently reduced, hence they are called MLCT transitions. As for LMCT bands they are usually very intense as they are dipole allowed.

The spectrum of a 0.1 mM solution of $[Fe(phen)_3]^{2+}$ is shown in Figure 5.27. The iron complex, $[Fe(phen)_3]^{2+}$, is sometimes known as ferroin and is formed between Fe(II) and the bidentate 1,10-phenanthroline (phen) ligand. This is commonly used in analytical determinations of Fe(II) due to its intense red colour and because

it is insensitive to Fe(III). (Fe(III) is commonly detected using red solutions formed with NCS^-.)

The red colour is due to the absorption band from 18 000 to 27 000 cm^{-1}, which has a molar absorptivity of ca. 11 500 dm^3 mol^{-1} cm^{-1} so blue and green light is strongly absorbed. As Fe(III) is a relatively low oxidation state, this feature can be assigned to a charge transfer transition from the metal to the 1,10-phenanthroline ligands. $[Fe(phen)_3]^{2+}$ is low-spin d^6, so the MLCT transitions will be from the full t_{2g} d orbitals, to the π^* antibonding orbitals of p character on the 1,10-phenanthroline. As this transition involves a d–p transition, they are allowed by the dipole orbital selection rule and are very intense.

Figure 5.27 demonstrates the type of spectrum observed if the concentration is too high for the bands of interest. With an absorbance of 3 or 4 there is very little light getting to the detector so the detector noise is ratioed against the reference channel. If this behaviour is observed and the peak position is required, the solution should be diluted appropriately.

In order to determine whether a LMCT or MLCT band is being observed, it is important to consider whether the metal can be transiently oxidised or reduced. If the metal is relatively easily reduced (e.g. $[MnO_4]^-$) or relatively easily oxidised (e.g. Fe(II)) then the charge transfer band is more likely to be in the visible part of the spectrum. If the metal ion is very stable to oxidation/reduction, then the charge transfer bands are much more likely to be observed in the UV part of the spectrum.

5.6.4 Optical electronegativities

Estimates of the position of the LMCT and MLCT bands can be made by using the concept of **optical electronegativity**. Optical electronegativities are similar to other measures of electronegativity, such as Pauling electronegativities, in measuring how strongly an atom attracts electrons towards itself. As charge transfer absorptions effectively involve transferring an electron from one atom to another the difference in their electronegativity values should correspond to the energy for that transition. Experimentally the relationship that has been found is

$$E_{exp} = 30\,000\,[\chi_{opt}(\text{ligand}) - \chi_{opt}(\text{metal})] \quad \textbf{Eqn 5.3}$$

where E_{exp} is the lowest, dipole-allowed, charge transfer absorption energy in cm^{-1} and χ_{opt}(ligand) and χ_{opt}(metal) are the optical electronegativities of the ligand and metal, respectively. (If eV is used, the 30 000 conversion factor becomes 3.72.) Equation 5.3 allows an estimate to be made of the energy, E_{exp}, of the first dipole-allowed absorption band for any complex, provided the values of χ_{opt} for the cation and anion are known. These have been tabulated for many common species (Table 5.4).

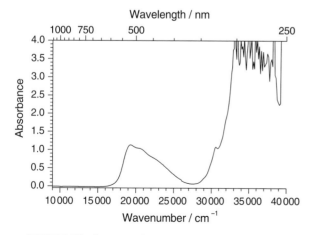

FIGURE 5.27 Electronic absorption spectrum of 0.10 mM $[Fe(phen)_3]^{2+}$ solution.

TABLE 5.4 Summary of χ_{opt} values for some metal ions and ligands

	χ_{opt}		χ_{opt}
V^{4+}	2.6	F^-	3.9
Cr^{3+}, Cr^{4+}	2.0, 2.6	Cl^-	3.0
Mn^{2+}, Mn^{3+}, Mn^{4+}	2.0, 2.2, 3.0	Br^-	2.8
Fe^{2+}, Fe^{3+}	1.75, 2.3	I^-	2.5
Co^{2+}, Co^{3+}	1.6, 1.8	NO_3^-	3.9
Ni^{2+}, Ni^{3+}	1.8, 3.0	OH^-	3.6
Cu^{2+}	2.2	CN^-	3.3
Zn^{2+}	1.1		
Mo^{3+}, Mo^{4+}, Mo^{6+}	1.7, 1.9, 2.1	SO_4^{2-}	3.6

For accurate analysis, E_{exp} needs to be corrected to take account of whether the destination orbital is t_{2g} or e_g and for different electron repulsion energies due to spin-pairing, for example, for d^4 and d^5 systems.

EXAMPLE 5.15

Application of optical electronegativities

Calculate the expected position of the lowest energy charge transfer absorption band (in cm^{-1}, eV, and nm) in the UV-vis spectra of the complex ions $[NiX_4]^{2-}$ (X = Cl, Br, and I).

ANSWER

Using optical electronegativity data in Table 5.4 and Eqn 5.3:

$[NiCl_4]^{2-}$ $E_{exp} = 30\,000(3.0 - 1.8)\ cm^{-1}$
           $= 36\,000\ cm^{-1}$ (4.46 eV, 278 nm)

$[NiBr_4]^{2-}$ $E_{exp} = 30\,000(2.8 - 1.8)\ cm^{-1}$
           $= 30\,000\ cm^{-1}$ (3.72 eV, 333 nm)

$[NiI_4]^{2-}$ $E_{exp} = 30\,000(2.5 - 1.8)\ cm^{-1}$
           $= 21\,000\ cm^{-1}$ (2.60 eV, 476 nm)

The observed experimental values are at $35\,000\ cm^{-1}$, $28\,000\ cm^{-1}$, and $19\,000\ cm^{-1}$, respectively, showing good agreement with theory. It should also be noted that the LMCT transition moves to a lower wavenumber/longer wavelength on going from Cl^- to Br^- and I^-.

SELF TEST

The first charge transfer absorption band observed in the UV-vis spectrum of the $[CoF_6]^{2-}$ anion is at $28\,300\ cm^{-1}$. Calculate an optical electronegativity for Co(IV). A new compound was isolated and proposed to contain either the $[CoCl_6]^{2-}$ or $[CoCl_6]^{3-}$ anion. Its first charge transfer band was at $36\,000\ cm^{-1}$. Determine which anion is present in this compound.

5.6.5 Intervalence (or metal to metal) charge transfer (IVCT) transitions

When two metals are in close proximity and in different oxidation states (i.e. a mixed-valence compound), absorption of light may cause the transient oxidation of one of the metal ions and reduction of the other. The extent of the interaction between the two sites determines whether an intervalence charge transfer (IVCT) band will be observed or not. If there is no interaction, because the valences are localised as in **Robin–Day** Class I mixed-valence compounds, no IVCT band is observed. For intermediate interactions, the oxidation states of the two metal sites can be thought to be distinct, but there is a low energy to their interconversion, and an intense IVCT band is observed. These belong to Class II in the Robin–Day classification. If there is a strong interaction (Class III), then an IVCT band will still be observed, but it is often better to consider the metal oxidation states as equivalent and intermediate, and possibly as one site, rather than two isolated ones.

EXAMPLE 5.16

What is the origin of the intense blue colour in Prussian blue, $Fe(III)_4[Fe(II)(CN)_6]_3 \cdot xH_2O$

ANSWER

Prussian blue is a good example of a compound with IVCT bands (Robin–Day Class II) and was one of the first modern synthetic pigments. The intense blue colour is due to an absorption in the orange part of the visible spectrum around 680 nm, where an electron moves from an iron(II) centre to an iron(III) centre mediated by an intervening CN^- ion. The extended network structure involves CN^- groups between the Fe(II) and Fe(III) centres.

SELF TEST

The iron(III)–ruthenium(II) analogue of Prussian blue, $Fe(III)_4[Ru(II)(CN)_6]_3 \cdot 18H_2O$, is dark purple with an IVCT band at 495 nm. What does this tell us about the extent of delocalisation in the two materials?

The stunning colours of many minerals and gemstones are often due to IVCT transitions. For example, blue aquamarines are beryls ($Be_3Al_2Si_6O_{18}$) with iron(II) and iron(III) impurities. If just iron(II) is present a pale blue colour is observed, and exclusively iron(III) yields a golden colour, but with both present an intense blue colour is observed. (Emeralds are green beryls, and the colour is caused by chromium dopants, but in this case d–d transitions rather than IVCT transitions are the source of the colour.) Vivianite, an iron(II,III) phosphate, is blue because of IVCT transitions. The mixed valence required for IVCT is not limited to having the same metal. For example, the blue colour of sapphire is due to IVCT transitions between low levels (ca. 0.02%) of Fe^{2+} and Ti^{4+} in the corundum (Al_2O_3) host. Ti^{4+} on its own in Al_2O_3 is colourless, and Fe^{2+} is very pale yellow, but when both are present the characteristic sapphire blue is observed. In contrast red rubies require about 1% of Cr doping in the corundum host, and the colour is due to d–d transitions in the violet and yellow-green regions. The colour and lustre of rubies is enhanced by the stimulation of 694 nm emission by shorter wavelengths (Section 5.13.5.2). This emission is also used in ruby lasers.

5.7 Intra-ligand transitions

The high-energy transitions in the spectrum of $[Fe(phen)_3]^{2+}$ (Figure 5.27) are intra-ligand transitions due to the π to π* transitions in the phen ligands themselves. These transitions are in the UV so phen is a white solid, but if the ligand is coloured, or becomes coloured after deprotonation, then these transitions will be observed in its colour and in the spectra. The intra-ligand bands may be shifted, but in general are less important than the d–d transitions which are much more diagnostic. However, as they may be intense, they can mask the weak d–d transitions.

5.8 Electronic transitions in other inorganic materials

The intense blue colour of lapis lazuli (lazurite) is due to a strong absorption at 600 nm (orange) in bent S_3^- molecular radical anions located within the complex aluminosilicate lazurite mineral, which results in an electronic excitation within the polysulfide anions, and not between them. The natural lapis lazuli when ground forms the intense blue ultramarine pigment. This has been valued since antiquity, and was especially sought after in the Renaissance and demanded a high price as the only source was in Afghanistan. Synthetic ultramarine was first produced in the 1820s and was much used by Impressionist painters. It was also used as the 'blueing' agent in laundry products such as Reckitt's Blue.

5.9 f–f electronic transitions in lanthanides

The UV-vis absorption spectra of Ln^{3+} compounds are characterised by numerous weak, sharp absorptions (Figure 5.28) that contrast with the generally stronger

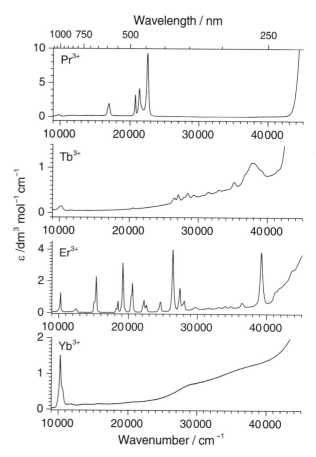

FIGURE 5.28 Electronic absorption spectra of f–f transitions in Pr^{3+}, Tb^{3+}, Er^{3+}, and Yb^{3+} aqueous solutions.

and broader absorptions seen for transition metal compounds in the preceding sections. The narrowness of the spectral features and their insensitivity to the nature of coordinated ligands are characteristic of contracted 4f orbitals that are shielded from the ligands. As a result the spectra are not very sensitive to different ligand fields and can be interpreted as 'atomic-like' where spin–orbit coupling rather than ligand fields predominate.

Absorption bands in Ln^{3+} compounds arise due to excitation from the ground electronic f-electron configuration into excited states. These are usually f–f transitions, but as they are forbidden by the orbital selection rule they are very weak. 4f–5d transitions are more intense and sensitive to the coordination environment, but are broader and at higher energy. In general only those of Ce^{3+} (ca. 32 000 cm^{-1}), Pr^{3+} (ca. 45 000 cm^{-1}), and Tb^{3+} (ca. 45 000 cm^{-1}) are observed; the others are too high in energy. Charge transfer transitions are also possible at high energy, but in aqueous solution only those for Eu^{3+} (ca. 36 000 cm^{-1}) and Yb^{3+} (ca. 42 000 cm^{-1}) are normally observed. As the position of the f–f absorptions in the visible region and their intensities vary little with respect to lanthanide(III) environment, the pale colours of many lanthanide compounds are characteristic of the element present regardless of the exact compound composition. For example, Pr^{3+} compounds are pale green, Nd^{3+} is lilac, Sm^{3+} yellow, and Er^{3+} compounds are rose pink.

Figure 5.28 shows f–f absorption spectra of Pr^{3+}, Tb^{3+}, Er^{3+}, and Yb^{3+} in aqueous solutions from the NIR to the UV. Rose pink Er^{3+} solutions have a very complex spectrum extending from the NIR to the UV. (See section 5.12, for analysis of complementary emission spectra.)

EXAMPLE 5.17

Why are Pr^{3+} solutions observed as pale green?

ANSWER

The weak absorptions for Pr^{3+} occur mainly between 21 000 and 24 000 cm^{-1} (420–480 nm) and at 17 000 cm^{-1} (590 nm) which correspond to the blue and orange parts of the visible spectrum. The remaining light has slightly more intensity in the green-yellow region giving Pr^{3+} its characteristic pale green colour.

SELF TEST

Why are the solutions of Tb^{3+} and Yb^{3+} essentially colourless?

Problems

5.1 Samples of sulfates, $MSO_4.nH_2O$, with M = Cr, Mn, Fe, Co, Ni, and Cu, were found unlabelled in the laboratory. Would it be possible to identify them by their colours alone?

5.2 The UV-vis spectrum of $[Ti(H_2O)_6]^{3+}$ in Figure 5.10 has a broad asymmetric peak at 20 000 cm^{-1} (500 nm). Use the spectrochemical series to predict the position of the d–d bands in the UV-vis spectrum of the $[Ti(NH_3)_6]^{3+}$ ion. The UV-vis spectra of $[Ti(NCS)_6]^{3-}$ and $[Ti(urea)_6]^{3+}$ show absorptions at 544 nm and 570 nm, respectively. Place the [NCS]$^-$ and urea ligands in the spectrochemical series.

5.3 Heating bright blue $CuSO_4.5H_2O$ to 700°C produces an almost colourless solid. Estimate the position of the sulfate anion in the spectrochemical series based on this observation.

5.4 Use the UV-vis spectral data for the selection of Cr(III) complexes in Table 5.5 to determine Δ_{oct} in each case and put the ligands in order of Δ_{oct}. Explain the order you obtain.

TABLE 5.5 UV-vis data (in cm^{-1}) for some octahedral chromium complexes

$[CrF_6]^{3-}$	14 900	22 700	34 400
$[CrCl_6]^{3-}$	13 200	18 700	
$[Cr(CN)_6]^{3-}$	26 700	32 600	
$[Cr(H_2O)_6]^{3+}$	17 400	24 600	37 800
$[Cr(NH_3)_6]^{3+}$	21 550	28 500	

5.5 Two spin-allowed transitions in $[Fe(CN)_6]^{4-}$ are at 31 000 and 37 040 cm^{-1}. Estimate Δ_{oct} for this complex.

5.6 Predict the differences in the LMCT spectra of the following groups of complexes.

(a) $[VO_4]^{3-}$, $[CrO_4]^{2-}$, $[MnO_4]^-$

(b) $[VO_4]^{3-}$, $[VS_4]^{3-}$, $[VSe_4]^{3-}$

(c) $[CrO_4]^{2-}$, $[MoO_4]^{2-}$

5.7 The charge transfer transition in $[IrCl_6]^{3-}$ is at 43 100 cm^{-1} and for $[IrBr_6]^{3-}$ at 36 800 cm^{-1}. Predict an optical electronegativity value for Ir(III) based on these data.

5.8 Explain why many complexes of Ln^{2+} and Ln^{4+} are intensely coloured.

ADVANCED TOPICS

In order to interpret the electronic absorption and emission spectra of lanthanides and transition metals in more detail it is necessary to build on the previous sections and introduce models of electronic structure that explicitly deal with electron–electron interactions caused by the coupling of the angular momentum associated with the orbital motion and the spin of the electrons. This will start from the familiar **orbital configurations** (e.g. p^2, d^2, f^2, etc.), before considering how these are split into **terms** (described using term symbols such as ^3F) and **levels** (e.g. $^3F_2$).

5.10 Configurations, terms, and levels

The important concepts required to interpret atomic and molecular electronic spectra are shown in Figure 5.29 for a d^2 atom or 'free' ion. The d^2 configuration, which contains 45 **microstates** (10 ways of organising the first electron, and 9 for the second (due to the Pauli principle), and then divided by 2 because of indistinguishabilty of electrons), is first split by **electron–electron repulsion** into terms (denoted using the symbol ^{2S+1}L), and then by

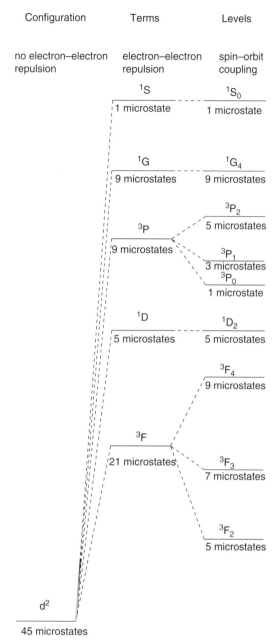

Configuration	Terms	Levels
no electron–electron repulsion	electron–electron repulsion	spin–orbit coupling

^1S — $^1S_0$ — 1 microstate — 1 microstate

^1G — $^1G_4$ — 9 microstates — 9 microstates

^3P — 9 microstates — $^3P_2$ 5 microstates — $^3P_1$ 3 microstates — $^3P_0$ 1 microstate

^1D — 5 microstates — $^1D_2$ 5 microstates

^3F — 21 microstates — $^3F_4$ 9 microstates — $^3F_3$ 7 microstates — $^3F_2$ 5 microstates

d^2

45 microstates

FIGURE 5.29 Schematic diagram of configuration, terms, and levels for a d^2 atom or free ion.

spin–orbit coupling into levels (described using the symbol $^{2S+1}L_J$). Atomic spectra involve transitions between the levels. However, for compounds and complexes, the size of the crystal field interaction compared to spin–orbit coupling determines whether terms or levels are required to interpret the data. For lanthanides, the crystal field effect is usually much smaller than the spin–orbit coupling, so the spectra are treated as 'atomic-like', but for transition metal compounds and complexes, the crystal field is usually much larger than the spin–orbit coupling, and so only terms are required to interpret the spectra.

Configurations such as $2p^2$, $3d^2$, and $4f^6$ tell us how many electrons there are in each orbital, but not how they are arranged, in respect of the orbital occupied and their spin orientation. This is why terms and levels are required. Electron–electron repulsion causes the orbital angular momentum of the individual electrons to **couple** with each other. Instead of the orbital angular momentum quantum number, l, for the individual electrons, a new quantum number, L, representing the total orbital angular momentum is used. For a two electron system such as d^2, each electron has an l value of 2, and these couple according to $l_1 + l_2, l_1 + l_2 - 1, ..., |l_1 - l_2|$ to give L values of 4, 3, 2, 1, 0. For each value of L there are $2L + 1$ values of M_L which span $+L$ to $-L$ in integer units and these are degenerate in the absence of a magnetic (Zeeman effect) or an electric (Stark effect) field. For one electron systems, lower case s, p, d, f labels are used to represent l; for multielectron systems, upper case S ($L = 0$), P ($L = 1$), D ($L = 2$), F ($L = 3$), G ($L = 4$), H ($L = 5$), I ($L = 6$) are employed. Therefore, for d^2 with L values of 4, 3, 2, 1, and 0, the terms will be G, F, D, P, and S. The spin angular momentum couples to yield the total spin angular momentum, represented by the S quantum number. For each value of S there are $2S + 1$ values of M_S which span from $+S$ to $-S$. The value of $2S + 1$, which is called the spin degeneracy or **spin multiplicity**, indicates how many M_S values exist. The spin multiplicity is always one more than the number of unpaired electrons.

EXAMPLE 5.18

Calculate the spin multiplicity, $2S + 1$, for atomic d^n ($n = 0$–10) configurations.

ANSWER

For the atomic d^n configurations there are six possibilities:

d^n	S	M_S	$2S + 1$	name
d^0, d^{10}	0	0	1	'singlet'
d^1, d^9	½	+½, −½	2	'doublet'
d^2, d^8	1	+1, 0, −1	3	triplet
d^3, d^7	3/2	+3/2, +½, −½, −3/2	4	'quartet'
d^4, d^6	2	+2, +1, 0, −1, −2	5	'quintet'
d^5	5/2	+5/2, +3/2, +½, −½, −3/2, −5/2	6	'sextet'

SELF TEST

Determine the spin multiplicity of the atomic f^n configurations.

Terms are represented with a term symbol, ^{2S+1}L, where L is replaced by its corresponding code letter.

The terms are further split into levels due to spin–orbit coupling (Figure 5.29). There are two ways of describing this and both give J, the total angular momentum quantum number. In the Russell–Saunders (or LS) coupling case it is assumed that the spin–orbit coupling is small. The individual orbital angular momenta couple and are summed into L, and the individual spin angular momenta couple and are summed to give S. The L and S then couple ($L + S, L + S - 1, \ldots, |L + S|$) to give J. If the spin–orbit coupling is large (i.e. for heavy atoms) then the jj coupling scheme is required where the individual orbital and spin angular momenta couple first (i.e. l and s) to give j, and the resultant j are coupled to give J. The level is labelled using the term followed by a subscript showing the J value, $^{2S+1}L_J$. Each level is $(2J + 1)$ degenerate. (The Online Resource Centre contains further details about how to calculate terms and levels.)

5.10.1 Hund's rules

Hund's rules are applied to determine the ground (i.e. lowest energy) term and level.

5.10.1.1 Hund's first rule

The term with highest S (or $2S + 1$) is lowest in energy. This corresponds to the term with the largest number of electrons with unpaired and parallel spins.

5.10.1.2 Hund's second rule

If there are several terms with the same S, the one with the highest L is lowest in energy.

EXAMPLE 5.19

Determine the term for total angular momentum $L = 3$ and total spin angular momentum $S = 1$.

ANSWER

$L = 3$ is represented with F, and the spin multiplicity $(2S + 1)$ for $S = 1$ is 3, so the term is ^{3}F (pronounced 'triplet F'). It comprises $(2L + 1) \times (2S + 1) = 7 \times 3 = 21$ term wavefunctions, all with the same energy.

SELF TEST

Determine the term for total orbital angular momentum $L = 4$ and total spin angular momentum $S = 0$.

5.10.1.3 Hund's third rule

Hund's third rule is used to identify the ordering of the levels within the terms. This states that for less than half-filled shells, the smallest value of J is lowest in energy, and that for more than half-filled shells, the largest value of J lies lowest in energy.

Hund's rules always work for determining the ground term, and the ordering of the p^n excited (i.e. higher energy) terms. However, the ordering of the excited terms for d^n does not follow this rule. The ordering of the d^2 terms is shown in Figure 5.29.

For all singlet terms ($S = 0, 2S + 1 = 1$) or those involving $L = 0$ (i.e. S) there is only one J value as either the orbital angular momentum, or the spin angular momentum are zero, so there is no coupling. (Care needs to be exercised when both S and S are being used. S is a variable, sum of the m_S, whereas S is the code when $L = 0$.)

EXAMPLE 5.20

Identify the ground term of the $3d^2$ configuration from the ^{1}G, ^{3}F, ^{1}D, ^{3}P, ^{1}S allowed terms.

ANSWER

The largest $2S + 1$ is 3, so the largest S value is 1, and one of the triplet ^{3}F and ^{3}P terms is the ground term. Of these, the ^{3}F has the larger L value. Hence the ground term is ^{3}F.

SELF TEST

The allowed terms for d^7 are ^{2}H, ^{2}G, ^{4}F, ^{2}F, ^{2}D, ^{2}D, ^{4}P, ^{2}P; identify the ground term.

5.10.2 Atomic terms for p^n, d^n, and f^n configurations

Full shells, that is, s^2, p^6, d^{10}, and f^{14}, have ^{1}S terms. The method for determining the terms for partially filled orbitals is given in the Online Resource Centre and is summarised below.

5.10.2.1 Terms arising from p^n configurations (in order of energy)

p^1, p^5	^{2}P
p^2, p^4	^{3}P, ^{1}D, ^{1}S
p^3	^{4}S, ^{2}D, ^{2}P

5.10.2.2 Terms arising from d^n configurations (in order of energy)

d^1, d^9	^{2}D
d^2, d^8	^{3}F, ^{1}D, ^{3}P, ^{1}G, ^{1}S

d³, d⁷	⁴F, ⁴P, ²G, ²H, ²P, ²D, ²F, ²D
d⁴, d⁶	⁵D, ³H, ³P, ³F, ³G, ¹I, ³D, ¹G, ¹S, ¹D, ¹F, ³F, ³P, ¹G, ¹D, ¹S.
d⁵	⁶S, ⁴G, ⁴P, ⁴D, ²I, ⁴F, ²D, ²F, ²H, ²G, ²F, ²S, ²D, ²G, ²P, ²D

5.10.2.3 Terms arising from fⁿ configurations

For f^n configurations the number of possible terms increases significantly; for f^2 there are 91 microstates and for f^7 there are 3432 and the derivation of all the resulting terms lies beyond the scope of this text.

5.10.3 Ordering of levels with terms

The ordering of the spin–orbit levels within a term is given by

$$\Delta E = \lambda \frac{\left[J(J+1) - L(L+1) - S(S+1)\right]}{2} \qquad \textbf{Eqn 5.4}$$

where ΔE is the energy shift for the level from that of the term.

EXAMPLE 5.21

Plot the ordering of the spin–orbit levels in the ³F term in d².

ANSWER

The ³F term has $L = 3$, $S = 1$ and therefore $J = 4, 3, 2$. As d² is less than half-filled, $J = 2$, lies lowest in energy. When $J = 4$, ΔE is $+3\lambda$; when $J = 3$, ΔE is -1λ; and when $J = 2$, ΔE is -4λ. This is shown in Figure 5.30.

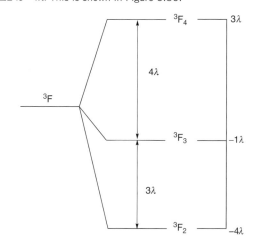

FIGURE 5.30 Splitting of ³F term into levels by spin–orbit coupling.

SELF TEST

Sketch the ordering and energy intervals of the levels in the ⁷F ground terms of Eu³⁺ and Tb³⁺.

Equation 5.4 and Figure 5.30 also demonstrate Landé's interval rule which states that 'the energy separation between two levels arising from the same term and characterised by integrally adjacent J levels is λ times the larger of the two J values'. $\lambda = \pm\zeta/2S$, where ζ (zeta) is the single electron spin–orbit coupling constant and λ is the term, or multi-electron, spin–orbit coupling constant. For less than half-filled shells $\lambda = +\zeta/2S$, and for more than half-filled shells $\lambda = -\zeta/2S$. This reverses the ordering of the levels when going from less than half-filled to more than half-filled and follows from Hund's third rule. If sufficient transitions are observed in the experimental data it is possible to determine the spin–orbit coupling constant, ζ.

5.11 Emission spectroscopy

When any element is promoted to an electronically excited state, it decays back to the ground state via a number of different pathways, which can be loosely described as either radiative (emission of light) or non-radiative. The rate of decay is dependent on the allowedness of the transition, and this is related to the selection rules. Transitions between states with the same spin ($\Delta S = 0$) and especially singlet states occur on a very fast timescale, typically nanoseconds or less. Processes involving triplet to singlet transitions can have longer lifetimes of microseconds, milliseconds or even seconds. For an atom or ion to be a good emitter, any non-radiative process must be much less favoured than the radiative pathway. For the majority of 3d, 4d, and 5d transition metal compounds the non-radiative pathway predominates, but in the case of the lanthanide elements the radiative process can become significant.

Radiative emission has traditionally been divided into fluorescence and phosphorescence. Fluorescence is radiative decay from an excited state of the same multiplicity as the ground state, with half-lives of the order of nanoseconds or less and is only observable while the sample is being irradiated. Phosphorescence is radiative decay from a state of different multiplicity from the ground state populated by inter-system crossing (ISC). The Jablonski diagram in Figure 5.31 summarises these processes. Phosphorescence is a spin-forbidden process and is often slow and still observable for a limited time once the exciting source is removed. These two processes are now often combined into a common term of luminescence, especially for inorganic systems. The fluorescence and phosphorescence will usually both be red-shifted from the excitation energy, but the extent is sample dependent. By careful gating of the detection it is possible to remove the prompt fluorescence and only detect the phosphorescence that has a longer lifetime.

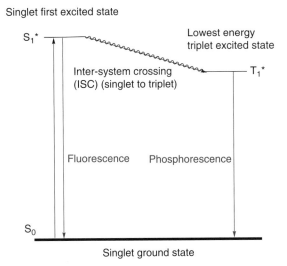

FIGURE 5.31 Simplified Jablonski diagram showing fluorescence and phosphorescence decay pathways. (Wavy line indicates non-radiative pathway.)

5.12 Lanthanide emission spectra— atomic spectra in solution

In general, atomic spectra are obtained from gas phase atoms or free ions. While gas phase spectra can be obtained for the lanthanides, the solution absorption spectra in Figure 5.28 contained sharp, very weak features characteristic of atomic spectra indicating it is possible to use the 'atomic' approach to understand their electronic spectra and structure in the more technologically useful solid or solution state. To a first approximation an atomic approach is justified because the 4f orbitals are buried deep within the atom and are little affected by the symmetry of their surroundings, so a spherical (atomic) approximation is reasonable. This should be contrasted with the transition metals where ligand field effects are significant.

Lanthanides are most commonly found in either the elemental or M^{3+} state. Many of the lanthanide M^{3+} ions fluoresce with narrow linewidths when exposed to UV light, with this fluorescence originating from transitions within the f sub-shell. The fluorescence properties of the lanthanides have led to their incorporation in the phosphors of domestic fluorescent tubes, the screens of colour televisions (old-fashioned cathode ray tubes), in some very widely used lasers (Nd-YAG), medical imaging, and as counterfeit protection on bank notes (europium is used on euro bank notes).

5.12.1 Ligand-enhanced lanthanide luminescence

While luminescence from lanthanide ions can sometimes be observed in aqueous solutions, due to the low absorption coefficient of the f–f transitions (forbidden by the $\Delta l = \pm 1$ selection rule) this is very weak, and is also quenched by a competing pathway involving ν_{O-H} vibrational overtones of coordinated H_2O. Complexation with particular ligands can increase the luminescence by several orders of magnitude. This emission is obtained by pumping the system using an allowed electronic transition (usually π to π^*) on the ligand to populate the higher energy f states which then decay to the ground state via emission of (visible) photons.

This is shown in the Jablonski diagram in Figure 5.32 where on the left-hand side the ligand is pumped from the ground singlet state, S_0, to excited singlet states, S_1^* and S_2^*. Most of the singlet states decay back to the ground singlet state, but a fraction go via ISC to an excited triplet state, T_1^* on the ligand. This can then decay to the excited lanthanide levels, and then via emission back to the ground state. The organic molecule is said to act as an **antenna**.

A great deal of research effort has gone into designing new ligands so that the energy relationships between the ligand singlet and triplet states and the lanthanide excited states are optimised for this ligand-enhanced luminescence pathway. These systems have found use in anion detection, cytometry, and biological imaging as well as in the detection of low levels (ppb) of lanthanides in water. A ligand that facilitates enhanced luminescence is 2,6-pyridinedicarboxylic acid (dipicolinic acid (H_2 dpa)) (Figure 5.33), which when deprotonated acts as a tridentate ligand to the

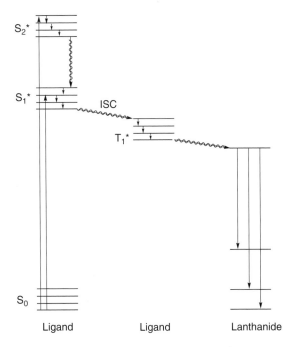

FIGURE 5.32 Jablonski diagram for photophysical processes in ligand-enhanced lanthanide luminescence. (The sub-levels on the ligand states are vibrational levels, those on the lanthanide are spin–orbit levels.)

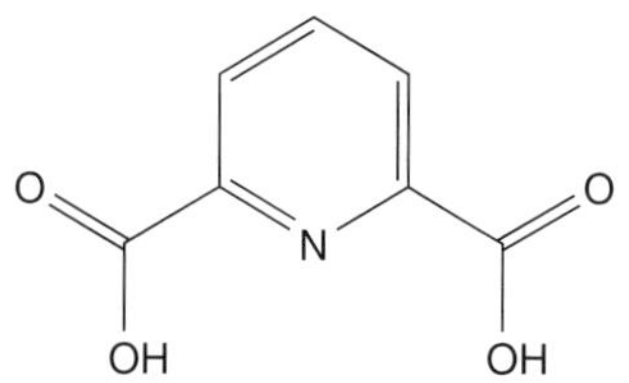

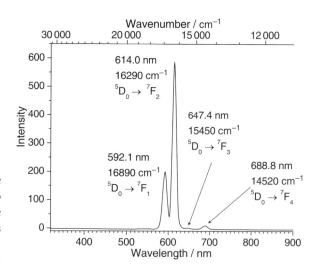

FIGURE 5.33 2,6-Pyridinedicarboxylic acid (dipicolinic acid (H_2 dpa)).

lanthanide excluding water from the coordination sphere (Figure 5.34). The ligand based transitions are the π to π* transitions in the pyridine aromatic system, which are fully allowed, and the lack of water coordination avoids quenching of the luminescence.

The lanthanide emissive state coupled to the ligand triplet state is $^5D_0$ for Eu^{3+} (ca. 17 270 cm^{-1} above the ground level) and $^5D_4$ for Tb^{3+} (ca. 20 500 cm^{-1} above the ground level). The spectra in Figures 5.35 (Eu^{3+}) and 5.36 (Tb^{3+}) show the ligand-enhanced luminescence due to transitions from these emissive states to the levels within the ^7F terms of Eu^{3+} ($4f^6$) and Tb^{3+} ($4f^8$) after excitation at 250 nm. The Eu^{3+} emission at 580–705 nm is orange/red, whereas the Tb^{3+} emission at 485–630 nm is green.

Figure 5.37 shows the relevant energy levels in Eu^{3+} and Tb^{3+} and how the assignments in Figures 5.35 and 5.36 were made. The spectra show the notation used to label the transitions, namely the higher energy level is put first, followed by the lower energy level, and the arrow indicates whether it is emission (as in this case) or absorption.

FIGURE 5.35 Emission spectrum of 12.5 μM Eu^{3+} dpa complex at 250 nm excitation.

EXAMPLE 5.22

Calculate, using Figures 5.35 and 5.37, the single electron spin–orbit coupling constant, ζ, for Eu^{3+}.

ANSWER

The separation between the levels in the ^7F terms can be used to determine the single electron spin–orbit coupling constant, ζ. The λ values are determined from the spectrum in Figure 5.35 and Figure 5.37 using the Landé interval rule, and $ζ = + 2Sλ$ (as f^6 is less than half full) for Eu^{3+}, to give the values below.

Observed transition/cm^{-1}	Assignment	$λ/cm^{-1}$	$ζ/cm^{-1}$
16 890 cm^{-1}	$^5D_0 \to {}^7F_1$		
16 290 cm^{-1}	$^5D_0 \to {}^7F_2$	300	1800
15 450 cm^{-1}	$^5D_0 \to {}^7F_3$	280	1680
14 520 cm^{-1}	$^5D_0 \to {}^7F_4$	233	1400

SELF TEST

Calculate the single electron spin–orbit coupling constant, ζ, for Tb^{3+} using the data in Figures 5.36 and 5.37.

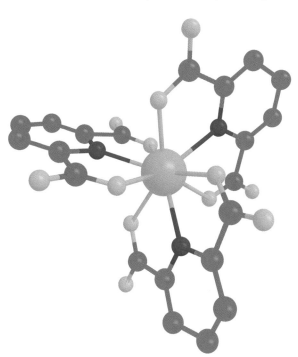

FIGURE 5.34 Structure of $[Ln(dpa)_3]^{3-}$ complexes.

The values of ζ are not the same for each transition because the Russell–Saunders approach is not completely accurate for the lanthanides, especially for high J values. However, these values compare favourably with those obtained using a sophisticated treatment with many

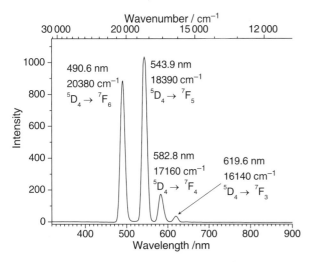

FIGURE 5.36 Emission spectrum of 12.5 μM Tb^{3+} dpa complex at 250 nm excitation.

more transitions which gave $\zeta = 1326$ cm^{-1} for Eu^{3+} and $\zeta = 1709$ cm^{-1} for Tb^{3+}.

Transition metal emission spectroscopy is discussed in Section 5.13.5.2, following the introduction of molecular term symbols.

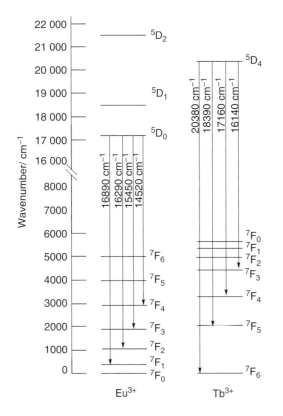

FIGURE 5.37 Schematic energy level diagram for Eu^{3+} and Tb^{3+} showing the principal luminescence transitions.

5.13 Advanced d–d spectroscopy

In the earlier sections, 0, 1, or 3 spin-allowed d–d transitions were predicted for octahedral and tetrahedral high-spin complexes of the first row transition series using a largely pictorial approach based on crystal field theory. The terms crystal field and ligand field are often used interchangeably, but they have subtle and important differences. Crystal field theory is predicated on the central metal and the ligands acting as point charges and the interaction is electrostatic or ionic. In contrast, in ligand field theory the d orbital splitting is interpreted by reference to chemical bonding effects and is, therefore, covalent in outlook. For transition metals the crystal field/ligand field effects are usually much greater than those of spin–orbit coupling, so d–d spectra are interpreted using terms, rather than levels. This section uses a more rigorous approach involving terms, so that a deeper insight into the electronic spectra and structure of transition metal complexes can be gained.

5.13.1 Octahedral and tetrahedral ground terms

In an octahedral d^1 configuration, such as that found in Ti(III), there are two configurations, a ground state, t_{2g}^1, and an excited state, e_g^1, with a separation of Δ_{oct} (Figure 5.12). For d^2, for example, V(III), there is one ground state, t_{2g}^2, and three excited states; two of these excited states are derived from the $t_{2g}^1 e_g^1$ configuration due to electron–electron repulsion, and one from the e_g^2 configuration. The terms determined in the atomic spectroscopy section were for atoms or free ions, that is, in a spherical field. In transition metal complexes the orbitals and terms use symmetry labels appropriate to the point group.

The free ion ground term for d^1 is ^2D, but the effect of an octahedral ligand field now needs to be considered. The t_{2g}^1 configuration consists of six possible arrangements, with three t_{2g} orbitals, and an up spin or down spin electron in each one (Figure 5.38).

As for the atomic case, the molecular term is constructed from the description of the spin and orbital degeneracies. The spin multiplicity is worked out in exactly the same way as for the atomic case. As there is only one electron for d^1, $\Sigma m_s = \frac{1}{2}$, so $S = \frac{1}{2}$, therefore $2S + 1 = 2$, that is, a doublet. As the total degeneracy was 6, this means that the orbital degeneracy must be 3.

FIGURE 5.38 Possible arrangements of electrons in a t_{2g}^1 configuration.

The only possible triply degenerate representations for an octahedral geometry (see O_h character table in Appendix 1) are T_{1g}, T_{2g}, T_{1u}, and T_{2u}. The d orbitals are even with respect to inversion in an octahedral environment, so it must be either a T_{1g} or a T_{2g} term. In this case it is T_{2g} as t_{2g} is the label used for the individual orbitals that contain the single electron. Therefore, the term for the t_{2g}^1 configuration is $^2T_{2g}$. A similar analysis gives the term for the e_g^1 excited state as ^2E_g. Therefore, the d^1 ^2D free ion term has split into $^2T_{2g}$ and ^2E_g terms in an octahedral environment.

For d^2 with a t_{2g}^2 ground configuration there are 15 arrangements. This can be calculated as follows. There are six ways of placing the first electron (three t_{2g} orbitals, up or down spin). The second electron cannot occupy the same orbital with the same spin as the first electron due to the Pauli exclusion principle, so there are now only five possibilities, giving $6 \times 5 = 30$ arrangements. Electrons are indistinguishable so the number of arrangements has been double counted so there are $^{30}/_2 = 15$ different microstates. The lowest energy term can be worked out in a similar way using Hund's rules so that the spins are parallel with the electrons in separate orbitals. There are three ways to achieve this (Figure 5.39).

Each of these arrangements has $S = \frac{1}{2} + \frac{1}{2} = 1$, so $2S + 1 = 3$, that is, a triplet. Each arrangement in Figure 5.39 has the same energy; as there are three of them, a triply degenerate representation is required, that is, a T term. This would often just be written as ^3T, but for completeness the trailing subscript is required. The subscript part of the label is much more difficult to work out for multi-electron configurations than the number of spin and orbital arrangements. Although it is often ignored or just looked up (see later discussion of Tanabe–Sugano diagrams in Section 5.13.5) it is important to appreciate its origin. In order to extract this information the effect of combining two t_{2g}^1 configurations needs to be considered. This is done by evaluating the direct product of the two t_{2g} electrons, $t_{2g} \times t_{2g}$. The direct product of a degenerate species with itself (i.e. t_{2g} in this case) is resolved into symmetric and antisymmetric components. In the table of direct products (Appendix 2), the antisymmetric components of the direct products are placed within [], and for $t_{2g} \times t_{2g}$ it is given as $a_{1g} + e_g + [t_{1g}] + t_{2g}$. Therefore, the T_{1g} term is antisymmetric, and the T_{2g} term is symmetric, as are the A_{1g} and E_g terms. For electronic terms, the total wavefunction must

be antisymmetric. To get a total wavefunction that is antisymmetric, the spin component must therefore be symmetric for T_{1g} and antisymmetric for T_{2g}. Spin singlet states are antisymmetric, while triplet spin states are symmetric, therefore the correct terms are $^3T_{1g}$ and $^1T_{2g}$. The total degeneracy of the $^3T_{1g}$ term is $3 \times 3 = 9$. The other 6 arrangements from the total of 15 for t_{2g}^2 are $^1A_{1g}$, ^1E_g, and $^1T_{2g}$ in which the spins are paired and these are spin-forbidden excited states. Determining the trailing subscript for more electrons becomes even more complicated. For example, while all spin quartet states are symmetric, the spin doublets are split between

---

EXAMPLE 5.23

Determine the ground term for octahedral d^3.

ANSWER

The lowest energy configuration for d^3 is t_{2g}^3 with all the electrons parallel and in different orbitals. As there is only one way to organise this it must be an A term. There are three unpaired electrons so the spin multiplicity ($2S + 1$) is 4, and the term is ^4A.

SELF TEST

Determine the ground terms for the other high-spin d^n configurations using the spin multiplicity and orbital degeneracy method outlined above (ignore the trailing subscripts). Check these against those in Table 5.6 which have the trailing subscripts for completeness. Repeat the process for tetrahedral terms and check against Table 5.6.

TABLE 5.6 Terms for ground and spin-allowed excited states for high-spin octahedral and tetrahedral d^n configurations

Octahedral			Tetrahedral		
d^n	ground term	spin-allowed excited terms	d^n	ground term	spin-allowed excited terms
1	$^2T_{2g}$	^2E_g	1	^2E	$^2T_2$
2	$^3T_{1g}$	$^3T_{2g}, {}^3T_{1g}, {}^3A_{2g}$	2	$^3A_2$	$^3T_2, {}^3T_1, {}^3T_1$
3	$^4A_{2g}$	$^4T_{2g}, {}^4T_{1g}, {}^4T_{1g}$	3	$^4T_1$	$^4T_2, {}^4T_1, {}^4A_2$
4	^5E_g	$^5T_{2g}$	4	$^5T_2$	^5E
5	$^6A_{1g}$		5	$^6A_1$	
6	$^5T_{2g}$	^5E_g	6	^5E	$^5T_2$
7	$^4T_{1g}$	$^4T_{2g}, {}^4T_{1g}, {}^4A_{2g}$	7	$^4A_2$	$^4T_2, {}^4T_1, {}^4T_1$
8	$^3A_{2g}$	$^3T_{2g}, {}^3T_{1g}, {}^3T_{1g}$	8	$^3T_1$	$^3T_2, {}^3T_1, {}^3A_2$
9	^2E_g	$^2T_{2g}$	9	$^2T_2$	^2E
10	$^1A_{1g}$		10	$^1A_1$	

FIGURE 5.39 Possible arrangements of electrons in the t_{2g}^2 ground term of a d^2 configuration.

symmetric and antisymmetric wavefunctions. While the trailing subscripts are given in the other examples below for completeness, it is not normally necessary to be able to calculate them as they are widely available (see the Tanabe–Sugano diagrams later in this section).

It should be noted from Table 5.6 in Example 5.23 that (i) the orbital part of the ground and excited terms are identical for d^n and d^{n+5} octahedral configurations and (ii) the terms for tetrahedral d^n are the same as those for octahedral d^{10-n} (apart from the inclusion of the g and u subscripts as there is no centre of symmetry in the T_d point group).

5.13.2 Excited state terms

The terms for the spin-allowed excited states are also given in Table 5.6. For the d^1 excited state, e_g^1, it is straightforward to determine the term as ^2E_g. For more than one electron the situation becomes a little more complicated, but for high-spin complexes it is still relatively easy to identify the spin and orbital parts of the term symbol. The t_{2g}^2 configuration represents the lowest energy way of occupying the d orbitals in an octahedral d^2 complex. There are two possible excited state arrangements: $t_{2g}^1 e_g^1$ and e_g^2, representing excitation of one and two electrons, respectively, without changing the spin.

There is only one way of achieving the highest energy e_g^2 arrangement while keeping the spins parallel (Figure 5.40): $\Sigma m_s = \frac{1}{2} + \frac{1}{2} = 1$, so $S = 1$ and $2S + 1 = 3$. There is only one orbital arrangement possible and this is represented by A. The term symbol for this arrangement is thus ^3A ($^3A_{2g}$ if the subscript is included).

There are six ways of arranging the electrons for the configuration $t_{2g}^1 e_g^1$ (Figure 5.17). There is no label for sixfold degeneracy in the O_h point group, so not all of these arrangements can have the same energy. Figure 5.18 showed that $(d_{xy})^1(d_{z^2})^1$ and $(d_{xy})^1(d_{x^2-y^2})^1$ configurations have different energies and that there are two pairs of three arrangements, so the six possibilities partition into two triply degenerate ^3T terms. Using the more rigorous approach, the direct product of $t_{2g} \times e_g$ gives $T_{1g} + T_{2g}$, so the terms are $^3T_{1g}$ and $^3T_{2g}$. These $^3T_{1g}$ and $^3T_{2g}$ terms arising from the $t_{2g}^1 e_g^1$ configuration are split by electron–electron repulsion. The lower energy $(d_{xy})^1(d_{z^2})^1$ configuration belongs to the $^3T_{2g}$ term as do $(d_{xz})^1(d_{x^2-y^2})^1$ and $(d_{yz})^1(d_{x^2-y^2})^1$. The higher energy $(d_{xy})^1(d_{x^2-y^2})^1$

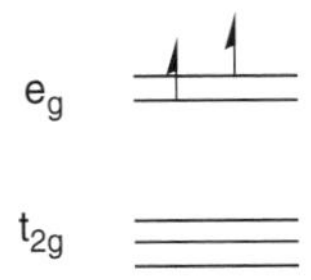

FIGURE 5.40 Arrangements of electrons in the e_g^2 excited term of a d^2 configuration.

TABLE 5.7 Terms derived from free ions in octahedral and tetrahedral fields

Free ion term	Orbital degeneracy	Terms in a cubic field*
S	1	A_{1g}
P	3	T_{1g}
D	5	$E_g + T_{2g}$
F	7	$A_{2g} + T_{1g} + T_{2g}$
G	9	$A_{1g} + E_g + T_{1g} + T_{2g}$
H	11	$E_g + T_{1g} + T_{1g} + T_{2g}$
I	13	$A_{1g} + A_{2g} + E_g + T_{1g} + T_{2g} + T_{2g}$

*In a tetrahedral field the g subscripts are not required as there is no inversion centre present.

configuration belongs to the $^3T_{1g}$ term together with $(d_{xz})^1(d_{z^2})^1$ and $(d_{yz})^1(d_{z^2})^1$. As a transition to the $^3A_{2g}$ term from the $^3T_{1g}$ ground term involves the excitation of two electrons, it is expected to be highest in energy, and is also likely to be weak (but not always). Therefore, for d^2, the ground term is $^3T_{1g}$ and the excited terms involved in spin-allowed transitions are $^3T_{2g}$, $^3T_{1g}$, and $^3A_{2g}$.

The free ion ground term for d^2 is ^3F. This has $2L+1$, that is, sevenfold, orbital degeneracy, but the total orbital degeneracy of the ground term ($^3T_{1g}$) and the three excited terms ($^3T_{1g}, {}^3T_{2g}, {}^3A_{2g}$) is $3 + 3 + 3 + 1 = 10$. Therefore, all of these cannot be derived from the ^3F free ion ground term. The complete list of allowed terms for d^2 is $^3F, {}^3P, {}^1G, {}^1D, {}^1S$ (see Section 5.10.2.2), and hence one of the triply degenerate excited terms ($^3T_{1g}$) must be derived from the ^3P excited term. Table 5.7 gives the terms derived from the free ion terms in cubic (octahedral, tetrahedral) fields.

EXAMPLE 5.24

Calculate the terms for the ground and spin-allowed excited states for octahedral high-spin d^4 complexes, and hence identify the number of spin-allowed transitions.

ANSWER

High-spin d^4 has a $t_{2g}^3 e_g^1$ ground configuration, and there is only one spin-allowed excited configuration, $t_{2g}^2 e_g^2$. Both of these are spin quintets. As there are two possibilities for the ground configuration the term is ^5E. For the excited configuration the term is ^5T as there are three ways of placing the two electrons in the t_{2g} orbitals. This is shown in Figure 5.41. Therefore, there is just one spin-allowed transition, $^5T_{2g} \leftarrow {}^5E_g$.

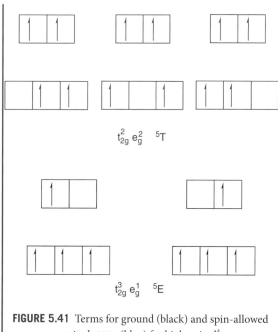

$t_{2g}^2 e_g^2$ ^5T

$t_{2g}^3 e_g^1$ ^5E

FIGURE 5.41 Terms for ground (black) and spin-allowed excited states (blue) for high-spin d^4.

SELF TEST

Calculate the terms for the ground and spin-allowed excited states for octahedral high-spin d^7 complexes, and hence identify the number of spin-allowed transitions.

5.13.3 Racah parameters

The electron–electron repulsion is usually expressed in terms of Racah A, B, and C parameters. The Racah A parameter describes the average of the total interelectronic repulsions. The Racah B parameter is related to the coulombic energies and the Racah C parameter to electron exchange energies between the d electrons. For d^2 the energies of the atomic terms are given in Table 5.8.

TABLE 5.8 Absolute and relative energies of the terms in a d^2 configuration

Term	Energy	Energy relative to ^3F
^1S	$A + 14B + 7C$	$22B + 7C$
^1G	$A + 4B + 2C$	$12B + 2C$
^3P	$A + 7B$	$15B$
^1D	$A - 3B + 2C$	$5B + 2C$
^3F	$A - 8B$	0

As it is the energies of terms relative to the ^3F ground term that are important, the absolute value of the Racah A parameter is not significant, and the differences are expressed in terms of Racah B and C parameters. (The absolute and relative energies for the other d^n configurations are available in the Online Resource Centre.)

For the two triplet terms, the energy separation is expressed only in terms of B, and the ^3P term lies $15B$ higher in energy than the ^3F ground term. In general, the energies of the excited states with the same multiplicity as the ground term will only involve multiples of B. The excited states of different multiplicities, will involve both B and C. The Racah B, C, and C/B values in the d^n 'free ions' in different oxidation states are given in Table 5.9. From the values it can be seen that it is usually assumed that $C \approx 4B$. For all of the 3d elements the Racah B parameter increases as one goes across the period, that is, the electron–electron repulsion increases as the electrons are held tighter to the nucleus. It also increases with oxidation state.

5.13.4 Orgel diagrams

The effect of electron–electron repulsion on the energies of the ground and excited terms can be represented by Orgel diagrams. For high-spin complexes there are two

TABLE 5.9 Racah B and C parameters (in cm^{-1}) and C/B ratios for 'free ion' first row transition metals in different oxidation states. (Data from M. Brorson and C. E. Schäffer, *Inorg. Chem.* **27**, 2522 (1988) and B.N. Figgis and M.A. Hitchman. (2000) *Ligand Field Theory and its Applications.* Wiley-VCH.)

	+2			+3			+4		
	B	C	C/B	B	C	C/B	B	C	C/B
Ti	714	2663	3.7						
V	760	2909	3.8	886	3520	4.0			
Cr	796	3298	4.1	933	3710	4.0	1038	4270	4.1
Mn	859	3527	4.1	950	4112	4.3	1088	4412	4.1
Fe	897	3877	4.3	1029	4269	4.2	1122	4755	4.2
Co	989	4214	4.3	1080	4560	4.2	1185	4950	4.2
Ni	1042	4604	4.4	1149	4826	4.2	1238	5210	4.2
Cu	1240	4712	3.8						

Orgel diagrams, one for systems with a free ion D ground term and a single excited state (e.g. d^1) and a second one where the free ion ground term is F, but with three spin-allowed excited states (e.g. d^2).

Before considering these in detail it is necessary to define what is meant by a weak and strong field in the context of crystal and ligand field theories. While these phrases are related to whether high-spin or low-spin configurations are observed, the strict definition within the context of crystal and ligand field theories is that in the strong-field limit the splitting of the d orbitals by the crystal/ligand field is considered first, and then the electron–electron repulsion is introduced. In contrast, in the weak-field approach the electron–electron repulsion in the free ion is perturbed by the (octahedral) ligand field to give the terms as given in Table 5.7. Orgel diagrams make use of the weak-field approach, and are limited to high-spin complexes.

The Orgel diagram for a d^1 configuration with a $^2T_{2g}$ ground term and an ^2E_g excited term shown in Figure 5.42 is essentially the same as that for d^1 as in Figure 5.12, but this diagram also includes the other d^n configurations that behave in a similar manner; this is why the spin multiplicity and g subscripts are not given. The centre of the diagram is when the crystal field splitting is zero, and corresponds to the atomic or free ion D term. As the crystal field increases either to the left or the right, the free ion D term is split into the molecular E_g and T_{2g} terms in an octahedral field and E and T_2 terms in a tetrahedral field. For d^1 and d^6 in an octahedral environment the T_{2g} term is lower in energy than the E_g term. This is reversed for octahedral d^4 and d^9. In all cases the value of Δ (Δ_{oct} or Δ_{tet}) can be read directly from the spectra from the peak corresponding to the transition or energy gap between the two terms.

Figure 5.42 also shows how diagrams intended for octahedral geometries can be used for tetrahedral

environments because d^n octahedral is equivalent to d^{10-n} tetrahedral. For octahedral usage, a subscript g needs to be added to each term, but is absent for the tetrahedral case as there is no inversion centre.

The remaining high-spin d^n configurations (d^2, d^3, d^7, and d^8) with an F ground term and an excited P term of the same spin multiplicity are also covered by one Orgel diagram, as shown in Figure 5.43. This has one ground term and three excited terms of the same multiplicity (derived from both F and P) in either a tetrahedral or an octahedral field. As this includes electron–electron repulsion, the relationships between the terms are more complicated as they include both Δ and B (the relevant equations are given in the Online Resource Centre). The relationships for d^2 and d^7 are the same, as are those for d^3 and d^8. The Orgel diagram in Figure 5.43 shows that terms of the same symmetry (i.e. T_1 in this case) cannot cross as they interact and repel each other, and this is a manifestation of the **non-crossing rule**. The left-hand side of the diagram explains why it is possible to obtain the value of Δ directly from the spectra of d^3 and d^8 octahedral complexes (d^2 and d^7 tetrahedral complexes) if the lowest energy transition can be identified. While there is an energy gap of Δ on the right-hand side of the diagram between the $T_2(F)$ and $A_2(F)$ terms, this cannot be accessed directly as it is the difference between transitions from the $T_1(F)$ ground term to the $T_2(F)$ and $A_2(F)$ excited terms, and can only be used if all three transitions are observed (and properly assigned). The assignment is made more complex because the $A_2(F)$ and $T_1(P)$ terms cross.

The Orgel diagrams in Figures 5.42 and 5.43 demonstrate the effect of electron–electron repulsion on the atomic terms,

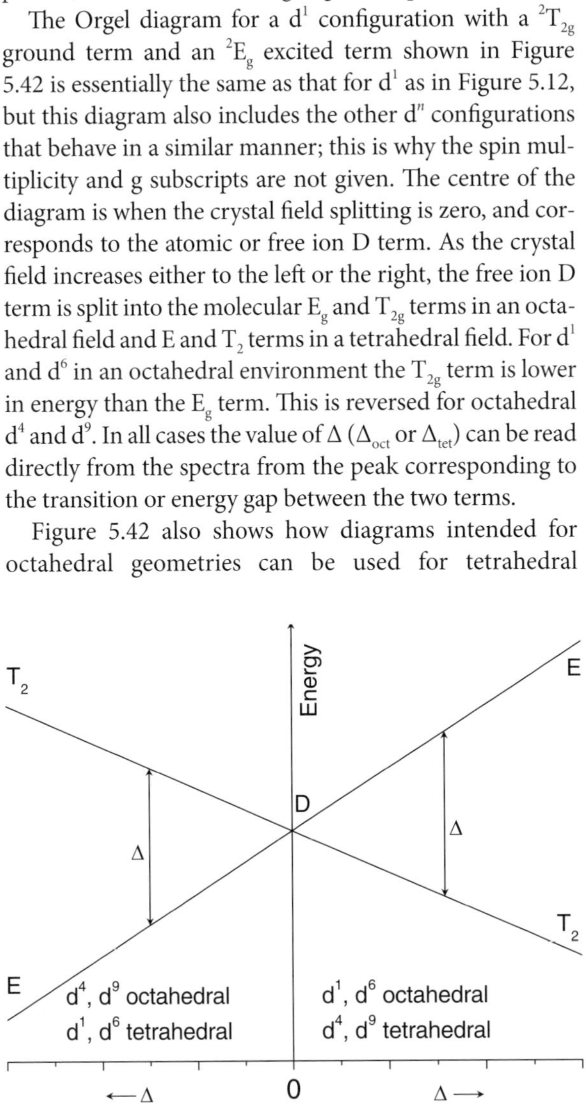

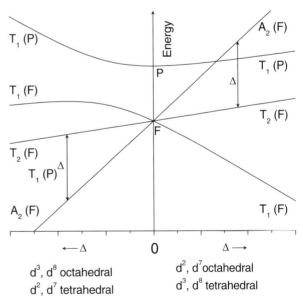

FIGURE 5.42 Orgel diagram for high-spin (weak-field) complexes with d^1, d^4, d^6, and d^9 configurations (for octahedral use, g subscripts need to be added to each term).

FIGURE 5.43 Orgel diagram for high-spin (weak-field) complexes with d^2, d^3, d^7, and d^8 configurations with a free ion F ground term (for octahedral use, g subscripts need to be added to each term).

and show how the relative energies of the terms vary with the crystal field splitting, but they are limited to the weak-field (i.e. high spin only) cases, and also only show spin-allowed transitions. They are usually presented in a generic form (so that only two are needed) and while they are not easily used to extract the value of Δ_{oct} or Δ_{tet} except for systems where the values can be read directly from the experimental spectra, they are very useful in giving a good qualitative interpretation, including for mixed ligand complexes.

5.13.5 Tanabe–Sugano diagrams

A much more complete picture is offered by Tanabe–Sugano diagrams. The main difference to the Orgel diagrams is that the ground term is now plotted as the *x*-axis, with all the excited terms plotted relative to this. They also include all the spin states, so that high-spin and low-spin spin-allowed transitions, as well as the spin-forbidden transitions can be identified. The *x*-axis is given as Δ/B. Older versions may have this as $10Dq/B$ (or Dq/B) as $10Dq = \Delta$. The *y*-axis is plotted as E/B (transition energy/Racah *B* parameter). Tanabe–Sugano diagrams are available in Appendix 4 (and in colour in the Online Resource Centre) for all d^n configurations, except d^1 and d^9 which are trivial, and where the Orgel diagram is sufficient.

While these diagrams look highly complex at first, once understood, they are a powerful and compact source of information. They allow for: (i) the identification of the ground and excited terms; (ii) the number of spin-allowed transitions; (iii) estimation of the ligand field splitting parameter and the Racah *B* parameter; (iv) identification of spin-forbidden transitions; and (v) the interpretation of additional features in the spectra.

5.13.5.1 d^3, chromium(III), vanadium(II)

The Orgel diagram for octahedral d^3 shown in Figure 5.43 indicates that it is possible to obtain the value of Δ_{oct} straight from the spectrum as it corresponds to the first d–d transition. However, this gives no information about the Racah *B* parameter which contains important information about bonding. The electronic absorption spectra of 0.05 M *trans*-$[CrCl_2(H_2O)_4]^+$ and 0.15 M $[Cr(H_2O)_6]^{3+}$ are shown in Figure 5.44 and the molar absorptivity is very similar (ca. 15 mol dm^{-3} cm^{-1}) for both complexes. *trans*-$[CrCl_2(H_2O)_4]^+$ is the green solution species formed when $CrCl_3.6H_2O$ is dissolved in water, and the blue/grey solution of $[Cr(H_2O)_6]^{3+}$ is formed after oxidation of $[Cr(H_2O)_6]^{2+}$ (produced by the action of 1 M H_2SO_4 on Cr metal).

The Tanabe–Sugano diagram for d^3 is shown in Figure 5.45. The quartet terms are the same as those in the left-hand side of the Orgel diagram in Figure 5.43, but Figure 5.45 now also includes some of the more important doublet terms, and these are shown in blue. (The complete set is given in Figure 2 in Appendix 4.)

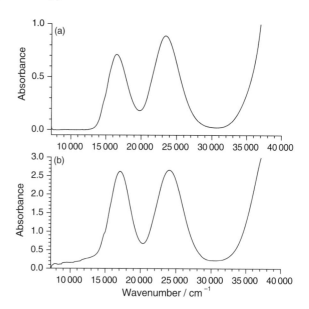

FIGURE 5.44 Electronic absorption spectra of (a) 0.05 M *trans*-$[CrCl_2(H_2O)_4]^+$ and (b) 0.15 M $[Cr(H_2O)_6]^{3+}$.

It is very important to note that the energy of the terms is plotted relative to the energy of the $^4A_{2g}$ ground term, and that, therefore, this ground term is lying on the *x*-axis. The free ion (atomic) terms are on the left-hand side. The ligand

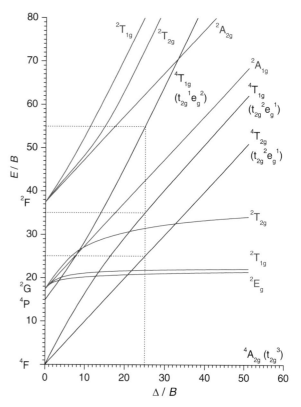

FIGURE 5.45 d^3 Tanabe–Sugano diagram with quartet terms in black, the more important doublet terms in blue, and the analysis for $[Cr(H_2O)_6]^{3+}$.

field, Δ, increases from the weak-field limit, at the left-hand side, to the strong-field limit on the right-hand side. The lines originating from the free ion terms show how these are split by the octahedral ligand field into the A_{1g}, A_{2g}, E_g, T_{1g}, and T_{2g} terms (Table 5.7), and also how their relative energies change. The ^4F ground term splits into a $^4A_{2g}$ ground term and $^4T_{2g}$, $^4T_{1g}$ excited terms in an octahedral field. The remaining excited triplet term ($^4T_{1g}$) is derived from the ^4P excited atomic term. Due to the non-crossing rule, terms of the same symmetry and spin multiplicity cannot cross (e.g. the pair of $^4T_{1g}$ terms), and diverge from each other. In addition many of the doublet lines appear bent because they are plotted with respect to the ground term, and there is a complex relationship between them.

The advantage of these diagrams is that it is very easy to identify the ground term (on the x-axis), the excited terms, and the number of spin-allowed transitions to excited terms with the same multiplicity as the ground term.

From the d^3 Tanabe–Sugano diagram (and earlier analysis) three spin-allowed transitions are expected from the $^4A_{2g}$ ground term and in terms of increasing energy they are: $^4T_{2g}$ (F) $\leftarrow$ $^4A_{2g}$ (F); $^4T_{1g}$ (F) $\leftarrow$ $^4A_{2g}$ (F); $^4T_{1g}$ (P) $\leftarrow$ $^4A_{2g}$ (F). These use the conventional spectroscopic notation with the higher energy term on the left-hand side, followed by the lower energy term on the right-hand side, with an arrow between them to indicate whether the process is either emission ($\rightarrow$) or absorption ($\leftarrow$). Only two peaks are observed in the experimental UV-vis spectra for most Cr(III) complexes (Figure 5.44) as the 'two-electron' transition ($^4T_{1g}$ (P) $\leftarrow$ $^4A_{2g}$ (F)) at high energy is masked by charge transfer bands. The first peak in the spectrum of $trans$-$[CrCl_2(H_2O)_4]^+$ is at $16\,510$ cm^{-1} and that in $[Cr(H_2O)_6]^{3+}$ is at $17\,080$ cm^{-1}, and these correspond to the values of Δ_{oct} in these two complexes. This ordering is as expected because Cl$^-$ is a weaker field ligand in the spectrochemical series than H_2O.

While it is straightforward to determine the value of Δ from the position of the first peak in the spectrum, to obtain a value of B, it is necessary to use the d^3 Tanabe–Sugano diagram shown in Figure 5.45.

EXAMPLE 5.25

Use the d^3 Tanabe-Sugano diagram (Figure 5.45) to determine the value of B for $[Cr(H_2O)_6]^{3+}$.

ANSWER

The basis of this process is to locate the value of Δ/B on the x-axis that is consistent with the relative energies of the observed transitions. For $[Cr(H_2O)_6]^{3+}$ the transitions in Figure 5.44(b) are at $17\,080$ and $24\,030$ cm^{-1}, giving a ratio of 1.41. From the d^3 Tanabe–Sugano diagram it is possible to measure the ratio of the $^4T_{1g}$ (F) $\leftarrow$ $^4A_{2g}$ (F) and $^4T_{2g}$ (F) $\leftarrow$ $^4A_{2g}$

(F) transition energies as 1.47 at $\Delta/B = 20$, 1.41 at $\Delta/B = 25$, and 1.35 at $\Delta/B = 30$. Therefore, a value of $\Delta/B = 25$ fits the data, and by drawing a vertical line on the diagram at $\Delta/B = 25$ in Figure 5.45 and reading across to the y-axis, the values of E/B for the two transitions can be determined as 25.0 and 35.0. The fact that Δ/B and E/B have the same value for the first of these indicates a gradient of 1 and that $\Delta = E$. This is confirmation that Δ can be obtained directly from the first spin-allowed transition in the spectrum in this case. In conjunction with the transition energies of $17\,080$ and $24\,030$ cm^{-1} this gives B values of 683 and 687 cm^{-1}, respectively, which can be taken as 685 cm^{-1}. Using a value of $\Delta/B = 25$ predicts the third spin-allowed transition will be at $E/B = 54.9$, that is, $37\,600$ cm^{-1}, which is obscured by the charge transfer transition in Figure 5.44(b).

An alternative approach is to use a Lever plot of the ratios of the spin-allowed transition energies, as shown in Figure 5.46. (This diagram can also be used for octahedral d^8, which is why the spin multiplicities are omitted.) From this it is immediately clear that with a ratio of 1.41 the only possible assignment of these two bands is to the two lower energy $^4T_{2g}$ (F) $\leftarrow$ $^4A_{2g}$ (F) and $^4T_{1g}$ (F) $\leftarrow$ $^4A_{2g}$ (F) transitions, as the other ratios are larger than this for all values of Δ/B. Using the ratio of 1.41 allows a more accurate Δ/B value of 24.2 to be obtained. The vertical line at 24.2 Δ/B in Figure 5.46 can be used to (i) estimate the three transition energies at 24.2, 34.1, and 53.5 E/B, and (ii) predict a

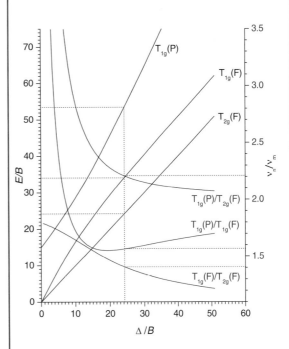

FIGURE 5.46 Lever plot of spin-allowed excited state terms (left-hand side) and ratios of transition energies (right-hand side) for the spin-allowed transitions in d^3 and d^8.

ratio of 1.57 for the $^4T_{1g}$ (P) ← $^4A_{2g}$ (F) and $^4T_{1g}$ (F) ← $^4A_{2g}$ (F) transition energies, and 2.21 for the $^4T_{1g}$ (P) ← $^4A_{2g}$ (F) and $^4T_{2g}$ (F) ← $^4A_{2g}$ (F) transition energies. Using the data from (i) and transition energies of 17 080 and 24 030 cm^{-1} results in values of B of 705 cm^{-1} and Δ of 17 080 cm^{-1}. Both methods predict the third spin-allowed transition will be at ca. 37 500 cm^{-1}, confirming that it will be masked by the intense charge transfer band at high wavenumber.

SELF TEST

The first two spin-allowed d–d transitions in *trans*-[CrCl$_2$(H$_2$O)$_4$]$^+$ (Figure 5.44a) are at 16 510 and 23 450 cm^{-1}. Using the d^3 Tanabe-Sugano diagram (Figure 5.45) and Lever plot (Figure 5.46) calculate Δ, B and predict the position of the third spin-allowed transition.

From the Tanabe–Sugano diagram in Figure 5.45 it can be seen that for Δ/B of 25 or 24.2, there are spin-forbidden transitions involving the $^2T_{1g}$ and ^2E_g terms that are close to, but slightly lower in energy than the spin-allowed transition to the $^4T_{2g}$ term. These spin-forbidden transitions are expected to be weak, but gain some intensity by mixing with the spin-allowed transitions in close proximity and are observed as the weak shoulders at 15 000 cm^{-1} in Figure 5.44.

5.13.5.2 Use of Tanabe–Sugano diagrams for understanding d^3 emission spectra

Many d^3 Cr(III) and Mn(IV) complexes are luminescent, and the d^3 Tanabe–Sugano diagram (Figure 5.45) can be used to understand the origin of these emissive

transitions. The classic example is [Cr(urea)$_6$]$^{3+}$ where the lowest energy d–d absorption band is at ca. 16 250 cm^{-1}, and irradiation of this results in both narrow phosphorescence at ca. 14 240 cm^{-1} and broad fluorescence at ca. 12 550 cm^{-1} (Figure 5.47).

From the Tanabe–Sugano diagram the lowest energy d–d transition is $^4T_{2g}$ (F) ← $^4A_{2g}$ (F) and the fluorescence is just the reverse process, that is, $^4T_{2g}$ (F) → $^4A_{2g}$ (F). The phosphorescence occurs because of the close proximity of the spin-forbidden $^2T_{1g}$ and ^2E_g states to the $^4T_{2g}$ term, which also gave rise to weak spin-forbidden transitions in absorption.

The fluorescence emission is red-shifted because there is a considerable increase in the metal–ligand bond length in the $^4T_{2g}$ excited term compared to the $^4A_{2g}$ ground term due to the population of the antibonding e_g orbitals (Figure 5.47). In contrast, the phosphorescence emission is very close to the absorption energy as there is little or no change in the equilibrium bond lengths for the $^2T_{1g}$ and ^2E_g excited states, as these only involve rearrangement of the t_{2g} electrons, a process known as a spin flip. This is also evident in the Tanabe–Sugano diagram where the ^2E_g and $^2T_{1g}$ terms are parallel to the $^4A_{2g}$ ground term. The 694.3 nm (14 400 cm^{-1}) ^2E_g → $^4A_{2g}$ emission is the basis of ruby lasers.

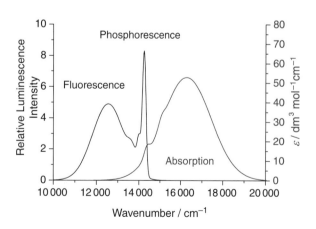

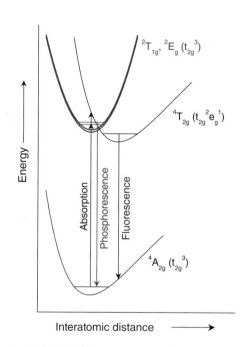

FIGURE 5.47 Absorption (298 K) and emission (78 K) spectra of [Cr(urea)$_6$]$^{3+}$ (left). Schematic potential energy diagram for absorption, fluorescence, and phosphorescence in octahedral d^3 complexes (right). (Data from G.B. Porter and H.L. Schläfer, *Z. Phys. Chem.* **37** 109 (1963).)

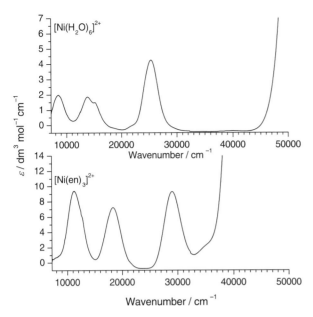

FIGURE 5.48 Electronic absorption spectra of $[Ni(H_2O)_6]^{2+}$ and $[Ni(en)_3]^{2+}$.

5.13.5.3 d^8, nickel(II)

As for d^3, the value of Δ_{oct} can be read directly from the first d–d transition in the spectra of d^8 complexes (Figure 5.48). But as this is not always observable, especially for nickel complexes with weak-field ligands, the use of the Tanabe–Sugano diagram (Figure 5.49) and the Lever plot of ratios of the term energies (Figure 5.46) is recommended as this also allows for the determination of B.

EXAMPLE 5.26

Use the d^8 Tanabe–Sugano diagram (Figure 5.49) to explain the splitting of the central feature at 14 500 cm⁻¹ in the electronic absorption spectrum of $[Ni(H_2O)_6]^{2+}$ (Figure 5.48).

ANSWER

The splitting of the central feature in the spectrum of $[Ni(H_2O)_6]^{2+}$ can be explained using the d^8 Tanabe–Sugano diagram where the spin-allowed $^3T_{1g}$ (F) ← $^3A_{2g}$(F) and the spin-forbidden ^1E_g ← $^3A_{2g}$ transitions lie in close proximity with Δ/B values of ca. 10, but are further apart as Δ decreases or increases. The ^1E_g ← $^3A_{2g}$ spin-forbidden transition, which would be expected to be much weaker than the spin-allowed transition, gains/steals intensity from the spin-allowed transition when the energies are very similar. It is also expected to be a sharp transition as the term line is close to horizontal.

SELF TEST

Use the d^8 Tanabe–Sugano diagram (Figure 5.49) and the three observed d–d transitions in $[Ni(H_2O)_6]^{2+}$ at 8550,

14 480, and 25 370 cm⁻¹ and in $[Ni(en)_3]^{2+}$ at 11 280, 18 350, and 29 040 cm⁻¹ (Figure 5.48) to calculate Δ_{oct} for $[Ni(H_2O)_6]^{2+}$, and for $[Ni(en)_3]^{2+}$. Using the ratio of the first and third transitions with either the d^8 Tanabe–Sugano diagram (Figure 5.49) or the Lever plot of transition energy ratios (Figure 5.46) calculate B for $[Ni(H_2O)_6]^{2+}$ and $[Ni(en)_3]^{2+}$. The d–d spectrum of $[Ni(NH_3)_6]^{2+}$ has peaks at 10 750, 17 500, and 28 200 cm⁻¹; use these to calculate Δ_{oct} and B for this complex. Comment on the values of Δ_{oct} and how the values of B compare to the free ion value of 1042 cm⁻¹ in Table 5.9.

The ^1E_g ← $^3A_{2g}$ spin-forbidden transition discussed in Example 5.26 is known as a spin-flip transition (Figure 5.50). This involves a change from a triplet to a singlet state, but as it only involves the change of spin of one electron in the e_g orbitals (marked in blue), the spatial distribution remains the same. This is why the line associated with the ^1E_g term is horizontal with respect to the $^3A_{2g}$ ground term in the Tanabe–Sugano diagram, and as a result these spectral features are usually relatively sharp.

This is a good example of how Tanabe–Sugano diagrams can also be used to explain the variation in

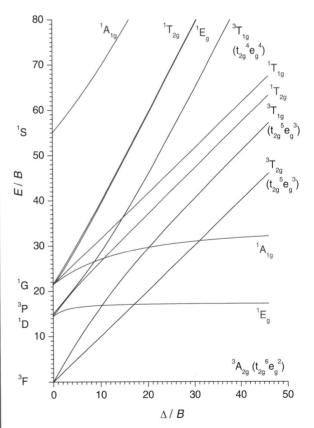

FIGURE 5.49 Tanabe–Sugano diagram for octahedral d^8. (Black lines indicate triplet terms, blue lines singlet terms.)

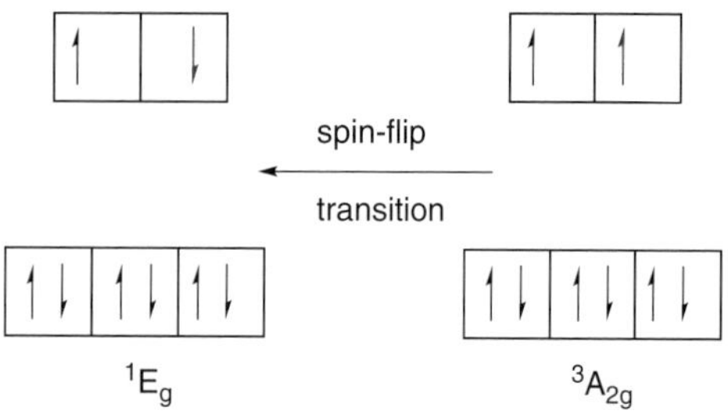

FIGURE 5.50 Spin-flip transition in Ni(II) d^8 complexes.

absorption band width observed in electronic absorption spectra. Vibrational motion of the ligands (vibronic coupling) with respect to the transition metal centre has the effect of varying Δ, and therefore, Δ/*B* with time. As the timescale of UV-vis spectroscopy is fast compared with the vibrational frequency it samples the complex over the vibrational positional range and the ion has a range of Δ/*B* values. This is shown in Figure 5.51 as a vertical rectangle rather than a line as used previously. This means that transitions can occur over a range of energies whose spread depends upon the gradient of the excited state term relative to the excited state. A steep gradient, such as those for the spin-allowed transitions to the triplet terms produces a large range of energies and thus a broad peak in the UV-vis spectrum, and a low gradient, as observed for the singlet terms, produces sharp bands in the spectra.

5.13.5.4 d^6, iron(II) spin-crossover complexes

d^4–d^7 octahedral complexes can display spin-crossover behaviour with temperature- or pressure-dependent spin states when the values of the pairing energy, *P*, and the crystal field splitting, Δ_{oct}, are similar to each other. Iron(II) complexes based on triazole and tetrazole

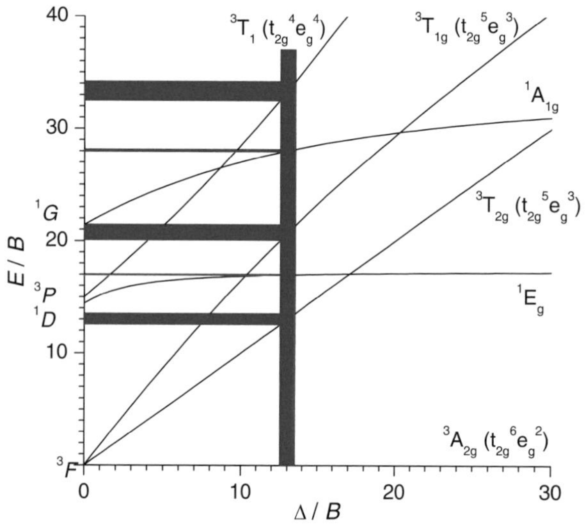

FIGURE 5.51 Expanded d^8 Tanabe–Sugano diagram showing effect of vibronic coupling on width of absorption features.

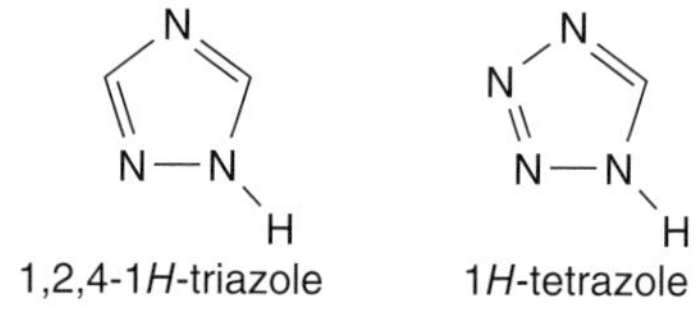

FIGURE 5.52 Structures of 1,2,4-1*H*-triazole and 1*H*-tetrazole.

ligands (Figure 5.52) are well suited to investigation by d–d spectroscopy as charge transfer bands do not interfere with the d–d transitions.

The triazole complexes have a very abrupt change of spin state around 373 K, whereas the 1-propyltetrazole complexes display an abrupt spin-state transition around 135 K. In both cases the low-spin isomer is formed at lower temperature and has a pink/purple colour, whereas the high-spin isomer is white (colourless). Figure 5.53 shows the UV-vis-NIR spectrum for $[Fe(ptz)_6](BF_4)_2$ (ptz = 1-propyltetrazole) at 293 and 20 K.

The 293 K spectrum of the high-spin colourless isomer has one spin-allowed transition in the NIR at 11 770 cm^{-1} yielding a value of Δ_{oct} of 11 770 cm^{-1}. The 20 K spectrum of the low-spin purple isomer has two transitions at 18 200 and 26 450 cm^{-1}, neither of which correspond to Δ_{oct}. As in Section 5.5.2.6, an estimate of 20 260 cm^{-1} for Δ_{oct} can be obtained by taking a quarter of the difference between the two values and adding this to the lower value. However, a more accurate value can be obtained using the d^6 Tanabe–Sugano diagram. The allowed terms for a d^6 free ion are ^5D, ^3H, ^3P, ^3F, ^3G, ^1I, ^3D, ^1G, ^1S, ^1D, ^1F, ^3F, ^3P, ^1G, ^1D, ^1S, and the total degeneracy is 210. As a result, the full d^6 Tanabe–Sugano diagram is very complex and busy, and not at all easy to use.

A simplified d^6 Tanabe–Sugano diagram is given in Figure 5.54 with the high-spin terms in black and low-spin terms in blue. (The version in Appendix 4 is more complete and shows the quintet, singlet, and low-lying triplet terms for comparison.)

This Tanabe–Sugano diagram is different from the others encountered so far as there appears to be a hiatus at Δ/*B* = 18.6, marked with a dashed line. To the left of this

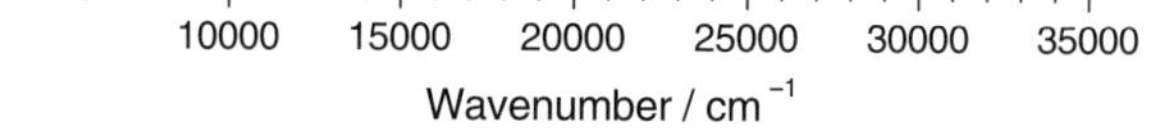

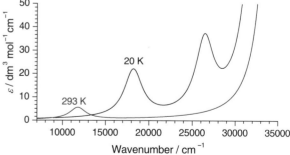

FIGURE 5.53 Electronic absorption spectra of $[Fe(ptz)_6](BF_4)_2$ (ptz = 1-propyltetrazole) at 293 and 20 K. (Data from A. Hauser, *J. Chem. Phys.* **94** 2741 (1991).)

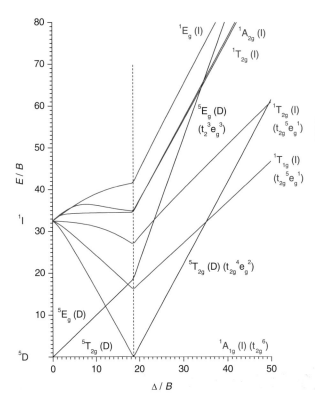

FIGURE 5.54 Simplified Tanabe–Sugano diagram for octahedral d^6.

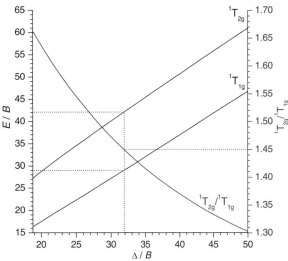

FIGURE 5.55 Lever plot of E/B (left) for the two lowest energy singlet excited states and the ratio of transition energies for these singlet states (right) versus Δ/B for low-spin d^6 complexes (see text for explanation of dashed lines).

point, high-spin $^5T_{2g}$ is the ground term. As the ligand field gets larger the energy difference between the low-spin $^1A_{1g}$ term and the high-spin $^5T_{2g}$ term decreases, so that by Δ/B = 18.6 the two are equienergetic, and after this the $^1A_{1g}$ term is now lower in energy and becomes the ground term on the x-axis. The relative energy of the $^5T_{2g}$ term increases rapidly after this point. Hence, the left-hand side of the diagram is for the high-spin isomer, and the right-hand side is for the low-spin isomer. Similar Tanabe–Sugano diagrams are also observed for d^4, d^5, and d^7 configurations.

For the high-spin isomer, the Tanabe–Sugano diagram (Figure 5.54) predicts one d–d transition, $^5E_g \leftarrow {}^5T_{2g}$, which gives a direct measure of Δ_{oct}, which in this case is 11 770 cm^{-1}. As there is only one data point for the high-spin isomer it is not straightforward to calculate B as the gradient of the line for the ^5E_g excited term is 1. While it is tempting to use the low-spin value determined in Example 5.27, B is affected by the extent of covalency in the bonding. The high-spin isomer is likely to be less co-valent than the low-spin isomer, and as a result the value of B will be larger for the high-spin complex than the low-spin complex.

For the low-spin isomer, the Tanabe–Sugano diagram shows that the two lowest energy spin-allowed transitions are $^1T_{1g} \leftarrow {}^1A_{1g}$ and $^1T_{2g} \leftarrow {}^1A_{1g}$. If the ratio of the two lowest energy spin-allowed transitions for the low-spin isomer is plotted against Δ/B (Figure 5.55), this can be used to quickly read off the appropriate Δ/B ratio.

EXAMPLE 5.27

Use the Lever plot of ratio of transition energies (Figure 5.55) to determine Δ_{oct} and B for low-spin $[Fe(ptz)_6](BF_4)_2$.

ANSWER

The two low-spin peaks at 18 200 and 26 450 cm^{-1} give a ratio of 1.45 and this yields a Δ/B value of 31.9 from the Lever plot. Having located the Δ/B value, the relevant lines from the Tanabe–Sugano diagram can then be used to determine B and then Δ. For Δ/B = 31.9 the E/B value for the $^1T_{1g} \leftarrow {}^1A_{1g}$ is 29.0, and the transition energy of 18 200 cm^{-1} indicates a value of B of 630 cm^{-1}. For the $^1T_{2g} \leftarrow {}^1A_{1g}$ transition the E/B value is 42.1, and the transition energy of 26 450 cm^{-1} also gives a value of B of 630 cm^{-1}. A value of 20 100 cm^{-1} is then obtained for Δ_{oct} by multiplying 31.9 x 630 cm^{-1}.

SELF TEST

Use Figure 5.54 and/or Figure 5.55 and the data in Figure 5.21 ($[Co(NH_3)_6]Cl_3$, 21 070, 29 460 cm^{-1}; $[CoCl(NH_3)_5]Cl_2$, 18 810, 27 540 cm^{-1}) to calculate Δ_{oct} and B for $[Co(NH_3)_6]$ Cl_3 and $[CoCl(NH_3)_5]Cl_2$. Compare these values with those in Example 5.9.

5.13.5.5 d^7, cobalt(II)

The electronic absorption spectrum of pink 0.2 M $[Co(H_2O)_6]^{2+}$ is shown in Figure 5.56(a), together with that of deep blue 0.02 M $[CoCl_4]^{2-}$ formed by tenfold dilution of the pink solution with 8 M HCl in Figure 5.56(b).

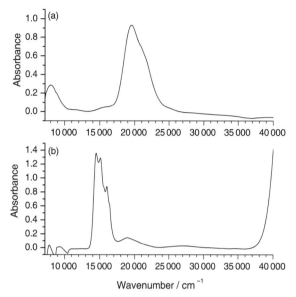

FIGURE 5.56 Electronic absorption spectra of (a) 0.2 M $[Co(H_2O)_6]^{2+}$ and (b) 0.02 M $[CoCl_4]^{2-}$.

EXAMPLE 5.28

Use the reported values of $\Delta_{oct} = 9200$ cm^{-1} and $B = 825$ cm^{-1} for $[Co(H_2O)_6]^{2+}$ and the octahedral d^7 Tanabe–Sugano diagram in Appendix 4 to predict the transition energies of the spin-allowed transitions and hence assign the peaks shown in Figure 5.56(a).

ANSWER

With a Δ/B ratio of 11.2 the three spin-allowed transitions for high-spin Co(II) are predicted to be: $^4T_{2g}$ (F) $\leftarrow$ $^4T_{1g}$ (F), 8100 cm^{-1}; $^4A_{2g}$ (F) $\leftarrow$ $^4T_{1g}$ (F), 17 300 cm^{-1}; $^4T_{1g}$ (P) $\leftarrow$ $^4T_{1g}$ (F), 19 300 cm^{-1}. The lowest energy peak in the spectrum can be readily assigned to the $^4T_{2g}$ (F) $\leftarrow$ $^4T_{1g}$ (F) transition. In this case the two electron $^4A_{2g}$ (F) $\leftarrow$ $^4T_{1g}$ (F) transition, is not the one highest in energy at this Δ/B ratio, and as it is also likely to be weak, the peak at 19 500 cm^{-1} is assigned to the $^4T_{1g}$ (P) $\leftarrow$ $^4T_{1g}$ (F) transition. The shoulder on the main peak could possibly be assigned to the remaining $^4A_{2g}$ (F) $\leftarrow$ $^4T_{1g}$ (F) transition, but it is on the wrong side (high rather than low energy). The consensus of opinion is that the structure on this peak is a result of some activation of spin-forbidden transitions and intensity stealing due to spin–orbit coupling, or a reduction in the symmetry from strict O_h. These assignments can be confirmed using the Lever plot of the d^7 transition energies also given in Appendix 4.

SELF TEST

Calculate ε_{max} for the peaks in Figure 5.56(b), and compare these with the values in Figure 5.56(a) and Table 5.1. The literature ε_{max} value for $[CoCl_4]^{2-}$ is 580 dm^3 mol^{-1} cm^{-1}.

What does your value indicate about the extent of reaction between $[Co(H_2O)_6]^{2+}$ and HCl to form $[CoCl_4]^{2-}$? As d^n tetrahedral configurations can be covered by d^{10-n} octahedral Tanabe–Sugano diagrams, use the d^3 Tanabe–Sugano diagram (Figure 5.45 and Appendix 4) and the reported values of Δ_{tet} (3120 cm^{-1}) and B (710 cm^{-1}) for $[CoCl_4]^{2-}$ to assign the features in Figure 5.56(b). It should be noted that for both the octahedral and tetrahedral Co(II) complexes the lowest energy transitions were outside the range of conventional spectrometers, and a detailed analysis using the Tanabe–Sugano diagrams was necessary to confirm the assignments.

5.13.5.6   d^5, manganese(II)

The d–d spectra of octahedral complexes considered so far have had broad peaks with molar absorptivities of 1–50 dm^3 mol^{-1} cm^{-1}. In the spectrum of a very pale pink saturated (ca. 4 M) solution of MnCl$_2$ (Figure 5.57), the peaks are sharp and weak with molar absorptivities of the order of 0.03–0.07 dm^3 mol^{-1} cm^{-1}.

$[Mn(H_2O)_6]^{2+}$ is high-spin d^5 and will therefore have a $^6A_{1g}$ ground term, as shown on the left of the octahedral d^5 Tanabe–Sugano diagram (Figure 5.58). There are no spin-allowed transitions as there are no sextet excited terms. Low-spin d^5 complexes, shown on the right-hand side of the Tanabe–Sugano diagram, are spin doublets with a $^2T_{2g}$ ground term. The spin-forbidden transitions observed in Figure 5.57 are to the spin-quartet excited states shown in blue in the Tanabe–Sugano diagram (Figure 5.58), as this only involves the change in spin of one, not two electrons. The widths of the various transitions observed reflect the gradient of the excited state energy line (Section 5.13.5.3 and Figure 5.51). Thus, transitions to the ($^4A_{1g}/^4E_g$), $^4T_{2g}$, and ^4E_g excited states are sharper features in the absorption spectrum as these are spin-flip transitions from the $^6A_{1g}$ ground state. The transition energy does not depend on the value of Δ/B as the lines of these excited terms are almost all horizontal in the Tanabe–Sugano diagram. In contrast, the lines associated with the other $^4T_{1g}$, $^4T_{2g}$, and $^4T_{1g}$ terms have considerable gradients.

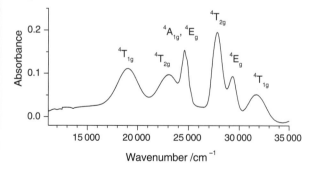

FIGURE 5.57 Electronic absorption spectrum of a saturated solution of MnCl$_2$.

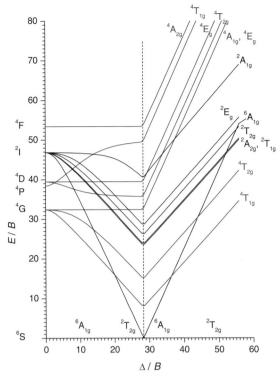

FIGURE 5.58 Simplified Tanabe–Sugano diagram for octahedral d^5.

If the lowest energy transitions at $19\,000$ and $23\,000$ cm^{-1} in Figure 5.57 are assumed to be due to the $^4T_{1g} \leftarrow {}^6A_{1g}$ and $^4T_{2g} \leftarrow {}^6A_{1g}$ transitions, respectively, their ratio of 1.21 can be used to obtain an estimate of $\Delta/B =$ 11.6 using the Lever plot of d^5 transition energy ratios in Appendix 4. This value can be used in conjunction with the Tanabe–Sugano diagram (Figure 5.58 and Appendix 4) to determine $\Delta_{oct} = 9200$ cm^{-1} and B $= 790$ cm^{-1}, which can then be used to assign the other peaks in Figure 5.57.

5.13.6 Summary of the analysis of UV-vis spectra from first row transition metal ions

Table 5.10 gives a summary of the above observations and the methods for determining the ligand field splitting parameter, Δ, from the experimental spectra of octahedral 3d transition metal complexes. If Tanabe–Sugano diagrams are used, the value of B is also usually determined as well.

5.14 Nephelauxetic effect

As part of the determination of Δ using the Tanabe–Sugano diagrams values of the Racah B parameter were also obtained. The summary in Table 5.11 shows that

TABLE 5.10 Summary of observations and deductions that can be obtained from d–d spectra of octahedral 3d transition metal complexes

d^n configuration	No. of spin-allowed transitions	Determination of Δ from spectrum	Comment
d^0	0	n/a	
d^1	1	Directly from spectrum for octahedral complexes.	May be split peak due to Jahn–Teller distortion of excited state. The higher energy component is most likely to be Δ. This is also the case for vanadyl complexes.
d^2	3	Δ is equivalent to the difference of the 1st and 3rd d–d bands.	Usually need to use Tanabe–Sugano diagram or Lever plot, as ordering of 2nd and 3rd transitions is Δ and B dependent. The highest energy transition may be masked by charge transfer bands. The $^3A_{2g}$ (F) $\leftarrow {}^3T_{1g}$ (F) transition is formally a two electron transition, and in this case it is expected to be weak
d^3	3	Δ is equivalent to the lowest energy d–d transition directly from spectrum.	Use Tanabe–Sugano diagram to determine B and confirm assignments. Although the two electron transition $^4T_{1g}$ (P) $\leftarrow {}^4A_{2g}$ (F) is expected to have appreciable intensity, it is often masked by CT bands.
d^4 high spin	1	Often only one band is observed, but this does not correspond to Δ due to ground state Jahn–Teller distortion.	Cr^{2+} very easily oxidised to Cr^{3+}
d^4 low spin		Require Tanabe–Sugano diagram.	Lowest spin-allowed transition expected at relatively high energy.
d^5 high spin	0	No spin-allowed transitions.	Δ and B can be estimated from spin-forbidden transitions (often very narrow) in conjunction with Tanabe–Sugano diagram. Fe^{3+} easily hydrolysed in aqueous solution.

TABLE 5.10 Summary of observations and deductions that can be obtained from d–d spectra of octahedral 3d transition metal complexes (*Continued*)

d^n configuration	No. of spin-allowed transitions	Determination of Δ from spectrum	Comment
d^5 low spin		Require Tanabe–Sugano diagram.	Rare, and in case of Fe^{3+}, the d–d bands are obscured by CT bands.
d^6 high spin	1	Directly from single peak in the spectrum.	May be split due to Jahn–Teller distortion of excited state, accurate value of Δ is dependent on orbital ordering.
d^6 low spin		Usually two spin-allowed bands observed. Δ can be estimated by adding a quarter of the difference of the two transition energies to the lower one.	Use Tanabe–Sugano diagram or Lever ratio plot to accurately determine Δ and B.
d^7 high spin	3	Δ is the difference of 1st and 2nd/3rd and d–d bands.	Usually need Tanabe–Sugano diagram and/or Lever plot as the lowest energy transition may be outside of spectral range, and the second one is often weak as a two electron transition, and there is the possibility of two of the excited states crossing.
d^7 low spin		Tanabe–Sugano diagram required.	Ground state Jahn–Teller distortion
d^8	3	Δ is equivalent to the lowest energy d–d transition.	The lowest energy transition may be at low energy; therefore the assignment should be confirmed using the Tanabe–Sugano diagram.
d^9	1	Directly from spectrum.	Often only one band is observed in the visible part of the spectrum, but this does not necessarily correspond to Δ due to ground state Jahn–Teller distortion.
d^{10}	0		No d–d transitions possible.

there is a variation of these compared to the free ion values. This variation can provide insight into the electronic structure and bonding.

The Racah B parameter reduces on going from the free ion values to the complex, and this can be expressed by the parameter β defined as:

$$\beta = B_{complex}/B_{free\ ion} \qquad \textbf{Eqn 5.5}$$

β is therefore always less than one. This indicates that there is less repulsion between the electrons in the complex, than in the free ion, and the smaller β becomes, the greater the reduction in B. This can be considered as being due to an increase in the size of the metal d orbitals. This expansion in the electron cloud gives rise to the phrase **nephelauxetic effect**, which is derived from the Greek for cloud expanding.

The reduction in B, the nephelauxetic effect, is due to covalency and arises from either, or both, of the following reasons. As the bonding becomes more covalent, the lone pairs of the electrons from the ligands penetrate the d orbitals more, with the effect that they screen the d electrons from its nucleus. The effective nuclear charge is reduced, and the d orbitals expand. Hence, B and β are expected to reduce with an increase in covalent bonding.

This is sometimes known as **central field covalency** and is regarded as the predominant effect. As the covalency increases in an octahedral complex (i.e. the d orbitals become more involved with the bonding), the metal e_g orbitals become progressively more σ antibonding. If there is also π bonding the t_{2g} orbitals may become either bonding (π acid ligands) or antibonding (π base ligands). While the σ bonding is the predominant effect, both result in some of the electron density of the metal being transferred to the ligands. Delocalisation of this sort can be considered to increase the distance and hence reduce the interaction between the electrons, resulting in a lower value of B. This is sometimes known as **symmetry restricted covalency**.

From the data in Table 5.11 it can be seen that an increase in oxidation state reduces B, indicating greater covalency in higher oxidation states. There is also a reduction in B on going from an octahedral to tetrahedral geometry, indicating greater covalency in tetrahedral complexes.

Using a larger data set, a nephelauxetic series for a given ion can be drawn up in terms of B or β:

$$\text{Free ion} > F^- > H_2O > NH_3 > en > \underline{N}CS^- > Cl^-$$
$$\approx CN^- > Br^- > N_3^- > I^- > S^{2-}$$

TABLE 5.11 Summary of Δ and *B* values from data in this chapter

Complex	Δ/cm⁻¹	B/cm⁻¹	Free ion B/cm⁻¹	β
[V(H₂O)₆]²⁺	11 800	620	760	0.82
[V(H₂O)₆]³⁺	18 600	610	886	0.69
[Cr(H₂O)₆]³⁺	17 080	705	933	0.76
trans-[CrCl₂(H₂O)₄]⁺	16 510	705	933	0.76
[Fe(ptz)₆]²⁺	20 100	630	897	0.70
[Mn(H₂O)₆]²⁺	9 200	790	859	0.92
[Co(H₂O)₆]²⁺	9 200	825	989	0.83
[CoCl₄]²⁻	3 120	710	989	0.72
[Co(NH₃)₆]Cl₃	22 900	620	1080	0.57
[CoCl(NH₃)₅]Cl₂	20 700	675	1080	0.63
[Ni(H₂O)₆]²⁺	8 550	930	1042	0.89
[Ni(NH₃)₆]²⁺	10 750	880	1042	0.84
[Ni(en)₃]²⁺	11 280	875	1042	0.83

In terms of the donor atom, the series is F > O > N > Cl > Br> I > S > Se > As. The order roughly correlates with the polarisability, as the series is related to the extent of covalent bonding present. β values close to 1 imply ionic bonding, and as β reduces the extent of covalent bonding increases. While the ordering of the nephelauxetic series appears at first glance to be similar to the spectrochemical series, there are significant differences in the ordering as the nephelauxetic series is dependent on covalent bonding, whereas the ordering of the spectrochemical series is dependent on the extent of π base and π acid character in the bonding. For example, Cl⁻ and CN⁻ are at the opposite ends of the spectrochemical series, but are very similar in the nephelauxetic series, and this indicates that the extent of covalency is very similar, but its origins are very different (Cl⁻ acts a π base, CN⁻ as a π acid).

If the ligands are kept constant, a nephelauxetic series in terms of the central metal can also be drawn up which is related to the polarising power of the metal:

$$Mn^{2+} \approx V^{2+} > Ni^{2+} \approx Co^{2+} > Mo^{3+} > Ru^{2+} \approx Cr^{3+}$$
$$> Fe^{3+} > Rh^{3+} \approx Ir^{3+} > Tc^{4+} > Co^{3+} > Ag^{3+} > Cu^{3+}$$
$$\approx Mn^{4+} > Pt^{4+} > Pd^{4+} > Ni^{4+}$$

For the same metal and ligand, β also decreases with a decrease in coordination number, indicating an increase in covalency.

Bibliography

A.B.P. Lever. (1984) *Inorganic Electronic Spectroscopy*. 2nd ed. Oxford: Elsevier.

B.N. Figgis and M.A. Hitchman. (2000) *Ligand Field Theory and its Applications*. New York: Wiley-VCH.

B.N. Figgis. (1966) *Introduction to Ligand Fields*. New York: Wiley.

S.F.A. Kettle. (1998) *Physical Inorganic Chemistry: A Coordination Chemistry Approach*. Oxford: Oxford University Press.

R.L. Carter. (1997) *Molecular Symmetry and Group Theory*. New York: Wiley.

S.F.A. Kettle. (1995) *Symmetry and Structure*. 2nd ed. Chichester: Wiley.

M. Gerloch and E. C. Constable. (1994) *Transition Metal Chemistry*. Weinheim: Wiley-VCH.

S. F. A. Kettle. (1969) *Coordination Compounds*. London: Nelson.

N. N. Greenwood and A. Earnshaw. (1984) *Chemistry of the Elements*. Oxford: Pergamon; 2nd ed. (1997) Oxford: Butterworth Heinemann.

M.T. Weller, T.L. Overton, J.A. Rourke, and F.A. Armstrong. (2014) *Inorganic Chemistry*, 6th ed. Oxford: Oxford University Press.

C.E. Housecroft and A.G. Sharpe. (2012) *Inorganic Chemistry*, 4th ed. Harlow: Pearson.

X-ray and photoelectron spectroscopy, electron microscopy, and energy dispersive analysis of X-rays

6.1 Introduction to photoelectron spectroscopy

Photoelectron spectroscopy (PES) employs electromagnetic radiation to excite electrons from a compound. The analysis of the kinetic energy distribution of the emitted photoelectrons provides information on the composition and electronic energy levels of the sample. The technique is often subdivided according to the energy of the exciting radiation. **X-ray photoelectron spectroscopy (XPS)** uses soft X-rays (with a photon energy of approximately 100–2000 eV) which excites and, therefore, provides information on core electron energy levels. **Ultraviolet photoelectron spectroscopy (UPS)** employs vacuum ultraviolet (UV) radiation (with photon energies in the range 10–50 eV) which excites and allows the study of outer or valence electrons; it is mainly used to study bonding electrons in molecular compounds and the band structures of solids.

Instrumentation for both experimental XPS and UPS is available commercially as laboratory-based instrumentation, though the cost is relatively high so, where available, it usually addresses the needs of a large number of users within, say, a university. Synchrotron radiation sources (Section 1.6.5) provide much more intense soft X-ray and hard UV radiation over the full range of energies needed to probe inorganic compounds (5–20 000 eV). This allows experimental data to be collected very quickly, permits studies of very small samples or allows investigations that map features systematically at a large number of points across an area of a material. Synchrotron radiation also allows high resolution studies to be carried out as the radiation spans a much wider and more complete energy range than laboratory-based instrumentation but such work remains a small minority of all photoelectron studies due to the expense, complexity, and limited availability of such specialised sources.

Figure 6.1(a) shows the energy transfer processes that are the basis of photoelectron spectroscopy, that is, the measurement of the kinetic energies of electrons

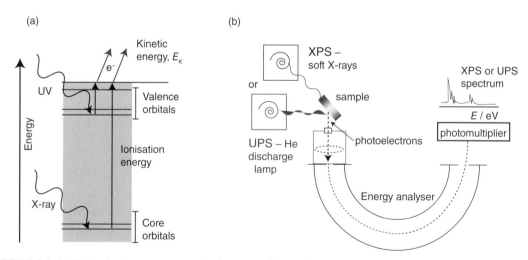

FIGURE 6.1 (a) The ionisation processes involved in UV and X-ray photoelectron spectroscopies. (b) Schematic of the photoelectron spectroscopy instrumentation which uses either soft X-rays (XPS) or He radiation (UPS) as the excitation source.

(photoelectrons) emitted by ionisation of a sample that is irradiated with high-energy monochromatic UV or X-ray electromagnetic radiation.

From the law of conservation of energy, the kinetic energy of the ejected photoelectron, E_K, can be related to its ionisation energy, E_i (also known as the binding energy, E_b), by the expression

$$E_K = E_{h\nu} - (E_i + \varphi) \qquad \textbf{Eqn 6.1}$$

where $E_{h\nu}$ is the energy of the incident radiation (of frequency ν) and φ is a work function term applicable to solids. **Koopmans' theorem** states that the ionisation energy is equal to the negative of the orbital energy, so the kinetic energy of the photoelectron can be used to determine the energy of the orbital from which it originated. The work function term, φ, arises in photoelectron spectroscopic studies of solids and results from the additional energy required to eject an electron from the surface of the material. There can also be a contribution to φ from the additional energy required to detect the photoelectron in the instrument analyser. φ is material dependent and factors that control its value include whether a metal, semiconductor or insulator is being investigated. These various work function terms can be calibrated in the photoelectron spectroscopy instrument so that E_i in Eqn 6.1 can be determined.

6.2 **X-ray photoelectron spectroscopy (XPS)**

The basic layout of the XPS instrument is shown in Figure 6.1(b). The standard laboratory source for the X-rays used in XPS is usually a magnesium or aluminium anode that is bombarded by a high-energy electron beam. This bombardment produces soft X-ray radiation with energies of 1.254 keV (Mg) and 1.486 keV (Al) due to the transition of a 2p electron into the vacancy in the 1s orbital caused by ejection of the core Mg or Al electron by the electron bombardment. These X-ray photons are directed onto the material being investigated and cause ionisations from core orbitals in the elements present. The ionisation energies are characteristic of the element and its oxidation state so their analysis provides information about which elements are present and their oxidation states. The linewidth is around 1–2 eV so XPS is not suitable for probing fine details of valence orbital energies but it can be used to study the band structures of solids. The mean free path of electrons in a solid is only about 1 nm, so XPS is most suitable for surface elemental analysis, and in this application it is commonly known as **electron spectroscopy for chemical analysis** (ESCA); see Section 6.3.1.

Figure 6.2 shows the simple XPS spectrum obtained from a copper foil with the XPS ionisation peaks observed labelled according to the orbital from which the electron originated. The Cu 3d orbitals have a very small binding energy, and the binding energy then increases on going from Cu 3p to Cu 3s, Cu 2p (two ionisations are resolved corresponding to the differing energies of the final $2p_{1/2}$ and $2p_{3/2}$ states of the copper atom), and Cu 2s ionisations. Note that in this case the excitation energy (from aluminium) was insufficient to photoionise the Cu 1s electrons. The spectrum also shows a number of so-called **Auger peaks**. These involve simultaneous electron transitions between orbitals. In this case an $n = 2$, denoted **L**, see Section 2.6.1, core hole was filled by an electron from an $n = 3$ level orbital (**M**) while another electron, called an **Auger electron**, was emitted from another $n = 3$ level orbital (**M**). Analysis of these Auger peaks is known as **Auger spectroscopy** (Section 6.3.2) and

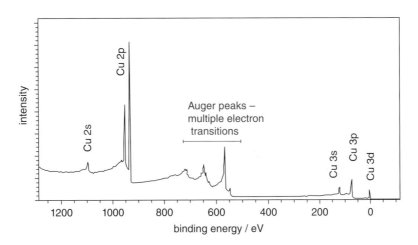

FIGURE 6.2 XPS spectrum of copper metal.

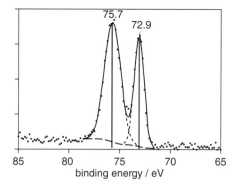

FIGURE 6.3 XPS spectrum in the range 65–85 eV of aluminium metal that has been exposed to air.

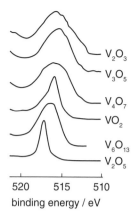

FIGURE 6.4 XPS spectra in vanadium 2p region of various vanadium oxides. (Data from M. Demeter, M. Neumann, and W. Reichelt, *Surf. Sci.* **41**, 454–456 (2000).)

ANSWER

There is a gradual decrease in the vanadium 2p electron binding energies going from V_2O_5 ($3d^0$) through V_6O_{13} (mixed $3d^0$, $3d^1$), VO_2 ($3d^1$), V_4O_7, V_3O_5 ($3d^1$-$3d^2$), and finally to V_2O_3 ($3d^2$). This reflects the increased shielding of the 2p electron as the number of the 3d electrons increases.

SELF TEST

Predict the vanadium 2p XPS spectrum of the oxide VO.

can provide additional information on the electronic energy levels of the atom.

The basis of the XPS technique is that the electron binding energy is sensitive to different chemical environments and, in particular, oxidation state. It is also possible to study multiple elements in a sample as the XPS spectrum will contain peaks from all the different elements present. In its simplest form XPS data can be used to determine chemical composition as for a characteristic transition to be observed that element must be present. The intensity of the absorption is proportional to the amount in the sample and this is exploited in the ESCA technique (Section 6.3.1). However, more detailed interpretation of the spectrum and the exact electron binding energies provides additional information on, for example, the oxidation state of the element.

Figure 6.3 shows the XPS spectrum of the surface of aluminium metal, that has been exposed to air, in the region where aluminium 2p electrons are ionised. Two peaks are seen which can be explained by the presence of two oxidation states of aluminium. The peak at 72.9 eV is that of aluminium metal while that at 75.7 eV is that from Al^{3+} in Al_2O_3, which has been formed on the surface. Note that the binding energy of a 2p electron in an Al^{3+} cation is greater than that in the atom because of the increased effective nuclear charge observed in the cation.

6.3 Techniques related to XPS

Techniques related to XPS that use X-rays to induce electronic transitions in solids and analyse the emitted photoelectrons include ESCA and Auger spectroscopy. These and closely related analysis techniques can also use high energy electron beams in electron microscopes, rather than X-rays, to induce similar photoelectron ionisations (see Sections 6.7 and 6.8).

6.3.1 Electron spectroscopy for chemical analysis (ESCA)

When applied to the investigation of the chemical composition of surfaces, particularly across an area of the material under investigation, XPS is normally referred to as **electron spectroscopy for chemical analysis (ESCA)**. As with XPS, ESCA typically uses Al K_α X-rays as the excitation radiation. The technique can be applied to a broad range of materials including metals, metal oxides, polymers and electronic thin film devices. ESCA is non-destructive and samples can be insulating, semiconducting,

EXAMPLE 6.1

Interpreting simple XPS spectra

Figure 6.4 shows the XPS spectra in the vanadium 2p region obtained from a series of vanadium oxides. Explain the trend in peak position seen in these spectra.

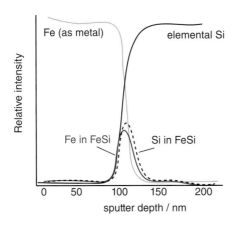

FIGURE 6.5 Depth profiling ESCA using ion milling on an iron metal thin film grown on silicon.

or metallic. Analysis of the energies of the emitted photoelectrons allows the elements and their quantities present in the sample to be determined; the sensitivity varies between 0.01 and 1 at. %, dependent upon the element and with greater sensitivity for heavier elements.

Due to short path lengths in the solid of the emitted photoelectrons the average depth of analysis for an ESCA measurement is only around 5 nm, while the X-ray beam can be focused to a surface area resolution as small as 10×10 μm². Scanning a micro-focused X-ray beam across the sample surface produces a compositional map of the material. Synchrotron radiation sources (Chapter 1, Section 1.6.5) provide very intense X-ray beams, allowing ESCA mapping at high spatial resolution to be undertaken rapidly.

Depth distribution information can be obtained by combining ESCA measurements with simultaneous ion milling (sputtering). Sputtering eats away at the surface exposing new material for analysis; this is particularly useful in the study of thin film structures and coatings. Figure 6.5 shows depth profiling ESCA data obtained from a sample manufactured by the deposition of a 100 nm thick, iron metal thin film on a silicon substrate. On the surface, a sputter depth of 0 nm, only iron metal peaks were observed in the ESCA XPS spectrum. At the interface, FeSi is formed and the iron and silicon atoms in this compound can be distinguished from the elemental forms by shifts (for both Fe and Si peaks) in their photoelectron emission energies in the ESCA spectra. Spectra obtained on further sputtering show only peaks from elemental silicon.

6.3.2 Auger spectroscopy with X-ray excitation

The Auger electrons emitted during excitation of a sample using an X-ray beam (Section 6.2) have low energies (around 40 eV–1 keV), meaning they can only travel

short distances before they escape from the surface to be detected. This means that their analysis, in Auger spectroscopy, is very sensitive to the surface of the sample. Auger spectroscopy typically provides information to depths of between 0.5 and 5 nm. Spatial resolutions as low as 7.5 nm can be achieved, though normally larger areas of a surface (1–1000 μm² in dimension) are analysed. Auger spectroscopy is more sensitive than ESCA for the investigation of light elements. Example applications include the determination of the thickness of graphene films and investigating the formation of metal oxides on metal surfaces by measuring the level and distribution of oxygen atoms in the outermost 5 nm layer. Most contemporary Auger spectroscopy is carried out using high-energy electron beams for excitation and, therefore, in association with scanning electron microscopy investigations (Section 6.7).

6.4 Ultraviolet photoelectron spectroscopy (UPS)

The source for UPS is typically a helium discharge lamp that emits He(I) radiation (from He atoms at 21.22 eV) or He(II) radiation (from He⁺ at 40.8 eV). In molecular species, UPS yields information about the energies and symmetries of molecular orbitals, and in solid-state materials their band structure. The technique can also be used to characterise small molecules adsorbed on surfaces. The linewidths in the UV photoelectron spectra are much narrower than in XPS, so the resolution is far greater allowing more detailed information on the nature of the bonding electrons to be extracted. For molecules, vibrational fine structure often provides important information on the bonding or antibonding character of the orbitals from which electrons are ejected (Figure 6.1). When the electron is removed from a non-bonding orbital the product species is formed in its vibrational ground state and a narrow line is usually observed. However, when the electron is removed from a bonding or antibonding orbital the resulting ion is formed in several different vibrational states and extensive fine structure is observed in the spectra. Bonding and antibonding orbitals can be distinguished by determining whether the vibrational frequencies in the resulting ion are higher (antibonding) or lower (bonding) than for the original molecule.

Another useful aid is the comparison of photoelectron intensities for a sample irradiated with He(I) and He(II) radiation. The higher energy source (He(II), 40.8 eV) preferentially ejects electrons from d or f orbitals, allowing these contributions to be distinguished from s and p orbitals, for which He(I) (21.22 eV) causes higher intensities.

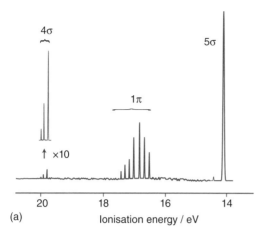

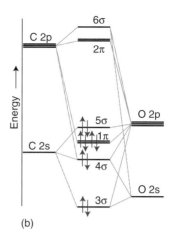

FIGURE 6.6 (a) Photoelectron spectrum of CO (He(I) excitation) with peaks assigned from (b) the molecular orbital diagram for CO.

6.4.1 Experimental considerations

The ionisation energy or binding energy, $E_{\text{binding/ionisation}}$, of the electron is determined using the same relationship as Eqn 6.1:

$$E_K = E_{h\nu} - E_{\text{binding/ionisation}}$$

where $E_{h\nu}$ is the excitation energy (21.218 eV for a He lamp) and E_K (electron) is the photoelectron energy measured at the detector. This expression assumes that the recoil kinetic energy of the residual ion formed from a molecule is negligible. As with XPS, corrections may need to be made to spectra from some solid samples to allow for the work function of ejecting the electron from the surface. A typical experimental UPS spectrum, therefore, consists of a series of bands, often with vibrational structure, of decreasing binding or ionisation energy.

Figure 6.1(b) shows a schematic of a UPS instrument. Monochromated He(I) or (II) radiation from the source impinges upon the sample causing ejection of photoelectrons. These are analysed using a variety of methods, including the spherical deflection analyser which allows only a specific photoelectron energy to travel a curved path between two charged hemispheres to the electron multiplier detector.

6.4.2 UPS spectrum of CO

Figure 6.6(a) shows the UPS spectrum obtained from CO and Figure 6.6(b) the molecular orbital diagram from this simple diatomic (note that ionisation from the 3σ orbital cannot be achieved with He(I) radiation). The 5σ is almost non-bonding as it shows very little vibrational fine structure. The 1π demonstrates vibrational structure with peak separations of around 0.190 eV

equivalent to 1530 cm^{-1}, which is significantly lower than the neutral molecule value of 2143 cm^{-1} indicating a strongly bonding orbital. The 4σ has an interval of 0.210 eV (1690 cm^{-1}) indicating a less bonding orbital with less bonding character than the 1π orbitals. The 5σ non-bonding orbital is the lone pair used by CO for σ bonding to metals. The UPS spectrum indicates that the 1π orbital is the most strongly bonding of the valence orbitals.

6.4.3 Applications of UPS

UPS has been used to investigate the bonding electron energies in a wide variety of molecular compounds, determine their relative energies and the degree of bonding or antibonding character. These experimental results can then be interpreted in relation to molecular orbital diagrams to provide a detailed picture of bonding in a compound. One particular application of UPS has been in the characterisation of organometallic compounds including, for example, sandwich compounds such as ferrocene $Fe(\eta^5\text{-}C_5H_5)_2$ and bent zirconocenes and titanocenes $M(\eta^5\text{-}C_5H_5)L_n$ with their applications in catalysis.

EXAMPLE 6.2

The series of compounds $M(BH_4)_4$, M = Ti, Zr, and U, all show similar UPS spectra, with a series of ionisation energies between 11 and 20 eV, except for the presence of an additional intense ionisation at 9.5 eV in $U(BH_4)_4$ that exhibits no vibrational structure. Propose the likely nature of this additional energy level in the uranium compound by consideration of its position in the periodic table of elements.

ANSWER

As an actinide uranium has 5f orbitals available for bonding which for U^{4+} in $U(BH_4)_4$ contain two electrons. It is the ionisation of these outer orbital electrons, which are non-bonding in nature in this compound, that gives rise to the 9.5 eV peak.

SELF TEST

Would the 4f electrons in $Hf(BH_4)_4$ be expected to be ionised below 20 eV and, therefore, be observable in its UPS spectrum?

6.5 X-ray absorption spectroscopy

XPS (Section 6.2) involves the study of the energies of the photoelectrons emitted when a compound is irradiated with X-ray radiation. Closely related to XPS is the study of the X-ray absorption spectrum itself and the interpretation of features in such spectra to yield chemical and structural information. X-ray absorption occurs at discrete energies when electrons are excited from core orbitals, such as the 1s, to either vacant valence orbitals or into the continuum, that is, when photoionisation takes place (see Figures 6.1(a) and 6.7).

The nomenclature used in X-ray absorption spectroscopy is such that if the excitation process involves a 1s electron, it is said to be a K-edge or K-shell process. If the principal quantum number, n, of the excited electron is 2,

then this is denoted as L-edge etc. At the L-edge there is the possibility of excitation from the 2s or 2p orbitals; a process involving the 2s electron gives rise to a so-called L_1 or L_I-edge. Due to spin–orbit coupling, producing the two energy levels termed $2p_{1/2}$ and $2p_{3/2}$, there are two edges close in energy arising through excitation from 2p orbitals and these are known as the $L_{2,3}$ or $L_{II,III}$-edges. Processes involving higher principal quantum number (3, 4, etc.) edges are labelled alphabetically, M, N, and so on. These core electron excitation processes are of high energy, 1000–50 000 eV (i.e. 1–50 keV), and generally require, except for the lightest elements with atomic numbers below sodium, X-ray radiation to perform them. The study of these excitation and ionisation processes is known as **X-ray absorption spectroscopy (XAS)**. The excitation energy for each element's K-edge is different and this gives the technique one of its major advantages: it is element specific. Furthermore, it is possible to obtain information from all the different elements present in the sample including, because of the technique's sensitivity, minor components.

6.5.1 X-ray absorption measurements: absorption and relaxation processes

Although excitations from many core shells are possible, the majority of experiments use K-edges or L-edges, and the discussion here focuses on K-edges. The X-ray absorption spectrum is conventionally divided into two parts, **X-ray absorption near edge structure (XANES)** and **extended X-ray absorption fine structure (EXAFS)**, as shown in Figure 6.7.

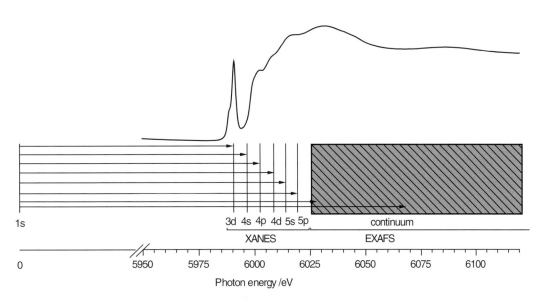

FIGURE 6.7 Schematic of a Cr K-edge X-ray absorption spectrum, showing the electronic transitions that can be involved and the XANES and EXAFS regions.

The basic instrumentation required for X-ray absorption measurements are a tuneable X-ray source, a monochromator to select an incident X-ray energy, and an X-ray detector. Because of the need to scan across an absorption edge, synchrotron X-ray sources, which produce intense X-radiation across a large energy range, are widely used to collect high quality X-ray absorption spectra. Further details of the instrumentation that is used to collect X-ray absorption spectra are given for the collection of EXAFS data in Section 6.5.4.1.

6.5.2 XANES

The XANES part of the spectrum comprises the features just prior to the edge, **pre-edge features** (Section 6.5.3), as well as the edge itself. The XANES edge position is sensitive to oxidation state, as the binding energy of the 1s electron is dependent upon the effective nuclear charge, Z_{eff}, it experiences. This is in turn related to the oxidation state; removing valence shell electrons, as occurs with increasing oxidation state, will increase Z_{eff}. The absorption edge energy-shifts are typically around +1 to +2 eV per oxidation number change for the first row transition elements; note that the K-edge for iron metal is at 7112 eV but such small edge shifts for, say, Fe^{2+} and Fe^{3+} compounds to approximately 7115 eV can be readily measured experimentally with good accuracy. However, changing the ligand element around a metal centre can also affect the absorption edge position, by withdrawing or donating electrons to the metal centre, so care has to be taken through the use of appropriate standards in determining oxidation state from XANES edges. Figure 6.8 illustrates the absorption K-edge position for a series of chromium compounds showing the gradual increase in energy with oxidation number.

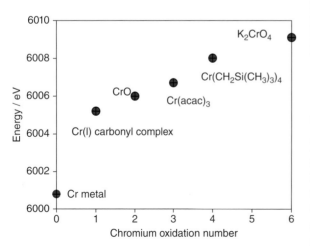

FIGURE 6.8 XANES Cr K-edge position determined in a series of compounds as a function of chromium oxidation number.

EXAMPLE 6.3

Phillips catalyst samples are prepared by depositing chromium from solution onto silica gel followed by heating in air at 600°C. The Cr edge XANES spectrum of the catalyst shows an absorption edge at 6009.2 eV. After reduction in CO at 673 K the edge is shifted to 6006.3 eV. What are the likely processes occurring in this system?

ANSWER

An edge at 6009.2 eV corresponds to the Cr(VI) oxidation state and this species is likely to be CrO_3 deposited on the SiO_2 surface. Figure 6.9 shows the proposed active chromium centre in this catalyst. The reduction process produces an absorption edge at an energy closest to that of Cr(II) in CrO or Cr(III) in Cr(acac)₃. So possibilities for this reduced phase are probably an oxide such as Cr_2O_3 or CrO on the silica surface.

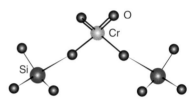

FIGURE 6.9 The proposed catalytic site in Phillips catalyst.

SELF TEST

The following data give the energy (eV) of the main pre-edge feature in Mn K-edge XAS as a function of oxidation state:

Mn(II) 6540.6	Mn(III) 6541.0	Mn(IV) 6541.5
Mn(V) 6542.1	Mn(VI) 6542.5	Mn(VII) 6543.8

An oxide containing manganese showed a pre-edge feature in its XAS spectrum consisting of peaks at 6540.6 eV (intensity 1) and 6540.9 (intensity 2). Explain the observed variation in the energy of the pre-edge feature and propose a formula for the manganese oxide.

6.5.3 Pre-edge features

Pre-edge features in the XANES spectrum are the result of transitions from the core level (1s for a K-edge) to a higher energy unoccupied orbital (e.g. 3d, 4p, 5p). The transitions are governed by the normal electronic absorption spectroscopy selection rules (Chapter 5, Section 5.4): dipole, orbital $\Delta l = \pm 1$, Laporte (if an inversion centre is present then there must be change in parity, u↔g), and spin ΔS. Pre-edge transitions are often observed with significant intensity in XAS and their intensity tends to be greatest for higher oxidation states, especially for the

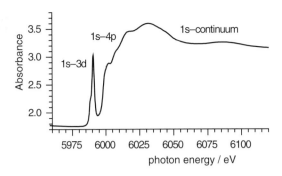

FIGURE 6.10 XANES region of the XAS spectrum of CrO_2Cl_2 showing clear pre-edge features.

lighter 3d elements because of the larger number of empty 3d orbitals. These 1s–3d pre-edge transitions in first row transition metals can be used to determine the metal co-ordination environment. For compounds with a centre of symmetry (e.g. octahedral and square planar) the pre-edge transitions are very weak as the 1s–3d transitions are orbitally forbidden. For tetrahedral compounds, which have no centre of symmetry, the 1s–3d pre-edge peaks are often found to be more intense. For square planar compounds there is usually a very diagnostic and intense peak at slightly higher energy, very close or on the absorption edge, which has been assigned to a $1s–4p_z$ transition. The X-ray absorption spectrum of CrO_2Cl_2 shown in Figure 6.10 has an intense pre-edge feature because it is a d^0 compound and is tetrahedral with extensive p–d mixing.

EXAMPLE 6.4

Which of the following would be expected to show strong pre-edge features in their K-edge XAS spectra?

$KMnF_3$ (perovskite structure with octahedral Mn as MnF_6), $K_2Ni(CN)_4$, K_3VO_4

ANSWER

The $[VO_4]^{3-}$ anion is a tetrahedral d^0 species and, therefore, strong pre-edge features would be expected from the 1s to the multiple, empty 3d orbitals. Octahedral Mn(II), d^5, would show no or only weak features as the transitions would be forbidden in a species with a centre of symmetry and only weak p–d mixing. $[Ni(CN)_4]^{2-}$ is square planar so a strong pre-edge or on-the-edge feature, due to a 1s to 4p transition, would be predicted.

SELF TEST

Mn^{3+} can be doped into $YInO_3$ to produce a bright blue pigment in which the Mn^{3+} ion partly replaces In^{3+} on a trigonal bipyramidal site rather than yttrium on a site with cubic symmetry. Would the location of Mn^{3+} be evident in the Mn K-edge XAS data?

6.5.4 EXAFS

The EXAFS part of the X-ray absorption spectrum involves the photoionisation of the atom, with the creation of a photoelectron and a hole in the core electron shell ('core hole'). As well as the abrupt strong absorption at the edge and the XANES features, the X-ray absorption spectrum contains a slowly attenuating but undulating tail after the edge. This is because the core electrons will still be ionised but the resultant ejected electron or **photoelectron** carries away the excess kinetic energy (see Eqn 6.1). These photoelectrons ejected from the atom whose absorption edge is being studied propagate as spherical waves whose wavelengths are dependent upon their kinetic energies. As the excitation energies increase, so does their kinetic energy, but their wavelengths are reduced (through the de Broglie relationship). If there are other atoms surrounding the photoelectron emitting atom, then the photoelectron is scattered by these, part of which results in a backscattered wave, which can then interfere with the outgoing wave. Figure 6.11 shows the cases for two atoms at different interatomic distances, scattering from one of which (blue broken line) results in constructive interference (peaks coincide with peaks), while the other (grey broken line) results in destructive interference (peaks coincide with troughs). As the photoelectron wavelength changes through the spectrum (at higher energies above the absorption edge the photoelectron will have higher energies and therefore a shorter associated wavelength) it carries away different amounts of excess kinetic energy. This pattern of constructive and destructive interference will also change depending on the separation of the photoelectron emitting atom and the backscattering one. These phenomena give rise to the **EXAFS** spectrum, which is a series of oscillations after the absorption edge caused by these interference effects.

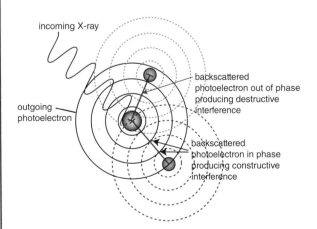

FIGURE 6.11 Effect of distance on backscattered photoelectron interference effects in EXAFS.

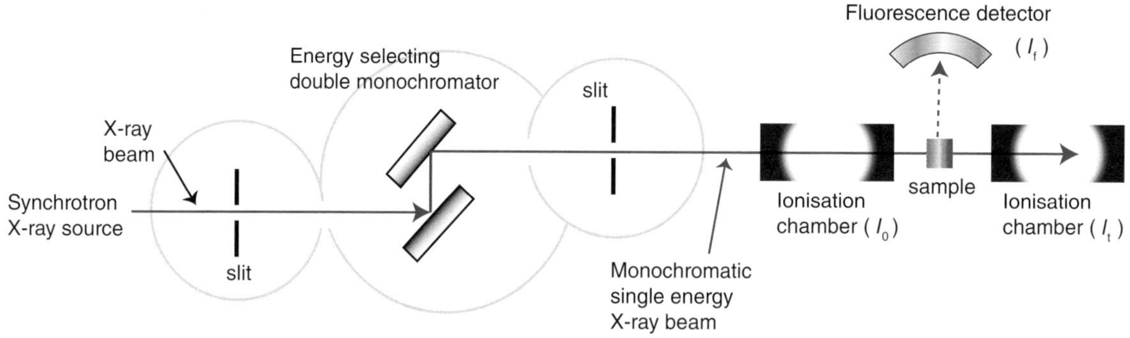

FIGURE 6.12 Schematic of instrument used to collect high-quality X-ray absorption spectra needed for EXAFS analysis.

6.5.4.1 EXAFS experimental requirements

In order to observe the EXAFS oscillations it is necessary to scan across the absorption edge and define the weak oscillations at energies above it. This requires a high intensity X-ray source with variable energies and, therefore, synchrotron sources (Section 1.6.5) are used. A typical set-up for a simple EXAFS experiment is shown in Figure 6.12.

X-rays enter from the synchrotron source and the beam size and resolution are defined by pre-slits. A monochromator constructed from two crystals of silicon select a specific wavelength (through their rotation and using Bragg's law) and, therefore, energy from the beam and this energy may be varied by rotating the monochromator angle. The incident X-ray beam intensity (I_0) is then measured using an ionisation chamber. The simplest way to obtain the X-ray absorption spectrum is to measure the transmitted intensity (I_t) after the beam has passed through the sample with a second ionisation chamber. The absorption is given by $-\ln(I_t/I_0)$. However, if the element of interest is very dilute in the sample, this will only produce a very small edge feature on a very high background with a poor signal-to-noise ratio, and as a result it is very difficult to extract accurately the weak EXAFS oscillations. In this case the fluorescence intensity (I_f) is measured (against a much lower background) and it is assumed that the absorption is directly proportional to this.

6.5.4.2 EXAFS data analysis

In order to extract the EXAFS oscillations from the raw X-ray absorption data the background is subtracted and the photoelectron energy, in eV, is converted to wave-vector (or wavenumber), k (in Å^{-1}), using the following equation:

$$k = \sqrt{[(2m_e/\hbar^2)(E - E_0)]}$$ **Eqn 6.2**

which can be simplified to

$$k(\text{Å}^{-1}) = \sqrt{0.2624628\left(E - E_0(\text{eV})\right)}$$ **Eqn 6.3**

where E is the measured photoelectron energy and E_0 the absorption edge position. The resulting graph contains the extracted oscillations as intensity, given the symbol $\chi(k)$, versus k. Much of the useful structural information is at high k values and to obtain a more equal spread of intensity throughout the k-range spectrum the intensity is weighted by k^2 or k^3 to give $k^2\chi(k)$ or $k^3\chi(k)$, respectively; see Figure 6.13(a).

The $\chi(k)$ versus k spectrum consists of superimposed damped sine waves and each interatomic distance surrounding the emitting atom gives rise to a different wave in the $k^3\chi(k)$ versus k spectrum (Figure 6.13a). These oscillations are controlled and weighted by structural parameters, such as atomic separation, number of backscattering centres and their ability to scatter the photoelectron. In essence the spectrum tells us what is there, how many there are, and how far away they are. The Fourier transformation (Chapter 1, Section 1.5) of this spectrum gives a pseudo-radial distribution function, a graph of backscattering intensity versus distance. Full analysis of EXAFS data involves fitting the $\chi(k)$ versus k spectrum using structural models that adjust and refine parameters in atomic 'shells'. Each shell represents a backscattering atom type, the number of this atom type and its distance from the photoelectron emitting atom. This fitting procedure, which is done computationally, allows the following information to be extracted from the EXAFS data:

(i) The interatomic distances between absorbing and surrounding atoms up to about 6 Å; the error of these distances is about ±1% in the best cases but typically a few per cent.

(ii) The likely atom types of the backscattering atoms based on knowledge of the sample composition. The

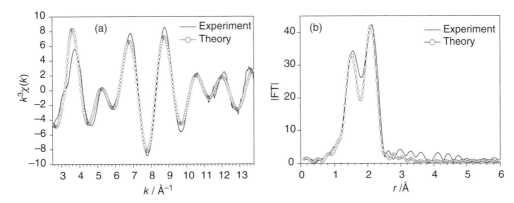

FIGURE 6.13 (a) The form of a typical EXAFS spectrum obtained from the Cr K-edge in solid CrO_2Cl_2 at 10 K with the intensity weighted by k^3. (b) The Fourier transformed EXAFS spectrum.

oscillation intensity is proportional to the number of electrons in the backscattering atom so a large atomic number neighbour, say selenium, can be distinguished from a low atomic number atom, say oxygen. However, it is not generally possible to distinguish, say, a backscattering sulfur atom from a near isoelectronic phosphorus atom.

(iii) The number of atoms or coordination number in each shell can be determined—though the accuracy is only of the order of ±0.5 atoms. Therefore, a four-coordinated metal can be easily distinguished from an octahedrally coordinated site.

While full analysis of the EXAFS spectrum requires detailed computational modelling to extract, for example, accurate distances, key structure features can usually be seen directly in the Fourier transform (FT). This is

illustrated in the FT of the chromium K-edge spectrum of CrO_2Cl_2 (Figure 6.13b), where distinct shells are observed at distances of 1.57(2) Å and 2.12(3) Å, which correspond to the Cr–O and Cr–Cl bond lengths, respectively (these can be compared to gas-phase electron diffraction values of 1.581 and 2.126 Å). The peak intensities in the FT reflect the atomic numbers (number of electrons) of the neighbouring atoms with the Cr–Cl peak being distinctly more intense than the Cr–O one. Note that the number of backscattering atoms is the same in each shell but intensity will be weighted on the number of that atom type in a specific shell. It should also be recognised that the backscattering is damped as a function of $1/r^2$ so peak intensities will decrease with higher r values. As the Cr–Cl shell is further out than Cr–O its peak intensity is less than that predicted on the basis of electron number alone.

EXAMPLE 6.5

Simple analysis of EXAFS data

Figure 6.14 shows the Fe K-edge EXAFS spectrum from $Fe(CO)_5$ and its FT. Interpret the form of this spectrum.

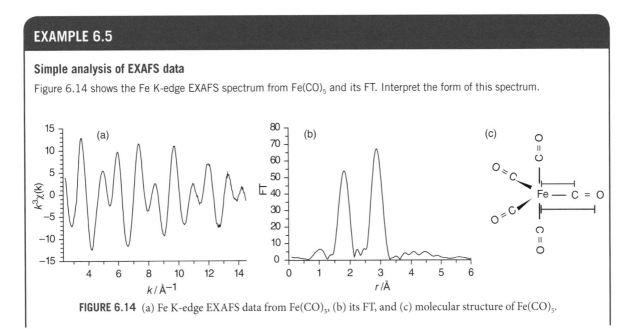

FIGURE 6.14 (a) Fe K-edge EXAFS data from $Fe(CO)_5$, (b) its FT, and (c) molecular structure of $Fe(CO)_5$.

ANSWER

Two shells of atoms are observed around the iron centre—one at around 1.9 Å and one near 2.9 Å. These correspond to the Fe–C and Fe . . . O distances in the carbonyl complex (Figure 6.14c). Note that the intensity of the oxygen peak is higher than might be expected based on the backscattering strengths of carbon and oxygen and

their distances from iron. This is due to a phenomenon known as **multiple scattering** where the photoelectron is scattered by more than one atom. In linear, or near linear, species, such as Fe–C–O, these multiple scattering pathways become a significant contributor to the EXAFS signal, which in $Fe(CO)_5$ enhance the intensity of the oxygen atom shell.

SELF TEST

The equilibrium between the dimer of the palladium complex (a) and its monomer (b) in Figure 6.15 is controlled by the level of dilution in the solvent NMP (*N*-methyl-2-pyrrolidone) with the monomer being favoured at high dilutions. What would be the main changes predicted in the FT of the Pd K-edge EXAFS spectrum as a solution of the monomer is concentrated?

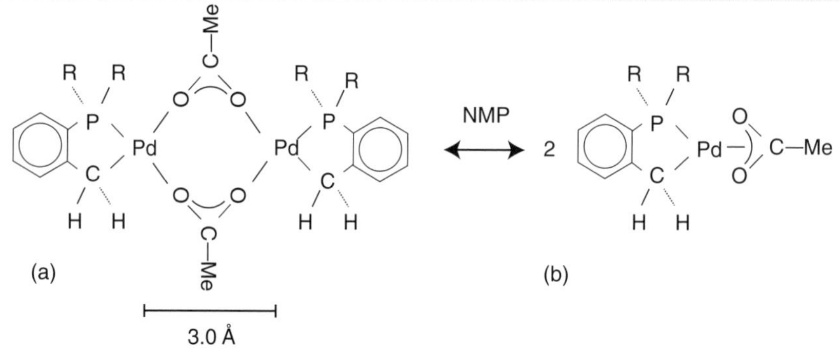

FIGURE 6.15 Equilibrium between the dimer of a palladium complex (a) and its monomer (b).

6.5.4.3 XANES and EXAFS in the structural characterisation of inorganic compounds

While XANES and EXAFS are not core, laboratory-based, analytical methods in the characterisation of an inorganic compound or material, these techniques are able to produce very useful compositional and structural information—often where other techniques are unable to do so. This section summarises some of these specific technique attributes.

Each element's K-edge (or any other edge) will occur at a unique energy. Therefore, it is possible to probe one element in the presence of other elements and obtain information on that element's oxidation state (XANES) and local environment (EXAFS).

XAS is a local structural probe providing information that is principally concerned with the first few coordination shells. This is very powerful in combination with the element specificity for studying the distribution of multiple elements in a complex system. For example, in the complex oxide spinels, AB_2O_4, where A and B represent two different sites in the structure, one with octahedral coordination and the other tetrahedral, the distribution of several different transition metals over these two

sites in the same sample, even at very low levels, can be ascertained.

No long-range order is needed, and hence many different types of samples, materials, phases can be studied including amorphous compounds such as glasses and polymers, nanomaterials, compounds in solution and even gases. This contrasts with diffraction methods which require single crystal or polycrystalline solid material (Chapter 2). It is also feasible to study systems *in situ* or *in operando* so that it is possible to identify the structural changes occurring during process chemistry and in catalytic cycles.

6.6 Introduction to electron microscopy and energy dispersive analysis of X-rays

Widely used electron microscopy methods that are applied to the characterisation of inorganic compounds and materials include **scanning electron microscopy (SEM)** and **transmission electron microscopy (TEM)**. These techniques visualise materials at distances down to 1 nm, and many of the recent advances made in nanochemistry have relied on their application.

An imaging electron microscope operates like a conventional optical microscope but instead of imaging photons, as in the visible microscope, electron microscopes use electrons. In these instruments, electron beams are accelerated through 1–200 kV and electric and magnetic fields are used to focus the electrons. In SEM the backscattered electrons from the sample are imaged while in TEM electrons passing through a very thin sample are collected and analysed.

Both SEM and TEM microscopes allow the elemental composition of the material under investigation to be determined through analysis of X-rays that are generated when the high energy electron beam impinges on the sample. These electrons cause core electron excitation and ionisation and subsequent production of characteristic X-rays akin to those produced in XPS. The analysis of the energies of these X-rays produced in the electron microscope provides very useful information on the chemical composition of the material under investigation. This information is available on areas of the sample where the electron beam is focused, which can be a few microns or less in size. The sample can be translated in the electron beam allowing maps of chemical composition to be constructed.

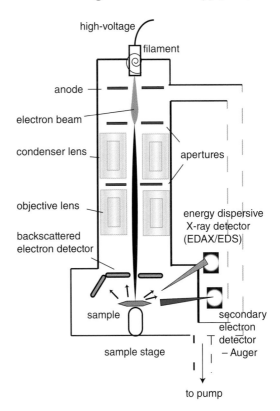

FIGURE 6.16 Schematic of a scanning electron microscope showing the key detector systems.

6.7 Scanning electron microscopy (SEM)

Figure 6.16 shows a schematic of a **scanning electron microscope** where an electron beam is scanned over the object and the reflected (scattered) beam is then imaged by the detector. An electron beam emitted from a tungsten filament or lanthanum hexaboride (LaB_6) cathode is accelerated using a potential between 0.2 keV and 50 keV, and focused using electric fields to a spot about 0.4 nm to 5 nm in diameter. By adjusting the focusing electric fields the beam can be scanned over a rectangular area of the sample surface, and the sample stage, which can be translated perpendicular to the electron beam, will accommodate samples up to several centimetres in size. The beam penetrates between 100 nm and 5 μm into the surface depending upon the energy of the electron beam and the sample density and atomic mass (dense samples containing high atomic mass elements will absorb the electron beam most rapidly). Three main processes can occur as the electron beam interacts with the sample: reflection of high-energy electrons by elastic scattering, emission of secondary electrons through inelastic scattering, and the production of X-rays (via excitation of core electrons and electron relaxation); each of these can be studied with specialised detectors.

Imaging of the backscattered electrons provides a picture of the sample under investigation as the electron beam is scanned across the sample; see, for example, Figure 6.17. Magnification in an SEM typically covers six orders of magnitude from about 10 to 500 000 times with a resolution at the highest magnifications of between 0.5 nm and 20 nm.

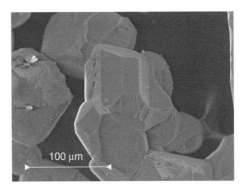

FIGURE 6.17 Scanning electron micrograph of copper zinc tin sulfide crystals. The scale bar shown indicates that the crystallites are around 50 μm in dimension.

6.7.1 Sample preparation and environmental SEM (ESEM)

SEM samples need to be conductive, otherwise electrons collect on the sample and interact with the electron beam itself, resulting in blurred images. Non-conductive samples are, therefore, usually coated with a thin layer of a conducting material, typically gold or graphitic carbon. Conventional SEM also requires that samples be imaged in a vacuum to allow the electron beam to be focused and not scattered by gas molecules. As a consequence, samples that produce a significant amount of vapour, for example, zeolites or hydrated or solvated salts, decompose in the scanning electron microscope. **Environmental SEM (ESEM)** allows samples to be observed in low-pressure gaseous environments (1–50 mbar) and high relative humidity (up to 100%). ESEM achieves this by separating a vacuum region around the electron beam from the sample chamber; it also uses a secondary-electron detector that can operate in the presence of water vapour. ESEM is especially useful for hydrated and bioinorganic materials as coating with conductive carbon or gold is no longer necessary.

6.7.2 EDAX and EDS

The analysis of the X-rays produced when a high energy electron beam impinges on the sample in an electron microscope is known as either **energy dispersive analysis of X-rays** (EDAX, sometimes EDX) or **energy dispersive spectroscopy** (EDS). These X-rays are generated when the high-energy electron ejects a core electron and outer shell electrons fall back into the vacant low-lying energy level. The process is analogous to that in XPS (Figure 6.1) but with the initial excitation caused by electrons rather than high energy X-rays. These X-rays are analysed using a **silicon drift detector.** In this type of detector, made of very pure silicon, the level of ionisation caused by an X-ray is proportional to, and thereby determines the energy of the original X-ray, typically within 100 eV. This allows the characteristic X-ray radiation from each element present in the sample to be determined. Specialist 'windowless' detector systems are needed for elements lighter than oxygen because of the absorption of their soft, characteristic X-ray energies by most detector window materials. The characteristic X-ray energies as a function of atomic number are shown in Figure 6.18(a). The EDAX spectrum, typically collected for X-ray energies between 0 and 20 keV, will contain a series of peaks with each one derived from the characteristic radiation of an element present; these will be K lines for low atomic number elements and L and M lines for the heavier elements. Representative EDAX data collected from $CaZrO_3$ (Figure 6.18b) shows characteristic peaks from Ca (K line at 3.7 keV), Zr (two peaks, the K line at 15.7 keV and L line at 2.1 keV) and O (K line at 0.53 keV).

6.7.2.1 Accuracy and sensitivity

The emitted X-ray intensities in the EDAX spectrum are proportional to the amount of that element present in the sample so the individual peak intensities reflect the sample composition. Measurement of peak intensities in the spectrum can, therefore, give a quantitative measure of composition which, with the use of standards to calibrate the instrument, can give accuracies for the atomic

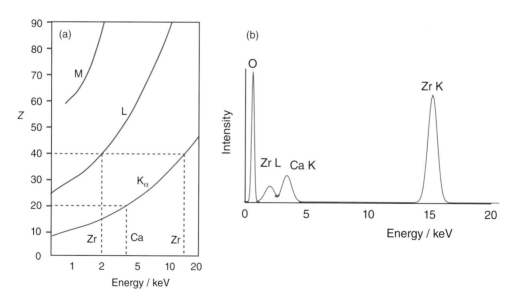

FIGURE 6.18 (a) EDAX element X-ray energies for various shells as a function of atomic number, Z. (b) EDAX spectrum from $CaZrO_3$.

composition of around ±2% for the major components. Accuracy is lower where elements with similar atomic numbers are present in the sample. For example, in the analysis of aluminium and silicon in the natural clay kaolinite, $Al_2Si_2O_5(OH)_4$, the characteristic K-line X-ray peaks from Si and Al partly overlap in the EDAX spectrum reducing how accurately their individual intensities can be measured.

EXAMPLE 6.6

The EDAX spectrum collected from a sample of a photovoltaic semiconducting material, copper zinc tin sulfide selenide, is shown in Figure 6.19 together with the experimentally derived atom percentages. Calculate the composition of this phase.

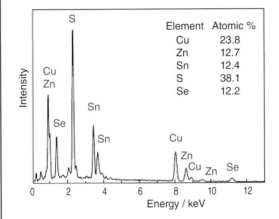

Element	Atomic %
Cu	23.8
Zn	12.7
Sn	12.4
S	38.1
Se	12.2

FIGURE 6.19 EDAX spectrum and compositional data from a phase in the Cu–Zn–Sn–S–Se system.

ANSWER

EDAX analysis produces the compositional data as atomic percentages so these may be used to directly obtain the atomic ratios. Dividing through the values obtained by the smallest value (Se at 12.2%) gives the following atomic ratios Cu 1.95: Zn 1.04: Sn 1.02: S 3.12: Se 1.00. Allowing for the typical experimental errors in EDAX, and assuming stoichiometry and integer relationships between the elements, this suggests a composition for this phase of Cu_2ZnSnS_3Se.

SELF TEST

Why would the copper and zinc atom percentages in this analysis have a higher error than those for tin and selenium?

6.7.2.2 Spatial resolution and X-ray mapping

The spatial resolution of the EDAX spectrum obtained is determined by the degree of penetration and spreading

of the electron beam in the specimen. The degree of penetration depends upon how absorbing the sample is to the electron beam which in turn is controlled by the average atomic number and hence the density of the material. In a typical material with a density of 2–3 g cm^{-3}, such as an aluminosilicate or sodium chloride, the nominal spatial resolution is a few microns under typical SEM operating conditions. By scanning the electron beam across the sample its EDAX spectrum can be obtained at multiple points and these data can be used to obtain a compositional map. Such elemental or X-ray maps are useful when studying multi-phase materials where crystallites of the different phases can be identified.

6.7.3 Auger spectroscopy in electron microscopy

Auger spectroscopy, using X-rays for sample excitation, was described previously in Section 6.3.2. Auger electrons can be detected and analysed in the scanning electron microscope using an additional detector placed close to the sample (Figure 6.16). The information available on the surface composition (to a depth of 5 nm) of the material that is available from Auger spectrum has been described previously (Section 6.3.2).

6.8 Transmission electron microscopy (TEM)

In TEM, the electron beam passes through a very thin sample being examined and is imaged on a phosphorescent screen. The instrumental set-up is essentially the same as that of the scanning electron microscope (Figure 6.16), but with the image produced from transmitted electrons below the sample. For the electron beam to pass through the sample and obtain an image, specimens have a thickness around 100 nm, but this value depends on the accelerating voltage of the microscope, which on modern instrumentation is typically in the range 100–500 kV, much higher than used for SEM. Very thin, <10 nm thick, specimens may be imaged using low voltage 30 kV TEM instrumentation. In TEM, an image is formed from the interaction of the electrons transmitted through the specimen; the image is magnified and focused onto an imaging plate. At these high operating voltages the wavelength of the electrons is very short allowing imaging of features on the scale of individual atoms or columns of atoms. At smaller magnifications, contrast in a TEM image is due to absorption of electrons in the material, due variations in the thickness and composition of the material. For materials of constant thickness such images, often termed **micrographs**, can be directly interpreted in terms of area of high and low electron density (see Section 6.8.2). At

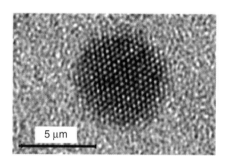

FIGURE 6.20 A bright field image of a 5 μm gold nanoparticle obtained using TEM.

higher magnifications and sample thicknesses complex wave interactions modulate the intensity of the image, requiring expert analysis of observed images. Chemical analysis of the sample composition can be undertaken using EDAX (Section 6.7.2).

6.8.1 TEM sample preparation and mounting

Sample preparation in TEM can be a lengthy procedure in order to get samples that are thin enough to transmit the electron beam (at most a few hundred nanometres thick), though sometimes the edges of particles are sufficiently thin to be imaged directly. Nanoparticles are often small enough to be studied directly and TEM has become an essential technique in their characterisation in recent years. High-quality samples are normally prepared by grinding and polishing the sample mounted in a polymer matrix to only a few tens of nanometres thick. The nanoparticle, finely powdered material (where the grain edges may be thin enough to transmit the beam directly) or polished thin section is usually supported on a carbon mesh grid.

6.8.2 Bright field image

The most common mode of operation for a TEM is the bright field imaging mode. In this mode the image is the shadow formed directly by absorption of electrons by the sample and is termed the **bright field image**. Thicker regions of the sample, or regions with higher electron densities associated with large atomic numbers, will appear dark in the image while regions with no sample or light elements in the beam will appear bright. The image is effectively a simple two-dimensional projection of the sample down the optical axis and provided the sample

FIGURE 6.21 SAED pattern obtained in a transmission electron microscope.

is oriented appropriately can be a projected image along one unit cell direction of the material (Figure 6.20).

6.8.3 Diffraction in TEM

By changing the magnetic lenses in the transmission electron microscope it is possible to obtain a diffraction pattern instead of a bright field image. Effectively the experiment is very similar to single crystal or powder X-ray diffraction (Chapter 2) but a beam of electrons is diffracted by a very small single crystal or multiple crystals. The wavelength of electrons accelerated in a 200 kV transmission electron microscope is 2.5 pm, which, while shorter than X-ray diffraction wavelengths (~100 pm), produces well-resolved diffraction patterns. Two types of diffraction experiment are normally performed: **selected area electron diffraction (SAED),** which uses a parallel beam of electrons, and **convergent beam electron diffraction (CBED),** which can reveal the full three-dimensional symmetry and structure of the material. The electron diffraction pattern (Figure 6.21) can be combined with direct bright field image of the sample allowing the structure solution and refinement of a crystal structure from a single crystallite of material.

Problems

6.1 Which technique would you use to determine the following information?

  (a) The composition of the product(s) obtained from the reaction:

$$CaO + ReO_2 \xrightarrow{O_2}$$

  (b) The silicon to oxygen distance in a silica glass.

  (c) The compounds formed on the surface of a sample of calcium exposed to air.

  (d) In a sample of Nd-doped yttrium aluminium garnet ($Y_3Al_5O_{12}$) whether the neodymium ions occupy cubic (8-fold coordination), octahedral (6-fold coordination), or tetrahedral (4-fold coordination) sites.

6.2 Figure 6.22 shows the FT of the Br K-edge EXAFS of molecular Br_2O. Explain the form of this spectrum and calculate a value for the Br–O–Br bond angle.

6.3 An XPS study of rhenium metal and its oxides showed binding energies (eV) for the rhenium 4f electrons as follows:

Re metal 42.3 ReO_2 45.0 ReO_3 47.2 Re_2O_7 49.5

Interpret these data.

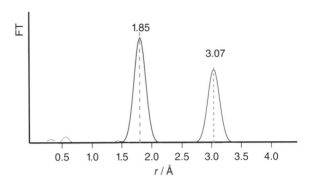

FIGURE 6.22 The Br K-edge FT-EXAFS of molecular Br_2O.

Heating a sample of ReO_2 deposited on platinum under hydrogen produces a material showing in its XPS spectrum a peak at 43.5 eV. Discuss a likely composition of the phase formed under these conditions.

6.4 XANES spectra of titanium(IV) compounds show a pre-edge feature whose position depends on the coordination of the titanium atom as follows: four-coordinate Ti 4969.6 eV, five-coordinate Ti 4970.5 eV, and six-coordinate Ti 4971.4 eV. Explain the origin of the pre-edge feature in these data and suggest a reason as to why its position depends upon the titanium coordination number.

Bibliography

P. van der Heide. (2011) *X-ray Photoelectron Spectroscopy: An Introduction to Principles and Practices.* Chichester, UK: Wiley.

A.M. Ellis, M. Feher, and T.G. Wright. (2011) *Electronic and Photoelectron Spectroscopy: Fundamentals and Case Studies.* Cambridge: Cambridge University Press.

J.H.D. Eland. (1984) *Photoelectron Spectroscopy: An Introduction to Ultraviolet Photoelectron Spectroscopy in the Gas Phase.* Oxford: Butterworth-Heinemann.

S. Calvin. (2013) *XAFS for Everyone.* Boca Raton, FL: Taylor & Francis.

N.A. Young. (2014) The application of synchrotron radiation and in particular X-ray absorption spectroscopy to matrix isolated species. *Coord. Chem. Rev.* **277** 224.

B.K. Teo. (1986) *EXAFS: Basic Principles and Data Analysis.* Berlin: Springer.

D. Koningsberger and R. Prins (Eds.) (1988) *X-ray Absorption: Principles, Applications, Techniques of EXAFS, SEXAFS and XANES.* New York: Wiley.

J. Goldstein, D.E. Newbury, D.C. Joy, C.E. Lyman, P. Echlin, E. Lifshin, L. Sawyer, and J.R. Michael. (2007) *Scanning Electron Microscopy and X-ray Microanalysis.* 3rd ed. Berlin: Springer.

L. Page (Ed.) (2015) *Scanning Electron Microscopy.* New York: Research Press.

R. Egerton. (2011) *Physical Principles of Electron Microscopy: An Introduction to TEM, SEM, and AEM.* Berlin: Springer.

M.T Weller, T. L. Overton, J. A, Rourke, and F.A. Armstrong. (2014) *Inorganic Chemistry.* 6th ed. Oxford: Oxford University Press.

Mass spectrometry, chemical, and thermal analysis techniques

7.1 Introduction: compositional and thermal analysis

The determination of the elemental composition of a compound is a fundamental stage in its characterisation and a wide suite of methods is available to the inorganic chemist to undertake this analysis. This chapter focuses initially on methods that can be used to determine the elemental composition of a compound and this analysis is normally destructive. The behaviours of compounds when they are heated, known as thermal analysis methods, are also covered here as these characterisation techniques often provide information on the sample's composition. Studying the behaviour of compounds as a function of temperature also yields information on their phase behaviour, such as changes in structure and melting points.

7.2 Mass spectrometry of inorganic compounds

In **mass spectrometry** (also referred to erroneously, particularly in older publications, as mass spectroscopy) gaseous ions are produced from a compound and then analysed to determine their mass-to-charge ratio. As one of the gaseous species produced in the mass spectrometer is often the **molecular ion**, which is simply the charged molecule of the compound under investigation, the **relative molecular mass (RMM)** of the compound is directly measured. Where the compound undergoes fragmentation into smaller charged species in the mass spectrometer, additional information on the molecule's configuration, that is, which atoms are bonded to each other, can be obtained. Normally the ions produced in a mass spectrometer are positively charged though negative ion mass spectrometry can provide additional

information, particularly on the stability of the species, as these ions fragment through different pathways. Mass spectrometry is a destructive analytical technique as the sample cannot be recovered for further analysis.

7.2.1 Experiment basis and instrumentation

The basic mass spectrometry instrument requires a method of producing gaseous charged species, their separation, using either electrostatic, magnetic, or radiofrequency fields, and their detection (Figure 7.1). A wide variety of methods is used both in terms of generating the gaseous ion species and their separation by mass/charge ratio. The choice of instrument configuration depends on the form of the sample to be analysed, the speed of the measurement demanded, and the accuracy required in the determination of the masses of the species. Mass spectrometry is typically a highly sensitive technique and for most characterisation experiments far less than a milligram of compound is required. Some modern instrumentation can detect analytes at concentrations in the attomolar (10^{-18} moles) range.

7.2.1.1 Ionisation methods

Mass spectrometry requires the conversion of a sample into gaseous ions without, in most applications, overly fragmenting the compound of interest. A variety of methods exists to produce the ions and these are applicable to gases, liquids, solutions, and some solids and in many cases the degree of ion fragmentation can be controlled. The majority of methods used for the study of complex inorganic compounds use 'soft' ionisation methods. These avoid a high degree of fragmentation, which can be prevalent in weakly bonded inorganic compounds, such as coordination complexes.

Electron ionisation (EI) relies on the interaction of a sample with high-energy electrons (70 eV) to produce positive ions. The exchange of energy between the analyte and the high energy electron causes ionisation and

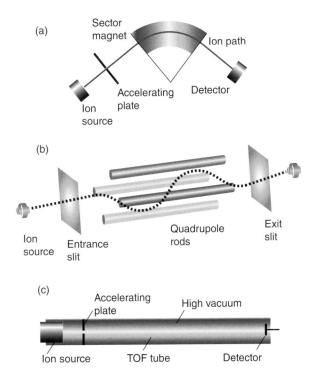

FIGURE 7.1 Schematics of mass spectrometers using various mass analysis methods: (a) a sector magnet; (b) a quadrupolar electric field; and (c) a time-of-flight (TOF) tube.

the emission of a further electron. Gases and volatile liquids can be directly introduced in a vacuum space where the electron to analyte molecule energy transfer process occurs. Solids and non-volatile liquids are heated and, thereby, vaporised close to the ionisation region (a process known as **direct insertion EI**). A disadvantage of EI, as a high energy process, is that it usually causes significant fragmentation of larger molecules. **Fast atom bombardment (FAB)** uses the impact of neutral atoms, such as argon or xenon, on a sample suspended in a matrix to vaporise and ionise the analyte. The high energy atom beam is produced by charge-exchange (neutralisation) of ions initially accelerated through a potential of 8–15 keV. Common analyte matrices include glycerol and *m*-nitrobenzyl alcohol. FAB induces less fragmentation than EI but matrix ions can appear in the mass spectrum. In **liquid secondary ion mass spectrometry (LSIMS),** high-energy ions, such as 35 keV Cs^+ ions, are used to bombard the sample.

Matrix-assisted laser desorption/ionisation (MALDI) uses a short UV-laser pulse to simultaneously vaporise and ionise the sample in a matrix. This technique is particularly effective with polymers and large atomic mass molecules. It has been applied to a variety of inorganic samples, such as supramolecular assemblies, synthetic inorganic polymers, fullerenes, metal clusters, large organometallic

complexes, and metallodendrimers. Despite the complexity of the molecules and materials to which the technique is applied, MALDI often gives clean, simple mass spectra with very few peaks.

In **electrospray ionisation (ESI)**, droplets of a solution containing the ionic species of interest are sprayed into a vacuum chamber; here solvent evaporation results in the generation of individually charged ions. ESI mass spectrometry is the method of choice for ionic compounds in solution as it avoids further sample preparation stages, which may change the nature of the species under investigation.

Inductively coupled plasma mass spectrometry (ICP-MS) employs a very high energy (inductively-coupled) plasma ion source, typically of argon ions, into which a solution of the sample has been nebulised. At the very high temperatures and energies of the plasma, the constituent molecules are ripped apart leaving just the individual atoms, which then form ions in the plasma. The detection of the metal (and some non-metal) ions in ICP-MS leads to applications in chemical analysis at very low concentrations (see Section 7.4.3).

Mass spectrometry is a core technique for the analysis of many molecular inorganic compounds, including (where MALDI is used) large molecules, complexes, and inorganic polymers. However, mass spectrometry, using the ionisation techniques described so far, is of limited use in the characterisation of ionic structures (such as metal oxides) or infinite, covalently bonded networks (e.g. SiO_2). For these materials, which do not contain molecules or form molecules or molecular ions in the gas phase, chemical characterisation using mass spectrometry is mainly limited to elemental analysis using ICP-MS (see Section 7.4.3). **Secondary ion mass spectrometry (SIMS)** can be used to analyse the surfaces of inorganic materials and thin films. In SIMS a high energy ion beam (typically oxygen, argon or caesium ions) is focused onto the specimen. This beam sputters and ionises atoms from the surface—the secondary ions. These secondary ions can then be separated and analysed. As the sputtering slowly removes the surface of the sample under investigation depth profiling of the surface layers of the material can be undertaken.

7.2.1.2 Ion separation and detection methods

Once the gaseous ions have been formed they can be accelerated, using an electric field, and, thereby, given a known velocity prior to separation. In the classical application of mass spectrometry the ion separation method employed a **magnetic field** to deflect the moving ions. Ions with a lower mass-to-charge ratio are deflected more in the magnetic field than heavier ions of the same charge. The magnetic field can be varied so that ions with different mass-to-charge ratios are sequentially directed on to the detector (Figure 7.1a). However, modern MS instrumentation uses alternative methods of ion separation that leads to greater resolution of the ion masses and much greater sensitivities.

A **quadrupole mass spectrometer** (Figure 7.1b) uses oscillating electric fields applied to four parallel rods to control the trajectories of the ions passing through the analyser. In a **time-of-flight (TOF)** mass spectrometer, the ions are accelerated by an electric field for a fixed time and then allowed to fly freely (Figure 7.1c); the lighter ions are accelerated to higher speeds than the heavier ions and strike the detector earlier. In an **ion cyclotron resonance (ICR)** mass spectrometer (often denoted FTICR, for Fourier transform-ICR), ions are collected in a small cyclotron cell inside a strong magnetic field. The ions circle round in the magnetic field, effectively behaving as an electric current. Because an accelerated current generates electromagnetic radiation, a signal is produced by a circulating ion in proportion to its mass; this can be detected and used to establish the mass-to-charge ratio for the ion.

7.2.1.3 Precision and accuracy in mass spectrometry

The precision of measurement of the mass of the ions varies according to the spectrometer analyser system being employed and the aims of the experiment. A mass spectrometer will provide the species mass to within at least one m_u (where m_u is the atomic mass constant, 1.66054×10^{-27} kg, also known as **1 dalton, 1 Da**). High resolution mass spectrometers can determine masses to within 10^{-4} Da and this allows **atom mass defects** to be determined; that is, while the mass of a ^{12}C atom is $12.0000 m_u$ by definition, the mass of ^{14}N is $14.00307 m_u$ and that of ^{16}O is $15.9949 m_u$. With a mass spectrometer of high resolution and precision (10^{-4} Da) molecules of nominally the same RMM (i.e. 28 g mol^{-1}, $28 m_u$), such as $^{12}C^{16}O$ (of mass $27.9949 m_u$) and $^{14}N_2$ (of mass $28.0061 m_u$), can be distinguished. High resolution mass spectrometry will separate species with different isotopic compositions for large complex ions

having masses up to $2000 m_u$. Knowledge of the natural abundances of the various isotopes that could be present in the sample allows rapid assignment of the molecular composition and structural fragments (Section 7.2.2.2).

7.2.2 Interpretation of mass spectral data from inorganic and organometallic compounds

7.2.2.1 A simple mass spectrum

In its simplest form the mass spectrum consists of the molecular ion or parent peak, and, possibly, a number of lighter ion species caused by the molecular ion fragmenting. The molecular ion peak provides the mass of the molecule or ion that is being studied (and hence its molar mass). The lighter species observed in the mass spectrum provide information about fragmentation pathways of a molecule. Their analysis can help determine the structure of a molecule and how the atoms are bonded to each other. Complex ions often lose one or more ligands in the mass spectrometer and peaks are observed that correspond to the molecular ion less these ligand masses. For example, the mass spectrum of PF_5 consists of a molecular ion peak at $126 m_u$ ($1 \times {}^{31}P + 5 \times {}^{19}F$) and a series of molecular fragments corresponding to the sequential loss of fluorine atoms (107, 88, 69, 50, and $31 m_u$ from PF_4, PF_3, PF_2, PF, and P, respectively). Note that sometimes mass spectra contain a peak that occurs at half the parent species' molecular mass; this can be ascribed to a doubly charged ion.

7.2.2.2 Isotopes in inorganic mass spectrometry

In inorganic chemistry many of the elements exist naturally as two or more isotopes and this complicates the form of the mass spectrum. Figure 7.2 shows a typical mass spectrum from a simple inorganic compound containing different isotopes, in this case PCl_3. To interpret a spectrum, it is helpful to detect a peak or peaks corresponding to the singly charged, intact parent molecular

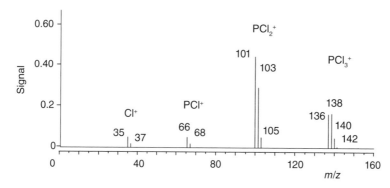

FIGURE 7.2 Positive ion mass spectrum obtained from PCl_3 with major mass fragments marked.

ion, usually the highest mass peak. For PCl$_3$ the molecular ion appears as a group of peaks representing the different masses of the isotopes present (see Example 7.1).

Where isotopes exist, multiple peaks are observed in the mass spectrum reflecting the isotope masses and abundances. Chlorine occurs naturally as 75.5% ^{35}Cl and 24.5% ^{37}Cl. Thus, for a molecule containing one chlorine atom, the mass spectrum will show two peaks, $2m_u$ apart and in an intensity ratio of approximately 3:1. Molecules with different isotopic compositions are known as **isotopomers** (see also Chapter 3, Section 3.10).

Different and more complex patterns of mass spectrum peaks are obtained from molecules that contain two or more atoms of an element that has isotopes. These patterns can be used to identify the number of atoms of the element in compounds of unknown composition. Peaks from multiply charged ions are usually easy to identify in this type of spectrum because the separation between the peaks from the different isotopomers is no longer m_u but fractions of that mass. For instance, in a doubly charged ion, isotopic peaks are ½m_u apart, in a triply charged ion they are ⅓m_u apart, and so on.

These species are formed by the fragmentation of PCl$_3$ in the mass spectrometer.

SELF TEST

Explain the form of the mass spectrum obtained from bromine, Br$_2$, shown in Figure 7.3 (bromine isotopes and natural abundances: ^{79}Br 50.5%, ^{81}Br 49.5%).

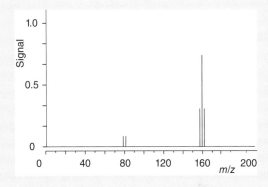

FIGURE 7.3 The mass spectrum obtained from Br$_2$.

7.2.2.3 Isotopes and inorganic geochemistry

Mass spectrometry not only has the ability to identify the various isotopes present in a compound (through the presence of isotopomers) but to measure their relative levels very precisely. Because the actual proportion of isotopes of an element can vary according to its geographic source, the determination of the proportions of different isotopes can be used to determine the source of a sample. This application is of particular interest in geology and mineralogy and for dating rocks and minerals which contain specific isotope patterns generated through radioactive decay.

7.2.2.4 Ultra-high resolution mass spectrometry and mass defect

Ultra-high resolution mass spectrometry can aid in the assignment of a peak in a mass spectrum using the so-called **mass defect** (or sometimes referred to as the **Kendrick mass defect**). Atomic masses are not exact integers, except for ^{12}C atom (12.0000m_u) by definition, due to the energy of combining the proton and neutrons together in the nucleus. Therefore, the masses of seemingly equivalent species such as ^{10}BH and ^{11}B (approximately 11m_u but actually 11.0208 and 11.0093m_u, respectively) and 2 × ^{16}O and 1 × ^{32}S (approximately 32m_u but actually 31.9898 and 31.9721m_u, respectively) can be separated at very high mass resolutions. The main applications of this method are in organometallic and bioinorganic compounds where large, complex fragments, which have the

EXAMPLE 7.1

Interpret fully the mass spectrum of PCl$_3$ shown in Figure 7.2, assigning all the peaks and their relative intensities to specific molecular fragments.

ANSWER

The parent intact molecular ion will be PCl$_3^+$ and we need to take note that chlorine exists as two isotopes ^{35}Cl (~75%) and ^{37}Cl (~25%). With three chlorine atoms four isotopomers are possible: P(^{35}Cl)$_3$, P(^{35}Cl)$_2$(^{37}Cl), P(^{35}Cl)(^{37}Cl)$_2$, and P(^{37}Cl)$_3$ with masses of 136m_u, 138m_u, 140m_u, and 142m_u, respectively. These isotopomers of PCl$_3$ can be assigned to the group of peaks between 136m_u and 142m_u. Note that the intensity of these peaks can be rationalised in terms of the isotopic abundances and probability of their presence within a PCl$_3$ molecule. Thus the probability of having P(^{37}Cl)$_3$ is (¼)^3 = ~0.016 (1.6%) as ^{37}Cl is only 25% naturally abundant, while the probability of P(^{37}Cl)(^{35}Cl)$_2$ is 3 × (¼)(¾)^2 = 0.422 (42.2%) as there are three ways of choosing which two chlorine sites are occupied by ^{35}Cl. So the relative intensities in this parent group are predicted to be close to 42.2:42.2:14.0:1.6, in accordance with the measured peak heights.

The groups of peaks at 101–105, 66–68, and 35–37m_u can be assigned on their masses to PCl$_2$, PCl, and Cl, respectively with the number of peaks and their intensities reflecting their isotopomers and natural abundances.

same m_u to the nearest integer, can be assigned to specific species. For example, in the analysis of vanadium porphyrins found in crude oil, a fragment peak could be assigned to $[C_{39}H_{36}N_4O_1S_1V_1+H]^+$ 660.21275m_u, rather than the similar mass $[C_{39}H_{36}N_4O_3V_1+H]^+$ and other possible species, helping identify that the parent vanadium compounds present contained thiophene residues.

7.2.3 Example applications in inorganic chemistry

One of the major uses of mass spectrometry in inorganic chemistry is the identification of a compound by comparison with previously collected spectra in a database. These databases contain hundreds of thousands of entries though the majority are organic and biological molecules. For the characterisation of an unknown compound, the analysis of a mass spectrum should allow the empirical formula to be determined using the molecular ion peak; then further analysis of the fragmentation pathways often allows the constituent parts of the molecule to be identified.

7.2.3.1 Characterisation of inorganic molecular compounds

This analysis of chemical composition and molecular structure is exemplified using the mass spectrum of an iron nitrosyl bromide complex, as shown in Figure 7.4. The three highest mass peaks are found at 390, 392, and 394m_u with intensity ratios of 1:2:1. This implies the molecule contains two bromine atoms (remembering that there are two bromine isotopes with natural abundances: ^{79}Br 50.5%, ^{81}Br 49.5%). The residual weight after allowing for two bromine atoms is $394 - (2 \times 81) = 232m_u$. Iron, nitrogen, and oxygen can be treated as monoisotopic with atomic masses of 55.85 (56 to the nearest mass integer), 14, and 16m_u and we can obtain the residual 232m_u with (2×56) + $(4 \times (14 + 16))m_u$, giving the formula $Fe_2(NO)_4Br_2$. The mass spectrum also shows a series of peaks that can be assigned to the successive loss of $4 \times NO$ groups (30m_u each) and a pair of equal intensity peaks at 191 and 193m_u, which can be assigned to $Fe_2{}^{79}Br$ and $Fe_2{}^{81}Br$, respectively. The presence of this fragment indicates that a group of atoms of this composition, that is, Fe–Br–Fe or Fe–Fe–Br, is present in the compound. However, this assignment requires some caution as molecular rearrangements can occur readily in the mass spectrometer between the formation of an ion and its detection. Putting this information together and using knowledge of chemical bonding the dimeric species shown in the insert of Figure 7.4 can be proposed for $[Fe_2(NO)_4Br_2]$.

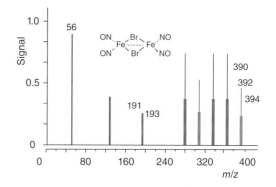

FIGURE 7.4 Mass spectrum obtained from the iron nitrosyl bromide complex $[Fe_2(NO)_4Br_2]$.

EXAMPLE 7.2

Interpreting mass spectra from metal carbonyls

Figure 7.5(a) shows the mass spectrum of $HCo(CO)_4$; assign the main peaks.

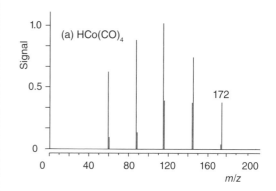

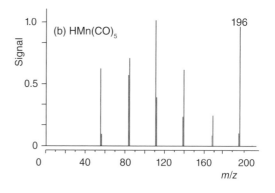

FIGURE 7.5 Mass spectra of (a) $HCo(CO)_4$ and (b) $HMn(CO)_5$.

ANSWER

Cobalt has a single stable isotope with an atomic mass of 59m_u so the parent peak would be expected to be at

59 + (4×28) + 1 = 172m_u. This parent peak is observed [HCo(CO)$_4$]$^+$ together with a weaker peak at 171m_u, which results from the loss of a hydrogen atom to give [Co(CO)$_4$]$^+$. A series of fragmentation pair peaks are observed resulting from the loss of increasing numbers of CO molecules (−28m_u, −56m_u, and so on). Note that the intensity ratio of [Co(CO)$_n$]$^+$ to [HCo(CO)$_n$]$^+$ increases as n decreases; this information can provide an understanding of the strength of bonds in, or rates of fragmentation of, the various fragments. This is because weaker bonds might be expected to break more easily giving more ion fragments from either side of this bond and associated stronger peaks in the mass spectrum. However, such interpretation should be undertaken with considerable caution as bond strengths are only one of several factors that can influence fragmentation pathways.

SELF TEST

Fully interpret the mass spectrum obtained from HMn(CO)$_5$ shown in Figure 7.5(b).

7.2.3.2 Secondary ion mass spectrometry

In SIMS, secondary ions, detected in the mass spectrometer, are produced by sputtering and ionising atoms from the surface of a solid. **Static SIMS (SSIMS)** is a technique that provides compositional analysis of just the topmost surface of a solid or the molecules adsorbed on the surface. Very low incident ion currents are used which remove less than a complete monolayer of the sample during the experiment. This means that the sputtered secondary ions are representative of the original surface, rather than surface that has already been eroded by previous ion impacts.

At higher ion impact and sputtering rates SIMS can be used for depth-profiling studies of surface layers of materials. The depth profile is obtained by recording sequential SIMS spectra as the surface is gradually eroded by the incident ion beam. A plot of the intensity of a given mass signal as a function of time is a direct reflection of the variation of its concentration with depth below the surface. Depth profiling SIMS offers high sensitivity, at parts per billion (ppb) concentrations of elements, and one important application is in the characterisation of semiconductors, where dopants are often present at very low concentrations.

SIMS can also be used to investigate the rates of reactions at surfaces and has been used to characterise ion diffusion in oxide ion conductors that are used in fuel cells. Exposure of an oxide ion conductor, such as $(Zr,Y)^{16}O_{2-y}$, to $^{18}O_2$ gas at 400°C results in exchange of the natural ^{16}O in the surface layers of the material with ^{18}O from the gas. By measuring, with SIMS, the depth profile of ^{18}O in the sample after a period of time the rate of oxide ion motion into and through the solid oxide can be determined.

7.2.4 Combined mass spectrometry analysis methods

7.2.4.1 Gas chromatography mass spectrometry (GC-MS), high performance liquid chromatography mass spectrometry (HPLC-MS), and thermal analysis mass spectrometry

A mass spectrometer can be added to other laboratory equipment to allow the analysis of any vapour or gaseous products being produced in that instrument. In **gas chromatography mass spectrometry (GC-MS)** the mass spectrometer is coupled to the outlet of a gas chromatograph so that as each compound is eluted from the column it is ionised and the mass spectrum recorded. This information plus the retention time on the gas chromatograph column allows the rapid analysis of complex mixtures. While perhaps of limited use directly in inorganic chemistry (compared with organic chemistry where it is very widely used to analyse the potential multiple products of a reaction) it has applications in organometallic chemistry and also in the analysis of the products of catalytic reactions at transition metal centres. In **high performance liquid chromatography mass spectrometry (HPLC-MS)** electrospray ionisation is normally used to convert the liquid emerging from the chromatography column into gas phase ions in the mass spectrometer.

Attaching a mass spectrometer to the gas outlet from a thermogravimetric analysis apparatus (Section 7.5.3) allows the products formed during heating and decomposition of a compound to be rapidly analysed. This **evolved gas analysis** is particularly helpful in identifying which weight losses are associated with which stage of a decomposition reaction. An example in the study of the decomposition of $CaC_2O_4.H_2O$ is given in Section 7.5.3.

7.2.4.2 Tandem mass spectrometry

The mass spectra of inorganic compounds can be highly complex, especially if the molecule is large, fragments through a variety of routes, and contains numerous elements with isotopes. The analysis of such spectra, as a result, can be highly problematic. One method of obtaining additional information from such spectra, and allowing full analysis and molecular structure resolution, is via **tandem mass spectrometry.** In this technique the fragment ion from one spectrometer is extracted and enters a second spectrometer and mass analyser. This second analyser then obtains a mass analysis of just the initial fragment and its own daughter fragmentation ions. By selecting a series of different fragment ions from the first mass spectrometer a detailed picture of the complex molecule and its fragmentation pathways can be extracted. Tandem mass spectrometry can also be used to obtain and study short-lived gas phase ion species and reaction intermediates. This is achieved by allowing the molecular fragment from the first spectrometer to react with a gas inside the second mass analyser.

7.3 **CHN, sulfur, and oxygen analysis**

Common components of inorganic and organometallic compounds are carbon, hydrogen, nitrogen (CHN), oxygen, and, sometimes, sulfur. The rapid analysis of these elements, in terms of atomic or weight per cent in the compound being analysed, is often a key stage in determining the overall compound composition and, subsequently, its structure. Data on C, H, N, O, and S contents are often combined with mass spectrometry data in determining a compound's formula. If the composition of a compound is known prior to CHN analysis then these analytical data can also be used to determine sample purity by matching the analysis against that expected for a pure phase.

The most common compositional analysis, **CHN analysis,** undertakes the automated determination of carbon, hydrogen, and nitrogen contents. Figure 7.6 shows a schematic of the instrument used. The sample is heated to 900°C in oxygen, and a mixture of carbon dioxide, carbon monoxide, water, nitrogen, and nitrogen oxides is produced. A stream of helium sweeps these gaseous products into a tube furnace at 750°C, where

copper reduces nitrogen oxides to nitrogen and removes excess oxygen; the copper oxide so-formed converts carbon monoxide to carbon dioxide. The resulting mixture of H_2O, CO_2, and N_2 is analysed by passing it through a series of three thermal conductivity detector pairs. The first cell of the first detector pair measures the total conductivity of the gas mixture, water is then removed in a trap and the thermal conductivity measured again. The difference in the two conductivity values corresponds to the amount of water in the gas and, thus, the hydrogen content of the sample. A second detector pair follows a carbon dioxide trap and yields the carbon content. The remaining nitrogen is measured at the third detector. The data obtained from this technique are reported as mass percentages of C, H, and N in the compound under analysis.

Oxygen content can also be analysed if the reaction tube is replaced with a quartz tube filled with carbon that has been coated with catalytic platinum. When the gaseous products are swept through this tube, the oxygen is converted to carbon monoxide, which is then converted to carbon dioxide by passage over hot copper oxide. The rest of the procedure is similar to that for CHN analysis alone and the oxygen content obtained from the augmented amount of CO_2 produced under these conditions. If sulfur is present in the compound its level can be measured if the sample is oxidised in a tube filled with copper oxide. Any water formed is removed by trapping in a cool tube and the amount of sulfur dioxide produced is measured at what would normally be the hydrogen detector in a CHN analyser.

CHN analysis is applicable to most inorganic molecular compounds, which burn to yield CO_2, H_2O, and N_2 quantitatively. For some inorganic solids and materials the combustion temperature may be insufficiently high to produce accurate analytical data. As an example, the C and N analysis of silicon aluminium carbide nitride ceramics would not be possible, as these refractory materials are very slow to react with oxygen at 900°C.

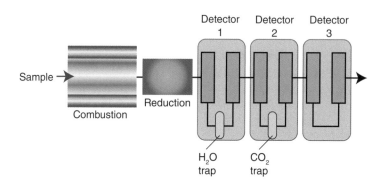

FIGURE 7.6 Schematic of a CHN analyser.

EXAMPLE 7.3

Interpreting CHNS analytical data

Reaction of $((CH_3)_3Si)_2NH$ with sulfur dioxide gave compound **A** whose CHN, sulfur, and oxygen analysis gave the following mass percentages for the elements present: C 21.03; H 7.65; N 8.17; S 18.72; and O 28.02. Determine the empirical formula of the compound, **A**.

ANSWER

The molar masses of C, H, N, S, O, and Si are 12.01, 1.008, 14.01, 32.06, 16.00, and 28.09 g mol^{-1}, respectively. As no other element is present, the residual percentage of silicon in the compound, **A**, is $100 - (21.03 + 7.65 + 8.17 + 18.72 + 28.02) = 16.40\%$. The amounts present are, therefore,

$$n(C) = \frac{(21.03\ \%)/100}{12.01\ \text{g mol}^{-1}} = 0.01751\ \text{mol g}^{-1}$$

$$n(H) = \frac{7.65/100}{1.008\ \text{g mol}^{-1}} = 0.0759\ \text{mol g}^{-1}$$

$$n(N) = \frac{8.17/100}{14.01\ \text{g mol}^{-1}} = 0.00583\ \text{mol g}^{-1}$$

$$n(S) = \frac{18.72/100}{32.06\ \text{g mol}^{-1}} = 0.00584\ \text{mol g}^{-1}$$

$$n(O) = \frac{28.02/100}{16.00\ \text{g mol}^{-1}} = 0.01751\ \text{mol g}^{-1}$$

$$n(Si) = \frac{16.40/100}{28.09\ \text{g mol}^{-1}} = 0.00584\ \text{mol g}^{-1}$$

These amounts are in the ratio 3:13:1:1:3:1 giving the empirical formula of $C_3H_{13}NSO_3Si$ for compound **A**. This compound can be formulated as $[NH_4][(CH_3)_3SiOSO_2]$, ammonium trimethylsilyl sulfite.

SELF TEST

The reaction forming compound **A** also gives compound **B** as a side product. CHNSO analysis of **B** gives 26.65% C, 6.66% H, 10.36% N, 23.71% S, and 11.84% O. Determine an empirical formula for **B**.

7.4 Atomic absorption spectroscopy, flame emission spectroscopy, and flame photometry

Compounds containing metals are important across much of inorganic chemistry but their analysis cannot be undertaken using CHN type instrumentation as they

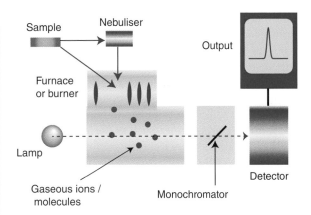

FIGURE 7.7 Schematic of the main elements of an atomic absorption spectrophotometer.

do not form volatile oxides. For the determination of the metal content of a material, X-ray absorption spectroscopy (Chapter 6) can be used, but much more widely employed, and of higher accuracy, are techniques that rely on light absorption and emission by gaseous metal atoms and ions.

7.4.1 Atomic absorption spectroscopy

The principles of **atomic absorption spectroscopy (AAS)**, also called **atomic absorption spectrometry,** are similar to those of ultraviolet–visible spectroscopy (Chapter 5) except that the absorbing species are free atoms, either neutral or charged. Their absorption spectra consist of sharply defined lines of known wavelength. By determining how strongly light is absorbed at these characteristic wavelengths (from a known quantity), the amount of that atom type in the sample can be calculated. Figure 7.7 shows the basic components of an atomic absorption spectrophotometer.

Initially the solution to be analysed needs to be converted to gas-phase species. **Flame atomisation** involves mixing the analyte solution with a fuel, typically acetylene (ethyne), in a 'nebuliser' to create an aerosol. The aerosol enters the burner where it reacts with air to produce flame temperatures of up to 2500 K, or with a nitrous oxide supply, temperatures of up to 3000 K are generated. Alternatively, an **electrothermal atomiser** can be used where the analyte solution or a solid is heated in a graphite furnace to similar temperatures; this thermal atomisation method has much increased sensitivity due to the ability to generate atoms quickly and keep them in the optical path of the spectrometer for longer. Other atomisation techniques include glow-discharge atomisation and, more rarely, hydride atomisation. In **glow-discharge atomisation,** ions of the analyte are produced by sputtering them from a cathode using a beam of argon cations produced in a glow discharge tube. The technique is used mainly for metals, as the cathode needs to be electronically conducting, though it is possible to mix non-conducting analytes with graphite to form a conducting cathode. The technique of **hydride atomisation** is limited to metallic elements that form gaseous hydrides, such as As, Sb, Sn, Pb, and Bi. The hydrides of these elements are formed through the reduction of the analyte solution using $NaBH_4$, before being carried into the AAS atomisation chamber using an inert gas.

In **line source AAS,** the incident light is generated from a lamp whose cathode is constructed from the element requiring analysis, so a different incident light lamp is required for each element that is to be analysed. If the element corresponding to that of the lamp element is present in the sample then the incident beam will be reduced in intensity at the detector because it stimulates an electronic transition in the sample and is partly absorbed. By determining the level of absorption, using the Beer–Lambert law (Chapter 5, Section 5.2), relative to that of a standard material a quantitative measurement can be made of the amount of the element present in the sample. In **continuum source AAS,** the lamp produces light across the wavelength range 190–900 nm and the absorption spectrum is analysed at the different characteristic wavelengths of interest, selected using a monochromator placed before the detector.

Almost every metallic element can be analysed using atomic absorption spectroscopy, although not all with high sensitivity or a usefully low detection limit. For example, the detection limit for Cd in a flame ioniser is 1 part per billion (1 ppb = 1 in 10^9), whereas that for Hg is only 500 ppb. Limits of detection using an electrothermal atomiser can be as low as 1 part in 10^{15} in the analyte. Analysis of some non-metallic species can be undertaken by means of AAS using indirect procedures. For example, phosphorus as $[PO_4]^{3-}$ can be analysed by reacting it quantitatively with $[MoO_4]^{2-}$ in acid conditions to form $H_3PMo_{12}O_{40}$. This complex can be extracted into an organic solvent and analysed for molybdenum (using a Mo cathode lamp), which in turn yields the original quantity of phosphorus. For the most accurate AAS analysis calibration standards prepared in a similar matrix to the sample are used.

Analytical data from AAS are normally reported in parts per million (ppm), or sometimes parts per billion (ppb). Parts per million is the same as milligrams per litre (mg/L), as one milligram of water is 1 ppm in 1 L or 1 kg of water. This can be converted to an atom per cent in the sample knowing the concentration of compound in the analyte solution.

EXAMPLE 7.4

Calculations based on ppm from AAS results

66.77 mg of a copper chloride was dissolved in 250 cm^3 of dilute acid. 10 cm^3 of this solution was then further diluted and made up to 250 cm^3 in a volumetric flask. Analysis of this diluted solution using AAS showed it contained 5.05 ppm Cu. Determine whether the compound was $CuCl_2$ or $CuCl$.

ANSWER

5.05 ppm copper is equivalent to 5.05 mg/L so the 250 cm^3 solution analysed contained 5.05/4 = 1.2625 mg of copper. This amount was also present in the 10 cm^3 that was diluted, so the 250 cm^3 of solution prepared initially contained $1.2625 \times 25 = 31.563$ mg of copper. Therefore, the percentage weight of copper in the copper chloride was 47.3%. This value corresponds to that of copper in $CuCl_2$ ($CuCl$ is 64.2% copper by mass).

SELF TEST

Calculate how much $MgSO_4.6H_2O$ needs to be dissolved in 1 litre of water to make a 10 ppm magnesium standard for AAS analysis.

7.4.2 Flame emission spectroscopy and flame photometry

In AAS the light absorbed by atoms or ions is analysed while in **flame emission spectroscopy** and **flame photometry** the intensity of light given out by ions excited in a high temperature flame is analysed. In essence the techniques are quantified flame tests and work best for species such as the Group 1 and Group 2 metal ions that impart characteristic colours to flames.

In flame photometry the analyte solution of the ions is nebulised and then introduced to a flame at a constant rate. The light given out is passed through a filter (so that only the emission line(s) specific to that element pass) and onto a photomultiplier detector. Once the instrument has been calibrated with solutions of known concentration the intensity of the light detected can be directly related to the concentration of an ion in the analyte solution. As with AAS the values are usually quoted in ppm.

Flame emission spectroscopy, sometimes also called **atomic emission spectroscopy (AES)** is a more sophisticated, and, therefore, more expensive, version of flame photometry where the emitted light is analysed via a monochromator, rather than an element-specific filter. As well as qualitative information, by analysis of the emission spectrum for the presence or absence of characteristic lines of each element, the instrument can be calibrated to give very accurate and highly sensitive analytical information. Over 60 elements can be analysed using this technique, including the alkali and alkaline earth metals and the majority of the transition metals, such as Fe, Mn, and Cu. Detectable limits are below 1 ppb for some Group 1 elements, for example, potassium and sodium, but more typically around 0.1–1 ppm. AES is often combined with ICP ionisation (Section 7.4.3) in ICP-AES, where the sample is broken down into atoms prior to spectral analysis; the ICP-AES technique is of very high sensitivity.

7.4.3 ICP-MS

The very high energy ionisation method of an inductively coupled plasma described in Section 7.2.1.1 causes molecules to break apart into their constituent atoms. These atoms then ionise in the plasma, normally forming singly charged ions (plasma temperatures can be adjusted to minimise a second ionisation). The ions produced under these conditions can then be analysed in a quadrupole mass spectrometer to determine which atomic ions are present (qualitative measurement) and their amounts (quantitative analysis).

The gas phase molecular species entering the ICP-MS instrument can be generated using a nebuliser (for liquid and solution samples) or by laser ablation (for solid samples). Unlike AAS which analyses a single element at a time, ICP-MS has the capability to scan all the elements in the mass spectrum, from lithium to uranium, simultaneously. This allows rapid sample processing.

Compared to AAS and flame emission techniques ICP-MS has much greater speed, precision, and sensitivity. A typical qualitative mass spectrum from a sample can be collected in less than a second and, with appropriate detectors, quantities as low as 1 part in 10^{15} (1 ppq, parts per quintillion) can be measured. Typically measurements of metal concentrations can be obtained over the range 100 ppm to 1 ppq. Because of this great sensitivity the applications of ICP-MS are mainly in trace metal analysis—for example, heavy metals in biological, pharmaceutical, and natural samples. Additional applications include in the electronics industry where very low levels (ppt, parts per trillion) of dopants in silicon semiconductor devices can be obtained by dissolution of the silicon in HF followed by ICP-MS analysis.

ICP-MS is also used widely in geochemistry for the radiometric dating of rocks where the relative abundances of different isotopes, in particular uranium and lead, obtained in the mass spectrum can be linked to the age of the specimen.

7.5 Thermal analysis

Thermal analysis concerns the change in a property of a sample induced by heating. The sample is usually a solid and the changes that occur include melting, phase transitions, sublimation, and decomposition. In terms of chemical analysis, the weight change that occurs during decomposition often provides useful information on the original composition of the compound—for example, water of crystallisation or solvent content.

7.5.1 Thermogravimetric analysis

The analysis of the change in the mass of a sample on heating is known as **thermogravimetric analysis (TGA)**. The measurements are carried out using a thermobalance, which consists of an electronic microbalance (weighing to microgram accuracy), a temperature-programmable furnace, and a computer-driven controller, which enables the sample to be simultaneously heated and weighed (Figure 7.8). Initially the sample is weighed into a sample holder which is then suspended from the balance within the furnace. The temperature of the furnace is usually increased linearly, but more complex heating schemes with, for example, different ramp rates and dwell times at specific temperatures are possible. Isothermal heating, which maintains a constant temperature for an extended period during a slow decomposition process, and cooling protocols can also be used.

The balance is situated within an enclosed system so that the gaseous atmosphere around the sample can be controlled. The gas may be inert or reactive, depending on the nature of the investigation, and can be static or

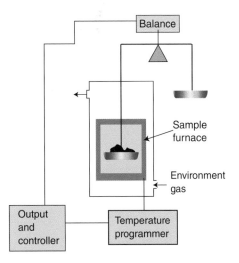

FIGURE 7.8 Schematic of the main elements of a thermogravimetric analysis instrument.

flow over the sample. A flowing gas has the advantage of carrying away any volatile or corrosive species and prevents the condensation of less volatile reaction products on the balance. In addition, any species produced during the experiment can be fed into a mass spectrometer for identification (Section 7.2.4.1). The reaction atmosphere, and whether it is static or flowing, can markedly change the reactions that occur and also the temperatures at which they occur. Thus, heating $(\eta^6\text{-}C_6H_6)Cr(CO)_3$ in CO on a thermogravimetric balance yields $Cr(CO)_6$, while heating the same compound in a benzene vapour produces $(\eta^6\text{-}C_6H_6)_2Cr$. Similarly heating $CaCO_3$ in flowing nitrogen causes complete decomposition to CaO by 750°C, but in flowing CO_2 the temperature has to be raised to 950°C to complete the decomposition.

TGA is most useful for characterising desorption, decomposition, dehydration, and reduction or oxidation processes. The technique can also identify the conditions needed to form of phases that have compositions not directly obtainable through room temperature reactions. As an example, the thermogravimetric curve for $CuSO_4.5H_2O$ from room temperature to 300°C shows three stepwise mass losses (Figure 7.9). These correspond to three stages in the dehydration, to form initially $CuSO_4.3H_2O$, then $CuSO_4.H_2O$, and finally $CuSO_4$. The conditions to synthesise a pure sample of $CuSO_4.3H_2O$ from $CuSO_4.5H_2O$ can be identified from this trace as heating at 100°C.

The percentage weight losses at each stage of the reaction correspond to the molar weight loss. The loss of two water molecules (RMM, 18 g mol^{-1}) from $CuSO_4.5H_2O$ (RMM, 249.69 g mol^{-1}) to give $CuSO_4.3H_2O$ has a weight loss equivalent to $[(2 \times 18)/249.69] \times 100 = 14.4\%$ of the original sample mass. Calculations based on the weight losses that occur during TGA can be used to determine the composition of the starting material.

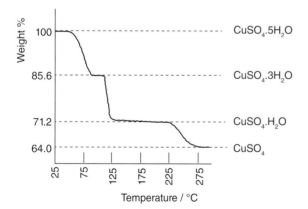

FIGURE 7.9 TGA trace showing the weight losses occurring on heating a sample of $CuSO_4.5H_2O$ to 300°C

EXAMPLE 7.5

Analysis of TGA data to obtain water of crystallisation

Heating a sample of hydrated iron(II) sulfate, $FeSO_4.nH_2O$, to 500°C resulted in a weight loss of 45.36%. Calculate a value for n assuming all the water is lost at 500°C.

ANSWER

The RMM of $FeSO_4.nH_2O$ is $151.90 + 18n$ g mol^{-1} and that of the anhydrous salt $FeSO_4$ 151.90 g mol^{-1}. The weight loss on heating to 500°C will be $18n$ g mol^{-1}. Therefore the percentage weight loss on dehydration is $[18n/(151.90 + 18n)] \times 100$ % which equals 45.36%. Solving this expression for n gives $n = 7$. So, hydrated iron(II) sulfate is $FeSO_4.7H_2O$.

SELF TEST

Heating hydrated iron(II) sulfate further to 900°C results in an additional weight loss to 25.2% of the original mass. Explain this observation in terms of the further decomposition of $FeSO_4$.

7.5.2 Differential thermal analysis and differential scanning calorimetry

In **differential thermal analysis (DTA)** the temperature of a sample is compared to that of an inert reference material while they are both subjected to the same heating procedure. Common inert reference samples for the analysis of inorganic compounds are alumina, Al_2O_3, and carborundum, SiC, neither of which undergoes a phase change or decomposition until very high temperatures. The temperature of the furnace is increased linearly and the difference in temperature between the sample and the reference is plotted against the furnace temperature. If an endothermic event takes place within the sample, the temperature of the sample lags behind that of the reference and a minimum is observed in the DTA curve. If an exothermic event takes place, the temperature of the sample rises above that of the reference and a maximum is observed on the curve. The area under the endotherm or exotherm (the resulting curve in each case) is related to the enthalpy change accompanying the thermal event.

DTA is most useful for the investigation of processes such as phase changes, where one form of a solid changes to another, and there is no weight change observed in the TGA experiment. Examples include the crystallisation of amorphous glasses and the transition from one structure type to another (e.g. the transition of TlI from a rock salt structure type to a CsCl structure type that occurs

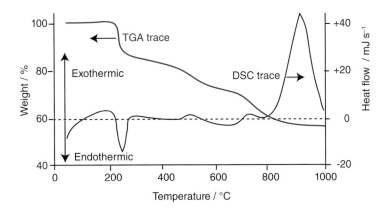

FIGURE 7.10 DSC (black) and TGA (blue) traces from $[Me_3NCMe_2(CH_2OH)][B_5O_6(OH)_4]$ from 25 to 1000°C. An initial endothermic loss of water from the decomposition of the $[B_5O_6(OH)_4]^-$ anion occurs at 200°C and is followed by decomposition of the organic amine cation and exothermic recrystallisation at higher temperatures. (Data from M.A. Beckett, P.N. Horton, M.B. Hursthouse, D.A. Knox, and J.L. Timmis, *Dalton Trans.* **39** 3944 (2010).)

on heating to 175°C). DTA can be combined with TGA by having separate sample and reference material crucibles (in Figure 7.8) on the sample side of the balance. Thermocouples are placed next to the sample and inert reference material and the difference in temperature between them measured; note that as the inert reference material neither gains nor loses weight during heating then the TGA measurement is unaffected.

More widely used these days is a technique closely related to DTA called **differential scanning calorimetry (DSC)**. In DSC, the sample and a reference material are maintained at the same temperature as each other (measured using separate thermocouples) throughout a heating procedure by using separate power supplies to the sample and reference holders. Any difference between the power supplied to the sample and the reference, to maintain them at same temperature, is recorded against the furnace temperature. Thermal events appear as deviations from a level DSC baseline as either endotherms or exotherms, depending on whether more or less power has to be supplied to the sample relative to the reference. Endothermic reactions are usually represented as negative deviations and exothermic events as positive deviations from the baseline (Figure 7.10, black trace).

The information obtained from DTA and DSC is very similar. The former can be used to higher temperatures although the quantitative data, such as the enthalpy of a phase change, obtained from DSC measurements are more reliable. Both DTA and DSC can be used to obtain a 'fingerprint' material in that a pure sample gives a characteristic heating curve. A DSC trace from an unknown sample can be compared with that of a pure reference material to determine sample purity and this leads to applications in quality control. Melting behaviour and exact melting points can be determined with high detail and accuracy in DSC analysis and as these depend on particle size and morphology DSC can provide a method to characterise these crystal properties. For example, crystalline silver metal has a melting point of 960°C, while DSC analysis of silver nanoparticles has shown them to have melting points as low as 100°C.

7.5.3 Combined thermal analysis and mass spectrometry

As noted previously (Section 7.2.4.1), the attachment of a mass spectrometer to the outlet from a DSC or TGA instrument provides a route to the identification of the gaseous decomposition products. Figure 7.11 shows the complete TGA/DSC/mass spectrometry data from the thermal decomposition analysis of calcium oxalate monohydrate $CaC_2O_4.H_2O$ (RMM 146.08 g mol^{-1}) and analysis of these data allow the complete decomposition pathway to be defined.

The initial weight loss near 200°C produces a molecular ion in the mass spectrum of $18m_u$ which corresponds to H_2O and the weight loss ($18/146.08 \times 100 = 12.3\%$) is consistent with one water molecule per compound formula unit. The process is endothermic as energy is input to release the water of crystallisation as gaseous water:

$$CaC_2O_4.H_2O(s) \rightarrow CaC_2O_4(s) + H_2O(g)$$

At 500°C, CO is released ($28m_u$ in the mass spectrum and a further weight loss of 19.2%):

$$CaC_2O_4(s) \rightarrow CaCO_3(s) + CO(g)$$

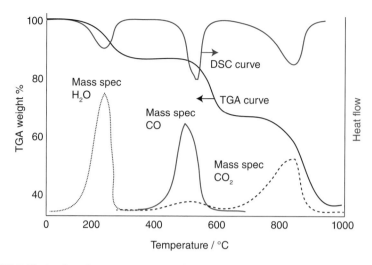

FIGURE 7.11 Combined TGA–DSC–MS analysis trace obtained from heating $CaC_2O_4.H_2O$.

Finally, above 800°C calcium carbonate decomposes losing CO_2 ($44m_u$ in the mass spectrum and a final weight loss of 30.1%) to form CaO.

7.5.4 Melting point apparatus and hot stage microscopes

While the melting point of a compound does not give direct information on its composition or structure it can provide useful characterisation of the sample purity. It also provides a reference point for anyone subsequently synthesising the same compound, so its value is often included in published journal data on new compounds.

In a typical melting point apparatus a few milligrams sample is placed inside a glass capillary tube. This part-filled tube is then placed in a heated metal block alongside a thermometer and the two are observed simultaneously during a gradual heating until melting is observed. For air sensitive samples the capillary can be filled in a nitrogen or argon filled glovebox and the end sealed with a plug of silicone grease before being removed to the melting point apparatus outside the glovebox.

The behaviour of a compound can also be observed during heating or cooling under an optical microscope. These microscopes frequently use transmitted polarised light and analysers. Under this illumination behaviours such as melting, crystallisation, and phase changes become very clear. For example, liquids are isotropic and when viewed between cross polarisers appear dark, while after crystallisation the individual crystal domains can rotate the plane of polarisation showing as bright areas in the optical microscope.

7.5.5 Other thermal analysis methods

Volume changes caused by physical or chemical processes can be measured using a **dilatometer**. These volume changes in a sample can be measured in a variety of ways. For example, the mechanical displacement can be determined using a rod directly mounted on the sample or from the change in capacitance of a pair of parallel plates, where one plate is again mounted directly onto the sample. More accurate measurements use light and either digital photographic imaging of the sample (optical dilatometer) or an interferometer (laser dilatometer).

While dilatometers are mainly employed to measure thermal expansion of materials and determine accurate thermal expansion coefficients, they can also be used to monitor the progress of a chemical reaction particularly where there is a large change in sample volume. This technique is known as **reaction dilatometry**. Examples include the study of polymerisation reactions that use inorganic catalysts, such as the Ziegler–Natta catalytic process that uses $TiCl_4$ or $TiCl_3$ supported on $MgCl_2$ as a catalyst. The polymerisation of propylene to polypropylene, over these catalysts, causes a large decrease in volume and, therefore, pressure in a sealed vessel. Measuring this change in volume or pressure over time enables the polymerisation reaction to be monitored and provide information on the kinetics of the process.

Temperature sweeping **dynamic mechanical analysis (DMA)** involves measuring the applied stress–strain behaviour of a material as a function of temperature. It is frequently used to investigate and determine the **glass transition temperature, T_g,** of a polymer or amorphous compound as the stiffness of a material drops dramatically above T_g.

Problems

7.1 Describe how you would determine the following:

  (a) The temperature at which a porous metal–organic framework (MOF)

    (i) loses solvent molecules from its pores;

    (ii) undergoes a collapse of the framework to form a dense solid.

  (b) The amount of the antibiotic drug cefazolin ($C_{14}H_{14}N_8O_4S_3$) that has been incorporated between the layers of the $Zn(OH)_2$ in the hybrid inorganic–organic composite $Zn(OH)_2{:}C_{14}H_{14}N_8O_4S_3$.

  (c) The liquid product of the reaction of $I_2(s)$ with $Cl_2(g)$.

  (d) The composition of the solid product from the reaction of CH_3NH_3Cl with $PdCl_2$ in aqueous 2 M HCl.

  (e) Accurate values for the zinc and copper content of a phase in the solid solution $Cu_{2-x}Zn_xSnS_4$ with $x \sim 0.95$.

  (f) The nitrogen to oxygen ratio at the surface (to a depth of 5 nm) of a series of silicon oxynitride thin films heated to various temperatures.

7.2 The mass spectrum of a sample of P_4S_{10} contaminated with a small level of oxygen showed a peak at ${\sim}332 m_u$ (D.W. Muenow and J.L. Margrave, *J. Inorg. Nucl. Chem.* **34** 89 (1972)). Which molecular phosphorus oxysulfide species, $P_nS_mO_p$, could give rise to this peak and how could the actual species producing this peak be identified using mass spectrometry?

7.3 How can SIMS be used to study the rate of nitrogen diffusion in silicon nitride at 1000°C?

7.4 A standard solution for analysis of iron using AAS was prepared by dissolving 0.1563 g of iron wire in acid and diluting to 250 cm³. After a further 100 times dilution the solution had a measured absorbance in the atomic absorbance spectrometer at 248.3 nm of 0.843. Calculate the concentration of iron in a natural water sample having an absorbance at 248.3 nm of 0.543.

7.5 Heating a 20 mg sample of black CrO_2 on a thermogravimetric balance to 300°C in $H_2(g)$ led to the formation of a green product with a 1.26% weight gain. Cooling the sample to room temperature and reheating to 300°C in oxygen gas yielded 20 mg of pure CrO_2. Heating a 20 mg sample of black CrO_2 on a thermogravimetric balance to 1000°C in $H_2(g)$ led to a 35.00% weight loss. Interpret these observations.

7.6 The thermal decomposition of metal borohydrides, MBH_4, can proceed to give two different end products. The two possible reactions are

$$MBH_4(s) \rightarrow M(s) + B(s) + 2H_2(g)$$

or

$$MBH_4 \rightarrow MH + B + \tfrac{3}{2}\,H_2(g)$$

  (a) Heating $LiBH_4$ to 800°C on a TGA apparatus results in a 13.9 % weight loss. Determine the reaction pathway.

  (b) Thermal analysis data for KBH_4 and KH are shown in Figure 7.12. Interpret these data.

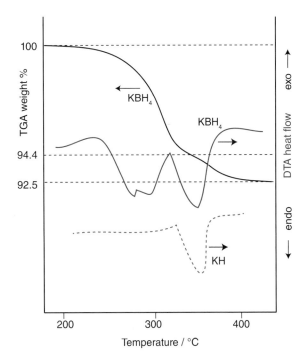

FIGURE 7.12 Thermal analysis data for KBH_4 and KH. (Data from J.A. Dilts and E.C. Ashby, *Inorg. Chem.* **11** 1230 (1992).)

Bibliography

S. Becker. (2007) *Inorganic Mass Spectrometry.* Chichester: Wiley.

M.P. Balogh. (2014) *The Mass Spectrometry Primer (Waters Series).* Wiley-Blackwell.

B. Welz and M. Sperling. (1999) *Atomic Absorption Spectrometry.* 3rd ed. New York: Wiley-VCH.

J.E. Cantle. (1982) *Atomic Absorption Spectrometry (Techniques and Instrumentation in Analytical Chemistry).* Amsterdam: Elsevier Science Ltd.

P.J. Haines (Editor). (2002) *Principles of Thermal Analysis and Calorimetry.* 1st ed. Cambridge: Royal Society of Chemistry.

P.J. Haines. (1995) *Thermal Methods of Analysis: Principles, Applications and Problems.* New York: Springer.

M.E. Brown. (2001) *Introduction to Thermal Analysis: Techniques and Applications.* 2nd ed. New York: Kluwer Academic Publishers.

M.T Weller, T. L. Overton, J.A. Rourke, and F.A. Armstrong. (2014) *Inorganic Chemistry.* 6th ed. Oxford: Oxford University Press.

Magnetism

8.1 Magnetometry

Many transition metal and lanthanide compounds, and a few non-transition metal compounds, have unpaired electrons. This results in an associated magnetic moment which can be studied using magnetometry. Compounds are classified as **diamagnetic** if they are repelled by a magnetic field and **paramagnetic** if they are attracted by a magnetic field. Most main group compounds are diamagnetic as the electrons are all paired, but the presence of unpaired d- and f-electrons in many transition metal and lanthanide compounds makes them paramagnetic. By measuring the size of the magnetic moment generated by the unpaired electrons, the number of these per transition metal centre can be determined. This in turn provides useful characterisation information such as the oxidation state of the metal, its d- or f-electron configuration and, in many cases, the likely coordination environment of the metal. Further analysis of the magnetic moment as a function of temperature often provides additional characterisation information, including greater detail on the energy levels occupied by the electrons and the nature of the interaction between the valence electrons and the ligands. Functional materials which change their electronic configuration with temperature (spin-crossover compounds) and those where the electron spins and associated magnetic moments interact and align throughout the crystal structure (cooperative magnetism) are also characterised in detail using magnetometry.

8.1.1 Magnetic moment instrumentation

The classic way of investigating the magnetic properties of a compound is to measure the weight of a sample with and without an applied magnetic field inside a **Gouy balance** (Figure 8.1a). The sample can be attracted into, or repelled by, the magnetic field depending on whether it contains unpaired electrons (paramagnetic) or not (diamagnetic), thereby changing the measured weight. In a **Gouy balance**

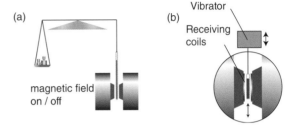

FIGURE 8.1 (a) A Gouy balance and (b) a vibrating sample magnetometer (VSM).

the sample is hung on one side of a balance by a fine thread so that only one end of the sample lies in the field of a strong electromagnet (which can be switched on or off through an applied current); the other end of the sample experiences only the Earth's magnetic field. The sample is weighed with the electromagnet field applied and subsequently with it turned off. From the change in apparent weight, the force acting on the sample, as a result of the application of the field to one part of it, can be determined. From this, with knowledge of various instrumental constants, the sample volume, and the relative molecular mass, the **molar susceptibility** (Section 8.1.2) can be derived.

The molar susceptibility measures how susceptible the sample is to alignment of the magnetic moments (or its magnetisation), derived from the presence of the unpaired electrons, in the magnetic field. Accordingly a sample with a large number of unpaired electrons has a higher susceptibility to alignment or magnetisation than one with fewer unpaired electrons. The effective magnetic moment of a d-metal ion present in a material may, therefore, be deduced from the magnetic molar susceptibility and used to infer the number of unpaired electrons and the spin state (Section 8.2).

In a **Faraday balance** a controllable magnetic field gradient is generated between two curved magnets rather than the simple inhomogeneous field that exists inside the Gouy balance. This yields more accurate susceptibility measurements and also allows the collection of magnetisation data as a function of the magnitude and direction of the applied field. Measurements can be carried out as the sample temperature is varied and the magnetic field can be applied or not during the cooling or heating ramps. This gives rise to 'field-cooled' and 'zero-field cooled' magnetism measurements which may differ, especially if permanent magnetism is induced in a sample, such as in a ferromagnet (see Section 8.3). In the related **Evans balance method** the sample under investigation is placed between a pair of suspended magnets which are torsionally balanced against a second pair of electromagnets. The introduction of the sample causes a deflection in the balance dependent on the magnetic properties of the sample. The balance equilibrium can be restored by supplying a current to a small coil situated between the pair of electromagnets. Measurement of this current allows a rapid measurement of magnetic susceptibility without the need for a high precision weighing balance.

The more modern **vibrating sample magnetometer (VSM)** (Figure 8.1b) is, again, a modified Gouy balance. The sample is placed in a uniform magnetic field, which partially aligns the electron spin directions in the sample and so induces a net magnetisation. As the part-magnetised sample is vibrated, it induces an electrical current in adjacent receiving coils. The signal has the same frequency as the sample vibration and its amplitude is proportional to the level of induced magnetisation. The vibrating sample may be cooled or heated allowing the study of magnetic properties as a function of temperature.

In the research laboratory measurements of magnetic properties are now made more routinely using a **superconducting quantum interference device** (SQUID, Figure 8.2). When the sample is raised and lowered in a magnetic field it produces an alternating current in the pick-up coil The current is determined by the value of the magnetic flux generated and, hence, the magnetic susceptibility of the sample. This is detected by the SQUID, which is a device that makes use of the quantisation of magnetic flux and the property of current loops in superconductors to produce a voltage that is proportional to the magnetisation of the sample. This generated voltage can measure extremely small magnetisations, as low as 10^{-14} T, so SQUID magnetometers may be used to study very weakly magnetic systems and very small samples.

The **Evans NMR method** can be used to determine the magnetic moment of a paramagnetic species in solution and, thus, the number of unpaired electrons. The technique uses the change in the NMR chemical shift of a dissolved species (Chapter 3) that results from adding the paramagnetic species to the solution. This observed shift in resonance frequency, Δv (in Hz), is related to the specific susceptibility, χ_g, of the solute ($cm^3 \, g^{-1}$) through the expression

$$\chi_g = \frac{-3\Delta v}{4\pi v m} + \chi_o + \frac{\chi_o(d_o - d_s)}{m} \qquad \textbf{Eqn 8.1}$$

where v is the spectrometer frequency (Hz), χ_0 the mass susceptibility of the solvent ($cm^3 \, g^{-1}$), m is the mass of substance dissolved per cm^3 of solution, while d_o and d_s are, respectively, the densities of the solvent and solution ($g \, cm^{-3}$). Values derived for the susceptibility of a transition metal species in solution may differ from those obtained for the solid phase using a magnetic balance because of a change in the metal coordination geometry, and, therefore, the number of unpaired electrons, on dissolution.

8.1.2 Initial analysis of experimental magnetisation data

For the simple analysis of magnetisation data from inorganic compounds, the figure of most interest is the **effective magnetic moment**, given the symbol μ_{eff} and, normally, expressed as a number of **Bohr magnetons (BM)**. The Bohr magneton, μ_B, is a fundamental quantity based on the magnetic moment associated with a single electron orbiting a nucleus; $\mu_B = e\hbar/2m_e = 9.274 \times 10^{-24} \, J \, T^{-1}$.

The raw experimental data provide the **susceptibility** of the sample, that is, how easily the sample was magnetised and the electron spins aligned (Figure 8.3). These raw data from the susceptibility measurement are obtained as the volume susceptibility (how much magnetisation was induced in a sample per unit volume) or, more normally, the molar susceptibility, χ_M. The SI units of χ_M are $m^3 mol^{-1}$ though literature data often use the cgs unit of $cm^3 \, mol^{-1}$. We also use cgs units in this text. Note that experimental data often need to be corrected for the presence of paired electrons, such as core electrons, in a compound and the vessel in which it is contained, typically diamagnetic glass. So, for example,

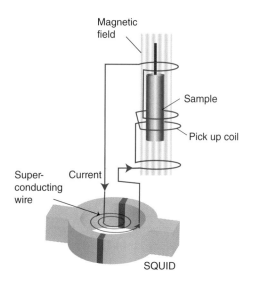

FIGURE 8.2 Schematic diagram of a SQUID magnetometer.

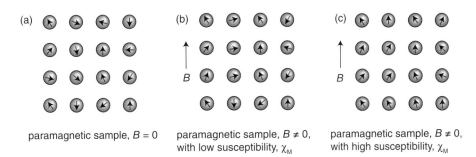

FIGURE 8.3 Illustration of magnetic susceptibility. In a paramagnetic sample with no applied magnetic field (a), the individual atomic magnetic moments are randomly orientated. Application of a magnetic field will cause the individual moments to align parallel to the applied field direction. The degree to which they do this, weakly as in (b) and or strongly as in (c), determines the magnitude of the susceptibility of the sample.

in analysing data from $FeSO_4.7H_2O$ a small diamagnetic correction for the sulfate anion, the core electrons on iron and the water molecules would be made. These diamagnetic corrections, often known as **Pascal's constants** or corrections, are effectively constant for a particular atom, ion, or molecule across all of their compounds. Hence tabulated values, which are additive, may be used to make this correction. Pascal's constants for many common inorganic cations and anions can be found in G.A. Bain and J.F. Berry, *J. Chem. Ed.* **85** 532 (2008).

χ_M can be converted to the effective magnetic moment, μ_{eff}, using the expression

$$\mu_{eff} = \sqrt{\frac{3k_B T \chi_M}{N_A \mu_0 \mu_B^2}}$$ **Eqn 8.2**

where k_B is the Boltzmann constant, T the temperature, μ_0 the vacuum permeability, χ_M the molar susceptibility, N_A the Avogadro constant, and μ_B is one Bohr magneton (9.274×10^{-24} J T^{-1}). Defined in this way, μ_{eff} is a dimensionless number which refers to the effective magnetic moment of a single molecule relative to the single electron value, μ_B. Rearranging Eqn 8.2 gives $\mu_{eff} \times \mu_B =$ constant $\times \sqrt{\chi_M}$, so μ_{eff} is often seen stated as being a number of BM or μ_B and this text keeps to this common terminology. Note also that the terminology 'effective magnetic moment' is strictly incorrect as μ_{eff} is a number, not a magnetic moment. Putting the correct physical constants into Eqn 8.2 gives

$$\mu_{eff} = 2.828 \sqrt{\chi_M T}$$ **Eqn 8.3**

where χ_M has the cgs unit of cm^3 mol^{-1} (the unit usually employed in susceptibility measurements). Using the experimental measurement of the molar susceptibility, χ_M, μ_{eff} is easily obtained at a specific temperature using this expression.

EXAMPLE 8.1

The experimental molar susceptibility, χ_M of $CrCl_3$ was measured as 6350×10^{-6} cm^3 mol^{-1} at 25°C. Calculate μ_{eff} for a Cr^{3+} ion in this compound.

ANSWER

Substituting these values into Eqn. 8.3 gives

$$\mu_{eff} = 2.828 \sqrt{6350 \times 10^{-6} \times 298}$$

$$\mu_{eff} = 3.89 \text{ BM} = 3.89 \mu_B$$

SELF TEST

Assuming the Cr^{3+} ions have the same μ_{eff} as in $CrCl_3$ calculate a molar susceptibility, in cm^3 mol^{-1}, for $Cr_2(SO_4)_3$ bearing in mind that one mole of this compound contains two moles of chromium ions.

χ_M is dependent on temperature and, therefore, experimental measurements as a function of temperature should obey the **Curie law** which arises from rearranging Eqn 8.3 to give

$$\chi_M = \frac{C}{T}$$ **Eqn 8.4**

where C is known as the **Curie constant**. C collects together the various physical constants in Eqn 8.2 and its value depends on μ_{eff} for the compound under investigation. Magnetic susceptibility data are often presented as a function of temperature or inverse temperature (Figure 8.4). Plotting the magnetic susceptibility, χ_M, against temperature gives the simple inverse Curie law

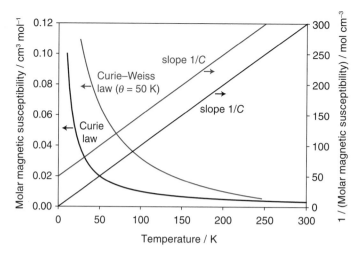

FIGURE 8.4 Variation of magnetic susceptibility (plots referred to the left-hand axis) and inverse susceptibility (plots referred to the right-hand axis) for systems obeying the Curie law (black) and the Curie–Weiss law (blue).

relationship; plotting the inverse susceptibility, $1/\chi_M$, versus temperature, or χ_M against $1/T$, gives a straight line graph whose gradient, $1/C$, gives the Curie constant, and hence μ_{eff}. This variable temperature analysis allows a more accurate determination of the magnetic moment, μ_{eff}, than a single temperature measurement.

This observed variation of χ_M as a function of temperature for a paramagnetic material can be easily understood. In a paramagnetic solid the electron spins on all the different metal centres present in the bulk sample will be orientated in different random directions, as shown in Figure 8.3(a). The magnetic susceptibility is a measure of how easy it is to partially align the electron spins, and their associated magnetic moment, parallel to the applied field, as shown in Figure 8.3(b) and (c). Any thermal energy available will act to randomise the electron spin orientations and, thereby, reduce the magnetic susceptibility. As a paramagnetic material is cooled, the disordering effect of thermal motion is reduced, more spins become aligned in an applied magnetic field, and the magnetic susceptibility increases. The magnetic susceptibility data from many compounds, especially those that display cooperative magnetic phenomena at low temperatures where the alignment of electron spins is no longer random (Section 8.3), often deviate from the Curie law. The materials display higher temperature susceptibility data that follow, instead, the **Curie–Weiss law**,

$$\chi_M = \frac{C}{T - \theta} \qquad \text{Eqn 8.5}$$

where θ may be positive or negative. θ (units in K) is known as the **Weiss constant**. The effect of this on the experimental magnetic susceptibility data is shown in Figure 8.4 and discussed further in Section 8.3 where the value and sign

of θ provide information on how the individual electron spin orientations interact with each other in a solid.

8.2 Interpretation of magnetic moment, μ_{eff}, values

In a free atom or ion that has unpaired electrons both the orbital and the spin angular momenta give rise to a magnetic moment and can contribute to the measured paramagnetism. When the atom or ion is part of a complex the orbital angular momentum can be **quenched**, or partly quenched as a result of the interactions of the electrons with their non-spherical environment. Where quenching of the orbital angular momentum occurs this leads to a simplified model, the spin-only formula, for the measured magnetic moment, μ_{eff}. This simple theory, that only the spin derived magnetic moment is observed, works well for many first row transition metal compounds at room temperature. Furthermore qualitative allowances can be made for whether a small orbital contribution would be expected to slightly modify the measured magnetic moment. A more general model for analysing the total magnetic moment, with both orbital contributions and spin–orbit coupling (Russell–Saunders coupling), for an ion is described in Section 8.2.4; these models are applicable to the heavier transition metals, lanthanide ions, and low temperature measurements.

8.2.1 The spin-only formula

Spin-only paramagnetism, where the orbital angular momentum is considered as being completely quenched, is characteristic of many first row transition metal ($3d^n$)

complexes at room temperature. In this case the spin-only magnetic moment, μ_{eff}, of a complex with total spin quantum number, S, is given by the expression

$$\mu_{eff} = 2\sqrt{S(S+1)} \ \mu_B \qquad \textbf{Eqn 8.6}$$

where μ_B is the Bohr magneton, $S = \frac{1}{2}n$, and n is the number of unpaired electrons each with spin $s = \frac{1}{2}$. This expression is often written as

$$\mu_{eff} = \sqrt{n(n+2)} \ \mu_B \qquad \textbf{Eqn 8.7}$$

Therefore, a measurement of the magnetic moment of a 3d transition metal complex can be used to determine the number of unpaired electrons, n, that it contains. Tables 8.1 and 8.2, columns 1–5, summarise the expected magnetic moment for different d^n electron configurations in the most common transition metal coordination geometries, that is, octahedral and tetrahedral.

Inspection of these data shows that a measurement of a magnetic moment can distinguish between high-spin and low-spin complexes, help determine oxidation states, and, sometimes, provide information on coordination

TABLE 8.1 Octahedral d^n electronic configuration effective magnetic moments

Electronic configuration	Ground state term symbol*	Example transition metal ions	n, S	Spin-only magnetic moment, μ_{eff}/μ_B	Orbital contribution expected?	Typical experimental values, μ_{eff}/μ_B
$d^1 \ t_{2g}^1 e_g^0$	$^2T_{2g}$	$Ti^{3+} \ V^{4+}$	1, ½	1.73	Yes	1.6–1.8
$d^2 \ t_{2g}^2 e_g^0$	$^3T_{1g}$	$V^{3+} \ Cr^{4+}$	2, 1	2.83	Yes	2.7–2.9
$d^3 \ t_{2g}^3 e_g^0$	$^4A_{2g}$	$V^{2+} \ Cr^{3+} \ Mn^{4+}$	3, 3⁄2	3.87	No	3.7–4.0
d^4 high spin $t_{2g}^3 e_g^1$	^5E_g	$Cr^{2+} \ Mn^{3+}$	4, 2	4.90	No	4.7–5.0
d^4 low spin $t_{2g}^4 e_g^0$	$^3T_{1g}$	$Cr^{2+} \ Mn^{3+}$	2, 1	2.83	Yes	3.2–3.3
d^5 high spin $t_{2g}^3 e_g^2$	$^6A_{1g}$	$Mn^{2+} \ Fe^{3+}$	5, 5⁄2	5.92	No	5.7–6.1
d^5 low spin $t_{2g}^5 e_g^0$	$^2T_{2g}$	$Mn^{2+} \ Fe^{3+}$	1, ½	1.73	Yes	1.8–2.5
d^6 high spin $t_{2g}^4 e_g^2$	$^5T_{2g}$	$Fe^{2+} \ Co^{3+}$	4, 2	4.90	Yes	5.1–5.7
d^6 low spin $t_{2g}^6 e_g^0$	$^1A_{1g}$	$Fe^{2+} \ Co^{3+}$	0, 0	0	No	0
d^7 high spin $t_{2g}^5 e_g^2$	$^4T_{1g}$	$Co^{2+} \ Ni^{3+}$	3, 3⁄2	3.87	Yes	4.3–4.7
d^7 low spin $t_{2g}^6 e_g^1$	^2E_g	$Co^{2+} \ Ni^{3+}$	1, ½	1.73	No	1.8–2.0
$d^8 \ t_{2g}^6 e_g^2$	$^3A_{2g}$	Ni^{2+}	2, 1	2.83	No	2.9–3.3
$d^9 \ t_{2g}^6 e_g^3$	^2E_g	Cu^{2+}	1, ½	1.73	No	1.7–2.2

*See Chapter 5, Section 5.13

TABLE 8.2 Tetrahedral d^n electronic configuration effective magnetic moments (high spin-only)

Electronic configuration	Ground state term symbol*	Example transition metal ion(s)	n, S	Spin-only magnetic moment, μ_{eff}/μ_B	Orbital contribution expected?	Typical experimental values, μ_{eff}/μ_B
$d^1 \ e^1 \ t_2^0$	^2E	$V^{4+} \ Cr^{5+} \ Mn^{6+}$	1, ½	1.73	No	1.7–1.8
$d^2 \ e^2 \ t_2^0$	$^3A_2$	$Cr^{4+} \ Mn^{5+} \ Fe^{6+}$	2, 1	2.83	No	2.6–2.8
$d^3 \ e^2 \ t_2^1$	$^4T_1$	Fe^{5+}	3, 3⁄2	3.87	Yes	3.6–3.7
$d^4 \ e^2 \ t_2^2$	$^5T_2$	Fe^{4+}	4, 2	4.92	Yes	4.9
$d^5 \ e^2 \ t_2^3$	$^6A_1$	$Mn^{2+} \ Fe^{3+} \ Co^{4+}$	5, 5⁄2	5.92	No	5.8–6.2
$d^6 \ e^3 \ t_2^3$	^5E	Fe^{2+}	4, 2	4.92	No	5.3–5.5
$d^7 \ e^4 \ t_2^3$	$^4A_2$	Co^{2+}	3, 3⁄2	3.87	No	4.2–4.8
$d^8 \ e^4 \ t_2^4$	$^3T_1$	Ni^{2+}	2, 1	2.83	Yes	3.7–4.8
$d^9 \ e^3 \ t_2^3$	$^2T_2$	Cu^{2+}	1, ½	1.73	Yes	1.8–2.0

*See Chapter 5, Section 5.13

geometry. For instance, the octahedral hexaaqua complex $[Mn(OH_2)_6]^{2+}$ present in $MnSO_4.7H_2O$ is paramagnetic with a measured magnetic moment of $5.9\mu_B$. As shown in Table 8.1, this value is consistent with five unpaired electrons ($n = 5$ and $S = 5/2$) and the high-spin $t_{2g}^3 e_g^2$ configuration. Magnetic measurements on octahedral d^6 complexes, for example, $[Fe(OH_2)_6]^{2+}$ and $[Fe(CN)_6]^{4-}$ in their compounds, easily distinguish between a high-spin $t_{2g}^4 e_g^2$ ($n = 4$, $S = 2$, predicted $\mu_{eff} = 4.90\mu_B$) (experimentally $FeSO_4.7H_2O$, $\mu_{eff} = 5.2\mu_B$) and a low-spin t_{2g}^6 ($n = 0$, $S = 0$, $\mu_{eff} = 0$) configuration (experimentally $K_4Fe(CN)_6$ $\mu_{eff} = 0$).

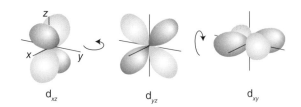

FIGURE 8.5 A set of degenerate d orbitals which allow an electron to rotate in three dimensions through the transformations arrowed and, thereby, generate orbital angular momentum.

EXAMPLE 8.2

Determining the electron configuration from an experimental magnetic moment

The experimental effective magnetic moment of an octahedral Co(II) complex is $3.91\mu_B$. What is its d-electron configuration?

ANSWER

We need to match the possible electron configurations of the complex with the observed magnetic moment. A Co(II) complex will be $3d^7$, In an octahedral crystal field the two possible configurations are $t_{2g}^5 e_g^2$ (high spin, $n = 3$, $S = \frac{3}{2}$) with three unpaired electrons, or $t_{2g}^6 e_g^1$ (low spin, $n = 1$, $S = \frac{1}{2}$) with one unpaired electron. The spin-only magnetic moments can be calculated as $3.87\mu_B$ and $1.73\mu_B$, respectively using the spin-only formula (see also Table 8.1). Therefore, the only assignment consistent with a measured value of $3.91\mu_B$ is the high-spin configuration, $t_{2g}^5 e_g^2$.

SELF TEST

The experimental magnetic moment of the complex $K_4[Mn(NCS)_6]$ is $6.06\mu_B$. What is the electron configuration of the manganese centre in this compound?

8.2.2 Qualitative treatment of orbital contributions and temperature-dependent paramagnetism

The calculated spin-only magnetic moments for 3d metal complexes, listed in Tables 8.1 and 8.2, can be compared with the experimental values, given in the final column of these tables. For most first row transition complexes (and some 4d complexes), the experimental values lie reasonably close to spin-only predictions. Deviations from the spin-only formula result from **orbital contributions** to the magnetic moment and these occur for

some 3d transition metal electron configurations. Orbital contributions are very commonly observed for transition metal complexes of the 4d and 5d blocks. The calculation of the orbital contribution to the magnetic moment and why this is important for the heavier transition metals is covered in more detail in Section 8.2.4. However, some simple predictions as to whether an orbital contribution might be present for a 3d transition metal complex can be made. Specifically, the orbital contribution is not fully quenched when the triply degenerate t_{2g} (octahedral) or t_2 (tetrahedral) electron energy levels contain one or two unpaired electrons (formally T ground states; see Chapter 5, Section 5.13). See also Tables 8.1 and 8.2 for the ground-state electronic term symbol for the most common electron configurations. In these configurations an applied magnetic field can cause the electrons in the t_{2g} (octahedral) or t_2 (tetrahedral) orbitals to circulate around the metal ion. This generates orbital angular momentum and a corresponding orbital contribution to the total magnetic moment (Figure 8.5).

Therefore, configurations such as octahedral d^6 high spin, $t_{2g}^4 e_g^2$, and tetrahedral d^4, $e^2 t_2^2$, are predicted to have an orbital contribution which may add to or subtract from the spin-only value. For less than half-filled electron shells, d^1 to d^4 ions, any orbital contribution is found to be small and would subtract from the spin-only value. However, the coupling between the spin and orbital angular momenta for these ions is found experimentally to almost cancel out the orbital contribution, and the measured moment for these electron configurations is very close to the spin-only value or very slightly less. For more than half-filled shells, d^6 to d^9 ions, the orbital contribution, where it exists, is larger and coupling between the spin and orbital contributions adds more significantly to the overall measured moment. For these electron configurations the deviations from the spin-only value are most obvious with measured values typically $0.5-1.0\mu_B$ higher. The size of the orbital contribution is also temperature dependent (Section 8.2.4 provides a more detailed explanation as to why). If variable temperature magnetic measurements are undertaken and a significant change in the measured value of μ_{eff} is observed then the electronic configuration of the metal can be determined as being a T term including t_2^1, t_2^2, t_2^4, or t_2^5.

EXAMPLE 8.3

Explain why $K_3[Fe(CN)_6]$ has a measured $\mu_{eff} = 2.3\mu_B$, which is temperature dependent, while FeF_3 shows an experimental $\mu_{eff} = 5.93\mu_B$, which is temperature independent.

ANSWER

The value for $K_3[Fe(CN)_6]$, which contains octahedrally coordinated Fe^{3+} d^5 in a low spin $t_{2g}^5 e_g^0$ electron configuration, is significantly higher than the predicted spin-only value for the one unpaired electron ($1.73\mu_B$). In this case there is a significant orbital contribution to the magnetic moment as expected for the one unpaired electron in the t_{2g} orbital set (Table 8.1). The level of orbital contribution will be temperature dependent. FeF_3 contains octahedral high-spin Fe^{3+}, electronic configuration d^5, $t_{2g}^3 e_g^2$, and has five unpaired electrons giving a predicted $\mu_{eff} = 5.92\mu_B$. No orbital contribution is expected for this configuration and the value will be temperature-independent; hence the spin-only prediction and experimental observation are very close.

SELF TEST

Would the spin-only formula be likely to predict the observed magnetic moment for the nickel centre in Cs_2NiF_6, which contains low spin $[NiF_6]^{2-}$ anions?

8.2.3 Spin-crossover complexes and temperature-independent paramagnetism (TIP)

The electronic state of a metal ion can change as a function of an external variable such as temperature or pressure. Most commonly this involves a transformation from a high-spin to low-spin electronic configuration, or *vice versa*, and a resultant change in the measured magnetic moment. Such complexes are referred to as **spin-crossover** complexes. Figure 8.6 shows the measured μ_{eff} of solid $[Fe(phen)_2(NCS)_2]$ (phen = 1,10-phenanthroline) as a function of temperature; the spin crossover from high-spin to low-spin Fe(II) is clearly visible at 175 K. The narrow temperature range over which this transition occurs indicates that the iron centres in each individual molecule transform their spin states cooperatively throughout the whole material.

In a few transition metal compounds, those containing the permanganate ion, $[MnO_4]^-$, are well-documented

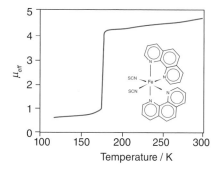

FIGURE 8.6 Variation of μ_{eff} of $Fe(phen)_2(NCS)_2$ as a function of temperature.

examples, a very small paramagnetic moment (typically $<0.1\mu_B$) is measurable. The spin-only formula predicts a zero value as the compound has no unpaired electron ($3d^0$ ground state for $[MnO_4]^-$). Similar behaviours are also seen for some actinide compounds, which also have no unpaired electron, such as those of $[UO_2]^{2+}$, and for low spin octahedral Co^{3+} (d^6) with a t_{2g}^6 electronic configuration. The magnitude of this magnetic moment is independent of sample temperature so is known as **temperature-independent paramagnetism (TIP)**. This paramagnetism derives from mixing in of higher energy electronic levels with the ground state that do provide a small magnetic moment due to their orbital angular momentum and spin–orbit coupling. Because these higher energy levels are not thermally populated and the amount of orbital mixing is independent of temperature the observed paramagnetic moment does not vary with temperature. In the case of $[Co(NH_3)_6]^{3+}$ an excited state with unpaired electrons, derived from the $t_{2g}^5 e_g^1$ configuration, lies 21 000 cm^{-1} above the ground state and mixes with it. As a result a small magnetic moment of $0.12\mu_B$ is seen for this ion.

8.2.4 Orbital contributions to magnetic moments

In the previous sections the orbital contribution to the magnetic moment (and its coupling to that derived from the electron spin) was, to a large extent, either neglected or treated in a highly qualitative manner. While this approach works reasonably well for the lighter elements, such as the 3d transition metals, the theory needs to be adapted and expanded for cases where the orbital contribution and its coupling to the spin value is significant. This includes the heavier, 4d and 5d transition metals and the lanthanides. This more detailed theory of spin–orbit coupling also allows a quantitative approach to the calculation of μ_{eff} values for the 3d transition metals where significant deviations from the spin-only values can occur.

8.2.4.1 Spin–orbit coupling and the van Vleck formula

In Chapter 5 various methods of treating the coupling of the orbital (L) and spin (S) angular momenta were discussed. Most commonly used in chemistry is **Russell–Saunders coupling** where the total angular momentum (J) is obtained by coupling together the possible combinations of L and S to give total angular momenta values of $J(= L+S), J-1, J-2, \ldots$ to $J = |L-S|$. Alternative schemes where the individual electron spin and orbital angular momenta are coupled together first can also be used. We will consider Russell–Saunders coupling further when interpreting magnetism data for lanthanide ions in Section 8.2.4.4. The strength of the spin–orbit coupling determines the separation of the energy levels corresponding to the total angular momentum quantum number $J (= L+S), J-1$, $J-2 \ldots |L-S|$, etc. The energy difference between the J and $J+1$ states is given by the expression

$$\Delta E = (J+1)\lambda \qquad \text{Eqn 8.8}$$

where λ is the many-electron spin–orbit coupling constant with units usually given in cm^{-1} (see Chapters 1 and 5). The use of the wavenumber unit, cm^{-1}, for λ arises because a common route to empirical values for this parameter is from electronic spectroscopy experiments (Chapter 5 Section 5.2), where data are often collected and analysed in cm^{-1} units. Note that spin–orbit coupling constants are sometimes given and tabulated using the one electron spin–coupling constant, ζ, where $\lambda = \pm \zeta/2S$, and the plus sign applies to less than half-filled shells and the minus sign to more than half filled. Amongst other factors the magnitude of λ is related to the atomic number and it ranges from less than 1 cm^{-1} for very light elements (Li ~ 0.2 cm^{-1}) to hundreds or thousands of wavenumbers for periods 5 and 6 of the periodic table (Cs ~ 400 cm^{-1}).

When the spin–orbit coupling constant, λ, is small, the spin and angular momenta can be considered as operating independently. In this case the **van Vleck formula,**

$$\mu_{eff} = \sqrt{4S(S+1)+L(L+1)} \qquad \text{Eqn 8.9}$$

can be applied to derive a value for the magnetic moment. In the free ions of the transition metals, that is, where they are not bound to ligands in a complex which may quench or remove the orbital contribution, μ_{eff} can be obtained from Eqn 8.9. If the orbital angular momentum is completely quenched, as nearly occurs in many first row transition metal complexes, L becomes zero. Putting $L = 0$ into the van Vleck formula produces the spin-only formula (Eqn 8.6, Section 8.2.1). Experimental values of μ_{eff} found for first row transition metal ions therefore normally lie between the spin-only value (where there is no,

or negligible, orbital contribution) and the value given by the van Vleck formula (where both the spin and angular momenta contribute independently but fully to μ_{eff}).

8.2.4.2 Spin–orbit coupling constants and orbital mixing in transition metal complexes

Table 8.3 summarises the values of free ion spin–orbit coupling constants for some transition metals. λ is positive for less than half-filled shells and negative for more than half-filled shells. Note that for half-filled shells, d^5 high spin, $L = 0$ so there is no orbital angular momentum to be coupled to the spin-only value though a value of λ can still be determined. The reported literature values for λ vary widely as there are numerous theoretical models and experimental approaches to deriving these values. However, trends in these values are clear, as well as the change in sign above a half-filled shell; values get larger on traversing the d-block and the much larger values found in 4d (and even larger in 5d) transition metals are evident.

The deviations of experimental values for μ_{eff} from the spin-only values found for the first row transition metals, first described in Section 8.2.2, can now be explained. Furthermore, in respect of compound characterisation, it becomes possible to interpret experimental values to get information about the electronic state and the coordination environment of a metal ion.

First of all we will consider electronic configurations where the orbital contribution in the ground state is completely quenched. In transition metal complexes, spin–orbit coupling allows electronic states with no orbital angular momentum to mix with those of higher energy that do. Thus, configurations which have no direct orbital contribution to the magnetic moment (Tables 8.1 and 8.2), such as octahedral high-spin d^5 (t$_{2g}$^{3}e$_g$^{2}) and d^3 (t$_{2g}$), can still demonstrate a small orbital contribution. The electronic ground states in octahedral fields that do not have a direct orbital contribution, as it is quenched, are the A and E states. For these states it is possible to modify the spin-only expression for μ_{eff} to account for a small orbital contribution derived from orbital mixing to give

$$\mu_{eff} = \mu_{eff} \text{ (spin-only)} \left(1 - \frac{\alpha\lambda}{\Delta_{oct}}\right) \qquad \text{Eqn 8.10}$$

where λ is the spin–orbit coupling constant of the free ion, Δ_{oct} is the crystal field splitting parameter for an octahedral field (Chapter 5, Section 5.5), and α is 4 or 2 for an A or E ground state, respectively. Remembering that λ is small and positive for d^1 to d^4 configurations and larger and negative for d^6 to d^9 configurations it can be seen that only a very small decrease in μ_{eff} (spin-only) is expected for the former but a more significant increase above μ_{eff} (spin-only) expected for the late transition

TABLE 8.3 Representative free ion λ values (cm^{-1}) for some 3d, 4d, and 5d transition metal ions[†]

Group number	4	5	6	7	8	9	10	11
Divalent metal ion	Ti^{2+}	V^{2+}	Cr^{2+}	Mn^{2+}	Fe^{2+}	Co^{2+}	Ni^{2+}	Cu^{2+}
$3d^n$ configuration	d^2	d^3	d^4	d^5	d^6	d^7	d^8	d^9
λ / cm^{-1}	60	57	58	60	−100	−171	−315	−830
Trivalent metal ion	Ti^{3+}	V^{3+}	Cr^{3+}	Mn^{3+}	Fe^{3+}	Co^{3+}	Ni^{3+}	Cu^{3+}
$3d^n$ configuration	d^1	d^2	d^3	d^4	d^5	d^6	d^7	
λ / cm^{-1}	152	105	92	89	92	−145	−235	−445
Metal ion	Zr^{3+}	Nb^{4+}	Mo^{5+}		Ru^{3+}	Rh^{4+}	Pd^{2+}	Ag^{2+}
$4d^n$ configuration	d^1	d^1	d^1		d^5	d^5	d^8	d^9
λ / cm^{-1}	500	750	900		1250	1700	−800	−1800
Metal ion		Ta^{3+}	W^{5+}	Re^{6+}		Ir^{2+}		
$5d^n$ configuration		d^2	d^1	d^1		d^7		
λ / cm^{-1}		700	2700	4200		−3500		

[†] Data mainly adapted from ζ values in B.N. Figgis and M.A. Hitchman. (2000) *Ligand Field Theory and Its Applications*. Wiley-VCH.

metal ions. The experimentally measured μ_{eff} value will be independent of temperature as the level of orbital mixing is also independent of temperature.

EXAMPLE 8.4

The experimental value of the magnetic moment of $NiSO_4.6H_2O$ is $3.20\mu_B$. Show that this value is consistent with Eqn 8.10. Using $\Delta_{oct} = 8550$ cm^{-1} in this complex, determine whether the ground state is an A or E term.

ANSWER

Ni^{2+} is d^8 and in $NiSO_4.6H_2O$ the metal ion is octahedrally coordinated by six water molecules. The spin-only value of μ_{eff} for a d^8, $t_{2g}^6e_g^2$ ion (Table 8.1) is $2.83\mu_B$ and we would only expect an orbital contribution through spin–orbit coupling to an excited T state. We can use the value (free ion) of λ from Table 8.3 and Eqn 8.10 to predict a value for μ_{eff} adjusted for an orbital contribution. For an A state, with $\alpha = 4$, we obtain $\mu_{eff} = 2.83[1 - (4 \times -315/8550)] = 3.25\mu_B$; for an E state, with $\alpha = 2$, the calculated μ_{eff} is $3.04\mu_B$. So the experimental value implies an A state; in fact the ground state of the $t_{2g}^6e_g^2$ ion is $^3A_{2g}$ (see Table 8.1).

SELF TEST

By consideration of the spectrochemical series (Section 5.5.3) and Eqn. 8.10, explain the following measured magnetic moments: $[CoI_4]^{2-}$ $4.77\mu_B$, $[CoBr_4]^{2-}$ $4.65\mu_B$, $[CoCl_4]^{2-}$ $4.59\mu_B$, and $[Co(NCS)_4]^{2-}$ $4.45\mu_B$.

8.2.4.3 Orbital contributions and Kotani plots

For T electronic ground states (see Section 5.13 for their derivation), the orbital angular momentum is not fully quenched (Figure 8.5). As we have seen in previous sections, for these T state systems of first row transition metal compounds, the measured μ_{eff} values often deviate significantly from the spin-only value and show a dependence on temperature. A rigorous analysis of these systems needs to take account of the combined contribution from the spin angular moment, the orbital angular moment, and spin–orbit coupling. The spin–orbit coupling produces energy levels whose separations are frequently of the order kT, and so temperature will have a direct effect on the population of the levels that arise in a magnetic field. This more rigorous analysis leads to complex expressions for μ_{eff}, which are a function of temperature and the spin–orbit coupling constant, λ. As the spin–orbit coupling constants are much larger for the second and third row transition metals, the theory is also applicable to the analysis of experimental magnetic moments of the compounds of these elements. These values also often differ markedly from spin-only values and show a pronounced dependence on temperature.

The theory, which analyses more fully the contributions to μ_{eff}, was first developed by Kotani, and can be applied to different electronic configurations and coordination geometries to obtain expressions for predicted μ_{eff} values. Plotting the derived expressions for μ_{eff} against $k_B T/\lambda$ (k_B is the Boltzmann constant) leads to **Kotani plots**. Figure 8.7 shows a Kotani plot for a d^1 ion in an octahedral field.

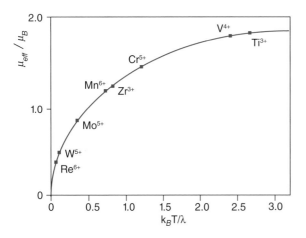

FIGURE 8.7 Kotani plot for an octahedral d^1, t_{2g}^1, configuration. The positions of some ions (at 298 K) with this electronic configuration are shown.

For low charge, early first row transition metal ions the spin–orbit coupling constants are relatively small, <100 cm^{-1} (Table 8.3), compared with $k_BT = 208$ cm^{-1} at room temperature, and values of μ_{eff} are close to the spin-only value ($\mu_{eff} = 1.73\mu_B$). While there is a temperature dependence of μ_{eff} for these ions, they only show a significant deviation from the calculated spin-only value well below room temperature. This observed magnetic behaviour for these ions can be rationalised in a simple picture. All the energy levels that could arise through spin–orbit coupling $J, J - 1, J - 2 \ldots$ and separated by $(J + 1)\lambda$ are occupied at room temperature, as λ is small. However, the orbital angular momentum generated is partially quenched in an octahedral complex (as the d orbitals are no longer degenerate in energy) and when combined with the spin magnetic moment, through spin–orbit coupling, results in only a very small decrease in the spin-only value. For higher charged d^1 species and for second and third row transition metal species λ is much larger and the μ_{eff} values are (for this d^1 less than half-filled shell configuration) normally significantly reduced below the spin-only value. They also show a very strong temperature dependence, as can be seen from the gradient of the curve at low k_BT/λ values in Figure 8.7.

EXAMPLE 8.5

The μ_{eff} values (in BM) obtained from Mo(OH)$_3$(acac)$_2$.3H$_2$O, (acac = acetylacetonate anion) as a function of temperature were as follows: 1.79 (298 K), 1.57 (222 K), 1.45 (186 K), 1.36 (160 K), and 1.26 (135 K) (P.C.H. Mitchell and R.J.P. Williams, *J. Chem. Soc.* 4570 (1962)). Interpret these data and explain how a spin–orbit coupling constant for this complex might be derived.

ANSWER

The complex contains Mo^{5+}, a d^1 electron configuration. We would expect it to have a magnetic moment much reduced through strong spin–orbit coupling and follow the behaviour seen in the Kotani plot (Figure 8.7). The room temperature experimental value is actually much closer to the theoretical spin-only value (1.73μ_B) than might be expected for a second row transition metal and implies that λ is fairly small in this compound (though covalency effects, see the text following this Example, will also moderate the decrease from the spin-only value). On cooling, the value of μ_{eff} decreases, as expected from the Kotani plot. The variable temperature data can be fitted computationally using various values of λ to see which gives the best agreement between theory and experiment; for this compound a value of 250 cm^{-1} was extracted. This is much lower than that given in Table 8.3 for Mo^{5+} (900 cm^{-1}) due to covalency effects.

SELF TEST

Estimate room temperature μ_{eff} values for the low spin, d^4 ions Cr^{2+}, Mn^{3+}, Ru^{4+}, and Re^{3+} by consideration of their spin–orbit coupling constants, λ, (58, 89, 350, and 630 cm^{-1} respectively) and the spin-only μ_{eff} value for a d^4 ion. Assume a similar form for the Kotani plot for a d^4 system to that shown for the d^1 system in Figure 8.7.

For more than half-filled shells, as λ is negative, the Kotani plots, with μ_{eff} plotted against $|k_BT/\lambda|$ as λ is negative, show an increase in μ_{eff} due to spin–orbit coupling. The magnitude of this increase is greater with increasing temperature and the values tend towards that predicted by the van Vleck formula (Eqn 8.9).

This analysis of the orbital contribution to experimentally observed magnetic moment has, so far, assumed that the transition metal species can be treated as an ion, and that the spin–orbit coupling constant taken as that of the free ion. The level of spin–orbit coupling depends on many factors, such as the nature of the orbitals involved (3d, 4d, 5d, 4f, etc.) and the degree of covalency in the bonding to the ligands. The e_g orbitals overlap with the ligand orbitals so should not be considered as purely d orbitals, as in the free ion, but they have a contribution from the ligand, normally p orbitals, mixed in. One way of interpreting magnetism data from these systems is to use a modified value for the spin–orbit coupling constant that is reduced from the free ion value to allow for these effects. Such modified spin–orbit coupling constants, typically around 0.7 times the free ion value, λ, are estimates often derived semi-empirically from fitting experimental magnetic susceptibility data. To some degree they can be considered a way of producing a fit between theory and experiment to

TABLE 8.4 Common lanthanide ion electronic configurations and derived μ_{eff} values.

Ion(s)	Electronic configuration	Ground state	Theory (0 K)	μ_{eff}/μ_B Experimental (298 K)
La^{3+}, Ce^{4+}	$4f^0$	$^1S_0$	0	0
Ce^{3+}, Pr^{4+}	$4f^1$	$^2F_{5/2}$	2.54	2.46–2.53
Pr^{3+}	$4f^2$	$^3H_4$	3.58	3.47–3.61
Nd^{3+}	$4f^3$	$^4I_{9/2}$	3.62	3.44–3.65
Pm^{3+}	$4f^4$	$^5I_4$	2.68	-
Sm^{3+}	$4f^5$	$^6H_{5/2}$	0.85 (1.55–1.65)*	1.54–1.65*
Eu^{3+}, Sm^{2+}	$4f^6$	$^7F_0$	0 (3.40–3.51)*	3.32–3.54*
Gd^{3+}, Tb^{4+}, Eu^{2+}	$4f^7$	$^8S_{7/2}$	7.94	7.9–8.0
Tb^{3+}	$4f^8$	$^7F_6$	9.72	9.69–9.81
Dy^{3+}	$4f^9$	$^6H_{15/2}$	10.65	10.0–10.6
Ho^{3+}	$4f^{10}$	$^5I_8$	10.61	10.4–10.7
Er^{3+}	$4f^{11}$	$^4I_{15/2}$	9.58	9.4–9.5
Tm^{3+}	$4f^{12}$	$^3H_6$	7.56	7.0–7.5
Yb^{3+}	$4f^{13}$	$^2F_{7/2}$	4.54	4.0–4.5
Lu^{3+}, Yb^{2+}	$4f^{14}$	$^1S_0$	0	0

*The calculated values are at 0 K for the ground state ions but for these ions a significant population of the higher J levels occurs at room temperature. This increases the experimental values and can be accounted for through a full theoretical treatment.

account for the effect of orbital mixing in the bond to the ligand. Modified spin–orbit coupling constants have the effect of flattening the curvature of Kotani plots and decreasing the reduction from the spin-only value, as we have seen in Example 8.5.

8.2.4.4 Russell–Saunders coupling and magnetism in lanthanide compounds

Russell–Saunders coupling is applicable to free ions and can be applied successfully to the magnetic properties of the majority of lanthanide ions derived from their partially filled f orbitals. In these species the interaction and overlap of the f orbitals with ligand orbitals is very weak and their role in modifying the spin–orbit coupling is very small, so they behave like the free ions. In Russell–Saunders coupling, the spin angular momentum of the 4f orbital electrons couples strongly with the orbital angular momentum. The magnetic moment μ_{eff} can, therefore, be expressed in terms of the total angular momentum quantum number, J:

$$\mu_{\text{eff}} = g_j \sqrt{J(J+1)}\,\mu_B \qquad \text{Eqn 8.11}$$

where the **Landé g_j factor** is

$$g_j = 1 + \frac{S(S+1) - L(L+1) + J(J+1)}{2J(J+1)} \qquad \text{Eqn 8.12}$$

and μ_B is the Bohr magneton.

Theoretical values for the magnetic moment of the ground states of the common Ln^{n+} ions are summarised in Table 8.4. In general these calculated values agree very well with experimental data. Therefore, an experimental measurement of the μ_{eff} value for a lanthanide compound can be used to assign an oxidation state to the lanthanide ion.

EXAMPLE 8.6

Heating Pr_2O_3 with BaO in a stream of oxygen gave a compound whose room temperature magnetic moment measured $\mu_{\text{eff}} = 2.50\mu_B$. Show that this value is consistent with the composition $BaPrO_3$.

ANSWER

$BaPrO_3$ should contain the Pr^{4+} ion which has a $4f^1$ configuration. The ground state level for this ion is $^2F_{5/2}$ with $S = \frac{1}{2}$, $L = 3$, and $J = \frac{5}{2}$:

$$g_j = 1 + \frac{\frac{1}{2}(\frac{1}{2}+1) - 3(3+1) + \frac{5}{2}(\frac{5}{2}+1)}{2 \times \frac{5}{2}(\frac{5}{2}+1)}$$

$$= 1 + \frac{3 - 48 + 35}{70} = \frac{6}{7}$$

Therefore, the predicted μ_{eff} is given by

$$\mu_{\text{eff}} = g_j \{J(J+1)\}^{1/2} \mu_B = \frac{6}{7}\left\{\frac{5}{2}(\frac{5}{2}+1)\right\}^{1/2} \mu_B = 2.54\mu_B$$

The experimental value is consistent with the Pr^{4+} oxidation state expected for $BaPrO_3$ (rather than that of the more common Pr^{3+} oxidation state; see the Self test).

SELF TEST

The measured magnetic moment of $Pr_2(SO_4)_3$ is $3.50\mu_B$. Show that this value is consistent with the presence of the Pr^{3+} ion (f^2, ground state level $^3H_4$, with $L = 5$, $S = 1$, and $J = 4$) in this compound.

Equation 8.11 can also be used to calculate μ_{eff} values for compounds of the heavier actinides where the 5f electrons are core, or free ion, like. Thus the measured μ_{eff} for Bk^{4+} in BkO_2 of $7.66\mu_B$ is in reasonably good agreement with the predicted value of $7.94\mu_B$ for an f^7 ion (Table 8.4).

8.3 Cooperative magnetism

In the solid state the individual magnetic centres are often close together and direct atomic neighbours, as in iron metal, or separated by only a single atom, typically oxygen in oxides. In such arrays cooperative properties can arise from interactions between electron spins and thus the magnetic moments on different atoms. Magnetic effects arising from cooperative phenomena can be very much larger than those arising from individual atoms and ions. The susceptibility and its variation with temperature in various material types that show **cooperative magnetism** are summarised in Figure 8.8.

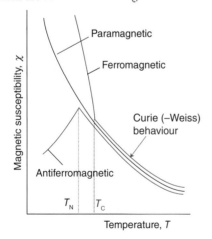

FIGURE 8.8 Magnetic susceptibility behaviours of compounds that show purely paramagnetic behaviour, order antiferromagnetically at T_N and order ferromagnetically at T_C.

At high enough temperatures all materials become paramagnetic with random orientation of the electron spins (Figure 8.9a) and show Curie or Curie–Weiss behaviour as a function of temperature, as discussed in Section 8.1.2 and shown in Figure 8.4. **Ferromagnetism** is exhibited by materials containing unpaired electrons in d or, more rarely, f orbitals that couple and align with unpaired electrons in similar orbitals on surrounding atoms. This coupling occurs below a temperature known as the **Curie temperature**, T_C. In a **ferromagnet,** the coupling between the individual electron spins results in the orientation of all the associated magnetic moments in the same direction (Figure 8.9b). The magnetic susceptibility of a ferromagnet is very large because it is easy to align the whole group or domain of magnetic moments if an external magnetic field is applied. If the sample is maintained below the Curie temperature (T_C) the magnetisation persists after the applied field is removed because the spins are locked together. At temperatures above T_C the disordering effect of thermal motion overcomes the ordering effect of the interaction and the material becomes paramagnetic. At these higher temperatures the material then shows the typical variation of the susceptibility with temperature following the Curie–Weiss law with θ positive in Eqn 8.5 (Figure 8.9b). The internal strong magnetic field, often greater than 10 T, that results from ferromagnetism has a marked effect on the nuclear energy levels, which can be probed using Mössbauer spectroscopy (Chapter 10, Section 10.1.3.3).

In an **antiferromagnetic** material, below the **Néel temperature,** T_N, neighbouring spins are locked into an antiparallel alignment (Figure 8.9c). The resulting collection of individual magnetic moments cancels out and the sample has a low magnetic moment and magnetic susceptibility (tending, in fact, to zero). Above T_N the magnetic susceptibility is that of a paramagnetic material, though the behaviour follows the Curie–Weiss law with θ negative in Eqn 8.5. In a **ferrimagnet** (Figure 8.9d), the individual spins on neighbouring atoms align in specific directions though these may not be all perfectly parallel or antiparallel and they may be of different magnitudes or they may be perfectly antiparallel but of different magnitudes; the overall effect is to produce a magnetic moment.

The magnetic structure of a ferromagnetic or antiferromagnetic material can be investigated using neutron diffraction (Section 2.16.3), as neutrons are scattered by the orientationally ordered electron spins. Where these spins are aligned and have long range order this scattering gives rise to magnetic diffraction peaks. Analysis of the d-spacings and intensities of these magnetic peaks allows determination of the magnetic structure of a material—a three-dimensional picture of the spin orientations.

FIGURE 8.9 Schematic of spin and magnetic moment orientations on neighbouring metal centres in (a) a paramagnetic compound, (b) a ferromagnet, (c) an antiferromagnet, and (d) a ferrimagnet.

EXAMPLE 8.7

Figure 8.10 shows the inverse molar magnetic susceptibility of an octahedral Co^{2+} complex with a molar volume of 385 g cm^{-3}. What information on the magnetic properties of the compound could be extracted from these data?

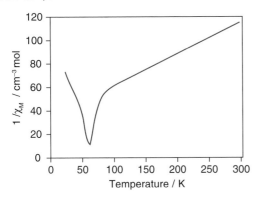

FIGURE 8.10 The inverse molar magnetic susceptibility of an octahedral Co^{2+} complex.

ANSWER

The data follow the Curie–Weiss law in the high temperature region between 75 K and 300 K. The room temperature value of $1/\chi_M$ or the gradient of this line can be used to determine μ_{eff}. Using the room temperature (298 K) value of $1/\chi_M$ (108 cm^{-3} mol) and Eqn 8.3, a value for μ_{eff} of $4.69\mu_B$ is obtained. The calculated spin-only magnetic moment for high-spin octahedral Co^{2+} d^7 $t_{2g}^5 e_g^2$ is $3.87\mu_B$ and the experimental value of $4.69\mu_B$ is consistent with the expected significant orbital contribution for the T state arising from the t_{2g}^5 electronic configuration. Extending the 75 K and 300 K data to the intercept on the temperature axis shows it follows Curie–Weiss behaviour and gives a value for the Weiss constant of around $\theta = -50$ K. The significant dip in the inverse susceptibility is associated with antiferromagnetic ordering and the minimum value occurs at the Néel temperature of 60 K. This antiferromagnetic ordering behaviour of the Co^{2+} centres is in accordance with the negative Weiss constant.

Problems

Problems 8.1–8.4 rely on the application of the spin-only formula and knowledge of coordination chemistry, geometry of complexes and crystal field theory, and the spectrochemical series. Later problems require the application of concepts from Section 8.2.4 onwards.

8.1 Predict spin-only magnetic moments for the following species:

(a) High-spin tetrahedral $[FeCl_4]^{2-}$

(b) Octahedral V^{3+} in $(NH_4)V(SO_4)_2.12H_2O$

(c) Low-spin octahedral Fe(III) in $K_3[Fe(CN)_6]$

(d) $[Fe(CN)_6]^{4-}$

(e) $[Ru(NH_3)_6]^{3+}$

(f) $[Cr(NH_3)_6]^{2+}$

8.2 The experimental magnetic moment of $[NiCl_4]^{2-}$ in a wide range of salts has been measured as $2.85\mu_B$. Use this information to determine whether the $[NiCl_4]^{2-}$ species is tetrahedral or square planar.

8.3 The measured magnetic moment of $[Co(tripyridylamine)_2]$ $(ClO_4)_2$ is $3.82\mu_B$ at 373 K and $1.75\mu_B$ at 4 K. Tripyridylamine is a neutral tridentate ligand. Explain how a change in the electron configuration on the cobalt centre at 200 K can explain these data.

8.4 A student mixes up samples of (a) $MnSO_4.7H_2O$ and (b) $CoSO_4.7H_2O$, both of which are pink in colour. How could magnetism measurements help determine which sample is which?

8.5 Explain the following experimental magnetic moments by considering any orbital contributions to the theoretical magnetic moments and their likely temperature dependence.

(a) VCl_4 μ_{eff} (77K) = $1.62\mu_B$, μ_{eff} (300 K) = $1.62\mu_B$

(b) $Cs_2[CoCl_4]$ μ_{eff} (77K) = $4.54\mu_B$, μ_{eff} (300 K) = $4.62\mu_B$

(c) $K_2[VCl_6]$ μ_{eff} (77K) = $1.42\mu_B$, μ_{eff} (300 K) = $1.83\mu_B$

8.6 Calculate a theoretical magnetic moment for the Bk^{3+} ion. Is this in agreement with the experimental value of $9.69\mu_B$ found for berkelium compounds?

8.7 Interpret the following observations on the magnetic susceptibility data of the cobalt chalcogenides CoS_2 and $CoSe_2$ in terms of cooperative magnetism phenomena.

For both CoS_2 and $CoSe_2$ the inverse susceptibility data obey the Curie–Weiss law above room temperature.

$CoSe_2$ shows a minimum in a plot of χ_M^{-1} versus temperature at 90 K and the Weiss constant determined from data above this temperature is −160 K.

CoS_2 shows a rapid increase in χ_M below 124 K and the Weiss constant is +220 K.

Bibliography

F.E. Mabbs and D.J. Machin. (2008) *Magnetism and Transition Metal Complexes*. New York: Dover Publications.

S.F.A. Kettle. (1969) *Coordination Compounds: Studies in Modern Chemistry*. London: Nelson.

J. Ribas Gispert. (2008) *Coordination Chemistry*. Weinheim: Wiley-VCH.

C. Benelli and D. Gatteschi. (2015) *Introduction to Molecular Magnetism: From Transition Metals to Lanthanides*. Weinheim: Wiley-VCH.

M.T Weller, T.L. Overton, J. A.Rourke, and F.A. Armstrong. (2014) *Inorganic Chemistry*. 6th ed. Oxford: Oxford University Press.

Electron paramagnetic resonance (EPR) spectroscopy

9.1 Introduction to EPR spectroscopy

Electron paramagnetic resonance (**EPR**) spectroscopy is a useful technique for characterising compounds that contain unpaired electrons (paramagnetic species), such as materials containing the d- and f-block elements and main-group compounds that exist as radicals. It provides molecular structural information about the species present, and can give a very detailed insight into the electronic structure. Just as the nuclear spin energy states can be split by a Zeeman interaction with an external magnetic field and studied using nuclear magnetic resonance (NMR) spectroscopy, so can the energies of the spin states of unpaired electrons in a paramagnetic sample. Transitions between these energy states can be observed using resonant absorption of the appropriate energy electromagnetic radiation; this is in the GHz or microwave region of the electromagnetic spectrum when using typical laboratory magnetic fields. This technique is called electron paramagnetic resonance (EPR) spectroscopy, and is also known as, **electron spin resonance** (**ESR**), and, more rarely, **electron magnetic resonance** (**EMR**) spectroscopy.

9.2 Experimental considerations for EPR spectroscopy

The basic requirements of an EPR spectrometer, which are shown in Figure 9.1, consist of a magnet providing a uniform field and a microwave radiation source.

The microwave source is usually connected to the sample cavity and the detector by waveguides. Microwave sources can only be tuned over a relatively limited frequency range so the frequency is kept constant and the magnetic field is varied by changing the current in the electromagnet, and this is known as continuous wave EPR (CW-EPR). While pulsed (Fourier transform, FT)

methods are becoming more common in EPR spectroscopy, they have not achieved the dominance that FT-NMR has. As CW-EPR and FT-EPR experiments provide different information, both techniques will continue to be required to fully characterise inorganic compounds. The detector records the level of microwave radiation absorbed by the sample by measuring the intensity of the signal that is reflected back to it as the magnetic field is varied. This employs phase sensitive detection and one consequence of this is that the output is usually observed not as an absorption spectrum but as its first derivative (Figure 9.2).

This usual format for EPR spectra makes it easier to identify the exact absorption frequency or magnetic field with greater accuracy (Figure 9.2). This is because the steepest gradient is readily identified in the first derivative curve while the exact top of an absorption maximum can be hard to identify for the complex and asymmetric peak shapes found in CW-EPR spectra.

The sample can be studied as either a solution or as a solid, but it must be dilute in terms of the proximity of unpaired electrons to each other, that is, they must be '**magnetically dilute**'. This avoids interaction of the electron spins between molecules that broaden the absorption

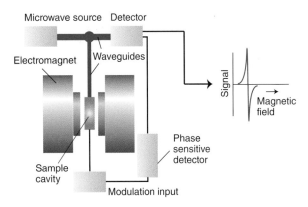

FIGURE 9.1 Schematic diagram of a CW-EPR spectrometer showing the main instrument components.

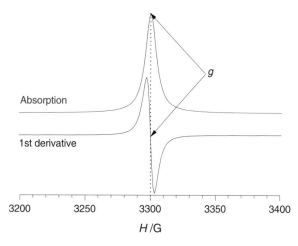

FIGURE 9.2 Typical first derivative EPR spectrum and its associated absorption spectrum. The peak position is marked with a dotted line and corresponds to the point where the first derivative crosses the baseline.

calibrated x-axis, but in the modern literature the x-axis is given in terms of magnetic field units of either G (gauss) or mT (millitesla).

While it is in principle possible to observe EPR spectra for radicals containing any number of unpaired electrons, in practice, the spectra of samples containing one unpaired electron are fairly easy to obtain and to interpret, those with an odd number of unpaired electrons ($S = \frac{3}{2}, \frac{5}{2}$, etc.) are easy to measure, but can be very challenging to understand, and those with an even number of unpaired electron ($S = 1, 2, 3$, etc.) may not even be observed.

It should be noted that EPR is a very sensitive way of detecting radical impurities either in the instrument, glassware (where iron ions can be present so high purity quartz tubes are used) or the samples themselves. Therefore, care is needed in sample preparation and spectral interpretation to avoid potential paramagnetic contaminants.

features. Solutions can be diluted appropriately to avoid any intermolecular interactions. For solid samples one method of diluting the systems is to crystallise a small amount of the material in a diamagnetic host that is ideally isostructural with the paramagnetic compound of interest. This then dilutes and isolates the molecules to be investigated. For example, many Cu(II) solid-state EPR spectra are obtained by doping the equivalent Zn(II) complexes in solution with a small level of the copper compound, molecules of which are then isolated in the solid when the compound is crystallised.

Fluid solutions yield some structural information, but much more detail (akin to that from diluted solids) can be obtained from frozen solutions wherein the orientations of the molecules with respect to the applied magnetic field are fixed. Even more information can be obtained from single crystal samples, in which the molecules have a fixed and known orientation relative to the applied magnetic field, but these are challenging experiments.

Although the microwave frequency can be determined accurately, the magnetic field at the sample is the largest source of uncertainty and EPR spectra need to be calibrated using a standard, stable, compound that has an unpaired electron, such as a free radical. The key parameter extracted from the analysis of EPR spectra is known as the g-value and these are well determined for the calibration standards. The most common one is DPPH (α,α′-diphenyl-β-picrylhydrazyl), whose g-value is 2.0036 ± 0.002, while more recently BDPA (1,3-bis-diphenylene-2-phenylallyl), with a g-value of 2.00264, has been used. The older literature often only displays a calibration marker (a g-value of either 2.0023 or 2.0036) together with a scale bar of the magnetic field with no

9.3 Information available from EPR spectra

There are two pieces of information within EPR spectra, the **g-value** or **g-factor**, and the **hyperfine coupling, A**, both of which are used to characterise inorganic compounds. The g-value is the measured peak position in the spectrum, and the hyperfine coupling constant, A, is the separation of the peaks within the multiplets. While this sounds analogous to chemical shifts and coupling constants in NMR spectroscopy, in EPR spectroscopy these are properties of the radical or molecule as a whole, because the unpaired electron can in principle be delocalised over the entire molecule. Therefore, only one coupled multiplet is observed for each compound, rather than a multiplet for each chemically equivalent nucleus. Because of the large number of potential couplings present in a large molecule the experimentally observed multiplet can be very complex. The g-value and the hyperfine coupling are both described mathematically by the **spin Hamiltonian**. This analysis is very complex and its discussion lies beyond the scope of this book; for details see the bibliography.

9.3.1 The g-value

The g-value of the free electron, g_e, is 2.002319. This is a fundamental constant related to the magnetic moment of the electron and is known to a very high level of precision. In a compound an unpaired electron also has a g-value that often varies from g_e providing structural information on the environment of that electron.

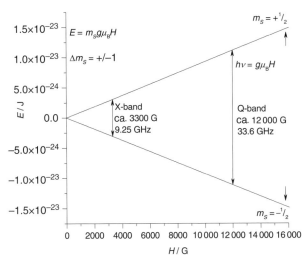

FIGURE 9.3 Splitting of the $m_s = \pm\frac{1}{2}$ levels in an applied magnetic field H.

The experimental g-value is determined from the spectra using the following relationship:

$$g = \frac{h\nu}{\mu_B H} = 714.46 \times \frac{\nu\ (\text{GHz})}{H\ (\text{G})} \qquad \textbf{Eqn 9.3}$$

Note that if the magnetic field units are in T, then the constant in Eqn 9.3 should be 0.071446, and 71.446 if mT are used. g is measured as the centre position of any multiplet, and its value is reported in dimensionless units.

The common frequency or field combinations for $g = 2$ are: X-band (ca. 9.2 GHz, 3300 G); Q-band (ca. 34 GHz, 12000 G); S-band (ca. 4 GHz, 1400 G); and W-band (ca. 94 GHz, 33000 G). X-, Q-, and S-band experiments make use of electromagnets, whereas W-band instruments require a superconducting magnet. For an X-band instrument operating at 9.2500 GHz, this gives a magnetic field of 3300.6 G for $g = 2.0023$.

Figure 9.3 shows the energy levels and transition processes involved in EPR spectroscopy for a single unpaired electron. The electron spin states, $m_s = +\frac{1}{2}$ and $m_s = -\frac{1}{2}$, are split by a Zeeman interaction in a magnetic field, and the electromagnetic radiation then flips the electron spin between the two states.

Figure 9.3 also shows the resonance conditions for the two common microwave frequencies used in EPR spectroscopy. X-band is the marine radar frequency of ca. 9.25 GHz (3.24 cm, 0.309 cm^{-1}) and Q-band at ca. 34 GHz (0.89 cm, 1.1 cm^{-1}) is the airport radar frequency. These are much higher than the frequencies used for NMR experiments (<1 GHz), but the magnetic fields are much smaller because the electron magnetic moment is much larger than the nuclear moment and hence at the same magnetic field the energy gap between the m_s levels is greater for an electron spin than for a nuclear spin. The magnetic fields for X-band experiments are ca. 3300 G (0.33 T) and ca. 12000 G (1.2 T) for Q-band. Although the SI unit for magnetic field, H, is T (tesla), the use of gauss (G) is so widespread in the EPR literature, that it is the unit employed here. 1 T = 10000 G, 10 G = 1 mT.

The energy of the two states is given by

$$E = m_s g \mu_B H \qquad \textbf{Eqn 9.1}$$

where H is the applied magnetic field (in G), μ_B is the Bohr magneton (9.2740 $\times$ 10^{-24} J T^{-1}, 4.6686 $\times$ 10^{-5} cm^{-1} G^{-1}), and g is the measurable g-value and relates the magnetic moment of an electron with its spin in the compound under investigation. As the selection rule is $\Delta m_s = \pm 1$, the energy separation is given by

$$h\nu = g\mu_B H \qquad \textbf{Eqn 9.2}$$

EXAMPLE 9.1

X- and Q-band magnetic fields

What is the magnetic field resonance condition for $g = 2.0023$ and a Q-band instrument operating at 33.600 GHz?

ANSWER

Substituting the appropriate values into Eqn 9.3:

$$2.0023 = 714.46 \times \frac{\nu\ (\text{GHz})}{H\ (\text{G})} = 714.46 \times \frac{33.600}{H}$$

giving $H = 11\,989$ G (1.1989 T)

SELF TEST

Calculate the separation, in magnetic field units of G, of two EPR absorptions with g-values of 2.0023 and 2.0071 using (i) an X-band instrument operating at 9.2500 GHz and (ii) a Q-band instrument operating at 33.600 GHz. Which instrument provides the greater resolution?

9.3.1.1 g-values of main group compounds

The experimental value of g obtained from compounds is often not the same as the free-electron g_e value (2.002319) due to the contribution from the orbital angular momentum via **spin–orbit coupling**, which mixes excited electronic states into the ground state. In the vast majority of organic radicals and non-transition metal inorganic examples, g is very close to 2.00 because in these cases the electron spin angular momentum is the main, if not only, contributor. Small differences in values of *g* are measurable and provide a means of fingerprinting the sample under investigation. The *g*-shift, Δg, can be represented as

$$\Delta g = g - g_e \qquad \text{Eqn 9.4}$$

and

$$g = g_e - \frac{a^2 n \lambda}{\Delta E} \qquad \Delta g = -\frac{a^2 n \lambda}{\Delta E} \qquad \text{Eqn 9.5}$$

where a^2 is a covalency parameter ($a^2 \leq 1$; 1 implies no covalency), *n* defines the extent of orbital mixing, and λ is the spin–orbit coupling constant ($\lambda = \pm \zeta/2S$, where ζ is the single-electron spin–orbit coupling constant). λ is positive for less than half-filled orbitals, and negative for more than half-filled orbitals (Hund's third rule; see Chapter 5, Section 5.10.1.3). ΔE is related to the energy gap between the SOMO (singly occupied molecular orbital) and LUMOs (lowest unoccupied molecular orbitals), which introduces orbital angular momentum via spin–orbit coupling. Essentially if the SOMO and excited state are related to each other by a rotation, this results in angular momentum about the rotation axis. Larger *g*-values than g_e (positive Δg) have resonances at lower magnetic field, whereas *g*-values smaller than g_e (negative Δg) have resonances at higher magnetic field.

As can be seen from Table 9.1, the ζ and hence λ values for the second period elements (Li to F) are very small, and this combined with the large ΔE values means that Δg is expected to be small for B, C, N, and O based radicals. For hydrocarbon-based organic radicals, $g \approx$ 2.002 to 2.003; for N/O-based radicals, $g \approx$ 2.003 to 2.006 as, although small, λ is now negative. For sulfur-based radicals, the *g*-shift is larger due to an increase in spin–orbit coupling and $g \approx$ 2.007 to 2.010. As the elements get

heavier, the spin–orbit coupling increases, and Δg will also increase. For example, the *g*-values for the Group 2 monohydride radicals decrease from 2.0021 for •BeH, through 2.0014 for •MgH, 1.9982 for •CaH, 1.9911 for •SrH, to 1.9825 for •BaH.

EXAMPLE 9.2

The *g*-value for the •KrF radical (formed from γ irradiation of KrF_2 at 77 K) is 2.0450. Predict the *g*-value for •XeF (formed from irradiation of XeF_4 or XeF_2).

ANSWER

λ is negative as the Xe 5p orbital is more than half-filled, and with $S = \frac{1}{2}$, $\lambda = -\zeta$. A simple ratio of the ζ values multiplied by the *g*-shift for •KrF ($\Delta_g = 0.0427$) in Table 9.1 gives $\Delta_g = 0.0746$ and a *g*-value of 2.0769 for •XeF. This is in good agreement with the experimental value of 2.0747.

SELF TEST

Predict the *g*-value for •ArF.

9.3.1.2 *g*-values of transition metal compounds

Large deviations of *g* from g_e arise in the EPR spectra of transition metal compounds because ΔE is smaller (the ground and excited states both involve d orbitals split by the crystal field) and the spin–orbit coupling constants are larger (see Table 8.3; the values given are λ and are oxidation state dependent). The orbital contribution depends on the electronic ground state of the transition metal, and to

---

TABLE 9.1 Table of single-electron spin–orbit coupling constants, ζ (in cm^{-1}), for neutral s- and p-block atoms. (Data from M. Montalto, A. Credi, L. Prodi, and M. T. Gandolfi. (2006) *Handbook of Photochemistry*. 3rd ed. CRC Press.)

Li	Be	B	C	N	O	F	Ne
0.23	2.0	10	32	78	154	269	520
Na	Mg	Al	Si	P	S	Cl	Ar
11.5	40.5	62	130	230	365	587	940
K	Ca	Ga	Ge	As	Se	Br	Kr
38	105	464	800	1202	1659	2460	3480
Rb	Sr	In	Sn	Sb	Te	I	Xe
160	390	1183	1855	2593	3384	5069	6080
Cs	Ba	Tl	Pb	Bi			
370	830	3410	5089	6831			

interpret their EPR spectra it is necessary to deal explicitly with spin–orbit split electronic states, which requires detailed analysis. The discussion that follows relates to orbital singlets (i.e. A or B terms). Δg can be either positive or negative for transition metal complexes because λ is positive for less than half-filled shells, and negative for more than half-filled shells. For VO^{2+}, V(IV), which has a d^1 configuration, λ is positive (250 cm^{-1}), so $g < g_e$ (Δg negative), and g is typically ≈ 1.9. For Cu(II), which is d^9, λ is negative (-830 cm^{-1}), so $g > g_e$ (Δg negative) and g is typically ≈ 2.1. For transition metal compounds, the central metal atom is usually the major source of g-shifts and the values of g can be found in the range 1–5.

9.3.1.3 Anisotropy in g-values

EPR spectra can be easily collected and interpreted from solutions, frozen solutions, and solids. In the case of fluid solutions where the paramagnetic species is freely tumbling, **isotropic g-values** are obtained, known as g_{iso}. g_{iso} represents the average of g_x, g_y, and g_z, the individual components of g, called the principal values and found along the different Cartesian directions. A typical EPR spectrum from a fluid solution is shown in Figure 9.4(a). This type of spectrum is also observed in the solid or frozen solution for cubic systems where $g_x = g_y = g_z$.

In frozen solutions or solid samples of non-cubic systems, g can be anisotropic and resolved into its g_x, g_y, and g_z components. The act of freezing fixes the molecules in all possible orientations with respect to the applied magnetic field. Therefore, the spectrum of a frozen sample represents the superposition of the spectra of all possible orientations (strictly a superposition not an average). The case for rhombic systems with $g_x \neq g_y \neq g_z$ is shown in Figure 9.4(d), which also shows how the g-values are measured in the data. This is known as a **powder spectrum** as it is also observed for magnetically dilute randomly oriented powders with **rhombic symmetry**.

A very common occurrence in EPR spectroscopy is when the radical or molecule has axial symmetry, that is, $g_x = g_y \neq g_z$ and then it is possible to resolve g into two components that are usually labelled $g_\perp$ (g_x, g_y) and $g_\parallel$ (g_z). g_z can be larger or smaller than g_x and g_y, so two different spectral motifs are possible, as shown in Figure 9.4(b) and (c). As the relative intensity of the $g_\perp$ (g_x, g_y) and $g_\parallel$ (g_z) features is a result of the superposition of all orientations, the $g_\perp$ features always have the larger intensity as there are many orientations in the xy plane compared to the unique z-axis for $g_\parallel$. For transition metal complexes the relative ordering of $g_\perp$ and $g_\parallel$ in axial geometries can be used to identify the SOMO, that is, the one

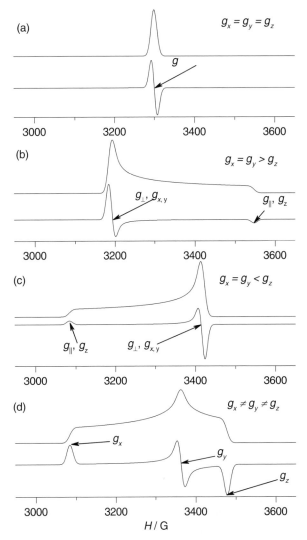

FIGURE 9.4 Representative absorption and first derivative EPR spectra: (a) isotropic ($g_x = g_y = g_z$), (b) axial ($g_x = g_y > g_z$), (c) axial ($g_x = g_y < g_z$), (d) rhombic ($g_x \neq g_y \neq g_z$).

with the unpaired electron in it, and hence the electronic ground term. For example, $g_\perp$ (g_x, g_y) $> g_\parallel$ (g_z) is characteristic of a d_{xy} SOMO in d^1 configurations and a d_{z^2} SOMO in a tetragonal compressed d^9 Jahn–Teller distorted octahedron, whereas $g_\perp$ (g_x, g_y) $< g_\parallel$ (g_z) is characteristic of a $d_{x^2-y^2}$ SOMO in a tetragonal elongated d^9 Jahn–Teller distorted octahedron. (The method of determining this is described in the Online Resource Centre)

All the spectra shown in Figure 9.4 have the same value of g_{iso} calculated using the following equation:

$$g_{iso} = \frac{g_x + g_y + g_z}{3} \qquad \textbf{Eqn 9.6}$$

g-values can be used for fingerprinting for main group radicals, and in conjunction with detailed analysis of the electronic states, the assignment of the ground term for transition metal complexes.

9.3.2 Hyperfine coupling constant, *A*

If the unpaired electron is in the vicinity of a nucleus with nuclear spin, $I > 0$, the magnetic field associated with this nucleus interacts with that from the electron and hyperfine coupling can be observed. This is analogous to the nuclear spin–nuclear spin coupling (*J*) observed in NMR spectra, and many of the same ideas and rules are employed. As the unpaired electron density can be distributed widely in a molecule, the hyperfine coupling in EPR may be associated with just one atom, or many across the whole molecule.

If the magnetic nucleus has $I = \frac{1}{2}$, then each m_S state interacts with the $m_I = +\frac{1}{2}$ and $m_I = -\frac{1}{2}$ states and this results in a slightly increased or decreased energy of the whole spin system, as shown in Figure 9.5.

It should be noted that the ordering of the energies of the coupled m_I states is reversed in the two m_S states. The selection rule is $\Delta m_S = \pm 1$ and $\Delta m_I = 0$, and there are now two values of the magnetic field that cause resonance separated by the hyperfine coupling constant, *A*. This is shown as the black lines in Figure 9.5, where the value of *A* is 200 G, together with the unobserved original single transition shown as a blue dashed line. Therefore, coupling of the unpaired electron to one $I = \frac{1}{2}$ nucleus gives an equal intensity doublet separated

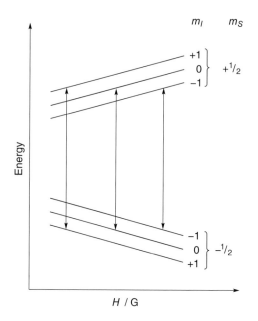

FIGURE 9.6 Effect of coupling of unpaired electron to one $I = 1$ nucleus.

by the hyperfine coupling, *A*. For coupling to an $I = 1$ nucleus, each m_S state is split by three m_I states (1, 0, –1) (Figure 9.6), which gives rise to three possible transitions and the experimental observation of a triplet of equal intensity (1:1:1).

For coupling to one $I = \frac{3}{2}$ nucleus a four line pattern of equal intensity (1:1:1:1) is observed. These patterns are just the application of the $2nI + 1$ rule that was also used in NMR spectroscopy (Section 3.7). Therefore, the same 'tree-splitting' diagrams and Pascal's triangles that are used to explain the coupling in NMR spectra can also be used to interpret EPR spectra. However, in contrast to NMR spectra, hyperfine coupling to quadrupolar nuclei (i.e. $I > \frac{1}{2}$) is observed much more frequently in EPR spectra, even for low symmetry environments.

Figure 9.7 shows the EPR hyperfine coupling patterns for different numbers (*n*) of equivalent $I = \frac{1}{2}$ and $I = 1$ nuclei. (A description of how to calculate the relative intensities when coupling to quadrupolar nuclei is given in the Online Resource Centre.) In both cases the simulation assumes $g = 2.0023$ with a 9.25 GHz operating frequency and a hyperfine coupling constant of 20 G. In each case there are $2nI + 1$ lines, whose intensities reflect the likelihood of various spin states, as with NMR spectra (Section 3.7). Where hyperfine coupling occurs to nuclei of different natural abundances then satellite peaks reflecting the isotopic composition and nuclear *I* values are observed—again paralleling the NMR spectra (Section 3.10).

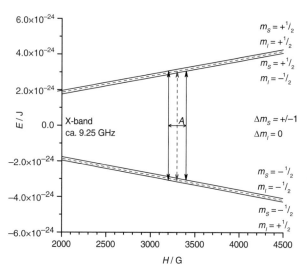

FIGURE 9.5 Effect of the interaction of an $I = \frac{1}{2}$ nucleus on the energy of the $m_S = \pm \frac{1}{2}$ levels.

$I = \frac{1}{2}$ $I = 1$

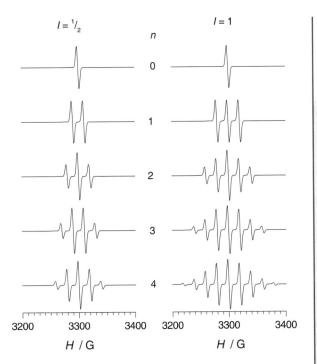

n

0

1

2

3

4

FIGURE 9.7 First derivative coupling patterns expected for coupling to *n* equivalent $I = \frac{1}{2}$ and $I = 1$ nuclei.

EXAMPLE 9.3

Interpret the X-band EPR spectrum shown in Figure 9.8.

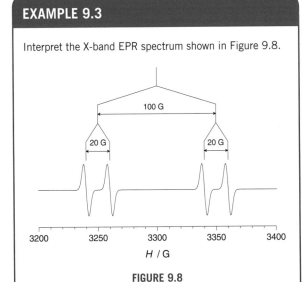

FIGURE 9.8

ANSWER

The spectrum consists of a doublet of doublets, indicating that the electron in the compound is coupled to two inequivalent $I = \frac{1}{2}$ nuclei, one with a hyperfine coupling constant of 100 G, the other with a hyperfine coupling constant of 20 G. The 100 G hyperfine coupling would give a doublet with peaks at 3250 and 3350 G, but these are

then split by the second $I = \frac{1}{2}$ nucleus with a hyperfine coupling constant of 20 G, so the observed peaks will be at 3240, 3260, 3340, and 3360 G, in a relative intensity of 1:1:1:1.

SELF TEST

Interpret the X-band EPR spectrum shown in Figure 9.9 obtained from a compound containing an element with a 9.5% abundant $I = \frac{3}{2}$ isotope; assume all other isotopes have $I = 0$.

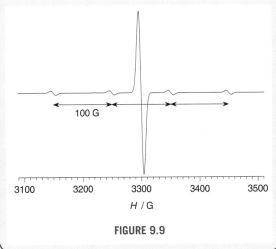

FIGURE 9.9

The hyperfine coupling constant is measured from the spectra in magnetic field units (i.e. G, T, mT). However, these are dependent on *g* and, as the hyperfine coupling constant is an interaction energy, for strict comparison between different compounds it needs to be converted to energy units using the following relationships:

$$A(\text{cm}^{-1}) = g\mu_B A(\text{G}) = g \times 4.6686 \times 10^{-5} \times A(\text{G})$$

Eqn 9.7

$$A(\text{MHz}) = gA(\text{G})/0.71446 \qquad \textbf{Eqn 9.8}$$

In the literature, hyperfine coupling constants may be listed in G (or mT), cm^{-1} (or $10^{-1}\ \text{cm}^{-1}$), or MHz.

For fluid solutions, an average or isotropic value, A_{iso}, is determined for the single g_{iso} value. For frozen solutions, *A* can be resolved into its *x*, *y*, and *z* components, or for axial symmetry its $A_\perp$ and $A_\parallel$ components, associated with each of the different *g*-values. Therefore, the spectra become complex very quickly, and detailed interpretation is normally only possible by simulating and fitting the data using a computer.

The value of *A* can be either negative or positive, and is often negative. The sign cannot be determined directly from EPR spectra, but the relative signs can be obtained by determining A_{iso} and the anisotropic A values and checking that

$$A_{iso} = \frac{A_x + A_y + A_z}{3} \qquad \textbf{Eqn 9.9}$$

but this must be carried out in energy units (cm^{-1}) or frequency units (MHz) and not magnetic field units (G, T).

For transition metal compounds, hyperfine refers to the coupling caused by interaction with the metal, and the term **superhyperfine** is used for coupling to magnetic nuclei in the ligands. In some circumstances, these are denoted by the different terms A and a, respectively.

9.3.2.1 Origin of hyperfine coupling

The hyperfine and superhyperfine coupling constants are a measure of the spin density of the unpaired electron on the different magnetic nuclei. Hence the magnitude of the hyperfine and superhyperfine coupling constant depends on the amount of interaction between the unpaired electron and the nuclei in the radical or molecule. The 'longer' the unpaired electron spends at any one nucleus, the greater the hyperfine coupling constant characteristic of that nucleus. This is a very powerful way of identifying how delocalised the unpaired electron is within the radical, and the extent of any covalency in the bonding. Therefore, hyperfine and superhyperfine couplings give an indication of where the unpaired electron is located, that is, is it metal centred or ligand centred, and which metal orbitals are involved. This can help identify the ground state.

In terms of characterising inorganic compounds the hyperfine and superhyperfine coupling helps to identify the species under investigation. The coupling patterns identify where the spin density of the unpaired electron is in the molecule, and the magnitude of the splitting identifies the extent of the spin density at each atom.

As the hyperfine coupling is an interaction between the unpaired electron and the nucleus, there must be a mechanism that mediates this. The only orbitals that penetrate the nucleus are the s orbitals (p, d, f orbitals have nodes at the nucleus), so it might be expected that only spin density in orbitals with s character at the magnetic nucleus will display hyperfine couplings. This is known as the **Fermi contact interaction** giving rise to the **Fermi contact term**, and if true would limit the usefulness of the technique, especially as in the majority of cases the unpaired electron is in a p or d orbital. The extent of the Fermi contact term can be increased if there is orbital mixing of the s and p or s and d orbitals when they have the same symmetry labels. Hyperfine coupling is observed for organic radicals with the unpaired electrons in p or π orbitals, and in these cases the hyperfine coupling is observed due to **spin polarisation**. That is, the spin of the unpaired electron in the p orbital is 'felt' throughout the rest of the molecule including by

electrons in s orbitals. Estimates of spin density in different orbitals can be made by comparing the observed hyperfine coupling with those tabulated for 100% spin density in that orbital. (See Online Resource Centre and Bibliography for specialised texts containing these data.)

9.4 Limitations of EPR spectroscopy

The key limitation is that the species under investigation must contain at least one unpaired electron. While there are numerous stable transition metal compounds with unpaired electrons, these are much less common in organic and main group chemistry. In principle it is possible to observe EPR spectra for any number of unpaired electrons, but in practice, the spectra of samples containing one unpaired electron ($S = \frac{1}{2}$) are fairly easy to obtain and to understand, those with an odd number of unpaired electrons greater than 1 ($S = \frac{3}{2}, \frac{5}{2}$) are easy to measure, but can be very challenging to understand. Those with an even number of unpaired electrons ($S = 1, 2$) can be challenging to observe using commercial spectrometers and electromagnets. As the vast majority of organic and main group radicals have $S = \frac{1}{2}$ this makes their study straightforward. In contrast for transition metal complexes, a much wider variation in the number of unpaired electrons can be found. A summary for the common transition metal ions and their ease of observation in EPR spectroscopy is given in Table 9.2. Of these

TABLE 9.2 Summary of EPR data for monometallic first row transition metal ions: d^n configurations, examples, and ease of detection

Configuration	Examples	S	Ground term (in O_h symmetry)	Ease of observation
d^1	Ti^{3+}, V^{4+}, Cr^{5+}	$\frac{1}{2}$	$^2T_{2g}$	Easy
d^2	Ti^{2+}, V^{3+}	1	$^3T_{1g}$	Very difficult
d^3	Cr^{3+}	$\frac{3}{2}$	$^4A_{2g}$	Easy
d^4 hs	Cr^{2+}	2	^5E_g (Jahn–Teller distorted)	Very difficult
d^5 hs	Mn^{2+}, Fe^{3+}	$\frac{5}{2}$	$^6A_{1g}$	Easy
d^5 ls	Fe^{3+}	$\frac{1}{2}$	$^2T_{2g}$	Usually difficult
d^6 hs	Fe^{2+}	2	$^5T_{2g}$	Very difficult
d^6 ls	Fe^{2+}	0	$^1A_{1g}$	Diamagnetic
d^7 hs	Co^{2+}, Ni^{3+}	$\frac{3}{2}$	$^4T_{2g}$	Very difficult
d^8	Ni^{2+}	1	$^3A_{2g}$	Difficult
d^9	Ni^+, Cu^{2+}	$\frac{1}{2}$	^2E_g (Jahn–Teller distorted)	Easy

the $S = \frac{1}{2}$ spin systems VO^{2+} and Cu^{2+} are some of the most studied, and examples based on these ions will be covered in subsequent sections.

9.5 Applications of EPR spectroscopy

The X-band EPR spectra in the following sections have been simulated using a constant microwave energy of 9.25 GHz together with the literature values for the g-shifts and hyperfine coupling constants. The effects of tumbling in the isotropic spectra have not been simulated, unless explicitly stated.

9.5.1 Main group examples

While there are stable organic and main group compounds with unpaired electrons, such as Frémy's salt (peroxylamine disulfonate $[ON(SO_3)_2]^{2-}$) and the $[S_3]^-$ anion found in ultramarine, in most cases the radical has to be prepared and then stabilised prior to characterisation. One way of producing radicals is through the use of energetic sources, such as pulse radiolysis with a ^{60}Co γ-ray source or UV photolysis. Once formed, the paramagnetic species needs to be stabilised for sufficient length of time for detection, and so they are usually generated in solid hosts or in frozen solutions/glasses of organic solvents at 77 K. Solidified Group 18 gases can be used as a matrix for stabilising the most reactive examples.

EXAMPLE 9.4

Interpret the isotropic X-band (9.25 GHz) EPR spectrum of the radical $^{\bullet}AlH_3^-$ shown in Figure 9.10 (^{27}Al, $I = \frac{5}{2}$, 100% abundant; ^1H, $I = \frac{1}{2}$, 99.99%).

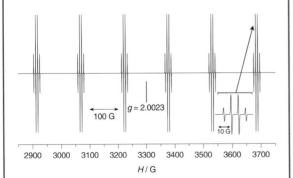

FIGURE 9.10 Isotropic X-band (9.25 GHz) EPR spectrum of $^{\bullet}AlH_3^-$. (Data from J.R.M. Giles and B.P. Roberts, *Chem. Commun.* 1167 (1981).)

ANSWER

Two different sets of hyperfine couplings can be observed in the spectrum. The six groups of lines of equal intensity are characteristic of coupling of the unpaired electron to aluminium, ($I = \frac{5}{2}$, m_I values $+\frac{5}{2}$, $+\frac{3}{2}$, $+\frac{1}{2}$, $-\frac{1}{2}$, $-\frac{3}{2}$, $-\frac{5}{2}$). Each of these is split into a binomial four-line pattern with a 1:3:3:1 distribution, characteristic of coupling to three $I = \frac{1}{2}$ (^1H) nuclei. As expected for main group radicals, the experimental value of g_{iso} of 2.0025 is very close to g_e. $A(Al)_{iso}$ is 154.2 G and $A(H)_{iso}$ is 7 G. The relatively large value of $A(Al)_{iso}$ indicates that the unpaired electron is largely located on the aluminium in a $3p_z$ orbital. It also suggests a non-planar geometry for $^{\bullet}AlH_3^-$ as this allows for mixing of the 3s and $3p_z$ orbitals which both have a_1 symmetry, resulting in an increase in s electron spin density and hence larger $A(Al)_{iso}$. In $^{\bullet}BH_3^-$, $A(B)_{iso}$ is 19.9 G indicating a much smaller 2s spin density, consistent with a planar structure and no 2s–2p mixing.

SELF TEST

Figure 9.11 shows the isotropic X-band (9.25 GHz) EPR spectrum of $^{\bullet}PF_2$ radicals produced by γ-irradiation of PF_3 in solid C_2F_6. Measure and explain the hyperfine coupling patterns.

How would the spectra of $^{\bullet}NF_2$ and $^{\bullet}AsF_2$ be different? (^{31}P, $I = \frac{1}{2}$, 100% abundant; ^{19}F, $I = \frac{1}{2}$, 100% abundant, ^{14}N, $I = 1$, 99.6%; ^{75}As, $I = \frac{3}{2}$, 100%)

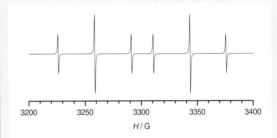

FIGURE 9.11 Isotropic X-band (9.25 GHz) EPR spectra of $^{\bullet}PF_2$. (Data from W. Nelson, G. Jackel, and W. Gordy, *J. Chem. Phys.* **52** 4572 (1970); A.J. Colussi, J.R. Morton, K.F. Preston, and R.W. Fessenden, *J. Chem. Phys.* **61** 1247 (1974).)

9.5.2 Transition metal examples

As highlighted earlier, electron configurations resulting in a spin doublet ($S = \frac{1}{2}$) are the easiest to study using EPR spectroscopy. For this reason the d^1 and d^9 configurations of the transition metals are the most often investigated, with V^{4+} the most prevalent d^1 system, usually occurring as VO^{2+}, and Cu^{2+} the most common d^9. In the sections below these ions will be considered, as well as other $S = \frac{1}{2}$ systems such as Cr(V) (as CrN^{2+}).

9.5.2.1 d^1 examples—vanadium(IV)

Of the V(IV) complexes, those containing the vanadyl unit VO^{2+} have been extensively studied by EPR spectroscopy as they are of interest to both coordination chemists and bioinorganic chemists. Figure 9.12 shows representative X-band EPR spectra of VO^{2+}.

Figure 9.12(a) shows an isotropic spectrum (g_{iso} = 1.96, A_{iso} = 108 G) expected for freely tumbling molecules in a fluid solution, and consists of an eight-line pattern of equal intensity due to coupling of the unpaired electron to the ^{51}V nucleus, which has $I = \frac{7}{2}$ and essentially 100% abundance. However, in practice, the spectrum in Figure 9.12(b) is usually observed for fluid solutions (at X-band frequency) because the tumbling rate of the molecules is not fast enough compared to the EPR timescale (ns) to generate a true isotropic spectrum. The area under each component is equal (this is a first derivative presentation,

so the spectrum must be integrated twice to obtain the area), but the peak separations may not be equal. The increasing separation from low to high field is called a 'second order effect', and these separations become more equal for spectra measured at higher microwave frequencies.

The spectrum in Figure 9.12(c) is for a frozen solution, and the short V=O bond in either OVL_4 or OVL_5 complexes means that the spectra are usually axial, with different $g_\perp$ (g_x, g_y) and $g_\parallel$ (g_z) features, and different $A_\perp$ (A_x, A_y) and $A_\parallel$ (A_z) values. (Simulated using g_z = 1.93, $g_{x,y}$ = 1.98, A_z = 170 G, and $A_{x,y}$ = 60 G.) The ordering of the $g_\perp$ (g_x, g_y) > $g_\parallel$ (g_z) can be used to identify that the SOMO is d_{xy} in this case. (See Online Resource Centre for further details.) The dotted lines between the bottom two spectra show how the m_I states in the solution and frozen solution are connected to give the isotropic (average) values. Although estimates of the g and A values can be made directly from the spectra, detailed calculations and simulation are required to obtain them accurately. Axial spectra are commonly observed for large metalloproteins, even in solution, as the tumbling rate is so slow.

9.5.2.2 d^1 examples—chromium(V)

Figure 9.13 shows the isotropic X-band (9.25 GHz) spectra for two Cr(V) complexes containing the CrN^{2+} unit, $[Cr(N)(NCS)_4]^{2-}$ and $[Cr(N)(CN)_4]^{2-}$, both of which have a d^1 configuration.

As with V(IV) the g-values for Cr(V) are close to, but slightly less than, g_e, with a bigger g-shift for $[Cr(N)(NCS)_4]^{2-}$ than $[Cr(N)(CN)_4]^{2-}$. These spectra indicate the accuracy with which g-shifts can be measured. Nitrogen superhyperfine coupling is observed for both cases, in addition to the weak ^{53}Cr hyperfine satellites (9.5% abundance) either side of the main feature. The presence of the nitrogen superhyperfine is direct evidence of delocalisation of the unpaired electron onto the ligands, indicating some covalent bonding in these compounds. In $[Cr(N)(CN)_4]^{2-}$ the 1:1:1 superhyperfine triplet coupling of 2.94 G (2.73×10^{-4} cm^{-1}) is due to just the nitrido (N≡Cr) ^{14}N atom with $I = 1$. In $[Cr(N)(NCS)_4]^{2-}$ the eleven-line (1:5:15:30:45:51:45:30:15:5:1) superhyperfine pattern (see Online Resource Centre for details of how to calculate this) with a coupling constant of 2.94 G (2.71×10^{-4} cm^{-1}) is due to coupling to the five equivalent ^{14}N nuclei (nitride plus four isothiocyanato ligands). The equivalence of the axial and equatorial N superhyperfine coupling is believed to be coincidental. There is a more marked difference in the chromium hyperfine coupling, which is significantly smaller in $[Cr(N)(CN)_4]^{2-}$ (24.5 G, 22.8×10^{-4} cm^{-1}) than $[Cr(N)(NCS)_4]^{2-}$ (29.3 G, 27.0×10^{-4} cm^{-1}). This indicates that the unpaired electron spends less time on the chromium in the cyano complex, implying much greater delocalisation of the unpaired electron onto the ligands and hence greater covalency in the cyano complex.

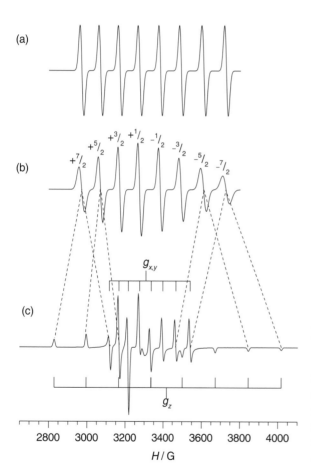

FIGURE 9.12 Representative X-band (9.25 GHz) EPR spectra of VO^{2+} species (a) isotropic freely tumbling, (b) isotropic restricted tumbling, and (c) as a frozen solution.

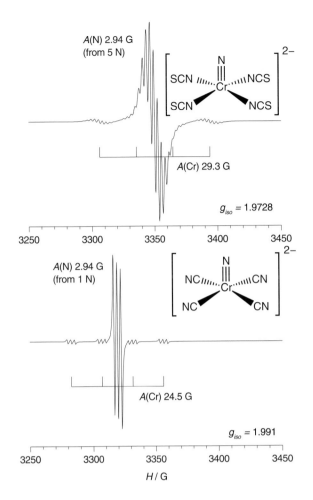

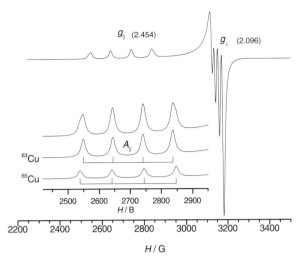

FIGURE 9.14 X-band (9.25 GHz) EPR spectrum of frozen solution of $[Cu(H_2O)_6]^{2+}$.

FIGURE 9.13 Isotropic X-band (9.25 GHz) EPR spectra of $[Cr(N)(NCS)_4]^{2-}$ and $[Cr(N)(CN)_4]^{2-}$. (Data from J. Bendix, T. Birk, and T. Weyhermüller, *Dalton Trans.* 2737 (2005); J. Bendix, K. Meyer, T. Weyhermüller, E. Bill, N. Metzler-Nolte, and K. Wieghardt, *Inorg. Chem.* **37** 1767 (1998).)

9.5.2.3 d⁹ examples—copper(II)

Many Cu(II) complexes yield typical axial EPR spectra. The two naturally occurring isotopes of copper, ^{63}Cu (69.17%) and ^{65}Cu (30.83%), both have nuclear spins of $^3/_2$ so that each m_S state will be split into four lines ($m_I = +^3/_2, +^1/_2, -^1/_2, -^3/_2$), giving four transition energies, but because the magnetic moments of these two isotopes are very similar, the observed hyperfine couplings are nearly coincident.

Figure 9.14 shows the frozen solution X-band (9.25 GHz) spectrum of $[Cu(H_2O)_6]^{2+}$. This displays the classic signature of an axial system due to the Jahn–Teller distortion expected for Cu(II). The assignment to $g_\perp$ (g_x, g_y) at 2.096 and $g_\parallel$ (g_z) at 2.454 is made on the basis that $g_\parallel$ (g_z) will be less intense than $g_\perp$ (g_x, g_y). The A(Cu) values are also anisotropic with $A_\parallel$ (96 G) being larger than $A_\perp$ (17.4 G). Although the $g_\perp$ region looks complex there are four lines as expected, but the unequal amplitude arises

because the linewidth is similar to the hyperfine splitting. In the $g_\parallel$ region the $A_\parallel$ is well resolved, but as shown in the insert it is actually made up of 1:1:1:1 quartets from hyperfine coupling to ^{63}Cu and ^{65}Cu which both have $I = ^3/_2$. In the majority of spectra the linewidths are sufficiently broad that these ^{63}Cu and ^{65}Cu components are not usually resolved.

The relative ordering of the $g_\perp$ and $g_\parallel$ values in the axial spectra of magnetically dilute powders or frozen solutions can be used to identify what type of Jahn–Teller distortion is present. $g_\parallel$ (g_z) > $g_\perp$ (g_x, g_y) with $g_\parallel$ (g_z) at a lower magnetic field than $g_\perp$ (g_x, g_y) is characteristic of a tetragonal elongation with a $d_{x^2-y^2}$ SOMO. A tetragonal compression with a d_{z^2} SOMO will have $g_\perp$ (g_x, g_y) > $g_\parallel$ (g_z), so $g_\perp$ (g_x, g_y) will occur at lower magnetic field than $g_\parallel$ (g_z), which will be close to g_e. (See Online Resource Centre for further details.) Therefore, the experimental data for $[Cu(H_2O)_6]^{2+}$ are consistent with a tetragonal elongation. This is by far the most common Jahn–Teller distortion for Cu(II) complexes, with only a few very rare examples of tetragonal compression, such as *trans*-$[CuCl_4(NH_3)_2]^{2-}$, which has $g_\perp = 2.22$ and $g_\parallel = 2.00$.

As well as identifying the type of Jahn–Teller distortion present, EPR spectra can also determine whether the Jahn–Teller distortion is static, dynamic, or temperature dependent.

Figure 9.15 shows the X-band (9.25 GHz) spectra of $[Cu(ptz)_6](BF_4)_2$ (ptz = 1-propyltetrazole) powder at 300 K and 77 K. The axial spectrum at 77 K is characteristic of a conventional tetragonal elongation, as $g_\parallel > g_\perp$. However, on warming to 300 K, an isotropic type spectrum is observed. This is characteristic of a change from a static Jahn–Teller distortion where the individual configurations are frozen in at low temperature to a dynamic Jahn–Teller distortion, where the long and short Cu–N bonds are interchanging rapidly on the EPR timescale (typically

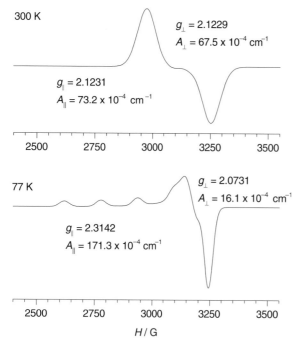

300 K

$g_\perp = 2.1229$
$A_\perp = 67.5 \times 10^{-4}$ cm^{-1}

$g_\parallel = 2.1231$
$A_\parallel = 73.2 \times 10^{-4}$ cm^{-1}

2500   2750   3000   3250   3500

77 K

$g_\perp = 2.0731$
$A_\perp = 16.1 \times 10^{-4}$ cm^{-1}

$g_\parallel = 2.3142$
$A_\parallel = 171.3 \times 10^{-4}$ cm^{-1}

2500   2750   3000   3250   3500

H / G

FIGURE 9.15 X-band EPR spectra of [Cu(ptz)$_6$](BF$_4$)$_2$ (ptz = 1-propyltetrazole) powder at 300 K and 77 K. (Data from V.S.X. Anthonisamy and R. Murugesan, *Chem. Phys. Lett.* **287** 353 (1998).)

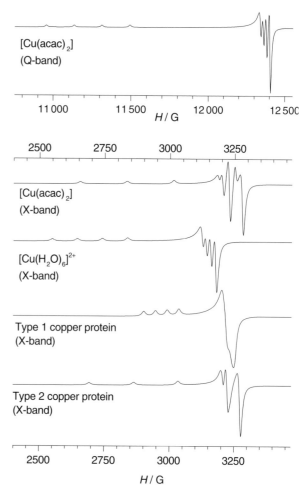

[Cu(acac)$_2$]
(Q-band)

11 000   11 500   12 000   12 500

H / G

2500   2750   3000   3250

[Cu(acac)$_2$]
(X-band)

[Cu(H$_2$O)$_6$]$^{2+}$
(X-band)

Type 1 copper protein
(X-band)

Type 2 copper protein
(X-band)

2500   2750   3000   3250

H / G

FIGURE 9.16 Frozen solution EPR spectra for a range of copper compounds.

nanoseconds). As expected the *g*-values are essentially the same. The loss of Cu hyperfine coupling is also characteristic of a dynamic Jahn–Teller distortion and arises because the linewidths increase to such an extent due to the dynamic processes that they are similar to the coupling constants. As with the majority of EPR experiments, detailed simulation making use of single-crystal data was required to obtain the *g* and *A* values given in Figure 9.15.

Copper(II), d^9, is also very important in bioinorganic chemistry, and there have been many studies of both metalloproteins and small molecule mimics. EPR has found widespread use to elucidate their geometric and electronic structure, including during the catalytic cycles.

Figure 9.16 shows the X-band (9.25 GHz) frozen solution spectra of two complexes, one octahedral, [Cu(H$_2$O)$_6$]$^{2+}$, and the other square planar, [Cu(acac)$_2$], where acac = acetylacetonate or pentane-2,4-dionate. For complexes such as [Cu(acac)$_2$], the Q-band experiment separates the $g_\parallel$ (2.26) and $g_\perp$ (2.050) components completely, which are overlapping in the X-band spectra, as the hyperfine coupling ($A_\parallel$ (179 G), $A_\perp$ (20.9 G)) remains the same. The $g_\perp$ region in the X-band spectra is affected by 'over-shoot' peaks, where the peak at high field has an unrealistically small *g*-value. Such peaks arise in powder or frozen solution spectra with large *g* and *A* anisotropies because the *A* hyperfine coupling pattern expands faster than the *g*-value decreases and, on going from the *x,y*($\perp$)

region to the *z*($\parallel$) region, a peak appears beyond the expected range. (Undershoot can also be observed.) This can make counting the number of peaks difficult and can lead to the incorrect assignment of rhombic rather than axial symmetry, and to spurious *g*-values. Detailed computational simulation is used to confirm the presence of over- or under-shoot, and carrying out the experiment at higher microwave frequency (e.g. Q-band) can also help in the analysis of spectra of this type (see Figure 9.16). Nitrogen superhyperfine coupling, especially on the $g_\parallel$ peaks, can also sometimes be observed with N-donor ligands.

Copper metalloproteins are conventionally divided into three types based on their spectroscopic and geometric properties. Type 1 copper proteins such as plastocyanin (which employ an outer sphere electron transfer) have a highly distorted monometallic copper coordination environment. This contains two histidine nitrogen atoms and one cysteine thiolate sulfur atom together with the copper in a near planar arrangement, with a thioether unit of a methionine at a longer distance out of the plane.

These proteins are also often known as blue copper proteins due to the intense absorption at ca. 600 nm giving the characteristic colour. Their EPR spectra (Figure 9.16) are also very characteristic with very small $A_\parallel(Cu)$ hyperfine couplings of ca. 45 G indicative of significant delocalisation of electron density from the copper to the ligands. Type 2 proteins, sometimes called normal copper proteins, include copper–zinc superoxide dismutase and have g and A values much more similar to conventional copper(II) complexes (Figure 9.16). The Type 3 copper proteins are bimetallic and are EPR silent because the two copper centres are antiferromagnetically coupled.

EXAMPLE 9.5

Fungal laccase contains both Type 1 and Type 2 copper centres. When 0.7 mM of F^- is added to 0.7 mM of the protein, the $g_\parallel$ peaks of the Type 2 spectral features are split into doublets, and when 14 mM of F^- is added, triplets are observed. The addition of 10 mM H_2O_2 to 0.3 mM of this protein also only modifies the Type 2 resonances. Discuss what information these observations yield on the behaviour of the two different copper sites.

ANSWER

The observation that only the $g_\parallel$ peaks of the Type 2 spectral features are split into doublets and triplets on the addition of F^-, and that these are also only affected by addition of H_2O_2, indicates that it is the Type 2 site that is directly involved in any reaction or catalytic process. This shows that the Type 2 site is subsitutionally labile. The Type 1 site signal is unaffected because it acts only as the electron transfer partner of the Type 2 catalytic site and its coordination is unaffected by the addition of ligands.

9.5.3 $S > \frac{1}{2}$ systems

For spin systems greater than $S = \frac{1}{2}$, two additional concepts, zero-field splitting and Kramers' doublets, need to be considered to fully interpret the spectra. m_S states for $S > \frac{1}{2}$ are no longer degenerate in the absence of a magnetic field (i.e. they do not converge at 0 G), and are therefore subject to what is known as **zero-field splitting**. This gives rise to many additional resonances in the EPR spectrum. As the magnetic field increases, the m_S states begin to diverge in energy and for even numbers of unpaired electrons this can result in large transition energies between the m_S states that cannot usually be probed with commercial EPR instrument frequency and magnet combinations. In contrast, for odd numbers of unpaired electrons, even though zero-field splitting still occurs, there are always pairs of levels and for at least one of these pairs, $m_S = \pm\frac{1}{2}$ and these are separated by a smaller energy (it behaves like the $S = \frac{1}{2}$ spin system described above), which is accessible to commercial EPR instrumentation. These are known as **Kramers' doublets**, and hence $S = \frac{3}{2}, \frac{5}{2}$ are known as Kramers' systems, and those with $S = 1$, 2 are known as non-Kramers' systems. While spectra of Kramers' systems are readily observable, their interpretation and analysis requires dealing explicitly with the zero-field splitting parameters (D and E). A special non-Kramers' system is when two d^1 or d^9 metals are in close proximity and the unpaired electrons can couple to give $S = 0$ and $S = 1$ states, the latter giving rise to very characteristic spectra. The detailed analysis is beyond this text but is covered in more advanced texts (see the Bibliography, especially Mabbs and Collison).

Cr^{3+} ($S = \frac{3}{2}$) has been studied using EPR spectroscopy, but of greatest biological interest is high spin iron(III) ($S = \frac{5}{2}$), and this has very characteristic EPR spectral motifs for a variety of different geometries.

Problems

9.1 Predict the sign of Δg for Mo^{5+}.

9.2 Sketch the first derivative spectrum of a radical containing one $I = \frac{1}{2}$ nuclei with a hyperfine coupling constant of 100 G, and one $I = \frac{3}{2}$ nuclei with a hyperfine coupling constant of 20 G. Assume that $g = 2.0023$ and the instrument operating frequency is 9.25 GHz.

Sketch the spectrum with the hyperfine coupling constants reversed.

9.3 For $g_{iso} = 2.00$ and $A_{iso} = 200$ G, calculate the hyperfine coupling constant in terms of wavenumber (cm^{-1}) and MHz. ($\mu_B = 9.27401 \times 10^{-24}$ J $T^{-1} = 4.66858 \times 10^{-5}$ cm^{-1} G^{-1}).

For $g_z = 1.93$ and $A_z = 170$ G, and $g_{x,y} = 1.98$ and $A_{x,y} = 60$ G, calculate the $A_\perp$ and $A_\parallel$ hyperfine coupling constants in cm^{-1} and MHz.

9.4 Sketch the isotropic X-band (9.2500 GHz) spectrum of $^{\bullet}GaH_3^-$ ($g = 2.0055$; $A(^{69}Ga)_{iso} = 420.2$ G, 60.4%; $A(^{71}Ga)_{iso} = 534.1$ G, 39.6%; and $A(H)_{iso} = 10$ G).

9.5 The K-band (24.0619 MHz) isotropic EPR spectra of the irradiation products of BF_3 are shown in Figure 9.17. Due to the complexity of the spectrum from natural abundance boron, samples enriched in ^{10}B and ^{11}B (assume complete enrichment) were also prepared. Use these data to identify the radical formed. Natural boron: ^{10}B, 20%, $I = 3$; ^{11}B, 80%, $I = \frac{3}{2}$.

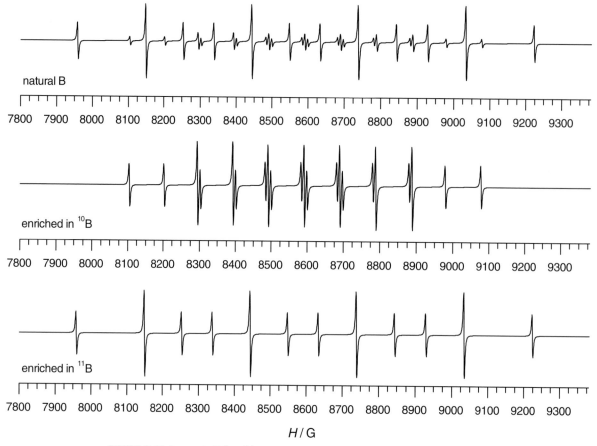

natural B

7800 7900 8000 8100 8200 8300 8400 8500 8600 8700 8800 8900 9000 9100 9200 9300

enriched in ^{10}B

7800 7900 8000 8100 8200 8300 8400 8500 8600 8700 8800 8900 9000 9100 9200 9300

enriched in ^{11}B

7800 7900 8000 8100 8200 8300 8400 8500 8600 8700 8800 8900 9000 9100 9200 9300

H / G

FIGURE 9.17 Isotropic K-band (24.0619 GHz) spectra of irradiation products of BF_3.

9.6 The isotropic X-band spectrum of bis(*O,O′*-diethyldithiophosphato)oxovanadium(IV) ([VO(dtp)$_2$], Figure 9.18) in ether solutions shows a 24-line pattern at room temperature, but this changes to a 16-line pattern on cooling. Account for these observations.

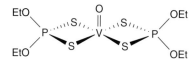

FIGURE 9.18 Structure of [VO(dtp)$_2$]. (Data from M. Sato, Y. Fujita, and T. Kwan, *Bull. Chem. Soc. Jpn.* **46** 3007 (1973); M. Sato, T. Katsu, Y. Fujita, T. Kwan, *Chem. Pharm. Bull.* **22** 1393 (1974).)

9.7 The isotropic X-band EPR spectrum of [Co(CO)$_2$(Ph$_2$C$_2$) (P(OMe)$_3$)] (prepared from the reduction of a dicobalt complex) consists of a 16-line pattern of approximately equal intensity with *g* = 2.061, and two hyperfine coupling constants of 45.4×10^{-4} cm^{-1} and 166.3×10^{-4} cm^{-1}. Explain these observations. (Data from L. V. Casagrande, T. Chen, P. H. Rieger, B. H. Robinson, J. Simpson, S. J. Visco, *Inorg. Chem.* **23** 2019 (1984).)

9.8 Reaction of a large excess of NaCN with [VCl$_2$Cp$_2$] resulted in the formation of a green compound. Elemental analysis gave the following: %C 61.75, %H 4.40, %N 12.0. The IR spectrum contained intense bands at 2116 and 2111 cm^{-1} with corresponding bands in the Raman spectrum at 2116 and 2110 cm^{-1}. The EPR spectrum in a water solution contained 8 lines of approximately equal intensity, with g_{iso} = 1.995 and A_{iso} = 60.8 G. When K^{13}CN was employed for the synthesis, each of the lines in the EPR spectrum was split into a 1:2:1 triplet with A_{iso} of 12.8 G.

Use these data to identify the complex formed (Data from J. Honzíček, J. Vinklárek, Z. Černošek, and I. Císařová, *Magn. Reson. Chem.* **45** 508 (2007)).

Bibliography

F.E. Mabbs and D. Collison. (1992) *Electron Paramagnetic Resonance of Transition Metal Compounds*. London: Elsevier.

M.C.R. Symons. (1978) *Chemical & Biochemical Aspects of Electron-Spin Resonance Spectroscopy*. New York: Van Nostrand Reinhold.

R.V. Parish. (1990) *NMR, NQR, EPR, and Mossbauer Spectroscopy in Inorganic Chemistry*. New York: Ellis Horwood.

F.E. Mabbs and D.J. Machin. (2008) *Magnetism and Transition Metal Complexes*. Mineola, NY: Dover Publications.

J.A. Weil and J.R. Bolton. (2007) *Electron Paramagnetic Resonance, Elementary Theory and Practical Applications*. Chichester: Wiley.

B.N. Figgis and M.A. Hitchman. (2000) *Ligand Field Theory and Its Applications*. New York: Wiley-VCH.

W. Weltner Jr. (1983) *Magnetic Atoms and Molecules*. New York: Scientific and Academic Editions.

A. Abragam and B. Bleaney. (2012) *Electron Paramagnetic Resonance of Transition Ions*. Oxford: Oxford University Press.

W.R. Hagen. (2009) *Biomolecular EPR Spectroscopy*. Boca Raton, FL: CRC Press.

H.M. Swartz, J.R. Bolton, and D.C. Borg. (1972) *Biological Applications of Electron Spin Resonance*. New York: Wiley.

V.Chechik, E. Carter, and D. Murphy. (2016) *Electron Paramagnetic Resonance*. Oxford: Oxford University Press.

10 Mössbauer spectroscopy and nuclear quadrupole resonance (NQR) spectroscopy

10.1 Mössbauer spectroscopy

10.1.1 Introduction to Mössbauer spectroscopy

The Mössbauer effect, Mössbauer spectroscopy, or recoil-less nuclear resonance absorption of γ-rays was discovered by Rudolf Mössbauer in 1958, who was awarded the Nobel Prize for physics in 1961. γ-rays are high-energy electromagnetic radiation in the keV to MeV range produced during radioactive decay involving transitions between nuclear energy levels (Chapter 1, Section 1.6). In Mössbauer spectroscopy, the absorption of γ-rays excites transitions between the nuclear energy levels, such as $I = \frac{1}{2}$ to $I = \frac{3}{2}$. Note that nuclear magnetic resonance (NMR) spectroscopy (Chapter 3) involves transitions between m_I energy levels within the ground-state nuclear energy levels, which are non-degenerate in the presence of a magnetic field. These NMR transitions occur in the MHz range (500 MHz ≈ 2×10^{-9} keV), that is, a billion times lower in energy than transitions between the nuclear levels with different values of I observed in Mössbauer spectroscopy.

As seen in earlier chapters, all spectroscopic techniques require a source of electromagnetic radiation that corresponds exactly to the difference between two energy levels in the sample under investigation, that is, resonant absorption. Laboratory-based Mössbauer experiments use γ-rays produced by the radioactive decay of the same isotope that is present in the sample under study. However, unless the isotope has a long half-life it is often produced in-situ by the decay of another radionuclide. With the advent of **synchrotron radiation** sources, it is also possible to use these as very high energy X-ray or γ-ray sources for Mössbauer spectroscopy.

10.1.2 Experimental requirements for Mössbauer spectroscopy

The requirement for an exact match between radiation source and γ-ray absorbing element in the sample means that the Mössbauer technique is element specific.

Thus an iron γ-ray source is needed to study iron atoms in the sample under investigation (^{57}Fe Mössbauer spectroscopy) and a tin γ-ray source to probe the tin atom environments in a separate experiment. This limits the number of elements to which the Mössbauer technique can be applied, as not all elements have the required isotopes with the energy level separations and appropriate excited state lifetimes, suitable radioactive sources, and sufficiently high natural abundances. The range of elements that can be studied using Mössbauer spectroscopy in described in Section 10.1.4.

For experiments investigating iron, which are by far the most common, the source contains a radioactive ^{57}Co source which decays by electron capture (EC) to generate an excited ^{57}Fe $I = \frac{5}{2}$ state. This can emit a variety of γ-rays with different energies to reach the ground state, as shown in Figure 10.1. Of

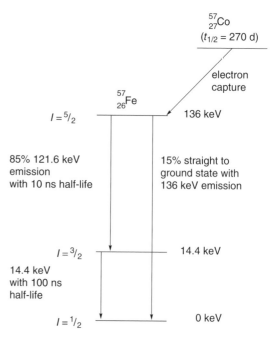

FIGURE 10.1 Schematic diagram of the decay process involving $^{57}_{27}$Co and $^{57}_{26}$Fe for Mössbauer spectroscopy.

FIGURE 10.2 Illustration of recoil occurring during the emission and absorption of γ-rays.

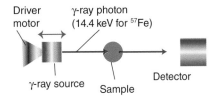

FIGURE 10.3 Schematic illustration of a Mössbauer spectrometer.

these, the 14.4 keV γ-rays associated with decay of the $I = \frac{3}{2}$ excited state to the $I = \frac{1}{2}$ ground state with a 100 ns half-life are ideally suited to ^{57}Fe Mössbauer spectroscopy.

The γ-ray emission and absorption involves high energy photons, and the emitting and absorbing atoms are subject to **recoil** due to the conservation of momentum. When the γ-ray is emitted from the atom shown in black on the left-hand side of Figure 10.2, it recoils as shown by the position of the atom shown in blue also on the left-hand side. Likewise, when the atom shown in black on the right absorbs the γ-ray, it is propelled forwards (but still called recoil), as shown by the final blue atom position.

The recoil energy, E_R, is given by

$$E_R = \frac{E_\gamma^2}{2mc^2} \qquad \textbf{Eqn 10.1}$$

where E_γ is the γ-ray energy and m is the effective mass of the nucleus under study. This recoil energy needs to be minimised in order for the resonant Mössbauer effect to occur. The effective mass of the absorbing and emitting atoms can be massively increased by embedding them in a tightly bound lattice and, thereby, the recoil energy, E_R, practically vanishes. As a result of this need for the absorbing atom to be fixed in a lattice, the Mössbauer effect can only be observed with solids and frozen solutions; the technique cannot be used for liquid or gas-phase samples.

The fundamental basis of the Mössbauer experiment, for the characterisation of inorganic solids, is that there are small changes in the nuclear energy levels which are dependent on the chemical environment of the nucleus. For resonant absorption of the emitted γ-ray it is necessary to effectively modify the γ-ray energy to observe absorption at different energies. For ^{57}Fe, the γ-ray energy is ca. 14.4 keV, and the absorption linewidth is ca. 5×10^{-9} eV, thus huge resolving powers of order of 10^{12} are required. This can be achieved using the **Doppler effect** where the γ-ray frequency and energy are very slightly increased for a radioactive source travelling towards the sample, and reduced in frequency and energy for a source travelling away from the sample. If the radioactive source is driven back and forth with a velocity of a few

mm s^{-1} (1 mm s^{-1} = 48.075 neV) this modulates the energy sufficiently to cover the whole spectral range needed to characterise iron containing samples. This is the origin of the units of mm s^{-1} for the x-axis, the energy axis, in Mössbauer spectra.

A schematic diagram of a simple Mössbauer spectrometer is shown in Figure 10.3. The γ-ray source is usually mounted on a solenoid/loudspeaker driver. The sample environment can be controlled in terms of temperature and pressure. The detector measures the extent of absorption/transmission of the γ-rays at each source velocity.

Figure 10.4 shows a typical Mössbauer spectrum with one peak at 0.0 mm s^{-1}. The source velocity is the x-axis, and the y-axis is the relative transmission (1–0), but per cent transmission (100–0%) is also very common. The experimental linewidth, Γ, is given as the **full width at half maximum** (FWHM), and is usually reported in units of mm s^{-1}. The spectra are calibrated using data from well-characterised, stable and easily handled materials. For ^{57}Fe there are two common calibration samples, iron foil at room temperature (RT) or sodium nitroprusside (Na$_2$[Fe(CN)$_5$(NO)].2H$_2$O), both of which are assigned isomer shift (δ) values of 0.00 mm s^{-1}. To convert between data reported with these two standards, sodium nitroprusside (SNP) is at −0.26 mm s^{-1} using RT iron foil for calibration. All the iron Mössbauer spectra in this book have been calibrated using iron foil at RT.

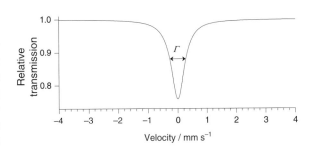

FIGURE 10.4 Typical form of a Mössbauer spectrum.

TABLE 10.1 Summary of hyperfine interactions in Mössbauer spectroscopy, and the chemical information they yield

Type of hyperfine interaction	Electric monopole	Electric quadrupole	Magnetic dipole
Mössbauer parameter	Isomer shift δ or IS	Quadrupole splitting ΔE_Q	Magnetic splitting ΔE_M
Characterisation	Oxidation state Spin state Bonding	Oxidation state Spin state Site symmetry	Magnetic properties

10.1.3 Information available from Mössbauer spectra

The information contained within a Mössbauer spectrum can be broken down into three components which result from the interaction between the nucleus and the surrounding electrons, known as **hyperfine interactions**. Table 10.1 summarises the three different types of hyperfine interaction and the chemical information derived from the sample.

10.1.3.1 Electric monopole interaction, isomer shift (IS), δ

The **isomer shift**, δ or IS, is a monopole (Coulomb) interaction that results from an interaction between the protons of the nucleus and electrons penetrating the nuclear field. (The isomer shift is also sometimes called the chemical shift.) The only electrons that penetrate the nucleus are s electrons. The effect arises because the size of the nucleus is different in the ground state and the excited state, denoted as ΔR. This leads to a difference in the energy between the $I = \frac{1}{2}$ and $I = \frac{3}{2}$ levels in the source and the absorber, as shown in Figure 10.5. Therefore, the isomer shift, δ or IS, measures the difference in the s electron density at the nucleus between the source and the absorber.

While Mössbauer spectroscopy is only able to measure variation in the s electron density at the nucleus, this information still has high chemical significance. The s electrons experience different levels of shielding from the nucleus by the p, d, and f electrons in different oxidation states, spin states, and bonding environments. For ^{57}Fe, $\Delta R/R$ is negative so an increase in s-electron density at the Fe nucleus is associated with a negative change in the isomer shift. The s electron density at the nucleus in all the s orbitals is affected by a number of factors, but it is thought to be the variation in the 3s and 4s electron density that contributes most to the isomer shift. The isomer shift in iron compounds normally increases initially from 0.0 mm s^{-1} for the iron foil calibrant (with a $4s^2 3d^6$ electronic configuration) as the 4s

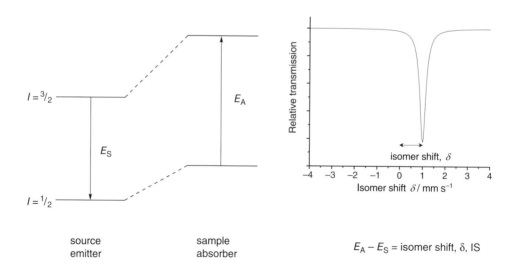

FIGURE 10.5 Origin of the isomer shift in Mössbauer spectra.

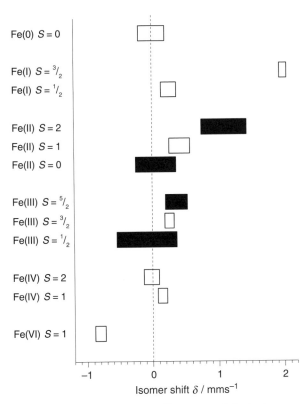

FIGURE 10.6 Representative ^{57}Fe isomer shifts for iron compounds in different oxidation states and spin states. The solid boxes represent the most commonly encountered examples.

EXAMPLE 10.1

Mössbauer isomer shifts

Predict a ^{57}Fe isomer shift for iron in the compound K_3FeO_4.

ANSWER

The compound contains Fe(V) with the $4s^n3d^n$ configuration $4s^03d^3$ and $S = \frac{3}{2}$. We would predict a level of s electron density at the nucleus intermediate in value between that of Fe(IV) and Fe(VI) predicted based on the effect of oxidation number. From Figure 10.6, a δ value in the range -0.35 to -0.6 mm s^{-1} would be predicted. Experimentally the value for K_3FeO_4 is at -0.55 mm s^{-1}.

SELF TEST

Predict isomer shift(s) for the main iron resonance(s) in the ^{57}Fe Mössbauer spectrum of Fe_3O_4.

electrons are the first to be lost in the formation of an iron compound. Therefore, for iron(II) compounds the isomer shift is a positive value. But as the iron oxidation state increases further, for example, iron(III), the nucleus attracts the remaining electrons more strongly, and so the remaining s electron density increases with oxidation state, and the isomer shift moves to smaller (more negative than iron(II)) values. In addition, an increase in electron donation by the ligands (more covalency in the bonding) also results in a decrease in the isomer shift. The isomer shift also decreases with coordination number, so that there is reduction of 0.2–0.3 mm s^{-1} on going from octahedral to tetrahedral geometry for the same ligands. A summary of ^{57}Fe isomer shifts for iron compounds and complexes is shown in Figure 10.6.

For ^{119}Sn, $\Delta R/R$ is positive, so an increase in the isomer shift is expected for higher oxidation state tin compounds compared to lower oxidation states. For ^{197}Au Mössbauer spectroscopy, $\Delta R/R$ is also positive and Au(I) isomer shifts are at ca. -1.5 to -1.0 mm s^{-1}, Au(III) at 0–1 mm s^{-1}, and Au(V) at >2 mm s^{-1}.

10.1.3.2 Electric quadrupole interaction, quadrupole splitting, ΔE_Q

The **electric quadrupole splitting**, ΔE_Q, results from an interaction between the nuclear magnetic moment, I, and an inhomogeneous (i.e. non-cubic) electric field. The non-spherical nature of the nucleus is given by the quadrupole moment, eQ, and is measured by the **electric field gradient (EFG)** in the z-direction. (See Section 10.2.3 in the NQR section for a more detailed discussion of quadrupolar nuclei.) In principle, this effect can be observed in all Mössbauer experiments as at least one of the nuclear states involved must possess a **nuclear quadrupole moment**, that is, $I > \frac{1}{2}$. In an inhomogeneous electric field (i.e. no longer having cubic symmetry, so that the z-direction is distinguishable from x- and y-directions) the degeneracy of the nuclear I states for a ($I > \frac{1}{2}$) nucleus is lifted according to the following equation:

$$E_Q = \frac{e^2qQ[3m_I^2 - I(I+1)]}{4I(2I-1)} \qquad \textbf{Eqn 10.2}$$

where e is the charge on the electron, q is the field gradient, and Q is the quadrupole moment (see Table 3.3 for values); e^2qQ is known as the **quadrupole coupling constant** (QCC).

For ^{57}Fe, the ground state is $I = \frac{1}{2}$ with m_I values of $+\frac{1}{2}$ and $-\frac{1}{2}$, and the excited state has $I = \frac{3}{2}$ with $m_I = +\frac{3}{2}, +\frac{1}{2}, -\frac{1}{2}, -\frac{3}{2}$. In a non-spherical charge density environment the $m_I = \pm\frac{1}{2}$ levels in the $I = \frac{1}{2}$ ground state will remain degenerate, but Eqn 10.2 shows that for the $I = \frac{3}{2}$ excited state, the levels are split by an inhomogeneous (non-spherical, non-cubic) electric field into $m_I = \pm\frac{3}{2}$ levels and $m_I = \pm\frac{1}{2}$ levels, as shown in Figure 10.7.

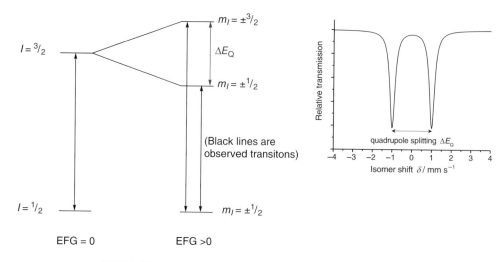

FIGURE 10.7 Origin of quadrupole splitting in Mössbauer spectra.

With the selection rule, $\Delta m_I = 0, \pm 1$, there are now two transitions possible from the ground to the excited state, that is, if there is an EFG at the nucleus, with a quadrupole moment ($I > \frac{1}{2}$), the degeneracy of the m_I levels will be lifted and this manifests itself as a doublet of resonances in the Mössbauer spectrum. The separation of the two spectral lines is measured (in mm s^{-1}) as the **quadrupole splitting**, ΔE_Q.

^{119}Sn also possesses an $I = \frac{1}{2}$ ground state and an $I = \frac{3}{2}$ excited state so Figure 10.7 is also appropriate for ^{119}Sn Mössbauer experiments. In the case of ^{197}Au, the ground state is $I = \frac{3}{2}$ and the excited state is $I = \frac{1}{2}$, so the energy level diagram needs to be inverted but a quadrupole doublet is also expected to be observed in ^{197}Au Mössbauer spectra if an EFG is present. If $I > \frac{3}{2}$, Eqn 10.2 shows that the separation of the m_I states split by an EFG are no longer equal (see Figure 10.16 in NQR section), and more complex spectra are observed.

An EFG around the nucleus under investigation can be present due to either a non-cubic lattice site (geometric) or a non-cubic valence electron distribution (electronic). The first of these is fairly easy to visualise. For example, if the nucleus under investigation is in a cubic symmetric environment where the x, y, and z directions are equivalent, such as in octahedral (O_h) or tetrahedral (T_d) symmetry, then the EFG is zero. However, for a lower site symmetry such as square planar (D_{4h}) (or pyramidal, C_{3v}, instead of T_d), the x, y, and z directions in the local environment are inequivalent and an EFG will be present (although its magnitude may be very small).

A non-cubic electron distribution can be illustrated by considering the d electron configurations of high-spin and low-spin octahedral iron compounds. Note that s electron orbitals are spherically symmetric around a nucleus, as are completely filled shells of electrons such as

p^6 and d^{10}. For Fe(II) it is the non-spherical distribution of the d electrons around the nucleus that would give rise to an EFG. In an octahedral field spherically symmetrical configurations such as high-spin d^5 ($t_{2g}^3 e_g^2$) do not give an EFG. Similarly complete and half-filled subshells (the t_{2g} and e_g sets in an octahedral field) such as low-spin d^6 ($t_{2g}^6 e_g^0$) and d^8 ($t_{2g}^6 e_g^2$) are spherically symmetric. The presence or not of an EFG can be readily determined from the term symbol for an electronic configuration (Tables 5.6 and 8.1 in Chapters 5 and 8, respectively). An A term indicates a spherical charge distribution, no EFG, and no valence electron induced quadrupole splitting, whereas E and T terms indicate a non-spherical charge distribution and hence a significant EFG. Such arguments demonstrate that high-spin, octahedral Fe(III) with the electronic configuration $t_{2g}^3 e_g^2$ and a $^6A_{1g}$ ground term will have no electronically derived EFG while low-spin octahedral Fe(III) with the electronic configuration $t_{2g}^5 e_g^0$ and a $^2T_{2g}$ ground term is expected to have an EFG and, hence, $\Delta E_Q \neq 0$. Similar arguments for octahedral Fe(II) would predict $\Delta E_Q = 0$ for low-spin d^6 ($t_{2g}^6 e_g^0$, $^1A_{1g}$) and $\Delta E_Q \neq 0$ for high-spin d^6 ($t_{2g}^4 e_g^2$, $^5T_{2g}$). ΔE_Q has a much stronger temperature dependence for low-spin Fe(III) due to the presence of close lying excited electronic states when slightly distorted, than low-spin Fe(II), where there is no valence contribution to the EFG. This can be used to differentiate between the two oxidation states, especially as the isomer shifts, δ, are very similar.

While it is convenient to discuss the isomer shift and quadrupole splitting separately, both need to be considered when interpreting most Mössbauer spectra. A summary of the combined effects of an isomer shift and quadrupole splitting on the Mössbauer spectrum is given in Figure 10.8.

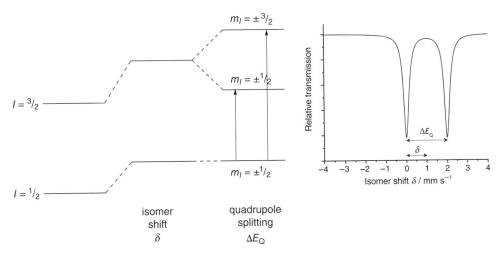

FIGURE 10.8 Combined effect of isomer shift and quadrupole splitting on a Mössbauer spectrum.

EXAMPLE 10.2

Identifying oxidation state and spin state in iron chemistry using isomer shifts and quadrupole splitting

Interpret the Mössbauer spectra of the four iron compounds, $K_3[Fe(CN)_6]$, $Fe_2(SO_4)_3.nH_2O$, $K_4[Fe(CN)_6].3H_2O$, and $(NH_4)_2Fe(SO_4)_2.6H_2O$, shown in Figure 10.9. In these spectra, the conventional way of presenting Mössbauer data is used, with the experimental data shown using + signs with the solid line representing the best fit to the data by varying δ, ΔE_Q, and Γ.

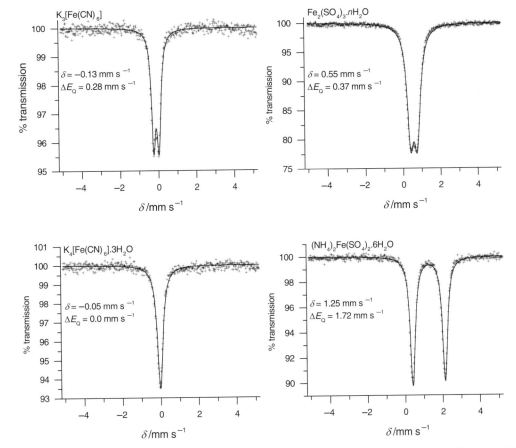

FIGURE 10.9 Illustrative ^{57}Fe Mössbauer spectra of high- and low-spin iron (II) and iron(III) complexes at 80 K.

ANSWER

For ^{57}Fe, increasing electron density at the nucleus results in a decrease in the isomer shift due to the negative $\Delta R/R$. Thus both an increase in oxidation state and increasing donation by the ligands give rise to a decrease in the isomer shift. The isomer shifts move to lower (more negative) values on going from Fe(II) to Fe(III), but also on going from high-spin to low-spin as the extent of covalency is greater in the low-spin than the high-spin complexes. Apart from high-spin Fe(II) $(NH_4)_2Fe(SO_4)_2.6H_2O$, with an isomer shift of 1.25 mm s^{-1} consistent with high-spin iron(II), the isomer shifts of the other three compounds are not that diagnostic as they are fairly similar, especially for the two complexes containing cyanide ligands. For these materials the quadrupole splitting, ΔE_Q, can be used to differentiate the spin

and oxidation states. Assuming that the complexes are octahedral there will be only a small geometric contribution to the EFG. A valence electron induced EFG is associated with an E or T ground state term. Therefore, for low-spin iron(II) with an $^1A_{1g}$ ground term, there are minimal geometric and electronic effects to the EFG, and so ΔE_Q is expected to be zero. In contrast, for low-spin iron(III) with a $^2T_{2g}$ ground term, there is now an electronic contribution to the EFG, and a modest ΔE_Q is observed. For the high-spin iron(II) $(^5T_{2g})$ an EFG is predicted and a large ΔE_Q is observed, but for high-spin iron(III) $(^6A_{1g})$ no valence electron EFG is expected. The small ΔE_Q observed for $Fe_2(SO_4)_3.nH_2O$ is due to geometric rather than electronic effects. The interpretation of the spectra in respect of the different oxidation states and spin states is summarised in Table 10.2.

TABLE 10.2 Summary of the assignments of the spectra for the high- and low-spin iron(II) and iron(III) complexes at 80 K in Figure 10.9.

		low-spin			high-spin			
			δ/mm s^{-1}	ΔE_Q/mm s^{-1}			δ/mm s^{-1}	ΔE_Q/mm s^{-1}
Fe(III)	$K_3[Fe(CN)_6]$	$^2T_{2g}$	−0.13	0.28	$Fe_2(SO_4)_3.nH_2O$	$^6A_{1g}$	+0.55	0.37
Fe(II)	$K_4[Fe(CN)_6].3H_2O$	$^1A_{1g}$	−0.05	0	$(NH_4)_2Fe(SO_4)_2.6H_2O$	$^5T_{2g}$	+1.25	1.72

SELF TEST

Figure 10.10 shows the ^{57}Fe Mössbauer spectrum of $Na_2[Fe(CN)_5(NO)].2H_2O$. Interpret this spectrum in terms of the geometric and electronic contributions to the EFG and hence ΔE_Q.

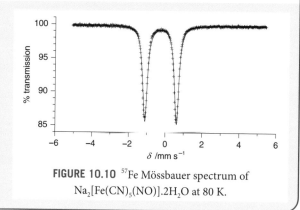

FIGURE 10.10 ^{57}Fe Mössbauer spectrum of $Na_2[Fe(CN)_5(NO)].2H_2O$ at 80 K.

10.1.3.3 Magnetic dipole interaction, ΔE_M

The Zeeman interaction between the nuclear magnetic dipole moment and a magnetic field at the nucleus, caused either by an external or an internal magnetic field, results in splitting of a Mössbauer spectrum into additional lines. An internal magnetic field is most often displayed by ferromagnetic materials (see Section 8.3). In a magnetic field the $m_I = \pm\frac{1}{2}$ levels of the $I = \frac{1}{2}$ ground state are no longer degenerate and are split into two components, $m_I = +\frac{1}{2}$ and $m_I = -\frac{1}{2}$, with the $m_I = +\frac{1}{2}$ lying lowest in energy for ^{57}Fe. This is the same process that is

occurring in NMR spectra in the presence of an external magnetic field. The $I = \frac{3}{2}$ excited state is split into four components, $m_I = +\frac{3}{2}, +\frac{1}{2}, -\frac{1}{2}, -\frac{3}{2}$, but in this case the lowest energy state is the $m_I = -\frac{3}{2}$ component because the sign of the splitting is reversed in the two states for ^{57}Fe. As the selection rule is still $\Delta m_I = 0, \pm 1$, six transitions are observed, as shown in Figure 10.11 for iron foil.

The separation between lines 1 and 2 gives the splitting in the excited state, while that between lines 2 and 4 gives the ground state splitting. For a randomly orientated sample such as a powder, the relative intensities of the six peaks are 3:2:1:1:2:3, as shown in Figure 10.11.

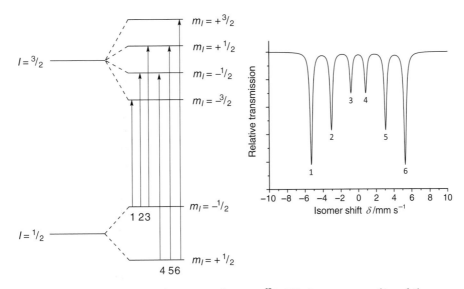

FIGURE 10.11 Origin of magnetic splitting in ^{57}Fe Mössbauer spectra of iron foil.

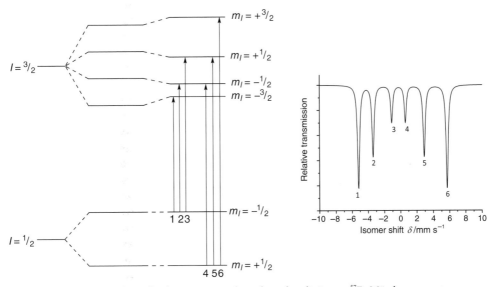

FIGURE 10.12 Effect of both magnetic and quadrupole splitting on ^{57}Fe Mössbauer spectra.

However, the relative intensity of the lines labelled 2 and 5 varies according to $(4\sin^2\theta/(1 + \cos^2\theta))$ (where θ is the angle between the magnetic direction and the γ-ray). If the magnetic field within the sample is orientated with respect to the γ-rays, their intensity varies from 0 to 4. If they are parallel, the intensities of lines 2 and 5 become zero; if perpendicular the relative intensity is 4.

If the EFG > 0 then quadrupole splitting also occurs and the situation becomes both more informative and more complex. The energy of $m_I = \pm\frac{3}{2}$ and $m_I = \pm\frac{1}{2}$ states of the $I = \frac{3}{2}$ excited state move in opposite directions depending on the sign of the quadrupole splitting. This results in a less symmetrical splitting pattern where the separation of the 1 and 2 lines is different from that of the 5 and 6 lines (see Figure 10.12) as these involve

both $m_I = \pm\frac{3}{2}$ and $m_I = \pm\frac{1}{2}$ states in the excited state. The separation of the 3 and 4 lines remains the same as these only involve $\pm\frac{1}{2}$ states in the excited state. (If the quadrupole splitting has the opposite sign, then lines 1 and 2 are further apart, whereas 5 and 6 are closer together.)

This is a very powerful way to study magnetic samples, in terms of their geometric, electronic, and magnetic structures. In addition, it is also a very valuable fingerprinting method for specific iron minerals as they each have characteristic magnetic and quadrupole splitting patterns. For example, a miniature Mössbauer spectrometer was included on the Mars Rover and Curiosity vehicles, and the identification of the mineral jarosite, $KFe_3(OH)_6(SO_4)_2$, implied the presence of water on Mars.

EXAMPLE 10.3

Effect of magnetic ordering—haematite, α-Fe₂O₃

α-Fe$_2$O$_3$ adopts the corundum structure of a hexagonal close-packed array of oxide ions with two thirds of the octahedral holes filled by iron(III). The ^{57}Fe Mössbauer spectrum of α-Fe$_2$O$_3$ at 371.3 K and 162.0 K, as well as small nanoparticles at 298 K, are shown in Figure 10.13. Interpret these spectra.

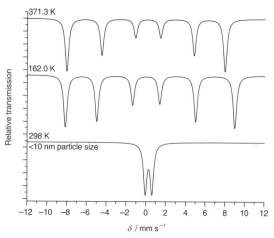

FIGURE 10.13 ^{57}Fe Mössbauer spectra of α-Fe$_2$O$_3$. (Data adapted from F. van der Woude, *Phys. Stat. Sol.* **17** 417 (1966) and W. Kündig, H. Bömmel, G. Constabaris, and R. H. Lindquist, *Phys. Rev.* **142** 327 (1966).)

ANSWER

In α-Fe$_2$O$_3$ (haematite) there is a magnetic transition at 260 K from weakly ferromagnetic (T > 260 K) to antiferromagnetic (see Section 8.3). It is not easy to read the values of isomer shifts, magnetic splitting, and quadrupole splitting directly off the 371.3 and 162.0 K spectra, but fitting of the data has shown that the isomer shift in both spectra is the same (0.38 mm s^{-1}), and the magnetic field B at the nucleus is 51.5 T. However, ΔE_Q changes from +0.12 mm s^{-1} at 371.3 K, above the transition temperature, to −0.22 mm s^{-1} at 162.0 K, below the transition temperature, and this change in ΔE_Q accounts for the shift in the position of the features in the two spectra. The spectrum of small (<10 nm) nanoparticles only displays a quadrupole splitting as the sample is **superparamagnetic** (where paramagnetic behaviour is observed in nanoparticles below the Curie temperature). If the sample size is increased, or the temperature reduced, the magnetic (Zeeman) splitting re-emerges. Hence, care should be taken in interpreting spectra if small particles are expected either in laboratory or field samples.

SELF TEST

Figure 10.14 shows the RT ^{57}Fe Mössbauer spectra of γ-Fe$_2$O$_3$ (maghemite) and Fe$_3$O$_4$ (magnetite). Assign the spectra to the correct iron oxide.

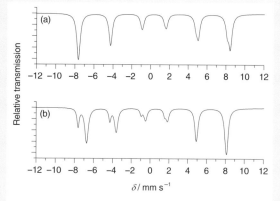

FIGURE 10.14 Room temperature ^{57}Fe Mössbauer spectra of γ-Fe$_2$O$_3$ and Fe$_3$O$_4$. (Data from R.J. Armstrong, A.H. Morrish, and G.A. Sawatzky, *Phys. Lett.* **23** 414 (1966); I.S. Lyubutin, C.R. Lin, Y.V. Korzhetskiy, T.V. Dmitrieva, and R.K. Chiang, *J. Appl. Phys.* **106** 034311 (2009).)

10.1.4 Elemental range in Mössbauer spectroscopy

While many elements have nuclear states that are in principle accessible for Mössbauer spectroscopy, in practice the range of elements that can be studied is much more limited as a suitable γ-ray source is required. As they are by definition radioactive isotopes, this poses serious limitations on the elements that are available for use in a laboratory spectrometer. ^{57}Fe is the most extensively studied; other elemental isotopes that have been commonly studied include ^{119}Sn, ^{121}Sb, ^{127}I, ^{129}I, ^{193}Ir, ^{197}Au, ^{151}Eu, and ^{237}Np. Table 10.3 summarises the more important nuclei that can be studied using Mössbauer spectroscopy as well as their important spectral parameters.

It can be seen from these data that some nuclei such as ^{57}Fe, ^{119}Sn, and ^{121}Sb make use of sources with reasonably long half-lives which mean the source can be used to collect data over several months or years. Others with short half-lives need to be prepared in a nuclear reactor immediately prior to use. As the recoil energy is proportional to the square of the γ-ray energy, those which require higher energy γ-rays have a smaller recoil-free fraction, and usually spectra have to be collected at low (typically liquid helium) temperatures. The theoretical linewidth is dependent on the transition energy

TABLE 10.3 Summary of the important parameters for the most common isotopes used with Mössbauer spectroscopy

Absorber isotope	Natural abundance (per cent)	Ground state I	Excited state I	E_γ (keV)	Source	Half-life of parent	Linewidth Γ, (FWHM, mm s^{-1})	$\Delta R/R$
^{57}Fe	2.1	½	³⁄₂	14.4	^{57}Co	270 d	0.192	negative
^{61}Ni	1.2	³⁄₂	⁵⁄₂	67.4	^{61}Co	99 m	0.77	negative
^{119}Sn	0.63	½	³⁄₂	23.8	^{119m}Sn	240 d	0.626	positive
^{121}Sb	2.1	⁵⁄₂	⁷⁄₂	37.2	^{121m}Sb	77 y	2.1	negative
^{127}I	100	⁵⁄₂	⁷⁄₂	57.6	^{127m}Te	105 d	2.54	negative
^{129}I	0.6	⁷⁄₂	⁵⁄₂	27.7	^{129m}Te	33 d	0.59	positive
^{99}Ru	12.8	³⁄₂	⁵⁄₂	89.4	^{99}Rh	16 d	0.147	positive
^{193}Ir	61.5	½	³⁄₂	73.0	^{193m}Os	30 h	0.60	positive
^{197}Au	100	³⁄₂	½	77.3	^{197m}Pt	18 h	1.87	positive
^{129}Xe	26.4	½	³⁄₂	39.6	^{129}I	17×10^7 y	6.85	positive
^{151}Eu	47.7	⁵⁄₂	⁷⁄₂	21.6	^{151}Gd	129 d	1.44	positive

and the excited state lifetime, and is derived from the Heisenberg uncertainty principle; experimentally the observed linewidths are greater than these values. The linewidth indicates how well resolved the peaks will be in the Mössbauer spectrum, and thus how sensitive the technique will be to changes in the isomer shift, and whether quadrupole or magnetic splitting will be clearly resolved or not.

A combination of the favourable experimental conditions of a relatively low energy γ-ray, giving small resonance linewidths (Γ), and a source with a reasonable lifetime, combined with a rich chemistry to explore, means that ^{57}Fe is the most commonly studied isotope. The fairly low natural abundance of ^{57}Fe at 2.1% can lead to long data collection times to get good quality spectra, especially in samples with low iron atomic percentage contents, such as biological systems. ^{57}Fe enriched samples can be prepared to improve the signal quality or reduce the data collection time. Additionally this also allows for complementary EPR experiments (Chapter 9). Although reasonable quality ^{57}Fe spectra can be recorded at RT, lower temperatures are usually employed to improve the data quality.

With the advent of synchrotron based Mössbauer experiments many more elements become accessible. Recent scientific literature reports include studies of ^{40}K, ^{151}Eu, ^{174}Yb, ^{61}Ni, ^{73}Ge, ^{119}Sn, ^{125}Te, ^{127}I, ^{149}Sm, and ^{189}Os and include the discovery of the existence of a new europium hydride, EuH$_{2+x}$, under high pressure from the presence of a Eu^{3+} resonance in the Mössbauer synchrotron spectrum.

10.1.5 Applications of Mössbauer spectroscopy to elements other than iron

The principles outlined above for ^{57}Fe Mössbauer spectroscopy can be applied to many other elements. However, as $\Delta R/R$ may be either positive or negative, interpretation of the isomer shift needs care. Many of the experiments exploit the ability of Mössbauer spectroscopy to determine the oxidation state of an element, so the technique is most widely employed for transition metals and post-transition metals where such characterisation information is particularly useful.

A number of transition metals other than iron have been studied using Mössbauer spectroscopy, including iridium, gold, ruthenium, tungsten, tantalum, and nickel. These elements are frequently components of catalysts and information on various stages of a catalytic cycle can often be obtained from Mössbauer spectroscopy. Although many of the lanthanides have isotopes amenable to Mössbauer spectroscopy, the emphasis has been on ^{151}Eu. $\Delta R/R$ is positive and it is possible to discriminate Eu^{2+} from Eu^{3+}. Amongst the actinides ^{237}Np is the most amenable to Mössbauer spectroscopy.

For ^{119}Sn, the isomer shift is sufficiently different for tin(II) and tin(IV) and identification of the oxidation state is usually possible. As tin(II) compounds formally have a lone pair of electrons, the stereochemical activity of this in producing highly asymmetric environments can be determined by analysis of ΔE_Q values. ^{119}Sn has $I = ½$ ground state and $I = ³⁄₂$ excited state, and there are clear parallels with iron spectral analysis. ^{121}Sb has an $I = ⁵⁄₂$ ground state with an $I = ⁷⁄₂$ excited state, and this results in more complex spectra with 8 or 12 overlapping lines. [121]Sb spectra can be used to distinguish antimony(III) and antimony(V) as the majority of antimony(III) complexes have a substantial quadrupole splitting due to the presence of a stereochemically active lone pair.

10.2 Nuclear quadrupole resonance (NQR) spectroscopy

10.2.1 Introduction to NQR spectroscopy

Nuclear quadrupole resonance (NQR) spectroscopy involves transitions within nuclear energy levels and, therefore, uses the concepts introduced in both NMR (Chapter 3) and Mössbauer spectroscopy (Section 10.1). The vast majority of NMR experiments probe a nucleus with an $I = \frac{1}{2}$ ground nuclear state, whereas for NQR the ground state must be quadrupolar ($I \geq 1$). In NQR spectroscopy the nuclear ground state degeneracy of the m_I levels is lifted by the presence of a local EFG at the nucleus. Transitions between the $\pm m_I$ energy levels occur, as with NMR, in the radiofrequency region. As the EFG at the nucleus is generated by the local electronic and chemical environment, the NQR spectrum can provide specific characterisation information on the number of different, and type of, local environments of the quadrupolar nucleus under investigation.

10.2.2 Experimental considerations for NQR spectroscopy

As the experimental requirements for NQR are essentially a magnet-free NMR experiment, similar technology to that used for NMR can be employed, that is, pulsed Fourier transform (FT) methods (Section 3.3). Pulsed FT techniques allow spectral averaging (where up to a million transients may be collected) and also allow for the use of indirect double resonance techniques that enables the NQR investigation of biologically relevant light nuclei (e.g. [2]H, [10]B, [11]B, ^{14}N, [17]O, [23]Na, [25]Mg, [27]Al) that are otherwise hard to observe, in the presence of those elements that have a stronger NQR signal.

10.2.3 Underlying NQR theory

A quadrupolar nucleus is one that has a non-spherical nuclear charge distribution, and this gives rise to a non-zero quadrupole moment, Q. Figure 10.15 shows that if Q is positive, the shape is elongated along the z-axis (rugby ball shaped) and denoted prolate. If Q is negative, the shape is oblate with the charge distribution contracted along the z-axis resulting in a flying saucer shape.

Quadrupole splitting was introduced in Section 10.1.3.2 to explain the number of peaks within Mössbauer spectra

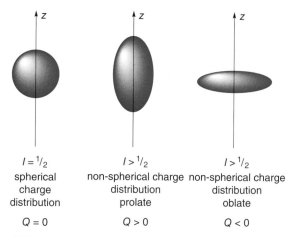

$I = \frac{1}{2}$	$I > \frac{1}{2}$	$I > \frac{1}{2}$
spherical charge distribution	non-spherical charge distribution prolate	non-spherical charge distribution oblate
$Q = 0$	$Q > 0$	$Q < 0$

FIGURE 10.15 Spherical and non-spherical nuclear charge distribution.

when the nucleus was in the presence of an EFG. In NQR spectroscopy exactly the same process removes the degeneracy of the $\pm m_I$ energy levels but in this case it is the transitions between the m_I levels of the nuclear ground state that are of interest. The energies of these $\pm m_I$ levels are given by

$$E_Q = \frac{e^2 q Q [3m_I^2 - I(I+1)]}{4I(2I-1)}$$

Eqn 10.3

where e is the charge on the electron, q is the field gradient, Q is the quadrupole moment (see Table 3.3 for values), and e^2qQ is known as the quadrupole coupling constant (QCC). The energy levels for the most commonly studied quadrupolar NQR nuclei with different I values are plotted in Figure 10.16. The calculations have assumed that both q and Q are positive, if the sign is changed, the energy levels are reversed, but the transition energies (and hence spectra) remain the same. These diagrams were generated using a common value of the QCC (e^2qQ) to show the relative splitting of the m_I states for the different values of I. The energy level with the largest values of m_I is always found at QCC/4. However, it should be noted that experimentally QCC is found to vary from ca. 10 MHz for ^{14}N to 2 GHz for [127]I. As the total spread of energies is very small, there is appreciable population of all the levels. As a result, transitions are expected from many levels, and because of the very small difference in their population, the intensities are effectively the same.

The NQR selection rule is $\Delta m_I = \pm 1$, so for half-integer spins, I, there are $(I + \frac{1}{2})$ energy levels and $(I - \frac{1}{2})$ allowed transitions. For integer spins, there are $(I + 1)m_I$ levels and I allowed transitions. Therefore, for nuclei with $I = 1$ and $I = \frac{3}{2}$ one transition is predicted, but for $I = \frac{5}{2}$ there are two transitions with relative energies in the ratio 1:2, and for $I = \frac{7}{2}$ there are three transitions with relative energies in the ratio 1:2:3.

This analysis assumes that the EFG is axial (i.e. $x = y \neq z$). If x and y are no longer equivalent, then an asymmetry

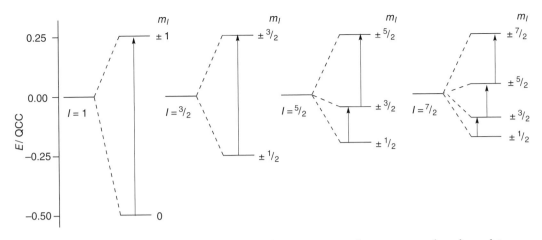

FIGURE 10.16 Energy levels, in terms of QCC, and NQR transitions for common quadrupolar nuclei.

parameter, η, is introduced. For half-integer spins the number of transitions is unaffected, but the energies change slightly. For $I = 1$, the $m_I = \pm 1$ states are no longer degenerate, and two transitions rather than one are observed.

The effect of a magnetic field on the energy levels shown in Figure 10.16 is analogous to that shown in Figure 10.11, in Mössbauer spectroscopy, and removes the degeneracy of the $\pm m_I$ levels, thus increasing the number of observed transitions. The application of a small external magnetic field (0.5 T) can be used to help assign the spectra, especially to determine the sign of EFG and the QCC.

10.2.4 Information determined from NQR spectra

Figure 10.16 shows that for $I = 1$ there is one transition at an energy corresponding to ¾QCC (QCC = e^2qQ), and for $I = \frac{3}{2}$ nuclei there is one NQR transition at ½QCC. For $I = \frac{5}{2}$ there will be two transitions at the energies $\frac{3}{20}$QCC and $\frac{6}{20}$QCC, and for $I = \frac{7}{2}$ there are three transitions at $\frac{1}{14}$QCC, $\frac{2}{14}$QCC, and $\frac{3}{14}$QCC. Therefore, for $I = 1$ and $I = \frac{3}{2}$ there will be one transition (assuming axial EFG) for each nucleus in a distinct environment, but the spectra of $I = \frac{5}{2}$ will contain two features per distinct environment, and for $I = \frac{7}{2}$ there will be three. The QCC for each environment can be readily obtained from the transition energies. Unlike NMR spectra, there are no coupling data present in the NQR spectra. Therefore, one of the principal benefits of NQR for characterising inorganic compounds is that the spectra can readily identify the number of environments in a sample. This is useful in identifying shape as well as the presence of different local sites in solid state compounds.

The separation of the energy levels is dependent on the EFG at the nucleus, and this can also be used to investigate the nature of bonding. In the case of the halogens, often studied in NQR spectroscopy, the halogen is usually covalently bound to only one other atom. This gives rise to a significant EFG as the electron density along

the bond is different from that perpendicular to it. In addition, the electron distribution along the bond, and hence the EFG, varies with the covalency or ionicity of the bonding. This gives rise to characteristic transition frequencies such as 54.247 MHz for $^{35}Cl_2$, 30–40 MHz for ^{35}Cl–C, and 15–20 MHz for the more ionic ^{35}Cl–Si bond. A more sophisticated treatment has been developed by Townes and Dailey for estimating the ionic character in the bonding, and this can be used for simple inter-halogens as well as transition metal complexes. In this analysis the ionic character of a bond, i, can be related to the ^{35}Cl NQR frequency ($\nu(^{35}Cl)$) in MHz using the expression:

$$i = 1 - \frac{\nu(^{35}Cl)}{54.873} \qquad \textbf{Eqn 10.4}$$

Values for the resonance frequency range from the purely covalent 54.247 MHz for $^{35}Cl_2$ to near zero MHz for the ionic alkali metal halides.

EXAMPLE 10.4

The ^{35}Cl NQR spectrum of $TiCl_4$ shows a resonance at 6.05 MHz. Calculate the iconicity, i, of the Ti–Cl bond.

ANSWER

Using Eqn 10.4 and the NQR resonance frequency of 6.05 MHz, a value of i of 0.89 is obtained.

SELF TEST

The ^{35}Cl NQR frequencies of $[TiCl_3Cp]$ and $[TiCl_2Cp_2]$ are 8.1 and 11.8 MHz, respectively. (Cp = $[C_5H_5]^-$,) Use these NQR data to comment on the effect of the cyclopentadienyl ligands (which are better π donor ligands than Cl^-) on the Ti–Cl bonding with respect to that in $TiCl_4$.

10.2.5 Range of elements for, and limitations of, NQR spectroscopy

The primary constraint on whether an element can be studied by NQR is that the nuclear ground state must be quadrupolar ($I \geq 1$). While many elements have isotopes that are quadrupolar (see Table 3.3), the most commonly studied elements are within the p- block such as the halogens (Cl, Br, I), Group 15 (N, As, Sb), Group 13 (B, Al), with much less usage in transition metal chemistry (Cu, Co, Mn). The halogens are widely studied because in the vast majority of their compounds, the bonding environment is very asymmetric, so that an EFG and hence an NQR spectrum is expected. For elements in more symmetric environments, the QCC may not be sufficient to allow a useful NQR spectrum to be obtained.

While the technology for NQR may be similar to that of NMR, it is very much still a specialist technique and only available in a few laboratories. However, in recent years ^{14}N NQR spectroscopy has been developed to identify different drugs and determine whether they are genuine or counterfeit. Many explosives contain nitrogen, and as each of these has very characteristic ^{14}N NQR frequencies, the technique is being developed for explosive detection both in the battlefield and at transportation centres.

As for Mössbauer spectroscopy the sample must be solid. This limitation is caused because the EFG is internal to the molecule and in liquids or gases tumbling takes place, and this averages the EFG to zero if the tumbling frequency is faster than the observation frequency. The technique is not very sensitive and even with FT spectrometers, a million transients may be required to obtain a good signal-to-noise ratio.

The frequency range over which transitions are expected can be quite large, so low resolution survey scans followed by higher resolution scans over the energies of interest are required. The linewidths can be very narrow (1–10 kHz) compared to the scan range (MHz) and this means that it is often possible to observe very subtle site effects in the solid state. Often more signals than expected from an isolated molecule are observed in the NQR spectrum in the crystalline solid due to variations in the local symmetry of the molecules.

10.2.6 Analysis of NQR data

The application of NQR spectroscopy to the characterisation of halogen environments can be illustrated using the phosphorus pentahalides. The ^{35}Cl NQR spectrum of molecular PCl_5 (sublimed and rapidly frozen at 77 K) contains two resonances at 29.258 and 33.571 MHz, which can be assigned to the axial and equatorial chlorine atoms, respectively.

In contrast, the ^{35}Cl NQR spectrum of solid crystallised PCl_5 displays two sets of peaks: one set at 28.424, 29.720,

30.478 MHz and another at 32.288, 32.396, 32.620 MHz. On the basis of the formal charges and the P–Cl distances in the ions, the lower frequency set belong to $[PCl_6]^-$, while the higher frequency set belong to $[PCl_4]^+$ as the solid is formed of an infinite array of $[PCl_4][PCl_6]$ ion pairs (with tetrahedral and octahedral geometry, respectively). While each of the chlorine atoms in the anion and cation might be expected to be equivalent, NQR is sufficiently sensitive to identify the different crystallographic chlorine atoms in the solid state. These sites are inequivalent as while all have a close phosphorus atom to which they are bonded (either in $[PCl_4]^+$ or $[PCl_6]^-$ ions) the chlorine atoms on neighbouring molecular ions form slightly different arrangements at longer distances and these marginally affect the local EFG.

EXAMPLE 10.5

Halogen atom distributions in $PCl_{5-n}F_n$ compounds

The ^{35}Cl NQR spectrum of PCl_4F (sublimed and rapidly frozen at 77 K) shows two resonances at 28.99 and 32.54 MHz with intensities in the ratio 1:3. Suggest a shape for this molecule.

ANSWER

PCl_4F would be predicted by the valence shell electronic pair repulsion (VSEPR) model to be trigonal bipyramidal. If the fluorine was located in an equatorial position, the relative intensity of the two peaks would be 1:1, reflecting two axial and two equatorial chlorine atoms, whereas a fluorine in an axial position would yield a relative intensity of 3:1. Therefore, on the basis of the NQR data the structure of PCl_4F has the fluorine atom in the axial position (Figure 10.17).

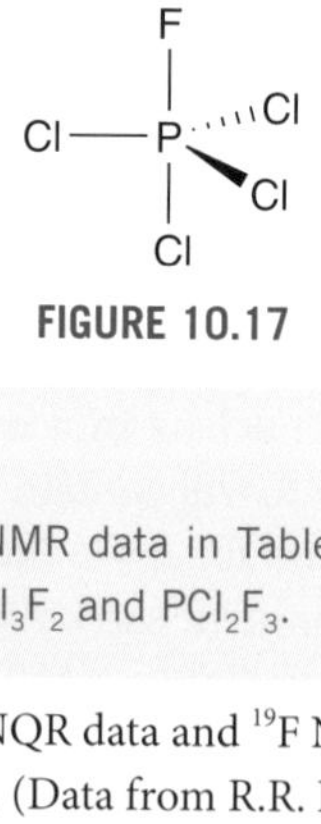

FIGURE 10.17

SELF TEST

Use the NQR and NMR data in Table 10.4 to determine the structures of PCl_3F_2 and PCl_2F_3.

TABLE 10.4 ^{35}Cl NQR data and ^{19}F NMR data for PCl_3F_2 and PCl_2F_3. (Data from R.R. Holmes, R.P. Carter, and G.E. Peterson, *Inorg. Chem.* **3** 1748 (1964); R.P. Carter and R.R. Holmes, *Inorg. Chem.* **4** 738 (1965); H. Chihara, N. Nakamura, and S. Seki, *Bull. Chem. Soc. Jpn.* **40** 50 (1967).)

Compound	^{35}Cl NQR/MHz	^{19}F NMR/ppm
PCl_3F_2	31.57	−123.0
PCl_2F_3	31.49	−67.4, +41.5

EXAMPLE 10.6

Copper NQR data of minerals

Copper has two isotopes, ^{63}Cu and ^{65}Cu, both of which are quadrupolar ($I = \frac{3}{2}$, Table 3.3). The mineral enargite, Cu_3AsS_4, which contains copper in distorted tetrahedral environments, gave the NQR spectrum in the 4–7 MHz range shown in Figure 10.18. Interpret these data.

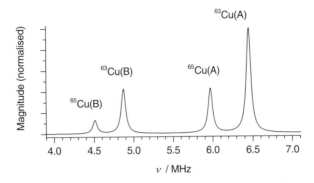

FIGURE 10.18 ^{63}Cu and ^{65}Cu NQR spectrum of Cu_3AsS_4 (enargite). (Data from J.A. Lehmann-Horn, D.G. Miljak, and T.J. Bastow, *Solid State Nucl. Magn. Reson.* **54** 8 (2013).)

ANSWER

The NQR spectrum contains two pairs of peaks due to two different copper environments in the solid state structure and also the isotopic abundance of the two copper isotopes (^{63}Cu and ^{65}Cu approximately 2.25:1). The ratio of the quadrupole moments (Table 3.3) and hence NQR frequencies for ^{63}Cu and ^{65}Cu in the same site is 1.078. Hence, assuming the most intense peak at 6.45 MHz is due to ^{63}Cu, the associated ^{65}Cu peak in the same environment is that at 5.98 MHz, not the peak of similar intensity at 4.87 MHz. Therefore, the peaks at 6.45 and 5.98 MHz are assigned to the ^{63}Cu and ^{65}Cu in one site (A), and peaks at 4.87 and 4.52 MHz to ^{63}Cu and ^{65}Cu in a second site (B), confirming the presence of two copper sites in a 2:1 ratio.

SELF TEST

The NQR spectrum of $[Co(NH_3)_6]^{3+}$ only has features from ^{14}N ($I = 1$) and none from ^{59}Co ($I = \frac{7}{2}$) and the ^{14}N quadrupole coupling constant in the complex is less than half that found for free NH_3. Explain these observations.

Problems

10.1 Variable temperature ^{57}Fe Mössbauer spectra for $[Fe(phen)_2(NCS)_2]$ (phen = 1,10-phenanthroline) are shown in Figure 10.19. Use the isomer shift and quadrupole splitting data in the 300 K and 77 K spectra and the magnetic susceptibility data in Figure 8.6 to explain the variable temperature behaviour.

10.2 Figure 10.20 shows pairs of ^{57}Fe Mössbauer spectra for iron(II) and iron(III) octahedral spin-crossover complexes recorded at different temperatures. Determine the isomer shift and quadrupole splitting for each spectrum, and use these to assign the spectra to the iron oxidation state and spin state.

10.3 The ^{57}Fe Mössbauer spectra and structures of some iron carbonyls are shown in Figure 10.21. Assign the spectra to the correct structure.

10.4 2-Acetylpyridinethiosemicarbazone can act as a tridentate ligand either in its neutral thione state (HATP), or after deprotonation of the thiol tautomer to give ATP^- (Figure 10.22). This ligand can form an iron(II) complex, $[Fe(HATP)_2]Cl_2$, and an iron(III) complex, $[Fe(APT)(HATP)]Cl_2$. Use the isomer shift and quadrupole splitting data in the ^{57}Fe Mössbauer spectra in Figure 10.22 to identify the oxidation states and spin states and hence assign the spectra

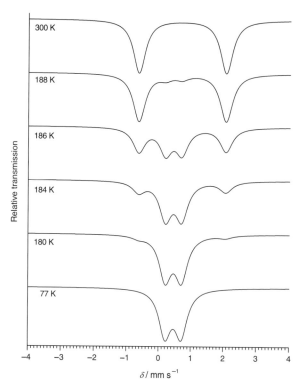

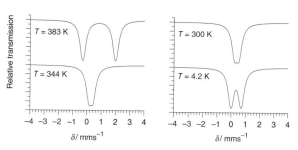

FIGURE 10.20 ^{57}Fe Mössbauer spectra of iron(II) and iron(III) octahedral spin-crossover complexes.

FIGURE 10.19 Variable temperature ^{57}Fe Mössbauer spectra of [Fe(phen)$_2$(NCS)$_2$] (phen = 1,10-phenanthroline). (Data from P. Gütlich, C.P. Kohler, H. Köppen, E. Meissner, E.W. Müller, and H. Spiering, *Trends in Mössbauer Spectroscopy*. University of Mainz (1983); I. Dézsi, B. Molnar, T. Tarnozci, and K. Tompa, *J. Inorg. Nucl. Chem.* **29** 2486 (1967).)

to the [Fe(HAPT)$_2$]Cl$_2$ and [Fe(APT)(HAPT)]Cl$_2$ complexes.

10.5 Ferredoxins are iron–sulfur proteins that mediate electron transfer in a variety of metabolic reactions. Electron

transfer is accompanied by iron redox chemistry in the Fe$_2$S$_2$ core. The ^{57}Fe Mössbauer spectra of the oxidised and reduced forms of *Scenedesmus* ferredoxin are shown in Figure 10.23. Use these data to account for the electronic and magnetic behaviour of ferredoxin in its oxidised and reduced form.

10.6 The ^{35}Cl NQR spectral frequencies (MHz) of some lanthanide trichlorides are summarised in Table 10.5. Interpret these data as far as possible (E.H. Carlson and H.S. Adams, *J. Chem. Phys.* **51** 388 (1969)).

10.7 The features at 31.7 MHz and 34.25 MHz due to ^{65}Cu and ^{63}Cu, respectively, in the NQR spectra of CuGeO$_3$ were both split by about 250 kHz. What does this say about the copper environment(s) in this material? (A.A. Gippius, E.N. Morozova, D.F. Khozeev, A.N. Vasil'ev, M. Baenitz, G. Dhalenne, A. Revcolevschi, *J. Phys.: Condens. Matter* **12** L71 (2000)).

10.8 The ^{35}Cl NQR frequency increased from 11.785 MHz for [TiCl$_2$Cp$_2$] to 11.930 MHz for [TiCl$_2$Cp$_2^*$]. (Cp = (C$_5$H$_5$)$^-$, Cp* = (C$_5$Me$_5$)$^-$) Explain this observation (J. Kubišta, M. Civiš, P. Španěl, and S. Civiš, *Analyst* **137** 1338 (2012)).

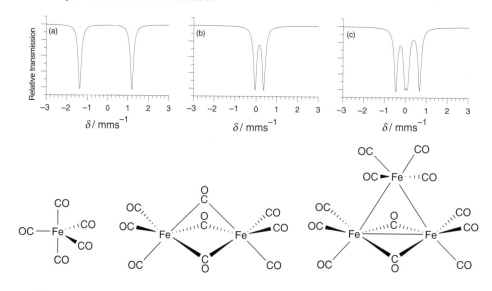

FIGURE 10.21 ^{57}Fe Mössbauer spectra and structures of iron carbonyls. (Data from R. Greatrex and N.N. Greenwood, *Disc. Farad. Soc.* **47** 126 (1969).)

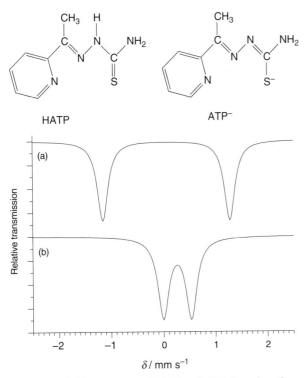

FIGURE 10.22 Structures of HATP and ATP⁻ ligands and
⁵⁷Fe Mössbauer spectra of [Fe(HAPT)₂]Cl₂ and [Fe(APT)
(HAPT)]Cl₂. (Data from R.H.U. Borges, A. Abras, and
H. Beraldo, *J. Braz. Chem. Soc.* **8** 33 (1997).)

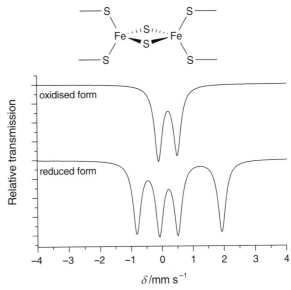

FIGURE 10.23 Structure of Fe_2S_2 core of ferredoxins and ⁵⁷Fe
Mössbauer spectra of *Scenedesmus* ferredoxin at 195 K. (Data
from C.E. Johnson, *J. Appl. Phys.* **42** 1325 (1971); K.K. Rao, R.
Cammack, D.O. Hall, and C.E. Johnson, *Biochem. J.* **122** 257
(1971).)

TABLE 10.5 ³⁵Cl NQR data (MHz) for some lanthanide trichlorides

$LaCl_3$	$CeCl_3$	$PrCl_3$	$NdCl_3$	$SmCl_3$	$GdCl_3$	$DyCl_3$	$HoCl_3$	$ErCl_3$	$YbCl_3$
4.167	4.387	4.567	4.729	5.033	5.315	4.203	4.277	4.424	4.738
							4.334	4.445	4.788

Bibliography

R.V. Parish. (1990) *NMR, NQR, EPR, and Mössbauer
Spectroscopy in Inorganic Chemistry*. New York: Ellis
Horwood.

P. Gütlich, E. Bill, and A. Trautwein. (2011) *Mössbauer
Spectroscopy and Transition Metal Chemistry*. Berlin: Springer
Verlag.

A.G. Maddock. (1997) *Mössbauer Spectroscopy: Principles
and Applications*. Chichester: Horwood Publishing.

D.P.E. Dickson and F. J. Berry. (Eds.) (1986) *Mössbauer
Spectroscopy*. Cambridge: Cambridge University Press.

N.N. Greenwood and T. C. Gibb. (1971) *Mössbauer
Spectroscopy*. London: Chapman and Hall.

M. Kalvius and P. Kienle. (Eds.) (2012) *The Rudolf Mössbauer
Story*. Berlin: Springer.

B.N. Figgis and M.A. Hitchman. (2000) *Ligand Field Theory
and Its Applications*. New York: Wiley-VCH.

S.J. Lippard and J.M. Berg. (1994) *Principles of Bioinorganic
Chemistry*. Mill Valley, CA: University Science Books.

Characterisation of inorganic compounds: example problems with multiple techniques

This chapter presents a series of problems, with worked answers, that exemplify the application of multiple techniques to the characterisation of inorganic compounds and demonstrates how these are used to identify reaction products and unknowns. Each of the previous chapters in the text covered a specific technique, or group of related techniques, and provided problems (in worked examples, self tests, and end of chapter problems) associated with that analytical method. Here the problems require the application of a combination of instrumental methods to elucidate the composition and structure of a compound. For example, the determination of the structure of a molecular, main group compound often needs information from several techniques, such as mass spectrometry, multinuclear nuclear magnetic resonance (NMR), and infrared (IR) spectroscopy. This type of problem also represents how much of inorganic chemistry research is undertaken, where a range of techniques is often needed to identify the composition and structure of an unknown. Many of the later, more challenging, example problems are taken from inorganic chemistry journals to demonstrate how characterisation methods are employed in the research laboratory.

The solution of these problems requires a knowledge and understanding of many fundamental concepts in inorganic chemistry. For example, typical coordination geometries of complexes, modes of bonding of ligands, valence shell electronic pair repulsion (VSEPR) theory, and reaction types and likely pathways. For each problem, information regarding the techniques applied in the question and the level of general chemistry knowledge assumed are summarised in an introductory sentence.

Atomic weights are given in the periodic table inside the front cover.

Problem 11.1 Multinuclear NMR spectroscopy and mass spectrometry

Required chemistry knowledge: transition metal coordination geometry and isomerism. Year 1 university level course

$RhCl_3$ reacts with three molar equivalents of PEt_3 to form a six-coordinate Rh(III) species, **A**. The mass spectrum of **A** has the highest parent mass peak at $568m_u$, among a group of peaks between 562 and $568m_u$. The ^{31}P NMR spectrum of **A** shows a doublet. Upon heating **A** at reflux, in the inert solvent $MeNO_2$, product **B** is generated and no other compound. The mass spectrum of **B** is almost identical to that of **A** and has the same group of peaks between 562 and $568m_u$. The ^{31}P NMR spectrum of **B** shows a doublet of doublets and a doublet of triplets (relative dd:dt integrals were 2:1). Use the data to identify **A** and **B** and account for the splitting patterns observed in the NMR spectra.

(^{31}P: $I = \frac{1}{2}$, 100%; ^{103}Rh: $I = \frac{1}{2}$, 100%; assume no coupling to other nuclei)

Answer

As the ^{31}P NMR spectrum of **A** shows one resonance, which is a doublet, there can only be one type of phosphorus environment in the compound. The doublet derives from the coupling of ^{31}P to a central ^{103}Rh ($I = \frac{1}{2}$) in the complex. The complex is six coordinate and contains three moles of PEt_3 (from the reaction stoichiometry). The complete compound stoichiometry of $[Rh(PEt_3)_3(Cl)_3]$ is consistent with a molecular mass of $568m_u$ where $3 \times {}^{37}Cl$ are present in the compound, and the other peaks in the group come from isotopomers with compositions ranging down to $3 \times {}^{35}Cl$, $6m_u$ lighter. A molecular structure with all phosphorus environments equivalent needs to be devised and this is the *fac* isomer of $[Rh(PEt_3)_3(Cl)_3]$, as shown in Figure 11.1(a).

Based on the NMR intensity data, compound **B** contains two different phosphorus environments in the ratio 2:1—and coupling between the phosphorus in the two different environments (here labelled P(2) and P(1), respectively) and to the rhodium atom will produce the splitting patterns observed. The single phosphorus environment will couple to rhodium to give a doublet which will then couple to the two other (non-equivalent to P(1)) phosphorus (P(2)) centres to produce triplets—giving a

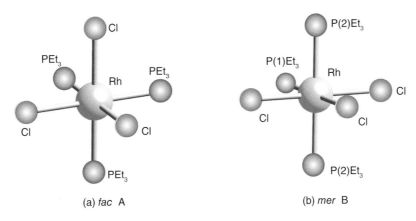

FIGURE 11.1 *fac* and *mer* isomers of $[Rh(PEt_3)_3(Cl)_3]$.

doublet of triplets. Similarly the 2-intensity peak will be formed from the two equivalent P(2) centres coupling to rhodium and then P(1). As **B** is an isomer of **A** (as there is no other reaction product) we can identify this as the *mer* isomer of $[Rh(PEt_3)_3(Cl)_3]$, as shown in Figure 11.1(b). This is consistent with the mass spectrometry data in that the isomer will have the same parent ion m_u and same isotopomers, as it also contains three chlorine atoms.

Problem 11.2 Powder X-ray diffraction and multinuclear NMR

Required chemistry knowledge: structures of simple solids, Lewis acids and bases. Year 2 university level course

(a) CsF reacts with PF_5 (in a 1:1 molar ratio) to give compound **C**. In solution the ^{19}F NMR spectrum of **C** is a doublet and the ^{31}P NMR spectrum is a septet. The powder X-ray diffraction (PXD) pattern ($\lambda = 1.54050$ Å) of **C** showed lines at the following positions (2θ in degrees):

18.733   21.665   30.827   36.318   37.993   44.156

Index the diffraction data obtained from solid **C**. Determine a lattice parameter and lattice type for this material. Describe a possible structure for **C** in solid form by comparison with the structure of CsF, which adopts the rock salt structure with $a = 6.12$ Å.

(b) Replacing PF_5 in the original reaction with AsF_5 gave product **Z**. Predict the following:

(i) the ^{19}F and ^{75}As NMR spectra of **Z** in solution;

(ii) any differences in the PXD pattern of **Z** compared to that of **C**, assuming both materials adopt the same structure type.

For a cubic unit cell, $\sin^2\theta = \dfrac{\lambda^2}{4a^2}(h^2+k^2+l^2)$

[^{31}P: $I = \frac{1}{2}$, 100%; ^{19}F: $I = \frac{1}{2}$, ^{75}As: $I = \frac{3}{2}$ 100%]

Answer

(a) The NMR data show that the phosphorus atom is surrounded by six equivalent fluorine atoms, which gives rise to the septet through spin–spin coupling (intensity ratios 1:6:15:20:15:6:1). The fluorine atoms are equivalent so only couple to the phosphorus centre, producing the observed doublet. These NMR data are consistent with the presence of the octahedral $[PF_6]^-$ anion in **C** and the compound is $CsPF_6$. The PXD data can be indexed following the procedure described in Chapter 2, Section 2.8, as shown in Table 11.1.

The systematic absences, with (h, k, l) all odd or all even, demonstrate a face-centred cubic lattice. The lattice parameter may be calculated using any reflection, but using a high angle line for improved precision:

$$0.13128 = [(1.54050)^2 \times 16]/4a^2 \text{ giving } a = 8.197\text{Å}$$

CsF has a rock salt structure and the simple F^- anion can be replaced in this structure with the complex octahedral anion $[PF_6]^-$. The larger lattice parameter of $CsPF_6$, compared to that of CsF, reflects the larger anion.

(b) **Z** is $Cs[AsF_6]$ containing the octahedral $[AsF_6]^-$ anion.

(i) The octahedral $[AsF_6]^-$ anion will have a single type of fluorine environment. In the ^{19}F NMR spectrum coupling to ^{75}As, $I = \frac{3}{2}$ 100%, will produce four resonances of equal intensity

TABLE 11.1 Indexation of the PXD data from compound **C**

2θ	$\sin^2\theta$	Ratio	$h^2 + k^2 + l^2$	(h, k, l)
18.733	0.02649	1	3	(1, 1, 1)
21.665	0.03532	1.33	4	(2, 0, 0)
30.827	0.07064	2.66	8	(2, 2, 0)
36.318	0.09713	3.66	11	(3, 1, 1)
37.993	0.10596	4	12	(2, 2, 2)
44.156	0.14128	5.33	16	(4, 0, 0)

representing interactions with the four possible m_I values, $\frac{3}{2}$, $\frac{1}{2}$, $-\frac{1}{2}$, $-\frac{3}{2}$. As the arsenic is in a high symmetry environment its NMR spectrum will be resolved despite being a quadrupolar nucleus. Coupling to six equivalent fluorine atoms will produce a septet.

(ii) $Cs[AsF_6]$ adopts the same basic structure type as CsF and $Cs[PF_6]$, with an *F*-centred lattice. As arsenic is below phosphorus in the periodic table it will have a larger atomic radius and have longer bonds to fluorine. The unit cell dimension of **Z** will be larger than that of **C** and through the Bragg equation, $2d\sin\theta = n\lambda$, the resultant larger *d*-spacings will mean the reflections (which will show the same set of systematic absences for an *F*-lattice) will shift to lower 2θ values in the PXD pattern.

Problem 11.3 Electronic spectroscopy and magnetism

Required chemistry knowledge: Transition metal coordination geometries and isomers; crystal field theory Year 1/2 university level course

Solutions of four different manganese compounds (**D**, **E**, **F**, and **G**), all at the identical concentration of 0.1 M, had been left unlabelled in the laboratory. The four compounds were known to be $K[MnCl_4]$, $K[MnO_4]$, $(NMe_4)_2[MnBr_4]$, and $[Mn(OH_2)_6]SO_4$. An enterprising student measured the molar extinction coefficient of the strongest absorption band in the ultraviolet–visible (UV-vis) spectrum. They then evaporated the solvents from the solutions to obtain a known molar quantity for the measurement (using a Gouy balance) of the molar magnetic susceptibility of each compound. Table 11.2 summarises the results obtained.

Identify compounds **D**, **E**, **F**, and **G**, explaining how the experimental data have been used to support the assignments.

Answer

We can predict the properties of the four manganese compounds and compare these data with the experimental results. Table 11.3 summarises this information derived as follows:

(i) The oxidation states, d-electron configurations, and ligand coordination geometries of the four possible compounds are summarised in Table 11.3, column 2.

(ii) The likely magnitude of the molar absorption coefficients can be predicted based on whether the electronic transitions are allowed or how many of the selection rules for the transitions need to be broken (Chapter 5, Table 5.1), as shown in Table 11.3, column 3.

TABLE 11.2 UV-vis spectroscopic and magnetism data for the manganese compounds **D**, **E**, **F**, and **G**.

Compound	Molar extinction coefficient ε_{max} / dm^3 mol^{-1} cm^{-1}	Experimental magnetic moment (μ_B) and any temperature dependence
D	1.8×10^4	0.05, none
E	0.06	5.88, none
F	14	5.94, none
G	320	4.82, varied

TABLE 11.3 Predicted properties for the manganese compounds

Compound	Oxidation state, coordination geometry and d-electron configuration	Predicted molar extinction coefficient range, $\varepsilon_{max}/dm^3\ mol^{-1}\ cm^{-1}$	Predicted spin only μ_{eff} (μ_B) and orbital contribution?
$K[MnCl_4]$	Mn^{3+} tetrahedral d^4, $e^2\ t_2^{\,2}$	250–1000	4.92 or slightly lower, yes
$K[MnO_4]$	Mn^{7+} tetrahedral d^0, $e^0\ t_2^{\,0}$	3000–50 000 based on LMCT	0; possible very small temperature-independent paramagnetism (TIP), 0.05
$(NMe_4)_2[MnBr_4]$	Mn^{2+} tetrahedral d^5, $e^2\ t_2^{\,3}$	1–20	5.92, no
$[Mn(OH_2)_6]SO_4$	Mn^{2+} octahedral d^5, $t_{2g}^{\,3}\ e_g^{\,2}$	<1	5.92, no

(iii) The spin-only formula (Chapter 8, Eqn 8.6) can be used to work out the expected magnetic moment, μ_{eff}, for each of the possible electronic configurations. In addition the likelihood of a modification to these values based on an orbital contribution can also be determined (Tables 8.1 and 8.2). These predictions are shown in Table 11.3, final column.

Comparison of the data, in Tables 11.2 and 11.3, of experimental and predicted molar absorption coefficients and magnetic moments, allows identification of the samples as follows. **D** is $K[MnO_4]$ with a zero μ_{eff} for the d^0 electron configuration (or possibly a very small temperature independent paramagnetism of ~$0.05\mu_B$) and very high ε_{max} as fully allowed ligand to metal charge transfer transitions can take place in this compound. **E** is $[Mn(OH_2)_6]SO_4$ with a μ_{eff} for the high-spin d^5 configuration of $5.94\mu_B$. **F** is $(NMe_4)_2[MnBr_4]$ again with five unpaired electrons but with a significantly higher absorption coefficient for the tetrahedral centre where the Laporte/parity selection rule does not apply. Finally, **G** is $K[MnCl_4]$ with a μ_{eff} close to the spin-only value for a high-spin d^4 configuration, though this is reduced slightly through an orbital contribution for the $e^2 t_2^{\,2}$ configuration and will have a temperature dependence (T ground state). This compound will also exhibit a reasonably strong absorption in the UV-vis spectrum for a tetrahedral complex without a centre of symmetry.

diamagnetic, and contains two different oxidation states of nickel. The visible spectra of the two complexes (measured as powders in the solid state) are shown in Figure 11.2. Interpret these data.

Answer

The average oxidation state of nickel in a compound of empirical formula $Ni(H_2NCH_2CH_2NH_2)_2Cl_3$ is Ni^{3+} with a d-electron configuration of d^7. Given that J contains two oxidation states of nickel, the most likely combination to give an average of Ni^{3+} would be $1 \times Ni^{2+}:1 \times Ni^{4+}$.

The magnetic moment in **H** is consistent with a spin-only value from one unpaired electron [μ(spin-only) $= \sqrt{n(n+2)}$, where n is the number of unpaired electrons, equal to $1.73\mu_B$]. This can be achieved with a low-spin octahedral configuration, $t_{2g}^{\,6}e_g^{\,1}$, in $[Ni(H_2NCH_2CH_2NH_2)_2Cl_2]^+Cl^-$; note that this would be expected to be Jahn–Teller distorted. The Jahn–Teller distortion does not remove the centre of symmetry in the complex and d–d transitions in this complex would be forbidden by both Laporte and parity selection rules consistent with the pale colour and weak absorbance of this complex.

As J is diamagnetic electronic configurations for Ni^{2+} (d^8) and Ni^{4+} (d^6) would have to leave no unpaired electron in either oxidation state. These can be achieved for square-planar d^8 (with an empty $d_{x^2-y^2}$ orbital) and

Problem 11.4 UV-vis spectroscopy and magnetism

Required chemistry knowledge: transition metal coordination geometries and isomers; mixed valence compounds. Year 2/3 university level course

Two complexes **H** and **J** of empirical formula $Ni(H_2NCH_2CH_2CH_2NH_2)_2Cl_3$ are known. **H** is pale yellow, exhibits $\mu_{eff} = 1.75\mu_B$, and electron paramagnetic resonance (EPR) data are consistent with an axially distorted octahedral geometry around nickel. **J** is deep blue,

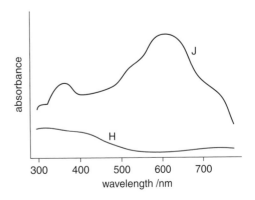

FIGURE 11.2 UV-vis spectra of compounds **H** and **J**.

low-spin octahedral d^6 (t_{2g}^6) coordinations. The compound formulation $[Ni(II)(H_2NCH_2CH_2NH_2)_2]^{2+}[Cl^-]_2 \cdot [Ni(IV)(H_2NCH_2CH_2NH_2)_2Cl_2]^{2+}[Cl^-]_2$, with these electron configurations, provides an explanation for the magnetism data and also the intense absorption in the visible region of the spectrum. Intervalence charge transfer between the Ni(II) and Ni(IV) (Section 5.6.5) gives rise to the strong absorption in the orange-red region of the visible spectrum and the resultant observed deep blue colour.

Problem 11.5 Powder X-ray diffraction and thermal analysis/mass spectrometry

Required chemistry knowledge: structures of simple transition metal oxides. Year 1/2 university level course

Heating a sample of an unknown first row transition metal sulfate hydrate, **Y**, to 1200°C resulted in two distinct mass losses during thermogravimetric analysis; one loss of 44.88% of the original weight occurred between 90 and 150°C, and another at 700°C of 28.5% of the original weight. Mass spectroscopic analysis of the gas released during the decomposition showed the gas liberated near 100°C had a molecular mass of $18m_u$ and that at 700°C one of $80m_u$. No other gas was evolved during the decomposition.

PXD data ($\lambda = 1.54050$ Å) obtained from the metal oxide residue left after the thermogravimetric analysis showed reflections at the following 2θ positions (degrees):

37.243   43.273   62.858   75.389   79.385   95.027

Table 11.4 contains a summary of lattice parameters (Å) and lattice types for several common first row transition metal oxides obtained from a database.

Use the PXD data to identify the metal oxide obtained following the thermogravimetric decomposition. Use the thermogravimetric analysis data to identify compound **Y**.

Answer

The PXD data can be indexed following the procedure described in Chapter 2, Section 2.8 and as summarised in Table 11.5.

The systematic absences, with (h, k, l) all odd or all even, demonstrate a face-centred cubic, **F**, lattice. The lattice parameter may be calculated using any reflection, but using a high angle line for improved precision:

$$0.5438 = [(1.54050)^2 \times 16]/4a^2$$
$$\text{giving } a = 4.178\text{Å}$$

Comparison of this lattice parameter and lattice type with the literature data given in Table 11.4 identifies this oxide product as NiO. **Y**, therefore, must be a nickel sulfate hydrate and as no oxygen was evolved during the decomposition then the oxidation state of the nickel must reflect that in the product NiO—that is, Ni^{2+}. The formula of **Y** can now be written as $NiSO_4 \cdot nH_2O$. The weight loss during the first stage of the decomposition $18m_u$ implies the evolution of water and this can be used to calculate n from the balanced reaction:

$$NiSO_4 \cdot nH_2O(s) \rightarrow NiSO_4(s) + nH_2O(g)$$

Using the relative molecular masses of $NiSO_4$ (154.75 g mol^{-1}), H_2O (18.0 g mol^{-1}), and $NiSO_4 \cdot nH_2O$ (154.75 +

TABLE 11.4 Lattice parameters and lattice types for some common cubic transition metal oxides

Metal oxide	MnO	Fe$_{0.92}$O	CoO	NiO	Cu$_2$O	ZnO
Lattice parameter (Å) and lattice type	4.446 **F**	4.332 **F**	4.263 **F**	4.176 **F**	4.174 **P**	4.489 **F**

TABLE 11.5 Indexing of the PXD data from compound **Y**

2θ	$\sin^2\theta$	Ratio	$h^2 + k^2 + l^2$	(h, k, l)
37.243	0.1020	1	3	(1, 1, 1)
43.273	0.1360	1.33	4	(2, 0, 0)
62.858	0.2719	2.66	8	(2, 2, 0)
75.389	0.3739	3.66	11	(3, 1, 1)
79.385	0.4079	4	12	(2, 2, 2)
95.027	0.5438	5.33	16	(4, 0, 0)

$18n$ g mol^{-1}), then the residual weight after the first stage decomposition is 55.12%:

$$\frac{\text{Mass of NiSO}_4}{\text{Mass of NiSO}_4.n\text{H}_2\text{O}} \times 100 = 55.12$$

$$= \frac{154.75}{154.75+18n} \times 100$$

Solving for n gives a value of 7 – so **Y** is NiSO$_4$.7H$_2$O. The second weight loss at 700°C is consistent with loss of SO$_3$(g) ($80m_u$) yielding NiO as the final product, as confirmed by the PXD analysis.

Problem 11.6 IR spectroscopy and chemical composition analysis

Required chemistry knowledge: transition metal carbonyl chemistry and coordination geometries; carbonyl stretching frequencies and bonding; structures of simple salts. Year 2 university level course

Chromium hexacarbonyl undergoes the following reaction chemistry:

$$[Cr(CO)_6] \xrightarrow{[NEt_4]Cl} J(+K\uparrow) \xrightarrow{Na} L + M$$

Atomic absorption spectroscopy, IR spectroscopy, and ^{13}C NMR spectroscopy data for compounds **J** to **M** were obtained as follows.

J is a salt; AAS spectroscopy: Cr 14.5%, Cl 9.9%; CHNO analysis gave C 43.6%, H 5.60%, N 3.92%, and O 22.4%; IR (v(CO)/cm^{-1}) 2058, 1911, 1872.
K is a gas with relative molecular mass from mass spectrometry of $28m_u$.
L is a salt; AAS spectroscopy: Na 19.3%, Cr 21.8%; IR (v(CO)/cm^{-1}) 1760, 1720.
M is a salt containing 21.5% Cl (from AAS) and CHN analysis gave C 58.0%, H 12.1%, and N 8.4%; its PXD pattern did not match that of NaCl.

(i) Use the analytical data to identify compounds **J** to **M**.
(ii) Use the IR spectroscopy data provided to determine the coordination geometry and point groups for the chromium complexes **J** and **L**.

Answers

(i) The analytical data for **J** yield an empirical formula of CrC$_{13}$NH$_{20}$ClO$_5$ and with the assumption that the [NEt$_4$]$^+$ cations remain intact and the CO

ligands either remain on, or are displaced from, the chromium centre this gives a molecular formula of [NEt$_4$] [Cr(CO)$_5$Cl]. Writing this as a salt gives [NEt$_4$]$^+$[Cr(CO)$_5$Cl]$^-$ containing a six-coordinate chromium centre. **K** is CO (with a mass of $28m_u$), the by-product of the first reaction evolved through displacement by Cl$^-$. **M** has the empirical formula C$_8$H$_{20}$NCl, which is equivalent to the salt [NEt$_4$]$^+$Cl$^-$. The sodium to chromium atomic ratio in **L** is Na$_2$: Cr and with no other product of this reaction the stoichiometry of the salt **L** is Na$_2$[Cr(CO)$_5$], containing the [Cr(CO)$_5$]$^{2-}$ anion.

(ii) The expected geometry of the chromium centre in **J** is octahedral with $5 \times$ CO and $1 \times$ Cl ligands. The point group of this species can be identified as C_{4v} (Chapter 4, Section 4.4.2) and it can be shown that this is consistent with the three bands seen in the CO stretching region of the infra-red spectrum. Following the procedure given in Chapter 4, Section 4.4.9, using the C_{4v} point group and the five CO stretching modes:

C_{4v} $h = 8$	E	$2C_4$	C_2	$2\sigma_v$	$2\sigma_d$
Γ_{CO}	5	1	1	3	1

On reduction Γ_{red}(CO stretch) $= 2A_1 + B_1 + E$, of which the $2A_1$ modes and the E mode will be IR active, giving rise to three bands in the CO stretching region of the IR spectrum, in agreement with the experimental data.

For **L**, which contains the [Cr(CO)$_5$]$^{2-}$ anion, the five-coordinate chromium centre could have either square-based pyramidal or trigonal bipyramidal coordination to the CO ligands. We have seen that square-based pyramidal coordination gives three IR active modes and only two are found for this compound. Therefore, undertaking analysis of the number of CO stretches for trigonal bipyramidal (D_{3h}) coordination:

D_{3h} $h = 12$	E	$2C_3$	$3C_2$	σ_h	$2S_3$	$3\sigma_v$
Γ_{CO}	5	2	1	3	0	3

On reduction $\Gamma_{CO} = 2A'_1 + A_2'' + E'$, of which the $A_2'' + E'$ symmetry modes are IR active. This predicts just two absorptions in the CO stretching region of the IR spectrum in agreement with the given data. Thus the [Cr(CO)$_5$]$^{2-}$ anion adopts trigonal bipyramidal geometry.

Problem 11.7 Elemental analysis, IR and Raman spectroscopy, EPR spectroscopy, and magnetic data

Required chemistry knowledge: transition metal coordination chemistry and complex geometries; crystal field theory and spectrochemical series; cyanide and carbonyl bonding to transition metals. Year 2 university level course

When yellow $[(Ph_3P)_2N]_2[Mn(CN)_6]$ was exposed to ambient sunlight, in a 1:1 solution of MeCN and CH_2Cl_2, a dark red solution was formed. Addition of Et_2O resulted in the crystallisation of burgundy red compound **A**. The CHN and AAS analysis of **A** gave 73.8% C, 4.89% H, 6.77% N, 10.0% P, and 4.44% Mn. The IR spectrum of **A** displayed a strong band at 2205 cm^{-1}, while the Raman spectrum showed a band at 2209 cm^{-1}. The room temperature EPR spectrum consisted of a sextet of equal intensity features centred at $g = 2.003$, with an isotropic hyperfine coupling constant of 71 G. The experimental magnetic moment of **A** was found to be $5.99\mu_B$. Determine a formula for **A** and establish the geometry and electronic ground state of the Mn-containing species.

Answer

The elemental analysis gives an empirical formula of $C_{76}H_{60}MnN_6P_4$ with an RMM of 1236.2 g mol^{-1}, and this corresponds to $[(Ph_3P)_2N]_2[Mn(CN)_4]$. Therefore, the spectroscopic properties are associated with the $[Mn(CN)_4]^{2-}$ anion, which contains Mn^{2+}. The vibrational spectra, both with an absorption at ~2200 cm^{-1}, are consistent with A_1 and T_2 ν_{CN} modes in coordinated cyanide anions. While it is normal for tetrahedral complexes of the first-row transition elements to be high-spin, with the strong field CN$^-$ ligand there is the possibility that this would result in a low-spin tetrahedral complex. So the EPR or magnetism data need to be used to distinguish these geometries.

Both the $^6A_1$ high-spin ($S = \frac{5}{2}$) and the low-spin $^2T_2$ low-spin ($S = \frac{1}{2}$) would be expected to show an EPR spectrum with a hyperfine splitting from interaction with Mn ($I = \frac{5}{2}$), so this does not distinguish whether high- or low-spin $[Mn(CN)_4]^{2-}$ has been prepared. However, the magnetic moment can be used to identify the spin state. For a high-spin complex, the spin-only magnetic moment is $5.92\mu_B$, in excellent agreement with the observed value of $5.99\mu_B$. (The spin-only value for low-spin $^2T_2$ is $1.73\mu_B$, though the observed value would be expected to be reduced slightly by an orbital contribution).

Therefore, all the spectroscopic and structural data are consistent with the presence of tetrahedral $[Mn(CN)_4]^{2-}$ which is high-spin with an $S = \frac{5}{2}$, $^6A_1$ ground state.

Despite the presence of strong-field CN$^-$ ligands, this tetrahedral complex is still high-spin (W.E. Buschmann, A.M. Arif, and J.S. Miller, *Angew. Chem. Int. Ed.* **37** 781 (1998)).

Problem 11.8 Crystallography, EPR, EXAFS, and technique timescales

Required chemistry knowledge: transition metal coordination chemistry and complex geometries; Jahn–Teller effect. Year 1/2 university level course

The single-crystal X-ray diffraction determined structure of bis[tris(2-pyridyl)methane]copper(II) dinitrate ($[Cu\{(C_5H_4N)_3CH\}_2](NO_3)_2$) shows six Cu–N bond lengths all of 2.103 Å at 295 K, and of 2.095 Å at 150 K. The EPR spectrum is isotropic at 293 K, but is axial with $g_\parallel > g_\perp$ at 150 K. The Cu K-edge EXAFS data at 298 K shows four nitrogen atoms at 2.04 Å and two nitrogen atoms at 2.25 Å. Account for these observations.

Answer

The single-crystal data at both 295 and 173 K imply that the coordination environment of the copper atom is a regular octahedron, rather than Jahn–Teller distorted as usually found for Cu(II). The EPR data indicate that the compound is isotropic, that is, regular octahedral, at 293 K. The spectrum becomes axial at 150 K, and the ordering of the g-values indicates that there is a conventional tetragonally-elongated Jahn–Teller distortion. The Cu K-edge EXAFS data at 298 K gives a Jahn–Teller distortion with four short and two long Cu–N bond lengths.

In order to rationalise these observations the timescales of the various characterisation methods need to be recognised. The interaction time of EXAFS is very short, ~10^{-15} s, so will give an 'instantaneous' picture of the molecule, and this indicates that there is a conventional Jahn–Teller distortion of the copper environment even at 298 K. The EPR data show that if there is a Jahn–Teller distortion at 298 K it is fast (dynamic) on the EPR timescale of nanoseconds, and the technique samples an averaged isotropic octahedral environment. However, at the lower temperature, 150 K, the rate of interconversion between elongated and shortened Cu–N distances slows sufficiently and individual complexes become frozen into a particular orientation. So at 150 K, EPR observes the axial distortion. The single crystal X-ray diffraction experiment sees an average of the dynamic processes at 298 K. At 150 K, where individual molecules are frozen with the axial distortion in different directions, SXD obtains data from the whole assembly across the crystal; this is orientationally averaged so again the Cu–N bonds all seem of the same length (T. Astley, P.J. Ellis, H.C. Freeman, M.A. Hitchman, F.R. Keene, and E.R.T. Tiekink, *J. Chem. Soc., Dalton Trans.* 595 (1995)).

Problem 11.9 Mössbauer spectroscopy, magnetic moments, and electronic absorption spectroscopy

Required chemistry knowledge: transition metal coordination chemistry and complex geometries; crystal field theory, spin-crossover complexes. Year 2/3 university level course

The ^{57}Fe Mössbauer spectrum of $[Fe(ptz)_6](BF_4)_2$ (ptz = 1-propyltetrazole) between 148 K and 97 K is shown in Figure 11.3. Over the same temperature range there is a reduction in the magnetic moment from ca. $5\mu_B$ to 0, and the complex turns from white to purple/pink. Account for the changes in the ^{57}Fe Mössbauer spectrum, magnetic susceptibility, and colour of the samples during the different processes.

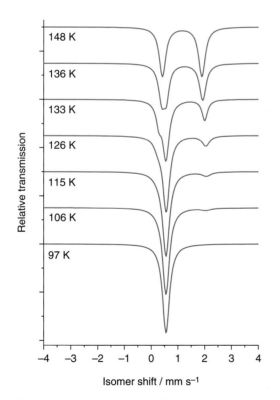

FIGURE 11.3 Variable temperature ^{57}Fe Mössbauer spectra of $[Fe(ptz)_6](BF_4)_2$.

Answer

At 148 K, the spectrum displays a quadrupole doublet characteristic of high-spin iron(II). At 97 K there is a single peak, characteristic of low-spin iron(II) in a highly symmetrical octahedral environment. In between these temperatures there is a mixture of the high-spin and low-spin spectral motifs. The proportions change very rapidly between 140 and 120 K. This is also reflected in the magnetic data with the change from high-spin iron(II) $t_{2g}^4 e_g^2$ with four unpaired electrons, μ_{eff}(spin-only) = $4.90\mu_B$ (and an expected small additional contribution for a T state with more than half-filled shell, so μ_{eff}(predicted) ~$5\mu_B$) to low-spin $t_{2g}^6 e_g^0$ with no unpaired electron and μ_{eff}(predicted) = 0. The electronic absorption spectra also support these electronic configurations, as discussed in Chapter 5, Section 5.13.5.4 (E. W. Müller, J. Ensling, H. Spiering, and P. Gütlich, *Inorg. Chem.* **22** 2074 (1983); S. Decurtins, P. Gütlich, C.P. Köhler, H. Spiering, and A. Hauser, *Chem. Phys. Lett.* **105** 1 (1984)).

Problem 11.10 Multinuclear NMR, IR and Raman spectroscopies, and main group chemistry

Required chemistry knowledge: VSEPR theory, main group and Group 18 chemistry, Lewis acid–base chemistry. Year 2 university level course

The reaction of a xenon fluoride **A**, which contained 63.3% xenon, with $N(CH_3)_4F$ at low temperature gave a colourless compound **B** and reaction of **A** with SbF_5 in anhydrous HF at 0°C gave pale green **C**. The ^{129}Xe and ^{19}F NMR spectra (with the spectrometer frequencies provided) and vibrational data for **A**, **B**, and **C** are given in Figure 11.4 and Table 11.6, respectively. Use these to identify **A**, **B**, and **C** and confirm whether the shapes of the xenon containing species are consistent with those predicted by the VSEPR model.

(^{19}F I = ½ 100%; ^{129}Xe I = ½ 26.4%; assume all other nuclei as non-NMR active)

Answer

Compound A

A xenon fluoride containing 63.3% xenon has 36.7% fluorine and, therefore, **A** is XeF_4 from the analytical data. The ^{129}Xe NMR spectrum of **A** consists of a binomial quintet with δ = 317 ppm indicating the presence of one xenon environment and coupling to four equivalent I = ½ nuclei; this is consistent with the formula XeF_4 with all fluorine sites equivalent. The ^1J(Xe-F) coupling constant is ca. 28 ppm, converting to Hz by multiplying

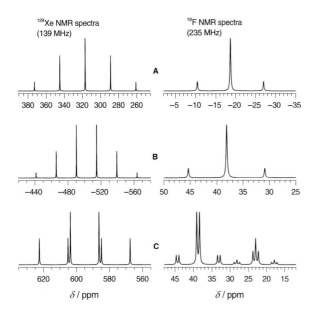

FIGURE 11.4 ^{129}Xe (139 MHz) and ^{19}F (235 MHz) NMR spectra of xenon fluoride species.

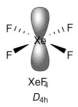

FIGURE 11.5

TABLE 11.6 Vibrational data in the Xe–F stretching region

Compound	IR	Raman
A	586	554 (p), 524 (dp)
B	465	502 (p), 423 (dp)
C	N/A	643 (p), 584(p)

by the spectrometer operating frequency gives ^1J(Xe-F) = 3880 Hz. The ^{19}F NMR spectrum only contains one resonance (which is split into a multiplet), therefore there is only one ^{19}F environment, indicating that (on the NMR timescale) they are all equivalent. While the ^{19}F NMR spectrum looks as though it contains a triplet, the intensity ratio is not that of a binomial triplet (i.e. 1:2:1), but is characteristic of coupling to dilute (non 100% abundant) spins. In this case ^{129}Xe has I = ½ with 26.4% abundance. Therefore, the central peak at −18.7 ppm comes from non-coupled ^{19}F nuclei and the two weaker satellites are a doublet due to the ^{19}F coupling with the 26.4% I = ½ ^{129}Xe nuclei. The ^1J(F-Xe) coupling will be the same as the ^1J(Xe-F) coupling (3880 Hz). As the ^{19}F NMR spectra were recorded at 235 MHz this translates to 16.5 ppm in agreement with the splitting observed. Therefore, both the ^{129}Xe and ^{19}F NMR spectra are consistent with **A** being XeF$_4$ and with all of the fluorine atoms being equivalent. Likely geometries for XeF$_4$ are tetrahedral or square planar.

VSEPR analysis of XeF$_4$ indicates that there are four bond pairs and two lone pairs, so the structure will be based on an octahedron with the two lone pairs *trans* to each other, with a D_{4h} point group (Figure 11.5).

The vibrational spectral data and group theory can be used to confirm this. $\Gamma_{Xe-F} = A_{1g} + B_{1g} + E_u$ for D_{4h} XeF$_4$. Of these A_{1g} and B_{1g} are Raman active, with the A_{1g} being polarised and the B_{1g} depolarised, and the E_u being IR active. As there is a centre of symmetry in D_{4h} the mutual exclusion principle applies. Therefore, the IR and Raman data are in agreement with XeF$_4$ being square planar. (If XeF$_4$ was tetrahedral $\Gamma_{Xe-F} = A_1 + T_2$, and although there would be two Raman active Xe–F stretching modes and one IR active stretching mode, the IR mode would be coincident with the weaker Raman band.)

Having identified that **A** is square-planar XeF$_4$, it can be postulated that it is acting as a fluoride ion acceptor with [N(CH$_3$)$_4$]F to form **B**, [N(CH$_3$)$_4$]$^+$[XeF$_5$]$^-$, and as a fluoride ion donor with SbF$_5$ to form **C** [XeF$_3$]$^+$[SbF$_6$]$^-$.

Compound B

For [XeF$_5$]$^-$ VSEPR predicts five electron bond pairs and two lone pairs. For seven electron pair compounds there are a number of possible geometries. The VSEPR predicted geometry is a mono-capped octahedron, but the mono-capped trigonal prism and pentagonal bipyramidal geometries are similar in energy. For main group compounds it is found that pentagonal bipyramidal is the preferred geometry (based on electronic rather than steric considerations). The two lone pairs will go *trans* to each other to give a pentagonal planar geometry with D_{5h} point group (Figure 11.6).

The NMR data for **B** indicate that there is one xenon environment and one fluorine environment. The binomial sextet in the ^{129}Xe spectrum at −527 ppm with ^1J(Xe-F) of 3400 Hz is consistent with coupling to five equivalent fluorine atoms. The singlet with ^{129}Xe satellites in the ^{19}F NMR spectrum is also consistent with a D_{5h} point group. However, it needs to be remembered that NMR spectra often display the effects of fluxionality, where the atoms are rapidly exchanging on the NMR timescale and appear to be equivalent. Therefore, NMR spectra on their

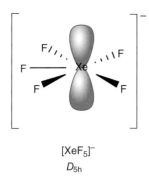

$[XeF_5]^-$

D_{5h}

FIGURE 11.6

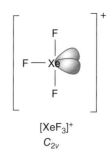

$[XeF_3]^+$

C_{2v}

FIGURE 11.7

own are not always sufficient to confirm the molecular shape, and another technique such as vibrational spectroscopy is required with a shorter observation timescale. Determining $\Gamma_{Xe\text{-}F}$ in the D_{5h} point group gives $\Gamma_{Xe\text{-}F} = A_1' + E_1' + E_2'$. Of these $A_1' + E_1'$ are Raman active (A_1' is polarised), and E_2' is IR active. Therefore, the vibrational data are consistent with $[XeF_5]^-$ as pentagonal planar. This was the first reported example of a pentagonal planar species (K.O. Christe, E.C. Curtis, D.A. Dixon, H.P. Mercier, J.C.P. Sanders, and G.J. Schrobilgen, *J. Am. Chem. Soc.* **113** 3351 (1991)).

Compound C

The ^{129}Xe and ^{19}F NMR spectra of **C** are more complex than those of **A** or **B**. The ^{129}Xe NMR spectrum could be a doublet of triplets at 595 ppm, due to the coupling of one xenon environment to two different fluorine environments, one with one F (2412 Hz) and the other with two F (2615 Hz). Alternatively it could be a pair of triplets, indicating two xenon environments, both coupling to two equivalent ^{19}F nuclei. However, with the triplet coupling constants being the same in both cases (2615 Hz) the most straightforward explanation is that it is a doublet of triplets. The ^{19}F NMR spectrum consists of a doublet at 38.7 ppm and a triplet at 23.0 ppm, both with ^{129}Xe satellites. This confirms that there are two different fluorine environments in **C**. The doublet is due to ^{19}F coupling with one other $I = \frac{1}{2}$ 100% abundant (^{19}F) nucleus, and the triplet is due to ^{19}F coupling with two equivalent $I = \frac{1}{2}$ (^{19}F) nuclei. While the ^2J(F-F) coupling of 174 Hz is the

same for both the triplet and the doublet, the separation of the ^{129}Xe satellites is different in the two multiplets. The ^1J(Xe-F) coupling constant for the doublet in the ^{19}F spectrum is 2615 Hz, which is the same as the triplet coupling in the ^{129}Xe spectrum. Likewise, the coupling constant of the ^{129}Xe triplet satellites in the ^{19}F spectrum (2412 Hz) is the same as the doublet coupling constant in the ^{129}Xe spectrum.

If compound **C** does contain $[XeF_3]^+$, the VSEPR predicted structure for this has three bonding electron pairs and two lone pairs and is, therefore, based on a trigonal bipyramidal geometry (Figure 11.7).

From VSEPR analysis the lone pairs will go in the equatorial position rather than the axial position, so $[XeF_3]^+$ is predicted to be 'T-shaped' with C_{2v} point group. This is consistent with both a doublet of triplets in the ^{129}Xe spectrum where the doublet coupling constant of 2412 Hz is associated with the ^1J(Xe-F$_{eq}$) and the triplet coupling constant of 2615 Hz is ^1J(Xe-F$_{ax}$). In the ^{19}F NMR spectrum the doublet is due to the two axial fluorine atoms coupling with the single equatorial fluorine, and the triplet is due to the single equatorial F coupling to the two axial fluorine atoms.

In this case there is only Raman data available, but this can be used to confirm whether the ion is 'T-shaped'. The z-axis is along the Xe–F$_{eq}$ bond, and the ion lies in the yz plane. $\Gamma_{Xe\text{-}F} = 2A_1 + B_2$, all of which are both IR and Raman active, and the presence of 2A$_1$ modes is consistent with the two polarised bands in the Raman spectrum of **C**. (If $[XeF_3]^+$ was planar $\Gamma_{Xe\text{-}F} = A_1' + E'$. Although these are both Raman active, only one (A_1') would be polarised).

Appendix 1

Character tables for some chemically important point groups

1.1 The non-axial groups

C_1	E
A	1

C_s	E	σ_h		
A′	1	1	x, y, R_z	x^2, y^2, z^2, xy
A″	1	−1	z, R_x, R_y	yz, xz

C_i	E	i		
A_g	1	1	R_x, R_y, R_z	$x^2, y^2, z^2, xy, yz, xz$
A_u	1	−1	$x, y, z,$	

1.2 The C_n groups

C_2	E	C_2		
A	1	1	z, R_z	x^2, y^2, z^2, xy
B	1	−1	x, y, R_x, R_y	yz, xz

1.3 The D_n groups

D_2	E	$C_2(z)$	$C_2(y)$	$C_2(x)$		
A	1	1	1	1		x^2, y^2, z^2
B_1	1	1	−1	−1	z, R_z	xy
B_2	1	−1	1	−1	y, R_y	xz
B_3	1	−1	−1	1	x, R_x	yz

D_3	E	$2C_3$	$3C_2$		
A_1	1	1	1		$x^2 + y^2, z^2$
A_2	1	1	−1	z, R_z	
E	2	−1	0	$(x, y), (R_x, Ry)$	$(x^2 - y^2, xy), (xz, yz)$

1.4 The C_{nv} groups

C_{2v}	E	C_2	$\sigma_v(xz)$	$\sigma_v'(yz)$		
A_1	1	1	1	1	z	x^2, y^2, z^2
A_2	1	1	−1	−1	R_z	xy
B_1	1	−1	1	−1	x, R_y	xz
B_2	1	−1	−1	1	y, R_x	yz

C_{3v}	E	$2C_3$	$3\sigma_v$		
A_1	1	1	1	z	$x^2 + y^2, z^2$
A_2	1	1	−1	R_z	
E	2	−1	0	$(x, y), (R_x, R_y)$	$(x^2 - y^2, xy), (xz, yz)$

C_{4v}	E	$2C_4$	C_2	$2\sigma_v$	$2\sigma_d$		
A_1	1	1	1	1	1	z	$x^2 + y^2, z^2$
A_2	1	1	1	−1	−1	R_z	
B_1	1	−1	1	1	−1		$x^2 - y^2$
B_2	1	−1	1	−1	1		xy
E	2	0	−2	0	0	$(x, y), (R_x, R_y)$	(xz, yz)

C_{5v}	E	$2C_5$	$2C_5^2$	$5\sigma_v$		
A_1	1	1	1	1	z	$x^2 + y^2, z^2$
A_2	1	1	1	−1	R_z	
E_1	2	$2\cos 72°$	$2\cos 144°$	0	$(x, y), (R_x, R_y)$	(xz, yz)
E_2	2	$2\cos 144°$	$2\cos 72°$	0		$(x^2 - y^2, xy)$

1.5 The C_{nh} groups

C_{2h}	E	C_2	i	σ_h		
A_g	1	1	1	1	R_z	x^2, y^2, z^2, xy
B_g	1	−1	1	−1	R_x, R_y	xz, yz
A_u	1	1	−1	−1	z	
B_u	1	−1	−1	1	x, y	

$C_{3\text{h}}$	E	C_3	C_3^2	σ_h	S_3	S_3^5		$\varepsilon = \exp(2\pi i/3)$
A′	1	1	1	1	1	1	R_z	$x^2 + y^2, z^2$
E′	$\left\{\begin{matrix}1\\1\end{matrix}\right.$	$\begin{matrix}\varepsilon\\\varepsilon^*\end{matrix}$	$\begin{matrix}\varepsilon^*\\\varepsilon\end{matrix}$	$\begin{matrix}1\\1\end{matrix}$	$\begin{matrix}\varepsilon\\\varepsilon^*\end{matrix}$	$\left.\begin{matrix}\varepsilon^*\\\varepsilon\end{matrix}\right\}$	(x, y)	$(x^2 - y^2, xy)$
A″	1	1	1	-1	-1	-1	z	
E″	$\left\{\begin{matrix}1\\1\end{matrix}\right.$	$\begin{matrix}\varepsilon\\\varepsilon^*\end{matrix}$	$\begin{matrix}\varepsilon^*\\\varepsilon\end{matrix}$	$\begin{matrix}-1\\-1\end{matrix}$	$\begin{matrix}-\varepsilon\\-\varepsilon^*\end{matrix}$	$\left.\begin{matrix}-\varepsilon^*\\-\varepsilon\end{matrix}\right\}$	(R_x, R_y)	(xz, yz)

1.6 The $D_{n\text{h}}$ groups

$D_{2\text{h}}$	E	$C_2(z)$	$C_2(y)$	$C_2(x)$	i	$\sigma(xy)$	$\sigma(xz)$	$\sigma(yz)$		
A_g	1	1	1	1	1	1	1	1		x^2, y^2, z^2
$B_{1\text{g}}$	1	1	-1	-1	1	1	-1	-1	R_z	xy
$B_{2\text{g}}$	1	-1	1	-1	1	-1	1	-1	R_y	xz
$B_{3\text{g}}$	1	-1	-1	1	1	-1	-1	1	R_x	yz
A_u	1	1	1	1	-1	-1	-1	-1		
$B_{1\text{u}}$	1	1	-1	-1	-1	-1	1	1	z	
$B_{2\text{u}}$	1	-1	1	-1	-1	1	-1	1	y	
$B_{3\text{u}}$	1	-1	-1	1	-1	1	1	-1	x	

$D_{3\text{h}}$	E	$2C_3$	$3C_2$	σ_h	$2S_3$	$3\sigma_\text{v}$		
$A_1′$	1	1	1	1	1	1		$x^2 + y^2, z^2$
$A_2′$	1	1	-1	1	1	-1	R_z	
E′	2	-1	0	2	-1	0	(x, y)	$(x^2 - y^2, xy)$
$A_1″$	1	1	1	-1	-1	-1		
$A_2″$	1	1	-1	-1	-1	1	z	
E″	2	-1	0	-2	1	0	(R_x, R_y)	(xz, yz)

$D_{4\text{h}}$	E	$2C_4$	C_2 (C_4^2)	$2C_2′$	$2C_2″$	i	$2S_4$	σ_h	$2\sigma_\text{v}$	$2\sigma_\text{d}$		
$A_{1\text{g}}$	1	1	1	1	1	1	1	1	1	1		$x^2 + y^2, z^2$
$A_{2\text{g}}$	1	1	1	-1	-1	1	1	1	-1	-1	R_z	
$B_{1\text{g}}$	1	-1	1	1	-1	1	-1	1	1	-1		$x^2 - y^2$
$B_{2\text{g}}$	1	-1	1	-1	1	1	-1	1	-1	1		xy
E_g	2	0	-2	0	0	2	0	-2	0	0	(R_x, R_y)	(xz, yz)
$A_{1\text{u}}$	1	1	1	1	1	-1	-1	-1	-1	-1		
$A_{2\text{u}}$	1	1	1	-1	-1	-1	-1	-1	1	1	z	
$B_{1\text{u}}$	1	-1	1	1	-1	-1	1	-1	-1	1		
$B_{2\text{u}}$	1	-1	1	-1	1	-1	1	-1	1	-1		
E_u	2	0	-2	0	0	-2	0	2	0	0	(x, y)	

D_{5h}	E	$2C_5$	$2C_5^2$	$5C_2$	σ_h	$2S_5$	$2S_5^3$	$5\sigma_v$		
A_1'	1	1	1	1	1	1	1	1		x^2+y^2, z^2
A_2'	1	1	1	−1	1	1	1	−1	R_z	
E_1'	2	2cos72°	2cos144°	0	2	2cos72°	2cos144°	0	(x, y)	
E_2'	2	2cos144°	2cos72°	0	2	2cos144°	2cos72°	0		(x^2-y^2, xy)
A_1''	1	1	1	1	−1	−1	−1	−1		
A_2''	1	1	1	−1	−1	−1	−1	1	z	
E_1''	2	2cos72°	2cos144°	0	−2	−2cos72°	−2cos144°	0	(R_x, R_y)	(xz, yz)
E_2''	2	2cos144°	2cos72°	0	−2	−2cos144°	−2cos72°	0		

D_{6h}	E	$2C_6$	$2C_3$	C_2	$3C_2'$	$3C_2''$	i	$2S_3$	$2S_6$	σ_h	$3\sigma_d$	$3\sigma_v$		
A_{1g}	1	1	1	1	1	1	1	1	1	1	1	1		x^2+y^2, z^2
A_{2g}	1	1	1	1	−1	−1	1	1	1	1	−1	−1	R_z	
B_{1g}	1	−1	1	−1	1	−1	1	−1	1	−1	1	−1		
B_{2g}	1	−1	1	−1	−1	1	1	−1	1	−1	−1	1		
E_{1g}	2	1	−1	−2	0	0	2	1	−1	−2	0	0	(R_x, R_y)	(xz, yz)
E_{2g}	2	−1	−1	2	0	0	2	−1	−1	2	0	0		(x^2-y^2, xy)
A_{1u}	1	1	1	1	1	1	−1	−1	−1	−1	−1	−1		
A_{2u}	1	1	1	1	−1	−1	−1	−1	−1	−1	1	1	z	
B_{1u}	1	−1	1	−1	1	−1	−1	1	−1	1	−1	1		
B_{2u}	1	−1	1	−1	−1	1	−1	1	−1	1	1	−1		
E_{1u}	2	1	−1	−2	0	0	−2	−1	1	2	0	0	(x, y)	
E_{2u}	2	−1	−1	2	0	0	−2	1	1	−2	0	0		

1.7 The D_{nd} groups

D_{2d}	E	$2S_4$	C_2	$2C_2'$	$2\sigma_d$		
A_1	1	1	1	1	1		x^2+y^2, z^2
A_2	1	1	1	−1	−1	R_z	
B_1	1	−1	1	1	−1		x^2-y^2
B_2	1	−1	1	−1	1	z	xy
E	2	0	−2	0	0	$(x, y), (R_x, R_y)$	(xz, yz)

D_{3d}	E	$2C_3$	$3C_2$	i	$2S_6$	$3\sigma_d$		
A_{1g}	1	1	1	1	1	1		x^2+y^2, z^2
A_{2g}	1	1	−1	1	1	−1	R_z	
E_g	2	−1	0	2	−1	0	(R_x, R_y)	$(x^2-y^2, xy), (xz, yz)$
A_{1u}	1	1	1	−1	−1	−1		
A_{2u}	1	1	−1	−1	−1	1	z	
E_u	2	−1	0	−2	1	0	(x, y)	

D_{4d}	E	$2S_8$	$2C_4$	$2S_8^3$	C_2	$4C_2'$	$4\sigma_d$		
A_1	1	1	1	1	1	1	1		x^2+y^2, z^2
A_2	1	1	1	1	1	−1	−1	R_z	
B_1	1	−1	1	−1	1	1	−1		
B_2	1	−1	1	−1	1	−1	1	z	
E_1	2	$\sqrt{2}$	0	$-\sqrt{2}$	−2	0	0	(x, y)	
E_2	2	0	−2	0	2	0	0		(x^2-y^2, xy)
E_3	2	$-\sqrt{2}$	0	$\sqrt{2}$	−2	0	0	(R_x, R_y)	(xz, yz)

D_{5d}	E	$2C_5$	$2C_5^2$	$5C_2$	i	$2S_{10}^3$	$2S_{10}$	$5\sigma_d$		
A_{1g}	1	1	1	1	1	1	1	1		x^2+y^2, z^2
A_{2g}	1	1	1	−1	1	1	1	−1	R_z	
E_{1g}	2	$2\cos72°$	$2\cos144°$	0	2	$2\cos72°$	$2\cos144°$	0	(R_x, R_y)	(xz, yz)
E_{2g}	2	$2\cos144°$	$2\cos72°$	0	2	$2\cos144°$	$2\cos72°$	0		(x^2-y^2, xy)
A_{1u}	1	1	1	1	−1	−1	−1	−1		
A_{2u}	1	1	1	−1	−1	−1	−1	1	z	
E_{1u}	2	$2\cos72°$	$2\cos144°$	0	−2	$-2\cos72°$	$-2\cos144°$	0	(x, y)	
E_{2u}	2	$2\cos144°$	$2\cos72°$	0	−2	$-2\cos144°$	$-2\cos72°$	0		

1.8 The S_n groups

S_4	E	S_4	C_2	S_4^3			
A	1	1	1	1		R_z	x^2+y^2, z^2
B	1	−1	1	−1		z	(x^2-y^2, xy)
E	$\left\{\begin{matrix}1\\1\end{matrix}\right.$	$\begin{matrix}i\\-i\end{matrix}$	$\begin{matrix}-1\\-1\end{matrix}$	$\left.\begin{matrix}-i\\i\end{matrix}\right\}$		$(x, y), (R_x, R_y)$	(xz, yz)

1.9 The cubic groups (T_d and O_h)

T_d	E	$8C_3$	$3C_2$	$6S_4$	$6\sigma_d$		
A_1	1	1	1	1	1		$x^2+y^2+z^2$
A_2	1	1	1	−1	−1		
E	2	−1	2	0	0		$(2z^2-x^2-y^2, x^2-y^2)$
T_1	3	0	−1	1	−1	(R_x, R_y, R_z)	
T_2	3	0	−1	−1	1	(x, y, z)	(xy, xz, yz)

O_h	E	$8C_3$	$6C_2$	$6C_4$	$3C_2$ (C_4^2)	i	$6S_4$	$8S_6$	$3\sigma_h$	$6\sigma_d$		
A_{1g}	1	1	1	1	1	1	1	1	1	1		$x^2 + y^2 + z^2$
A_{2g}	1	1	−1	−1	1	1	−1	1	1	−1		
E_g	2	−1	0	0	2	2	0	−1	2	0		$(2z^2 − x^2 − y^2, x^2 − y^2)$
T_{1g}	3	0	−1	1	−1	3	1	0	−1	−1	(R_x, R_y, R_z)	
T_{2g}	3	0	1	−1	−1	3	−1	0	−1	1		(xz, yz, xy)
A_{1u}	1	1	1	1	1	−1	−1	−1	−1	−1		
A_{2u}	1	1	−1	−1	1	−1	1	−1	−1	1		
E_u	2	−1	0	0	2	−2	0	1	−2	0		
T_{1u}	3	0	−1	1	−1	−3	−1	0	1	1	(x, y, z)	
T_{2u}	3	0	1	−1	−1	−3	1	0	1	−1		

1.10 The $C_{\infty v}$ and $D_{\infty h}$ groups for linear molecules

$C_{\infty v}$	E	$2C_\infty^\Phi$	...	$\infty\sigma_v$		
$\Sigma^+ \equiv A_1$	1	1	...	1	z	$x^2 + y^2, z^2$
$\Sigma^- \equiv A_2$	1	1	...	−1	R_z	
$\Pi \equiv E_1$	2	$2\cos\Phi$	...	0	$(x, y), (R_x, R_y)$	(xz, yz)
$\Delta \equiv E_2$	2	$2\cos 2\Phi$	...	0		$(x^2 − y^2, xy)$
$\Phi \equiv E_3$	2	$2\cos 3\Phi$	...	0		

$D_{\infty h}$	E	$2C_\infty^\Phi$	...	$\infty\sigma_v$	i	$2S_\infty^\Phi$	...	∞C_2		
$\Sigma_g^+ \equiv A_{1g}$	1	1	...	1	1	1	...	1		$x^2 + y^2, z^2$
$\Sigma_g^- \equiv A_{2g}$	1	1	...	−1	1	1	...	−1	R_z	
$\Pi_g \equiv E_{1g}$	2	$2\cos\Phi$	...	0	2	$−2\cos\Phi$	...	0	(R_x, R_y)	(xz, yz)
$\Delta_g \equiv E_{2g}$	2	$2\cos 2\Phi$	...	0	2	$2\cos 2\Phi$	...	0		$(x^2 − y^2, xy)$
$\Phi_g \equiv E_{3g}$	2	$2\cos 3\Phi$	...	0	2	$−2\cos 3\Phi$	...	0		
...	...	...	...	...	...	...	...	...		
$\Sigma_u^+ \equiv A_{1u}$	1	1	...	1	−1	−1	...	−1	z	
$\Sigma_u^- \equiv A_{2u}$	1	1	...	−1	−1	−1	...	1		
$\Pi_u \equiv E_{1u}$	2	$2\cos\Phi$	...	0	−2	$2\cos\Phi$	...	0	(x, y)	
$\Delta_u \equiv E_{2u}$	2	$2\cos 2\Phi$	...	0	−2	$−2\cos 2\Phi$	...	0		
$\Phi_u \equiv E_{3u}$	2	$2\cos 3\Phi$	...	0	−2	$2\cos 3\Phi$	...	0		
...	...	...	...	...	...	...	...	...		

Appendix 2

Table of direct products of irreducible representations

(Adapted from P.W. Atkins, M.S. Child, and C.S.G. Phillips. (1970) *Tables for Group Theory*. Oxford University Press and J.A. Salthouse and M.J. Ware. (1972) *Point Group Character Tables and Related Data*. Cambridge University Press.)

Direct products can be used to determine the symmetry properties of overtone and combination bands, as well as being at the heart of selection rules.

General rules

$A \times A = A$ $B \times B = A^{\S}$ $A \times B = B$

$A \times E = E$ $A \times E_1 = E_1$ $A \times E_2 = E_2$ $A_1 \times E = E$ $A_2 \times E = E$

$B \times E = E$ $B \times E_1 = E_2$ $B \times E_2 = E_1$ $B_1 \times E = E$ $B_2 \times E = E$

$A \times T = T$

Superscripts and subscripts

$' \times ' = '$ $'' \times '' = '$ $' \times '' = ''$

$_g \times _g = _g$ $_u \times _u = _g$ $_u \times _g = _u$

$_1 \times _1 = _1$ $_2 \times _2 = _1$ $_1 \times _2 = _2{}^{\dagger}$

Doubly degenerate representations‡

For C_3, C_{3h}, C_{3v}, D_3, D_{3h}, D_{3d}, C_6, C_{6h}, C_{6v}, D_6, D_{6h}, S_6, O, O_h, T, T_d, and T_h:

$E \times E = A_1 + [A_2] + E$ $E_1 \times E_1 = E_2 \times E_2 = A_1 + [A_2] + E_2$ $E_1 \times E_2 = B_1 + B_2 + E_1$

For C_4, C_{4v}, C_{4h}, D_{2d}, D_4, D_{4h}, and S_4:

$E \times E = A_1 + [A_2] + B_1 + B_2$

(If no subscripts on A, B, or E, read as $A_1 = A_2 = A$, etc.)

---

§ Except in the groups D_2 and D_{2h}: $B_1 \times B_2 = B_3$, $B_2 \times B_3 = B_1$, $B_3 \times B_1 = B_2$.

† Except in the groups D_2 and D_{2h}: $_1 \times _1 = _2 \times _2 = _3 \times _3 =$ no subscript; $_1 \times _2 = _3$, $_2 \times _3 = _1$, $_1 \times _3 = _2$.

‡ The direct product of a degenerate species with itself may be resolved into a symmetric direct product and an antisymmetric direct product. In vibrational spectroscopy, the symmetry species of the overtones of a degenerate fundamental are obtained from the symmetric direct products. In the determination of electronic terms, the symmetric and anti-symmetric direct products for orbital angular momentum are taken with the appropriate spin functions to ensure that the total wave functions are antisymmetric. The antisymmetric component of the direct products are placed within [ ] in the table above.

Triply degenerate representations

For T_d, O, and O_h:

$$E \times T_1 = E \times T_2 = T_1 + T_2$$
$$T_1 \times T_1 = T_2 \times T_2 = A_1 + E + [T_1] + T_2$$
$$T_1 \times T_2 = A_2 + E + T_1 + T_2$$

Linear molecules ($C_{\infty v}$, $D_{\infty h}$)

$$\Sigma^+ \times \Sigma^+ = \Sigma^- \times \Sigma^- = \Sigma^+ \qquad \Sigma^+ \times \Sigma^- = \Sigma^-$$
$$\Sigma^+ \times \Pi = \Sigma^- \times \Pi = \Pi \qquad \Sigma^+ \times \Delta = \Sigma^- \times \Delta = \Delta$$
$$\Pi \times \Pi = \Sigma^+ + [\Sigma^-] + \Delta \qquad \Delta \times \Delta = \Sigma^+ + [\Sigma^-] + \Gamma \qquad \Pi \times \Delta = \Pi + \Phi$$

Appendix 3

Selected correlation tables or descent of symmetry tables for some chemically important point groups

(Adapted from P.W. Atkins, M.S. Child, and C.S.G. Phillips. (1970) *Tables for Group Theory*. Oxford University Press; J.A. Salthouse and M.J. Ware. (1972) *Point Group Character Tables and Related Data*. Cambridge University Press; G. Davidson, (1991) *Group Theory for Chemists*. Macmillan; and R.L. Carter. (1998) *Molecular Symmetry and Group Theory*. Wiley.)

These tables show the relationship between the irreducible representations of the higher symmetry point group, and those of the lower symmetry sub-groups. They are useful for working out the irreducible representations after either ligand substitution or isotopic substitution. The top left corner contains the higher symmetry point group, and below this are listed the irreducible representations present within this point group. The remainder of the top line contains the lower symmetry sub-groups associated with this higher symmetry one. If there is more than one way of achieving the lower symmetry, for example, on going from C_{2v} to C_s either of the $\sigma(xz)$ or $\sigma(yz)$ planes can be maintained, then both possibilities are listed. The remainder of the table shows how the irreducible representations in the higher symmetry point group are transformed in the lower symmetry sub-groups.

C_{2v}	C_2	C_s $\sigma(xz)$	C_s $\sigma(yz)$
A_1	A	A'	A'
A_2	A	A''	A''
B_1	B	A'	A''
B_2	B	A''	A'

C_{3v}	C_3	C_s
A_1	A	A'
A_2	A	A''
E	E	A' + A''

C_{4v}	C_4	C_{2v} σ_v	C_{2v} σ_d	C_2	C_s σ_v	C_s σ_d
A_1	A	A_1	A_1	A	A'	A'
A_2	A	A_2	A_2	A	A''	A''
B_1	B	A_1	A_2	A	A'	A''
B_2	B	A_2	A_1	A	A''	A'
E	E	$B_1 + B_2$	$B_1 + B_2$	2B	A' + A''	A' + A''

C_{2h}	C_2	C_s	C_i
A_g	A	A'	A_g
B_g	B	A''	A_g
A_u	A	A''	A_u
B_u	B	A'	A_u

C_{3h}	C_3	C_s
A'	A	A'
E'	E	2A'
A''	A	A''
E''	E	2A''

D_{2h}	D_2	C_{2v} $C_{2(z)}$	C_{2v} $C_{2(y)}$	C_{2v} $C_{2(x)}$	C_{2h} $C_{2(z)}$	C_{2h} $C_{2(y)}$	C_{2h} $C_{2(x)}$	C_2 $C_{2(z)}$	C_2 $C_{2(y)}$	C_2 $C_{2(x)}$	C_s $\sigma(xy)$	C_s $\sigma(xz)$	C_s $\sigma(yz)$
A_g	A	A_1	A_1	A_1	A_g	A_g	A_g	A	A	A	A'	A'	A'
B_{1g}	B_1	A_2	B_2	B_1	A_g	B_g	B_g	A	B	B	A'	A''	A''
B_{2g}	B_2	B_1	A_2	B_2	B_g	A_g	B_g	B	A	B	A''	A'	A''
B_{3g}	B_3	B_2	B_1	A_2	B_g	B_g	A_g	B	B	A	A''	A''	A'
A_u	A	A_2	A_2	A_2	A_u	A_u	A_u	A	A	A	A''	A''	A''
B_{1u}	B_1	A_1	B_1	B_2	A_u	B_u	B_u	A	B	B	A''	A'	A'
B_{2u}	B_2	B_2	A_1	B_1	B_u	A_u	B_u	B	A	B	A'	A''	A'
B_{3u}	B_3	B_1	B_2	A_1	B_u	B_u	A_u	B	B	A	A'	A'	A''

D_{3h}	D_3	C_{3v}	C_{3h}	C_{2v} $\sigma_h \to \sigma_v(yz)$	C_s σ_h	C_s σ_v
A_1'	A_1	A_1	A'	A_1	A'	A'
A_2'	A_2	A_2	A'	B_2	A'	A''
E'	E	E	E'	$A_1 + B_2$	$2A'$	$A' + A''$
A_1''	A_1	A_2	A''	A_2	A''	A''
A_2''	A_2	A_1	A''	B_1	A''	A'
E''	E	E	E''	$A_2 + B_1$	$2A''$	$A' + A''$

D_{4h}	D_4	C_{4v}	C_{4h}	C_4	D_{2h} C_2'	D_{2h} C_2''	D_{2d} $C_2' \to C_2'$	D_{2d} $C_2'' \to C_2'$	S_4	D_2 C_2'	D_2 C_2''
A_{1g}	A_1	A_1	A_g	A	A_g	A_g	A_1	A_1	A	A	A
A_{2g}	A_2	A_2	A_g	A	B_{1g}	B_{1g}	A_2	A_2	A	B_1	B_1
B_{1g}	B_1	B_1	B_g	B	A_g	B_{1g}	B_1	B_2	B	A	B_1
B_{2g}	B_2	B_2	B_g	B	B_{1g}	A_g	B_2	B_1	B	B_1	A
E_g	E	E	E_g	E	$B_{2g} + B_{3g}$	$B_{2g} + B_{3g}$	E	E	E	$B_2 + B_3$	$B_2 + B_3$
A_{1u}	A_1	A_2	A_u	A	A_u	A_u	B_1	B_1	B	A	A
A_{2u}	A_2	A_1	A_u	A	B_{1u}	B_{1u}	B_2	B_2	B	B_1	B_1
B_{1u}	B_1	B_2	B_u	B	A_u	B_{1u}	A_1	A_2	A	A	B_1
B_{2u}	B_2	B_1	B_u	B	B_{1u}	A_u	A_2	A_1	A	B_1	A
E_u	E	E	E_u	E	$B_{2u} + B_{3u}$	$B_{2u} + B_{3u}$	E	E	E	$B_2 + B_3$	$B_2 + B_3$

D_{4h}	C_{2v} C_2, σ_v	C_{2v} C_2, σ_d	C_{2v} C_2'	C_{2v} C_2''	C_{2h} C_2	C_{2h} C_2'	C_{2h} C_2''	C_s σ_h	C_s σ_v	C_s σ_d
A_{1g}	A_1	A_1	A_1	A_1	A_g	A_g	A_g	A'	A'	A'
A_{2g}	A_2	A_2	B_1	B_1	A_g	B_g	B_g	A'	A''	A''
B_{1g}	A_1	A_2	A_1	B_1	A_g	A_g	B_g	A'	A'	A''
B_{2g}	A_2	A_1	B_1	A_1	A_g	B_g	A_g	A'	A''	A'
E_g	$B_1 + B_2$	$B_1 + B_2$	$A_2 + B_2$	$A_2 + B_2$	$2B_{2g}$	$A_g + B_g$	$A_g + B_g$	$2A''$	$A' + A''$	$A' + A''$
A_{1u}	A_2	A_2	A_2	A_2	A_u	A_u	A_u	A''	A''	A''
A_{2u}	A_1	A_1	B_2	B_2	A_u	B_u	B_u	A''	A'	A'
B_{1u}	A_2	A_1	A_2	B_2	A_u	A_u	B_u	A''	A''	A'
B_{2u}	A_1	A_2	B_2	A_2	A_u	B_u	A_u	A''	A'	A''
E_u	$B_1 + B_2$	$B_1 + B_2$	$A_1 + B_1$	$A_1 + B_1$	$2B_{2u}$	$A_u + B_u$	$A_u + B_u$	$2A'$	$A' + A''$	$A' + A''$

T_d	C_{3v}	C_{2v}	D_{2d}
A_1	A_1	A_1	A_1
A_2	A_2	A_2	B_1
E	E	$A_1 + A_2$	$A_1 + B_1$
T_1	$A_2 + E$	$A_2 + B_1 + B_2$	$A_2 + E$
T_2	$A_2 + E$	$A_1 + B_1 + B_2$	$B_2 + E$

O_h	C_{4v} ML_5L'	D_{4h} trans-ML_4L_2'	C_{2v} cis-ML_4L_2'	C_{3v} fac-ML_3L_3'	C_{2v} mer-ML_3L_3'	D_{3d}
A_{1g}	A_1	A_{1g}	A_1	A_1	A_1	A_{1g}
A_{2g}	B_1	B_{1g}	B_1	A_2	A_1	A_{2g}
E_g	$A_1 + B_1$	$A_{1g} + B_{1g}$	$A_1 + B_1$	E	$2A_1$	E_g
T_{1g}	$A_2 + E$	$A_{2g} + E_g$	$A_2 + B_1 + B_2$	$A_2 + E$	$A_2 + B_1 + B_2$	$A_{2g} + E_g$
T_{2g}	$B_2 + E$	$B_{2g} + E_g$	$A_1 + A_2 + B_2$	$A_1 + E$	$A_2 + B_1 + B_2$	$A_{1g} + E_g$
A_{1u}	A_2	A_{1u}	A_2	A_2	A_2	A_{1u}
A_{2u}	B_2	B_{1u}	B_2	A_1	A_2	A_{2u}
E_u	$A_2 + B_2$	$A_{1u} + B_{1u}$	$A_2 + B_2$	E	$2A_2$	E_u
T_{1u}	$A_1 + E$	$A_{2u} + E_u$	$A_1 + B_1 + B_2$	$A_1 + E$	$A_1 + B_2 + B_1$	$A_{2u} + E_u$
T_{2u}	$B_1 + E$	$B_{2u} + E_u$	$A_1 + A_2 + B_1$	$A_2 + E$	$A_1 + B_1 + B_2$	$A_{1u} + E_u$

$C_{\infty v}$	C_{2v}
$\Sigma^+ \equiv A_1$	A_1
$\Sigma^- \equiv A_2$	A_2
$\Pi \equiv E_1$	$B_1 + B_2$
$\Delta \equiv E_2$	$A_1 + A_2$

$D_{\infty h}$	D_{2h}
$\Sigma_g^+ \equiv A_{1g}$	A_g
$\Sigma_g^- \equiv A_{2g}$	B_{1g}
$\Pi_g \equiv E_{1g}$	$B_{2g} + B_{3g}$
$\Delta_g \equiv E_{2g}$	$A_g + B_{1g}$
$\Sigma_u^+ \equiv A_{1u}$	B_{1u}
$\Sigma_u^- \equiv A_{2u}$	A_u
$\Pi_u \equiv E_{1u}$	$B_{2u} + B_{3u}$
$\Delta_u \equiv E_{2u}$	$A_u + B_{1u}$

Appendix 4

Tanabe–Sugano diagrams and Lever plots

Tanabe–Sugano diagrams

Tanabe–Sugano diagrams for $3d^n$ configurations were introduced in Chapter 5, Section 5.13.5, as a means of identifying the ground and excited terms, the number of spin-allowed transitions, estimation of the ligand field splitting parameter and the Racah B parameter, identification of spin-forbidden transitions, and the interpretation of additional features in the spectra. A complete set of Tanabe–Sugano diagrams for d^2 to d^8 electron configurations in octahedral environments is collected together in this appendix. For use with $3d^n$ tetrahedral geometries, the d^{10-n} octahedral diagram can be used, but the g subscripts need to be removed.

The atomic/free ion terms lie on the y-axis, and as the crystal or ligand field is increased these split into the terms appropriate for the coordination environment (see Chapter 5, Table 5.7 for the terms derived from the free ions in an octahedral environment). The ground term lies on the x-axis, and the energies of the other terms are plotted relative to this. For d^4, d^5, d^6, and d^7 configurations there is the possibility of either the high-spin or low-spin isomer being the ground term. In the diagrams below the transition between these isomers is marked by a dashed line, with the high-spin ground term on the left-hand side and the low-spin ground term on the right-hand side.

The excited terms with the same spin multiplicity as the ground term are marked by black lines, and the excited terms with different spin multiplicities by blue lines. Spin-allowed transitions can take place between the terms represented by the black lines, and spin-forbidden transitions can occur from the ground term to the terms associated with the blue lines. These are more likely when the spin multiplicity changes by two as this only involves changing the spin of one electron.

These diagrams were originally introduced by Tanabe and Sugano, and are reproduced in many inorganic textbooks. Unfortunately, there was one T term missing from the d^5 ^2I and d^6 ^1I excited terms in the original diagrams, which are necessary to interpret some spectra. The Tanabe–Sugano diagrams in this text are based on more recently published ones (B.N. Figgis and M.A. Hitchman. (2000) *Ligand Field Theory and its Applications*. Wiley-VCH) as these have the full set of terms for each excited state. However, there do appear to be a number of errors in these diagrams, especially in the labelling of the atomic/free ion terms. The corrected Tanabe–Sugano diagrams in the main text and below were generated using the data kindly made available by Prof. Robert Lancashire (Department of Chemistry, The University of the West Indies, Mona, Jamaica). (There are some excellent tools for data analysis and interpretation available from his website at wwwchem.uwimona.edu.jm/courses/Tanabe-Sugano/TSintro.html)

For d^2, d^3, d^7, and d^8 all of the atomic and octahedral terms are present in the diagrams. For d^4, d^5, and d^6, some of the higher energy terms have been excluded to increase clarity, and these are given in brackets, together with a mark on the y-axis to indicate their location. In addition, for d^4, d^5, and d^6 configurations there are atomic terms that occur at values of E/B greater than 80.

The values of B and C in the captions are the atomic/free ion values used to calculate the relative energies. Whilst these values correspond to those given in Figgis and Hitchman, it appears that different values were actually used to generate their diagrams. The diagrams can be used with other values of B, and due to the nephelauxetic effect (see Chapter 5, Section 5.14), the value of B extracted from the analysis of experimental spectra of molecular complexes will always be lower than the atomic/free ion value.

Larger versions of these diagrams, together with full colour versions including all of the terms, are available from the Online Resource Centre.

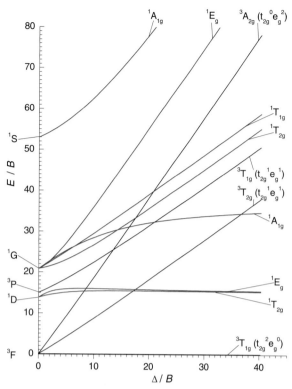

FIGURE 1 Tanabe–Sugano diagram for octahedral d^2
($B = 886$ cm^{-1}, $C = 3916$ cm^{-1}).

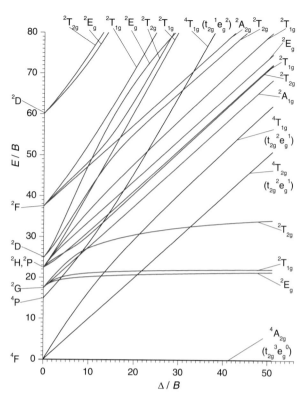

FIGURE 2 Tanabe–Sugano diagram for octahedral d^3
($B = 933$ cm^{-1}, $C = 4199$ cm^{-1}).

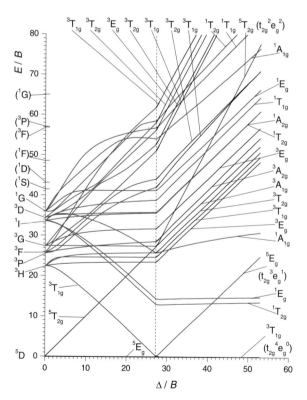

FIGURE 3 Tanabe–Sugano diagram for octahedral d^4
($B = 796$ cm^{-1}, $C = 3662$ cm^{-1}).

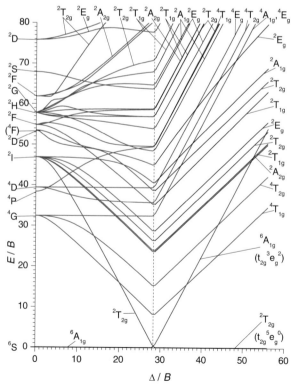

FIGURE 4 Tanabe–Sugano diagram for octahedral d^5
($B = 859$ cm^{-1}, $C = 3848$ cm^{-1}).

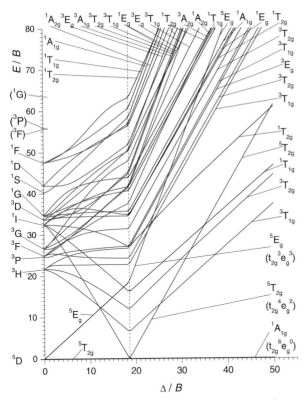

FIGURE 5 Tanabe–Sugano diagram for octahedral d^6 ($B = 1080$ cm^{-1}, $C = 4774$ cm^{-1}).

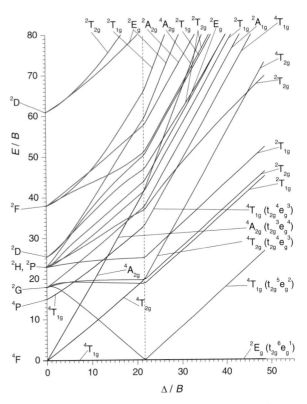

FIGURE 6 Tanabe–Sugano diagram for octahedral d^7 ($B = 986$ cm^{-1}, $C = 4565$ cm^{-1}).

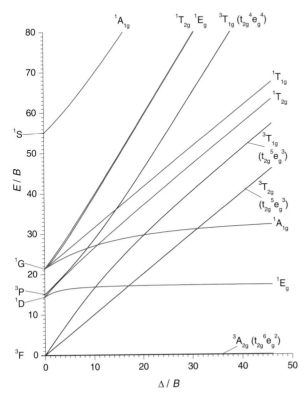

FIGURE 7 Tanabe–Sugano diagram for octahedral d^8 ($B = 1042$ cm^{-1}, $C = 4908$ cm^{-1}).

Lever plots

Tanabe–Sugano diagrams can be used to determine Δ and B if there are sufficient experimental data using an iterative approach of identifying the correct Δ/B ratio, or to the predict the energy of transitions from values of Δ and B. However, a quicker and more accurate method uses an approach introduced by Lever (A.B.P. Lever, *J. Chem. Educ.* **45** 711 (1968)), as described in Chapter 5, Section 5.13.5.1. The process involves the use of plots of the ratios of the calculated spin-allowed transition energies to determine the value of Δ/B. This can then be used with the Tanabe–Sugano diagram to identify the value of B and hence Δ. A selection of the most useful Lever plots of transition energy ratios is given below. The left-hand y-axis is the same as the parent Tanabe–Sugano diagram and gives the relative energy of the excited terms with the same multiplicity as the ground term and these are shown in black in the diagrams. The right-hand y-axis is the ratio of the transition energies, and these are shown in blue in the diagrams.

The diagram for octahedral d^3 and d^8 is identical because they both have the same A_{2g} ground term and T_{2g}, T_{1g}, and T_{1g} excited terms, although the spin multiplicity is different (and hence not included in the diagram).

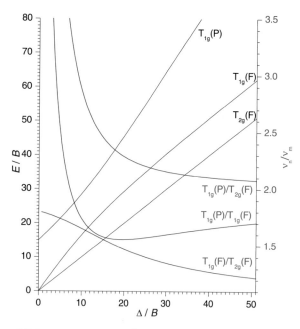

FIGURE 8 Lever plot for d^3 and d^8 octahedral configurations with A_{2g} ground terms.

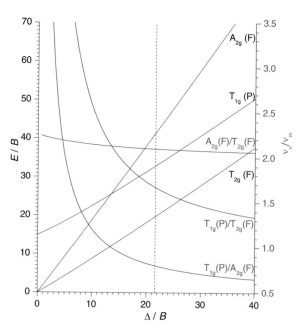

FIGURE 9 Lever plot for d^2 and high-spin d^7 octahedral configurations with T_{1g} ground terms. (The dashed line marks the limit of high-spin d^7.)

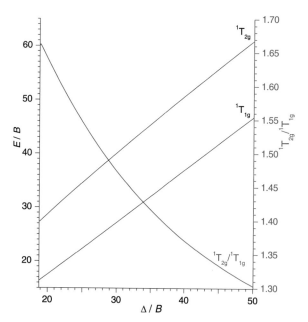

FIGURE 10 Lever plot for low-spin d^6 octahedral configurations with $^1A_{1g}$ ground term.

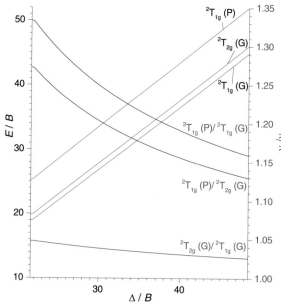

FIGURE 11 Lever plot for low-spin d^7 octahedral configurations with ^2E_g ground term.

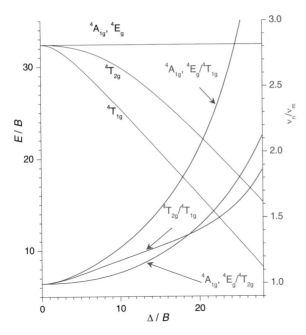

FIGURE 12 Lever plot for selected spin-forbidden transitions in high-spin d^5 octahedral configurations with $^6A_{1g}$ ground term.

Likewise, the same diagram can be used for d^2/d^7 as they both have T_{2g} ground terms and T_{2g}, T_{1g}, and A_{2g} excited terms (although it only applies to high spin d^7 and the limit is marked with a dashed line). Similar plots are also given for the spin-allowed transitions in octahedral low-spin d^6, and octahedral low-spin d^7 as well as the spin-forbidden transitions in high-spin d^5. Larger versions of the Lever Plots are available from the Online Resource Centre.

Index

Formula Index